STRINGS MICHIGAN THEORY

University of Michigan

STRINGS MICHIGAN THEORY

July 10-15, 2000

University of Michigan
Ann Arbor

International Advisory Committee

Sir Michael Atiyah, Edinburgh
L. Brink, Chalmers
A. Chamseddine, AUB
M. Cvetič, Pennsylvania
A. Dabholkar, TIFR
S. Deser, Brandeis
M.K. Gaillard, LBL
M. Green, Cambridge
D. Gross, ITP
J. Harvey, Chicago
T. Inami, Chuo
J. Maldacena, Harvard
H. Nicolai, Potsdam
D. Olive, Swansea
A. Polyakov, Princeton
E. Rabinovici, Hebrew University
P. Ramond, Florida
J. Schwarz, Caltech
N. Seiberg, IAS
A. Sen, MRI
S. Shenker, Stanford
G. Veneziano, CERN
S. Weinberg, Texas
E. Witten, IAS & Caltech

Local Organizing Committee

R. Akhoury
M. Brhlik
D. Chung
M. Duff
M. Einhorn
L. Everett
K. Freese
G. Kane
J. Liu
J.X. Lu
E. Myers
L. Pando Zayas
S. Rigolin
Y. Tomozawa
O. Yakovlev
E. Yao

Conference Secretaries

T. Wells
A. Yerks

Proceedings of the 2000 International Superstrings Conference, University of Michigan

STRINGS MICHIGAN THEORY

University of Michigan

University of Michigan, Ann Arbor July 10–15, 2000

edited by

Michael J Duff
James T Liu
Jianxin Lu
University of Michigan

World Scientific
Singapore • New Jersey • London • Hong Kong

Published by

World Scientific Publishing Co. Pte. Ltd.
P O Box 128, Farrer Road, Singapore 912805
USA office: Suite 1B, 1060 Main Street, River Edge, NJ 07661
UK office: 57 Shelton Street, Covent Garden, London WC2H 9HE

British Library Cataloguing-in-Publication Data
A catalogue record for this book is available from the British Library.

Cover illustration: Pencil drawing by Michael Duff of the University of Michigan Union Building.

PROCEEDINGS OF THE INTERNATIONAL SUPERSTRINGS CONFERENCE

ISBN 981-02-4509-2

Printed in Singapore by Uto-Print

PREFACE

The Strings 2000 Conference was held at the University of Michigan from July 10 through July 15, 2000. This conference brought together nearly 300 participants from around the world, and provided a successful forum for presenting the latest advances in the field. With the many recent developments in M-theory, a wide ranging array of topics were covered by the speakers at the conference, demonstrating the overall vitality of the field.

The conference ended on a lighter note, with the presentation of the "Top Ten Problems in Fundamental Physics" (for the next millennium). All conference participants were asked to provide a candidate problem, and the top ten were drawn from that list by a panel consisting of M. Duff, D. Gross and E. Witten. These problems are presented at the end of this volume.

We wish to acknowledge the generous financial support of the College of Literature, Science and the Arts Research Office and the Office of the Vice President for Research at the University of Michigan, as well as the National Science Foundation and the Department of Energy. We also acknowledge the Clay Mathematics Institute for the *CMI Super String 2000 Scholar* program.

This conference would not have been possible without the effort and dedication of a large number of people involved in both its organizing and management. We are grateful to our conference secretaries, T. Wells and A. Yerks, for ensuring that everything went smoothly, both in the planning stages and during the actual week of the conference. It is a pleasure to acknowledge the particle theory graduate students—A. Batrachenko, E. Krasteva, A. Mecke, H. Sati, H. Wang, L. Wang and S. Wen—for their enthusiasm and help in running the conference. We also wish to thank the other members of the local organizing committee—R. Akhoury, M. Brhlik, D. Chung, M. Einhorn, L. Everett, K. Freese, G. Kane, E. Myers, L. Pando Zayas, S. Rigolin, Y. Tomozawa, O. Yakovlev and E. Yao—for their efforts in making the conference a success. We appreciate the help of C. Sandler in making the Science Learning Center computing facilities available to participants during the week of the conference. Finally, much of the computing support for the conference would not have been possible without the hard work of M. Einhorn and E. Myers.

The Editors

CONTENTS

GAUGE FIELDS, SCALARS, WARPED GEOMETRY, AND STRINGS

EVA SILVERSTEIN
Department of Physics and SLAC
Stanford University
Stanford, CA 94305/94309, USA

We review results on several interesting phenomena in warped compactifications of M theory, as presented at Strings 2000. The behavior of gauge fields in dimensional reduction from $d+1$ to d dimensions in various backgrounds is explained from the point of view of the holographic duals (and a point raised in the question session at the conference is addressed). We summarize the role of additional fields (in particular scalar fields) in 5d warped geometries in making it possible for Poincare-invariant domain wall solutions to exist to a nontrivial order in a controlled approximation scheme without fine-tuning of parameters in the $5d$ action (and comment on the status of the singularities arising in the general relativistic description of these solutions). Finally, we discuss briefly the emergence of excitations of wrapped branes in warped geometries whose effective thickness, as measured along the Poincare slices in the geometry, grows as the energy increases.

1. Introduction

Generic general-relativistic spacetime backgrounds with d-dimensional Poincare invariance have a metric of the form

$$e^{2A(y)}\eta_{\mu\nu}dx^\mu dx^\nu + H_{IJ}(y)dy^I dy^J \tag{1}$$

where $\mu, \nu = 0, \ldots, d-1$. Canonical examples in M theory with a nontrivial warp factor $e^{2A(y)}$ include heterotic compactifications with $(0,2)$ worldsheet supersymmetry,[1] compactified Horava-Witten theory,[2] AdS_{d+1} and its relevant deformations, linear dilaton theories such as for example the NS5-brane solution[3] and the conifold singularity,[4] and no doubt many more solutions yet to be discovered with less supersymmetry.

Many warped backgrounds have non-gravitational holographic duals,[5] but most have no known equivalent "boundary theory." Almost all of these backgrounds have curvature singularities and/or strong coupling at some finite proper distance from a generic points on the component of the geometry parameterized by y^I, so that general relativity breaks down in this region of the background.

It is important and interesting to understand as precisely as possible the physics of this type of background, in particular to see if any new phenomena emerge from the warped shape of the spacetime. In this talk I will review results on three aspects of this physics:

(1) The behavior of gauge fields

(2) The role of for example scalar fields in making possible, to the leading order in a controlled approximation scheme, solutions with Poincare invariance even after some nontrivial quantum corrections to the vacuum energy have been included, and (3) The behavior of massive states coming from wrapped branes in this sort of geometry: in particular one finds a new corner of the theory where excitations can be seen to grow in size as they grow in energy as a consequence of the warping in the metric (1).

2. Gauge Fields

If we focus on cases where the warping occurs along a single direction y, the low energy effective action (to the extent that it is reliable) takes the form

$$S = \int d^dx dy \sqrt{g}\Big(a(\phi)R + b(\phi)(\nabla\phi)^2 + c(\phi)F^2 - \Lambda(\phi)\Big). \tag{2}$$

If the integral over y of the Einstein term is finite one obtains a finite d-dimensional Planck scale and "trapped gravity.[6]" This holds also for any d-dimensional graviphoton that arises from the $d+1$-dimensional graviton, as demonstrated recently in Ref. 7. On the other hand if there is a $d+1$-dimensional vector potential, the dimensional reduction of its kinetic term might give a divergence independent of what is going on with the graviton kinetic term. In particular in the $\int d^dx dy F^2$ term there are two powers of the inverse $d+1$-dimensional metric as opposed to the single power of $g^{\mu\nu}$ in the dimensional reduction of the Einstein term.

The coefficient of the d-dimensional gauge kinetic term $\frac{1}{e_d^2}\int d^dx trF^2$ is the inverse effective gauge coupling (charge) squared of the d-dimensional gauge theory. If the effective gauge coupling in d-dimensions is zero, then one might naively infer that this symmetry behaves like a global symmetry rather than a gauge symmetry after dimensional reduction. This would be very surprising since the black hole no-hair theorems, at least in contexts where they have been studied, indicate that information about global charges is lost in processes in which the black hole absorbs particles which carried this charge. On the other hand in this context, where the symmetry is a bona fide gauge symmetry in $d+1$ dimensions, the charges must be conserved.

In Ref. 8 we found that in several known examples of warped geometries in which $\frac{1}{e_d^2}$ diverges, this divergence is either indicative of a conventional screening effect or the effective theory (2) breaks down at some $y < \infty$ where new behavior takes over that also has a conventional behavior in the IR of the d-dimensional description of the physics.

The simplest example is the cutoff AdS_5 background studied by Randall and Sundrum.[6] In this background, $A(y) = -|y|/L$ where L is the curvature radius of AdS. The calculation giving the effective charge in $4d$ is

$$\frac{1}{e_4^2} \sim \frac{1}{e_5^2}\int_0^{y_0} dy e^{-4y/L}(e^{2y/L})^2 \sim y_0 \tag{3}$$

where we have introduced an IR momentum cutoff p_0 through the relation $y_0 \sim Llog(p_0L)$ following from the metric. The first factor in the integrand comes from $\sqrt{g}$, and the last from the two powers of the inverse metric involved in forming trF^2.

So we are finding

$$\frac{1}{e_4^2} \sim \frac{1}{e_5^2} Llog(p_0L) \tag{4}$$

This logarithmic IR divergence is the result one would expect from screening of the charge from the $4d$ description (as first pointed out to us by E. Witten, and as first noted in papers of Pomaral[9]). This interpretation is confirmed by an explicit calculation of the electrostatic potential arising from a point source of charge at $y = 0$:

$$A_0(p, y=0) = -\frac{QK_1(pL)}{2K_0(pL)} \rightarrow_{p\to 0} \frac{Q}{p^2 logp} \tag{5}$$

Similarly one finds a generalization of the screening effect to higher dimensions and higher-form fields. In dimension d, for a q-form field strength, as a function of IR momentum cutoff p_0, we find

$$\begin{aligned} \frac{1}{e_q^2} \propto \qquad & e^{\frac{R}{L}(2q-d)} \sim \frac{1}{p_0^{2q-d}} \qquad 2q \neq d \, . \\ & R \sim log(p_0) \qquad 2q = d \end{aligned} \tag{6}$$

At the conference, M. Duff asked about the consistency of this result with the possibility of dualizing $q+1$-form field strengths to $5-(q+1)$-form field strengths. For example a scalar field η with a 1-form field strength would be dual to a 3-form potential field C with a 4-form field strength. The latter, from (6), gives a mode with zero charge upon dimensional reduction; whereas a scalar field, like gravity, is left with nontrivial interactions after dimensional reduction.

I believe the answer to this is as follows (this result was obtained in collaboration with M. Schulz). The equation for dualizing a form, for example $d\eta = *dC$, is a linear differential equation which locally has a solution. There is no guarantee, however, that this solution is nonsingular everywhere.

Consider the equation for a scalar field in the background (1). A massless mode in $4d$ satisfies the equation

$$\eta'' + 4\eta' A' = 0 \tag{7}$$

where primes denote differentiation with respect to y. One obvious solution is the zero mode, $\eta = \eta(x)$ independent of y. This is the mode which gets "trapped" upon dimensional reduction, with a finite kinetic term. As noted by Duff in his question, this cannot be the solution dual to the three-form potential.

There is another solution to (7), which becomes singular at the AdS horizon in the RS geometry. Integrating (7), this solution satisfies

$$\eta' = \hat{\eta}(x) e^{-4A(y)} \tag{8}$$

For the RS geometry, this yields

$$\eta \sim \hat{\eta}(x) e^{+4y/L} \tag{9}$$

The dimensional reduction of the kinetic term for this scalar goes like

$$\int_0^{y_0} dy e^{-4y/L}(e^{4y/L})^2 \sim e^{4y_0/L} \tag{10}$$

from one power of $\sqrt{g}$ and two powers of $e^{+4y/L}$ from the solution (9) appearing in the quadratic action for η.

This is the same divergence which arises for the dimensional reduction of the standard zero-mode solution for the three-form potential C, which goes like

$$\int_0^{y_0} dy e^{-4y/L}(e^{2y/L})^4 \tag{11}$$

from one power of $\sqrt{g}$ and four powers of the inverse metric required to form the square of the four-form field strength. So the second solution (9) for the scalar is evidently the one dual to the zero-mode of the three-form potential.

Consider the N-NS 5-brane solution of type II string theory.[3] It has a string-frame metric and dilaton

$$\begin{aligned} ds^2 &= dx_6^2 + dr^2 + l_s^2 N d\Omega_3^2 \\ \phi &= \alpha r \end{aligned} \tag{12}$$

with $\alpha = 1/l_s\sqrt{N}$.

The string-frame ten-dimensional action is

$$\int d^6x dr d\Omega_3 \left[e^{-2\phi}(R + (\partial\phi)^2) + K_{RR}^2 \right] \tag{13}$$

where K is the field strength for the RR U(1) gauge field of type IIA string theory, or the field strength for the 2-form RR gauge potential of type IIB string theory.

In the dimensional reduction, $6d$ gravity survives (the $6d$ Planck scale ending up finite because of the coupling of the Einstein term to $e^{-2\phi}$ in string frame). On the other hand the RR gauge field effectively propagates in seven flat dimensions according to the metric (12) since its kinetic term is independent of ϕ. This is not a screening phenomenon in $6d$.

The resolution is that the breakdown of the solution (12) down the throat due to strong coupling effects is important. The IIA NS5-brane is fundamentally an M5-brane at a point on the eleventh circle of M theory.[10,11] This means that deep in the IR region of the solution the RR gauge symmetry is spontaneously broken, and this Higgs mechanism wards off the more exotic possibility of a conserved global symmetry persisting in $6d$.

These examples provide further evidence for the robustness of the arguments against global symmetries in quantum gravity. However, it is interesting to keep one's eyes out for more exotic examples that might arise in which a naive divergence in the dimensionally reduced gauge kinetic term persists into the infrared in a way that cannot be understood from screening or the Higgs mechanism. In such a background, the black-hole no-hair theorems would need to be analyzed carefully.

3. Scalars

Let us focus now on the physics of gravity plus scalars, with action in a $5d$ bulk and on a $4d$ δ-function localized brane given by

$$S = \int d^5x\sqrt{-G}\left[R - \tfrac{4}{3}(\nabla\phi)^2 - \Lambda e^{a\phi}\right] + \int d^4x\sqrt{-g}(-f(\phi)) \tag{14}$$

We have taken the bulk Λ to be zero to a leading approximation, having in mind bulk supersymmetry which is only broken at the level of interactions with the brane.

In contrast to the Randall-Sundrum system, this one has $4d$ Poincare-invariant solutions for *generic* brane tension $f(\phi)$.[12] The solutions in the bulk are:

$$\phi(y) = \pm\frac{3}{4}log|\frac{4}{3}y + c| + d \tag{15}$$

$$A(y) = \frac{1}{4}log|\frac{4}{3}y + c| + \tilde{d} \tag{16}$$

At $y = -\frac{3}{4}c$, this configuration has curvature singularities: general relativity breaks down near these points.

Einstein's equations at the wall at $y = 0$ (which boil down to Israel matching conditions there) are solved by adjusting the values of integration constants c, d, *not* by tuning parameters like $f(\phi)$ in the Lagrangian. So quantum corrections to $f(\phi)$ will not ruin the fact that there is a flat solution. This is perhaps encouraging, since in M theory all indications are that one does not have the freedom to tune parameters in the theory (there being no arbitrary dimensionless couplings put in); we at most have the freedom to choose among different solutions of the basic equations (still to be determined!) of the theory.

It is therefore an interesting goal to exhibit as a first step a background of M theory with near-vanishing cosmological term after nontrivial quantum corrections are included. The next (and perhaps most difficult) step is to understand why we live in such a background as opposed to one of the millions of others with large spacetime curvatures or otherwise unrealistic low-energy physics.

In our examples (14)(15)(16), we must first understand the physics of the singularities that appear in the general-relativistic description of the system. Progress was made on this front in two directions. Firstly, Horowitz, Low, and Zee found stringy cosmological solutions with again effectively zero cosmological term independent of the parameters in the Lagrangian.[13] In these solutions, there are again curvature singularities but they are null, so that there is no issue of additional boundary conditions needed for modes emanating from the brane toward the singularities.

Secondly, the static solutions above are in some ways analogous to the types of backgrounds that have been intensely studied recently as deformations of the AdS/CFT correspondence to confining theories.[14] In those backgrounds, a singularity in the general relativistic approximation corresponds to the presence of a mass gap beyond which there are no excitations of the field theory. Just as the presence of a mass gap is generic to asymptotically free quantum field theory, the presence of a singularity in the $5d$ gravity dual in the (bad) general relativistic approximation is generic. That is not to say that all singular backgrounds of this form are resolved by quantum gravity effects. Indeed, in this AdS/CFT context the field theory side can only vouch for a discrete subset of the continuum of apparent singular solutions in the GR description, since at finite N the field theory has a finite number of vacua. Luckily there are of order $e^{\sqrt{N}}$ such vacua.

So the upshot, in this context, is this. Couple a large-N gauge theory to the standard model in the way that is determined by the addition of a thin domain wall to cut off the UV end of the gravity dual to the gauge theory. Generically, one can choose a vacuum of the large-N theory to zero the $4d$ cosmological term in the full system, up to corrections that are parametrically suppressed relative to the TeV^4 contribution expected from standard model loops. These corrections could also introduce instabilities in the system, which is a problem endemic to these models as well as those of Ref. 6 (not to mention any string compactification with approximate moduli).

Is there any reason this vacuum is preferred? Aside from the partially anthropic arguments presented in Refs. 15,16 (where another mechanism for finding solutions of M theory with near-vanishing cosmological term was obtained) I know of no argument for this at present. One advantage of our setup in the context of the Brown-Teitelboim style analysis of Ref. 15 is that the "discretuum" of different vacua in our case all have manifestly similar "standard model" physics since this resides on the brane and the bulk is where the cancellation mechanism arises. As emphasized in Ref. 17, this is an important consideration in entertaining this sort of anthropic explanation for the history.

However, it is worth emphasizing that the actual solutions (15)(16) considered in Ref. 12 do not asymptote to AdS space in bulk, but instead tend toward flat space far from the singularity. Therefore they are not dual to a quantum field theory in toto. It is possible that the regime near the singularity does have an effective quantum field theory dual, since the warp factor decreases there.[18] In any case it is very interesting to try to understand the physics of warped geometries which do not have a quantum field theory dual. Related to this is the question of whether any backgrounds of this type can have sensible physics in which effective quantum field theory breaks down so that long-distance quantities like the cosmological constant might be affected naturally by high-energy excitations in the theory.

I cannot resist adding here more general comments on naked singularities in string theory. At such singularities, general relativity breaks down. To me this is one of the most interesting features in a spacetime, since it is an opportunity to learn about physics of M theory that goes beyond the long-wavelength general relativistic approximation. The resolution of singularities on the Coulomb branch of gauge theories in Seiberg-Witten theory,[19] in ADE limits of compactifications of type II string theory on K3,[20] at type II conifold points[21] and their heterotic cousins[22,23,24] (just to name a few examples), involved in a detailed way the physics of non-perturbative excitations of the theory and nontrivial information about its strong-coupling behavior. Without these many resolutions of general-relativistic naked singularities, the web of dualities relating different limits of M theory would not exist. One of the immediate applications of such resolutions has also been the analysis of controlled topology-changing transitions[4,25,24] in M theory.

For the most general solutions described in this section, it remains to be seen whether or not the singularities have a (perhaps discretized) resolution, and if so whether a standard 4d effective field theory arises at long distances. (In the above examples, on the order of a decade passed between the original nakedly singular solutions being written down (for the conifold for example in Ref. 26) and the eventual quantum resolution; it is not clear in our case when all the necessary ingredients will be available to answer this question.) This is an interesting open question.

4. Wrapped Branes

One of the features of quantum gravity emphasized recently for example in Refs. 27,28 is the fact that at the highest energies, excitations (namely black holes) grow in size as they grow in energy (mass). This is in stark contrast to the size=1/momentum uncertainty relation of elementary excitations in quantum field theory. This makes it conceivable that such high-energy excitations could naturally affect long-distance parameters.

It turns out that warped compactifications constitute another context in which some excitations grow in size as they grow in energy. In general given a metric (1), there will be wrapped branes on the compact component of the geometry whose mass $m_0(y)$ will depend on y since the volume of the cycle on which the brane is wrapped varies as a function of y. As measured along the Poincare slices, the energy will be (taking into account the warp factor as well as the variation of the cycle volume)

$$E(y) \sim \sqrt{g_{00}(y)}m_0(y) \tag{17}$$

On the other hand, because of the warp factor, the thickness of the object is rescaled from its proper thickness r_0 because of the warp factor:

$$R \equiv \delta x_{||} \sim \frac{1}{\sqrt{g_{ii}}}r_0 \tag{18}$$

In Ref. 29 concrete examples were exhibited in which both R and E grow in the same direction, so that

$$E = TR^q \tag{19}$$

for some power $q > 0$. The spectrum of excitations of these wrapped branes was worked out in a limit where a Kaluza-Klein analysis was valid.

Any situation in which there is this sort of growth of size with energy suggests that interesting nonlocal effects might emerge from the dynamics of these excitations. This is under investigation.[30]

Acknowledgments

The work in §2 was done in collaboration with some combinations of N. Kaloper, and L. Susskind. The work reviewed in §3 was done in collaboration with S. Kachru and M. Schulz, and concurrently by N. Arkani-Hamed, S. Dimopoulos, N. Kaloper, and R. Sundrum. M. Duff asked an interesting question about the material in §2. I thank all of the above, as well as those mentioned in the corresponding papers, for many interesting discussions on these and related topics. Finally I would like to thank the organizers of Strings 2000 for a stimulating conference. This work was supported by a DOE OJI grant, by the A.P. Sloan Foundation, and by the DOE under contract DE-AC03-76SF00515.

References

1. A. Strominger, *Superstrings With Torsion*, *Nucl. Phys.* **B274**, 253 (1986).
2. E. Witten, *Strong Coupling Expansion Of Calabi-Yau Compactification*, *Nucl. Phys.* **B471**, 135 (1996) [hep-th/9602070].
3. A. Strominger, *Heterotic Solitons*, *Nucl. Phys.* **B343**, 167 (1990); C.G. Callan, J.A. Harvey and A. Strominger, *Supersymmetric string solitons*, hep-th/9112030.
4. E. Witten, *Phases of $N = 2$ theories in two dimensions*, *Nucl. Phys.* **B403**, 159 (1993) [hep-th/9301042].
5. J. Maldacena, *The large N limit of superconformal field theories and supergravity*, *Adv. Theor. Math. Phys.* **2**, 231 (1998) [hep-th/9711200]; E. Witten, *Anti-de Sitter space and holography*, *Adv. Theor. Math. Phys.* **2**, 253 (1998) [hep-th/9802150]; S.S. Gubser, I.R. Klebanov and A.M. Polyakov, *Gauge theory correlators from non-critical string theory*, *Phys. Lett.* **B428**, 105 (1998) [hep-th/9802109].
6. L. Randall and R. Sundrum, *An Alternative to Compactification*, *Phys. Rev. Lett.* **83**, 4690 (1999) [hep-th/9906064].
7. M.J. Duff, J.T. Liu and W.A. Sabra, *Localization of supergravity on the brane*, hep-th/0009212; M. Cvetic, H. Lu and C.N. Pope, *Brane-world Kaluza-Klein reductions and branes on the brane*, hep-th/0009183.
8. N. Kaloper, E. Silverstein, and L. Susskind, *Gauge symmetry and localized gravity in M theory*, hep-th/0006192.
9. A. Pomarol, *Grand unified theories without the desert*, *Phys. Rev. Lett.* **85**, 4004 (2000) [hep-ph/0005293].
10. N. Itzhaki, J. Maldacena, J. Sonnenschein and S. Yankielowicz, *Supergravity and the Large N Limit of Theories with Sixteen Supercharges*, *Phys. Rev.* **D58**, 046004 (1998) [hep-th/9802042].
11. S. Minwalla and N. Seiberg, *Comments on the IIA NS5-brane*, *JHEP* **9906**, 007 (1999) [hep-th/9904142].
12. S. Kachru, M. Schulz, E. Silverstein, *Phys. Rev.* **D62**, 045021 (2000) [hep-th/0001206]; N. Arkani-Hamed, S. Dimopoulos, N. Kaloper and R. Sundrum, *A small cosmological constant from a large extra dimension*, *Phys. Lett.* **B480**, 193 (2000) [hep-th/0001197].
13. G. Horowitz, I. Low, and A. Zee, *Self-tuning in an outgoing brane wave model*, *Phys. Rev.* **D62**, 086005 (2000) [hep-th/0004206].
14. L. Girardello, M. Petrini, M. Porrati and A. Zaffaroni, *Novel local CFT and exact results on perturbations of $N = 4$ super Yang-Mills from AdS dynamics*, *JHEP* **9812**, 022 (1998) [hep-th/9810126]; D.Z. Freedman, S.S. Gubser, K. Pilch and N.P. Warner, *Renormalization group flows from holography supersymmetry and a c-theorem*, hep-

th/9904017; J. Polchinski and M.J. Strassler, *The string dual of a confining four-dimensional gauge theory*, hep-th/0003136.

15. R. Bousso and J. Polchinski, *Quantization of four-form fluxes and dynamical neutralization of the cosmological constant*, *JHEP* **0006**, 006 (2000) [hep-th/0004134].
16. J.L. Feng, J. March-Russell, S. Sethi and F. Wilczek, *Saltatory relaxation of the cosmological constant*, hep-th/0005276.
17. T. Banks, M. Dine and L. Motl, *On anthropic solutions of the cosmological constant problem*, hep-th/0007206.
18. J. Maldacena and C. Nunez, *Supergravity description of field theories on curved manifolds and a no go theorem*, hep-th/0007018.
19. N. Seiberg and E. Witten, *Electric-magnetic duality, monopole condensation, and confinement in N=2 supersymmetric Yang-Mills theory*, *Nucl. Phys.* **B426**, 19 (1994) [hep-th/9407087].
20. E. Witten, *String theory dynamics in various dimensions*, *Nucl. Phys.* **B443**, 85 (1995) [hep-th/9503124].
21. A. Strominger, *Massless black holes and conifolds in string theory*, *Nucl. Phys.* **B451**, 96 (1995) [hep-th/9504090].
22. E. Witten, *Small Instantons in String Theory*, *Nucl. Phys.* **B460**, 541 (1996) [hep-th/9511030].
23. S. Kachru, N. Seiberg and E. Silverstein, *SUSY Gauge Dynamics and Singularities of 4d N=1 String Vacua*, *Nucl. Phys.* **B480**, 170 (1996) [hep-th/9605036].
24. S. Kachru and E. Silverstein, *Chirality-changing phase transitions in 4d string vacua*, *Nucl. Phys.* **B504**, 272 (1997) [hep-th/9704185].
25. P.S. Aspinwall, B.R. Greene and D.R. Morrison, *Calabi-Yau moduli space, mirror manifolds and spacetime topology change in string theory*, *Nucl. Phys.* **B416**, 414 (1994) [hep-th/9309097]; J. Distler and S. Kachru, *(0,2) Landau-Ginzburg theory*, *Nucl. Phys.* **B413**, 213 (1994) [hep-th/9309110]; B.R. Greene, D.R. Morrison and A. Strominger, *Black hole condensation and the unification of string vacua*, *Nucl. Phys.* **B451**, 109 (1995) [hep-th/9504145].
26. P. Candelas and X.C. de la Ossa, *Comments On Conifolds*, *Nucl. Phys.* **B342**, 246 (1990).
27. J. McGreevy, L. Susskind, N. Toumbas, *Invasion of the giant gravitons from anti-de Sitter space*, *JHEP* **0006**, 008 (2000) [hep-th/0003075].
28. T. Banks, hep-th/0007146.
29. E. Silverstein, *Extended States from Warped Compactifications of M Theory*.
30. A. Adams and E. Silverstein, work in progress.

FUNDAMENTAL STRINGS AS NONCOMMUTATIVE SOLITONS*

FINN LARSEN
Enrico Fermi Institute, University of Chicago
5640 S. Ellis Ave., Chicago, IL 60637, USA
E-mail: flarsen@theory.uchicago.edu

The interpretation of closed fundamental strings as solitons in open string field theory is reviewed. Noncommutativity is introduced to facilitate an explicit construction. The tension is computed exactly and the correct spectrum is recovered at long wave length.

1. Introduction

It has recently been realized that strings and branes can be interpreted as noncommutative solitons in string field theory. The purpose of this talk is to review this development and provide a pedagogical introduction to the subject. The discussion is an extended version of the actual talk at STRINGS 2000 which in turn was based on Ref. 1. Section 5 reports partially on results obtained after the conference.

String field theory is designed to mimic many aspects of ordinary quantum field theories. It is therefore natural to develop intuition about string field theory by applying standard field theory techniques in this less familiar context. This motivates the construction of classical soliton solutions in string field theory. In this talk the specific goal is to discuss a class of solitons related to the closed fundamental string. It turns out useful to introduce noncommutativity in the field theory, as a tool that facilitates an explicit construction. The result of the computation will be a soliton with tension exactly equal to the tension of the fundamental string; and it has the same classical fluctuation spectrum as well. These facts provide strong circumstantial evidence that the soliton can be *identified* with the fundamental string. In the course of the talk some open questions raised by this interpretation will be discussed. For definiteness the bosonic string theory is considered but virtually identical results apply to the superstring case.

2. Tachyon Condensation and a First Look at Solitons

Open string theory by definition supports open string excitations. Although it is not usually stressed, the ends of the open strings may be situated anywhere in spacetime. Recalling that D-branes are defined as defects where open strings can end, the open string vacuum is thus characterized by a space-filling D-brane. In this terminology the 26-dimensional perturbative vacuum of open bosonic string theory is interpreted as a $D25$-brane. The spectrum of the open strings follows

*Talk presented at STRINGS 2000, July 10-15, 2000, University of Michigan at Ann Arbor.

from standard string theory computations; the result is that the lightest mode is a tachyon, *i.e.* it has negative mass-squared. This means the potential of the tachyon field has negative second derivative in the perturbative vacuum. As is well-known (*e.g.* from the Higgs phenomenon in the standard model) this kind of tachyon signals an instability, there is a true vacuum where the "tachyon" field has acquired an expectation value. The driving force behind the developments in the last few years was an important insight by Sen,[2] asserting that the true vacuum after tachyon condensation is in fact the standard perturbative closed string vacuum, *i.e.* the vacuum without the $D25$ brane and thus without the open strings. A consequence of this physical picture is that the energy liberated by the condensation of the tachyon precisely cancels the tension of the $D25$-brane.

We would like to develop a quantitative description of tachyon condensation in string theory. For fields with nonvanishing mass, such as the tachyon field, a constant field does not satisfy the equations of motion. It is therefore clear that tachyon condensation inherently involves off-shell properties. This is the reason that standard perturbative string theory is insufficient to analyze the problem, one must apply string field theory. The level truncation approximation to the cubic string field theory has provided convincing evidence that the energy liberated by tachyon condensation indeed equals that of the $D25$-brane.[3] This supports Sen's identification of the nonperturbative vacuum.

The interest in this talk is the more detailed question of *excitations* of the non-perturbative closed string vacuum; in particular, those adequately described as classical solitons. For definiteness the focus will be on the fundamental string excitations. Similar considerations apply to other soliton excitations, principally the lower-dimensional D-branes. The D-branes are in fact understood more precisely in this set-up; they are the topic of J. Harvey's talk at this conference.

The description of the fundamental string in open string theory is qualitatively as follows.[4,5,6] Consider first the situation *before* tachyon condensation. Then the open strings are described as some gauge field theory on the world-volume of the $D25$-brane. In this framework fundamental strings appear as electric flux-tubes. *After* tachyon condensation all open string degrees of freedom are removed from the spectrum, and in particular the gauge field no longer exists. It is therefore not so obvious how to describe the electric flux-tubes after tachyon condensation. This is precisely the problem of interest because, whatever the appropriate description of electric flux-tubes after tachyon condensation, these are the fundamental strings. The whole process of tachyon condensation is reminiscent of the confinement of quarks: the open string degrees of freedom cannot propagate in the closed string vacuum but instead manifest themselves as collective excitations, such as the fundamental strings.

3. From Noncommutativity to a Quantitative Description

We would like to turn these comments into a quantitative field theory description.

A suitable starting point for the discussion is the Born-Infeld Lagrangian[5]

$$S_I = -\int d^{26}x \; V(t) \; \sqrt{-\det[g_{\mu\nu} + 2\pi\alpha' F_{\mu\nu}]} \; . \tag{1}$$

One immediate problem is that the tachyon potential $V(t)$ is unknown (except for a few qualitative features discussed in the previous section). A more pressing concern is that (1) is justified only for *constant* fields. In the complete Lagrangian the tachyon has kinetic terms, and there are numerous higher derivative terms that generally couple the tachyon and the gauge field. All these terms are determined *in principle* by string field theory. In practice they are unfortunately difficult to compute accurately. For example, an exact determination of these terms require that infinitely many massive fields of the string field are taken into account.[a] The unknown derivative terms are important for fields varying over distances of order string scale. They are a serious problem for our purposes because the fundamental string solution we seek by definition varies over the string scale.

This is the point where noncommutativity turns out useful. Recall that a B-field can be incorporated in string theory by replacing the standard "closed string" metric $g_{\mu\nu}$ and coupling constant g_s with the "open string" quantities[9,10]

$$G_{\mu\nu} = g_{\mu\nu} - (2\pi\alpha')^2 (Bg^{-1}B)_{\mu\nu} \; , \tag{2}$$

$$G_s = g_s \left(\frac{\det G}{\det(g + 2\pi\alpha' B)} \right)^{\frac{1}{2}} \; . \tag{3}$$

More importantly, one must also replace the standard multiplication of fields with the noncommutative star-product

$$A \star B = \exp\left(\frac{i}{2} \theta^{\mu\nu} \partial_\mu \partial_{\nu'} \right) A(x^\mu) B(x'^\nu) \; , \tag{4}$$

where

$$\theta^{\mu\nu} = -(2\pi\alpha')^2 \left(\frac{1}{g + 2\pi\alpha' B} B \frac{1}{g - 2\pi\alpha' B} \right)^{\mu\nu} \; . \tag{5}$$

The reason this is useful for the present problem is that the noncommutativity parameter θ provides a new scale in the theory. The linear extent of the fundamental string solitons we seek are of order string scale $l_s = \sqrt{\alpha'}$ when measured in the *closed string metric*; but for large θ/α' this corresponds to a distance $\sqrt{\theta}$ when measured in the *open string metric*. By way of comparison, the complications of open string field theory we want to control are of string scale with respect to the open string metric; so these are negligible on the scale of the soliton, as long as we take the limit $\theta/\alpha' \to \infty$. For related discussions see Refs. 11,12.

Before proceeding with the main line of development it is helpful to discuss the limit of large noncommutativity in more detail. Many workers (including Seiberg

[a] In the recent works[7] it was noted that the tachyon can be decoupled from the massive fields in the BSFT formalism of string field theory.[8]

and Witten[10]) consider D-branes in background B-fields and take the low energy decoupling limit

$$\alpha' \sim \epsilon^{\frac{1}{2}} \to 0 \quad ; \quad g_{ij} \sim \epsilon \to 0 \ , \tag{6}$$

for i, j in the noncommutative directions. In this limit string theory reduces to noncommutative Yang-Mills theory (with some specific matter content). The decoupling limit (6) clearly implies large noncommutativity

$$\theta/\alpha' \sim \epsilon^{-\frac{1}{2}} \to \infty \ , \tag{7}$$

but it is *not* the limit we are considering. We take $\theta/\alpha' \to \infty$ *without* taking the low energy limit. This is important for our purposes because we want to keep string excitations. There are two dimensionless parameters in the problem $\alpha' E^2$ and θ/α'. The decoupling limit (6) takes $\theta/\alpha' \to \infty$ with θ,E fixed, and thus $\alpha' E^2 \cdot \theta/\alpha'$ fixed. In contract, we simply take $\theta/\alpha' \to \infty$ with $\alpha' E^2$ kept fixed. Our limit is that of noncommutative string field theory (NCSFT). As far as we are aware this limit has not been considered prior to Ref. 1.

We now return to the quest for a description of the fundamental string as a soliton by applying the limit of large noncommutativity to the Born-Infeld type Lagrangian (1). The string solution is going to be along some spatial direction, say x^1, as well as time x^0; we take large non-commutativity in all other directions. This introduces the open string metric G^{ij}, the open string coupling constant, and the star product in (1), yielding

$$S_I = -\frac{g_s}{G_s} \int d^{26}x \ V(t) \ \sqrt{-\det[G_{\mu\nu} + 2\pi\alpha' F_{\mu\nu}]} \ . \tag{8}$$

The overall factor arises from the replacement $g_s \to G_s$ in the D-brane potential $V(t) \propto 1/g_s$. The progress at this point is that it is now justified to ignore derivatives along all transverse directions. Thus, for solutions independent of x^0, x^1, we can simply use (8) as it stands.

It is convenient for our purposes to consider the conjugate Hamiltonian

$$H = -\int d^{25}x \ \left[\sqrt{V(t)^2 + E^2/(2\pi\alpha')^2} + \lambda \partial_1 E\right] - \lambda p \ , \tag{9}$$

where the electric field E is the canonically conjugate of the gauge field A_1, and λ is a Lagrange multiplier imposing a restriction.to the flux sector with quantum number p. The equations of motions become

$$\frac{V(t)V'(t)}{\sqrt{V(t)^2 + E^2/(2\pi\alpha')^2}} = 0 \ , \tag{10}$$

$$\frac{E}{\sqrt{V(t)^2 + E^2/(2\pi\alpha')^2}} + \lambda(2\pi\alpha')^2 = 0 \ . \tag{11}$$

These equations are quite simple. In fact, they appear too simple to allow nontrivial localized solutions. For example, the equations are obviously solved for tachyons with $V'(t) = E = \lambda = 0$ but this is only possible at the localized extrema of the potential and these solutions are therefore constant in spacetime. They are quite different from the *localized* solitons we seek.

At this point noncommutativity comes to the rescue again.[13] It is instructive to consider the equation

$$\phi * \phi = \phi \ . \tag{12}$$

The constant function $\phi = 1$ is the only solution, if $*$ is treated as the ordinary multiplication; but the noncommutative *-product involves infinitely many derivatives so the equation is actually a differential equation which may have nontrivial solutions. Indeed, there are many solutions;[13] the simplest is the Gaussian

$$\phi_0 = 2^{12} e^{-r^2/\theta} \ , \tag{13}$$

where $r^2 = (x^2)^2 + \cdots + (x^{25})^2$. Solutions to the equation $\phi * \phi = \phi$ are useful because functionals act on such functions in a simple way. For any functional f that can be expanded as a power series we have

$$f(a\phi) = \sum_{k=0}^{\infty} c_k a^k \phi^k = f(0) + [f(a) - f(0)]\phi \ , \tag{14}$$

where a is an ordinary number. As a result of this property the equations of motion become algebraic for *ansätze* built on solutions to $\phi * \phi = \phi$. It is therefore straightforward to find nontrivial solutions to the equations of motion.

4. The String Solutions

Let us consider some simple string solutions obtained this way. The simplest possibility is to take

$$t = t_* \phi_0 \ , \tag{15}$$

and other fields vanishing. Here t_* is chosen as the field at the perturbative extremum of the potential so that[13]

$$V'(t_* \phi_0) = V'(t_*)\phi_0 = 0 \ . \tag{16}$$

At large distances $\phi_0 \to 0$ so $t \to 0$ in the solution (15). In our conventions this corresponds to the nonperturbative vacuum. The solution (15) is interpreted as a D-string and discussed in more detail in Ref. 1.

It is simple to verify that

$$t = t_* \phi_0 \ ; \quad E = p\phi_0 \ , \tag{17}$$

satisfies (10-11) as well. Again, the solution is essentially the Gaussian (13); it is therefore fully localized and asymptotes the closed string vacuum at infinity. The tension of the string soliton is determined from (9) as

$$T = \frac{1}{2\pi\alpha'} \sqrt{\frac{1}{g_s^2} + p^2} \ . \tag{18}$$

This result suggests that the solitonic string can be identified with the $(p, 1)$ string, *i.e.* the bound state of p fundamental strings and a D-string. Repeating the computation starting with other solutions to $\phi * \phi = \phi$ we find more general string solutions which can be interpreted as (p, q) strings with $q > 1$. The main goal is to find a solution with precisely the tension of the fundamental string, without any D-branes present. This seems to require a separate consideration. A candidate string solution with the correct tension $T = \frac{1}{2\pi\alpha'}$ is

$$t = 0 \quad , \quad E = p\phi_0 \ . \tag{19}$$

Note that in each case discussed above the tension agrees *exactly* with the one known from perturbative string theory, even though the theory may not be supersymmetric.

The identification of (19) with the fundamental string is not entirely unproblematic. One issue is that the fundamental string tension is independent of g_s. This is puzzling because the action (1) depends on the coupling only through an overall factor $S \propto V(t) \propto 1/g_s$. The key feature that makes this possible is that $V(t) = 0$ in (19); this invalidates a simple scaling argument for the energy. The situation is similar to that of a massless particle with $V(t)$ playing the role of mass: the Lagrangian degenerates but the Hamiltonian presents no subtleties. Even though the tension computation is thus technically sound there is a cause for concern: $V(t) \to 0$ suggests that the effective coupling of the problem diverges, making quantum corrections important. Our understanding is that the correct loop counting parameter actually stays well-behaved so that quantum corrections are under control; however, this point deserves closer scrutiny.

Let us consider another issue. The solution (19) has the correct tension and electric flux to be identified with the fundamental string. The problem is that many other solutions have the same properties. Roughly speaking the equations of motion do not constrain the transverse profile of the solution at all. Specifically, if ϕ_k denotes a complete basis of solutions to (12) then the configurations

$$E = \sum_{k=0}^{\infty} a_k \phi_k \quad ; \quad a_k \in \mathbb{R} \ , \tag{20}$$

all satisfy the classical equations of motion and could potentially be interpreted as the fundamental string.

General quantum properties of the underlying gauge theory improve the situation somewhat by requiring the a_k integral. Even so, countably infinite candidate fundamental strings remain. Fortunately this is not the end of the story: there is no conserved quantity preventing these configurations from mixing quantum mechanically. In fact standard maximally supersymmetric D-brane dynamics would provide a unique quantum ground state. Unfortunately the situation is more involved here and the quantum problem cannot be analyzed precisely, but it may again have a unique ground state. The quantum problem deserves a better understanding.

In this talk we consider only the classical problem. Then the profile of the fundamental string is undetermined, it can spread out without violating any conservation laws. This is probably the correct result for infinite flux-tubes. To show flux confinement we need to add electric sources in bulk and show their flux escapes as a tube, rather than spreading out like a Coulomb field. It is possible that this can be understood already classically, as in Ref. 14, but we have not done so.

5. The Operator Formalism and Finite Noncommutativity

The presentation of the solutions above can be streamlined and the results significantly strengthened by introducing the operator formalism. The idea is to exploit the analogy between noncommutative geometry and the more familiar noncommutativity in quantum mechanics. The precise map between the two problems associates an operator $\hat{A}$ to each function A on the noncommutative space, and maps the noncommutative product $A * B$ to the more familiar operator product $\hat{A}\hat{B}$ in Hilbert space. In this way problems in noncommutative geometry translate into standard exercises in quantum mechanics.

Consider as an example the key equation $\phi * \phi = \phi$. In the operator formalism it reads $\hat{\phi}^2 = \hat{\phi}$ and therefore its solutions are simply the projection operators in Hilbert space. It is now clear that the equation has numerous solutions, indeed infinitely many. The soliton solutions presented in section 4 are thus essentially projection operators.

The operator formalism makes a huge symmetry manifest. Indeed, physics is left invariant under unitary transformations of operators and states

$$|\psi\rangle \to U|\psi\rangle \;\; ; \;\; \langle\psi| \to \langle\psi|U^\dagger \;\; ; \;\; \mathcal{O} \to U\mathcal{O}U^\dagger \; , \tag{21}$$

where

$$UU^\dagger = U^\dagger U = I \; . \tag{22}$$

These symmetries form the group $U(\mathcal{H})$. This points to a potential embarrassment because it shows that any soliton solution in the theory has infinitely many "images" under $U(\mathcal{H})$. In many situations we want a unique soliton, to be identified with its counterpart in closed string theory. The crucial observation is that the $U(\mathcal{H})$ is in fact a *gauge symmetry*.[15] The images under $U(\mathcal{H})$ are therefore not interpreted as distinct, but as gauge equivalent representations of a single physical state. In superstring theory the gauge symmetry is further enhanced by an infinite discrete group which removes certain tensionless solitons with no reasonable physical interpretation.[16]

A key step in the discussion of section 3 was taking the limit $\theta/\alpha' \to \infty$ in order to justify neglecting derivative terms. We are now in a position to present an alternative argument, valid at any θ.[17] The objectionable derivative terms are in fact all *gauge covariant* derivatives under the $U(\mathcal{H})$ symmetry. We can therefore imagine adjusting the gauge fields in the solution precisely such that the gauge covariant derivatives vanish identically, removing the need for neglecting them. That this is always possible relies on solution generating transformations of the form (21) but with (22) replaced by

$$UU^\dagger = I \;\; , \;\; U^\dagger U = I - P \; , \tag{23}$$

where P is some projection operator. We can therefore repeat the construction of solitons for finite θ.[17] This result is not surprising: in the closed string vacuum different values of the B-field are in fact gauge equivalent and it was therefore expected that vacua with different values of B are related. The argument above shows how this works in the open string variables by representing the noncommutative solitons as "almost" gauge equivalent to vacuum, at any θ.

6. Fluctuations

The noncommutative Born-Infeld type action (8) also describes long-wave length fluctuations depending on the commutative directions x^0, x^1. Allowing for these, the Hamiltonian (9) is replaced by

$$H = \int d^{25}x \left[\sqrt{E^\alpha M_{\alpha\beta} E^\beta + V(t)^2 \det(I+F)} + A_0 \nabla_\alpha E^\alpha \right] , \tag{24}$$

where

$$M_{\alpha\beta} = \delta_{\alpha\beta} - F_{\alpha\gamma} F^\gamma{}_\beta \ . \tag{25}$$

Using $t = 0 \Rightarrow V(t) = 0$ (closed string vacuum), $A_0 = 0$ (gauge condition), $F_{ij} = 0, F_{1i} = A_i'$ (derivatives negligible in NC directions) we find

$$H = \int d^{25}x \sqrt{(E^1)^2 (1 + (\vec{A}')^2) + \vec{E}^2 + (\vec{E} \cdot \vec{A}')^2} \ , \tag{26}$$

where prime denotes the spatial derivative along the string and the vector notation refers to the transverse coordinates. The *ansatz* for a fluctuating string is

$$\begin{aligned} \phi &= \phi_0(x^i - f^i(x^0, x^1)) \ , & (27) \\ E^1 &= p\phi_0 \ , & (28) \\ \vec{E} &= \vec{e}\phi_0 \ , & (29) \\ \vec{A}' &= \vec{a}' \phi_0 \ . & (30) \end{aligned}$$

We would like to find the effective action in $D = 1+1$ dimensions controlling the functions f^i. The Hamiltonian reduction procedure accomplishes this, with the result

$$H = \int dx^1 \sqrt{1 + \vec{\pi}^2 + (\vec{f}')^2 + (\vec{\pi} \cdot \vec{f}')^2} \ . \tag{31}$$

It is easy to compute the corresponding Lagrangian. In static gauge $X^\mu = (x^0, x^1, f^i)$ one finds

$$L = -\int d^2x \sqrt{(\dot{\vec{X}})^2 (\vec{X}')^2 - (\dot{\vec{X}} \cdot \vec{X}')^2} \ . \tag{32}$$

This shows that *the effective action of long wave length fluctuations is the Nambu-Goto action with the correct tension!* The spectrum of the soliton is therefore precisely the same as for a fundamental string. If we take the action (32) seriously and quantize it we find *very* light excitations propagating along the string, including the graviton and even the closed string tachyon. These would appear here as collective excitations in open string field theory, a fascinating result. There is a standard objection against this line of reasoning: it is only the lightest objects in a theory that can be quantized and solitons must therefore usually be treated classically. This objection fails here because, unlike in many superficially similar computations, the fundamental string soliton is indeed the lightest excitation of the closed string vacuum, and therefore subject to quantization. It is nevertheless unjustified to trust the present computation beyond the long-wave length approximation: we are using here a crude effective action that only takes into account constant fields. It is justified to neglect derivatives in the transverse directions, because they are noncommutative, but fluctuations in the spatial and temporal directions remain. The effective action therefore applies only for long wave lengths satisfying $\sqrt{\alpha'}F'_{\mu\nu} \ll F_{\mu\nu}$. It is reasonable to expect that, in a better description, the fluctuations of the closed strings can justifiably be interpreted in terms of open string variables. The discussion of θ-dependence in section 5 may be a step in this direction.

7. Summary

We conclude with a brief summary.

- The starting point was the process of tachyon condensation in open string field theory, from the standard perturbative vacuum to the nonperturbative vacuum, identified with the closed string vacuum. Our interest was in soliton excitations of the nonperturbative vacuum.

- Noncommutativity was introduced in the system because it generates a new scale. In the limit of large noncommutativity it is justified to ignore higher derivatives and use a simple effective field theory description of the open string field theory. An argument for ignoring the derivatives even at finite θ was presented in section 5.

- An explicit construction of solitons identified with (p, q) strings follow. Solitons with the exact tension and long wave length fluctuations of fundamental strings were similarly constructed. These are identified with the closed fundamental strings expected in the nonperturbative vacuum.

- Several problems remain in the identification of noncommutative solitons and fundamental strings: the quantum properties of the effective gauge dynamics need better understanding, confinement must be elucidated already in the classical description, and the spectrum of fluctuations must be computed beyond the long wave length approximation.

Acknowledgments

I am grateful to J. Harvey, P. Kraus, and E. Martinec for collaboration yielding the results reported here, among others. This work was supported by DOE grant DE-FG0290ER-40560 and by a Robert R. McCormick fellowship.

References

1. J.A. Harvey, P. Kraus, F. Larsen and E.J. Martinec, *D-branes and strings as non-commutative solitons*, *JHEP* **0007**, 042 (2000) [hep-th/0005031].
2. A. Sen, *Descent Relations Among Bosonic D-branes*, *Int. J. Mod. Phys.* **A14**, 4061 (1999) [hep-th/9902105].
3. V.A. Kostelecky and S. Samuel, *On a Nonperturbative Vacuum for the Open Bosonic String*, *Nucl. Phys.* **B336**, 263 (1990); A. Sen and B. Zwiebach, *Tachyon Condensation in String Field Theory*, *JHEP* **0003**, 002 (2000) [hep-th/9912249]; N. Moeller and W. Taylor, *Level truncation and the tachyon in open bosonic string field theory*, *Nucl. Phys.* **B583**, 105 (2000) [hep-th/0002237].
4. P. Yi, *Membranes from Five-Branes and Fundamental Strings from Dp-Branes*, *Nucl. Phys.* **B550**, 214 (1999) [hep-th/9901159]; O. Bergman, K. Hori and P. Yi, *Confinement on the brane*, *Nucl. Phys.* **B580**, 289 (2000) [hep-th/0002223].
5. A. Sen, *Universality of the tachyon potential*, *JHEP* **9912**, 027 (1999) [hep-th/9911116].
6. J.A. Harvey and P. Kraus, *D-branes as unstable lumps in bosonic open string field theory*, *JHEP* **0004**, 012 (2000) [hep-th/0002117].
7. A.A. Gerasimov and S.L. Shatashvili, *On exact tachyon potential in open string field theory*, *JHEP* **0010**, 034 (2000) [hep-th/0009103]; D. Kutasov, M. Marino and G. Moore, *Some exact results on tachyon condensation in string field theory*, *JHEP* **0010**, 045 (2000) [hep-th/0009148]; *Remarks on Tachyon Condensation in Superstring Field Theory*, hep-th/0010108.
8. E. Witten, *On background independent open string field theory*, *Phys. Rev.* **D46**, 5467 (1992) [hep-th/9208027]; S.L. Shatashvili, *Comment on the background independent open string theory*, *Phys. Lett.* **B311**, 83 (1993) [hep-th/9303143].
9. A. Connes, M.R. Douglas and A. Schwarz, *Noncommutative geometry and matrix theory: Compactification on tori*, *JHEP* **9802**, 003 (1998) [hep-th/9711162]; V. Schomerus, *D-branes and deformation quantization*, *JHEP* **9906**, 030 (1999) [hep-th/9903205].
10. N. Seiberg and E. Witten, *String theory and noncommutative geometry*, *JHEP* **9909**, 032 (1999) [hep-th/9908142].
11. K. Dasgupta, S. Mukhi and G. Rajesh, *Noncommutative tachyons*, *JHEP* **0006**, 022 (2000) [hep-th/0005006].
12. E. Witten, *Noncommutative tachyons and string field theory*, hep-th/0006071.
13. R. Gopakumar, S. Minwalla and A. Strominger, *Noncommutative solitons*, *JHEP* **0005**, 020 (2000) [hep-th/0003160].
14. J. Kogut and L. Susskind, *Vacuum Polarization And The Absence Of Free Quarks In Four-Dimensions*, *Phys. Rev.* **D9**, 3501 (1974).
15. L. Cornalba, *D-brane physics and noncommutative Yang-Mills theory*, hep-th/9909081; N. Ishibashi, *A relation between commutative and noncommutative descriptions of D-branes*, hep-th/9909176.
16. J.A. Harvey, P. Kraus and F. Larsen, *Tensionless branes and discrete gauge symmetry*, hep-th/0008064.
17. J.A. Harvey, P. Kraus and F. Larsen, *Exact noncommutative solitons*, hep-th/0010060.

VERTEX OPERATORS FOR THE SUPERMEMBRANE AND BACKGROUND FIELD MATRIX THEORY

JAN PLEFKA
Albert-Einstein-Institut, Max-Planck-Institut für Gravitationsphysik
Am Mühlenberg 1, D–14476 Golm, Germany

We derive the vertex operators that are expected to govern the emission of the massless $d = 11$ supermultiplet from the supermembrane in the light cone gauge. Our results immediately imply the linear coupling of matrix theory to an arbitrary supergravity background to all orders in anticommuting coordinates. Finally we address the definition of n-point tree level and one-loop scattering amplitudes. The resulting 3-point tree level amplitudes turn out to agree with $d = 11$ supergravity and are completely fixed by supersymmetry and the existence of a normalizable ground state.

1. Introduction

One of the most pressing question in string theory to date is as to what the precise microscopic degrees of freedom of M theory are. On the face of it the 11 dimensional supermembrane[1] appears as a natural candidate for M theory, as it sits atop of the main contenders for a unified theory of quantum gravity: supergravity,[2] superstrings[3] and matrix theory,[4] which all are obtained through certain limits of the membrane model. This is obvious for the case of eleven dimensional supergravity, where one simply discards all internal excitations of the membrane, being left with a first quantized description of supergravity in form of the 11 dimensional superparticle. Similarly one reaches type IIA superstrings at the kinematic level through a procedure called double-dimensional reduction.[5] And finally matrix theory, a proposed candidate for light cone M theory, emerges as a finite N regularization of the light cone supermembrane. Despite these features the study of the fundamental supermembrane has not received much attention caused by the tremendous difficulties one encounters once one turns to a quantization of the model. Opposed to the particle and the string the membrane is a nonlinear interacting field theory, with a priori no well defined perturbative scheme in form of a sum over worldvolumes. Moreover, the model possesses a continuous spectrum, which, following the insights from the matrix theory proposal, should be interpreted as a second quantized feature. In order to make progress we believe that a starting point for a quantum treatment of the supermembrane should be to clarify what the sensible quantities or operators are whose expectation values one would like to compute. We here want to push forward the concept of membrane vertex operators, which are expected to govern the emission of the massless d=11 supermultiplet from the membrane worldvolume. With these we are able to *define* scattering amplitudes in membrane theory, thus making the supermembrane more "computable". Moreover, these operators immediately translate into the corresponding objects in matrix theory, thus yielding the linear order background field coupling of matrix theory to all orders in fermions. We shall show that the resulting 3-point amplitudes agree with d=11 supergravity, and comment at ongoing work to compute the membrane four graviton amplitude, which yields the famous $\mathcal{R}^4$ quantum correction to the d=11 supergravity action.

In analogy to superparticle and superstring theory the supermembrane vertex operators are naturally defined as the linear coupling of the supergravity background fields $g_{\mu\nu}, C_{\mu\nu\rho}$ and ψ^α_μ to the embedding coordinates $X^\mu(\xi)$ and $\theta_\alpha(\xi)$, where ξ parameterizes the three dimensional membrane world-volume. In principle these operators are deducible from the known background field action of the supermembrane in superspace[1]

$$S = \int d^3\xi \sqrt{-g[Z(\xi)]} + \tfrac{1}{6}\epsilon^{ijk}\,\pi_i^A\pi_j^B\pi_k^C\,B_{CBA}[Z(\xi)] \tag{1}$$

where E_M^A denotes the super-dreiundvierzig-bein, B_{CBA} the super three-form and we have $Z^A = (X^\mu(\xi), \theta^\alpha(\xi))$, $\pi_i^A = \frac{\partial Z^M}{\partial \xi^i}\,E_M^A$ and $g_{ij} = \pi_i^r\,\pi_j^s\,\eta_{rs}$. Obtaining the linearized action in components from (1), however, is a highly nontrivial task. To date this has only been achieved up to second order in anticommuting coordinates θ [6] through a process called gauge completion,[7] i.e.

$$\begin{aligned}
E^r_\mu &= e^r_\mu + 2\,\bar\theta\Gamma^r\psi_\mu + \bar\theta\Gamma^r\Big[-\tfrac{1}{4}\,\hat\omega_\mu\cdot\Gamma + T_\mu\cdot\hat F\Big]\theta + \ldots + O(\theta^{32})\\
E^\alpha_\mu &= \psi^a_\mu - \tfrac{1}{4}\,\hat\omega_\mu{}^{rs}\,(\Gamma_{rs}\theta)^\alpha + (T_\mu\cdot\hat F\,\theta)^\alpha + \ldots + O(\theta^{32})\\
B_{\mu\nu\rho} &= C_{\mu\nu\rho} - 6\,\bar\theta\Gamma_{[\mu\nu}\psi_{\rho]} - 3\,\bar\theta\Gamma_{[\mu\nu}\Big[-\tfrac{1}{4}\,\hat\omega_{\mu]}\cdot\Gamma + T_{\mu]}\cdot\hat F\Big]\theta\ldots + O(\theta^{32})
\end{aligned}$$

Note that these expansions in principle extend all the way up to order 32 in θ's. We hence see that obtaining the complete expressions for the *covariant* vertex operators via superspace seems to be beyond reach.[a] This motivates the transition to a light cone description of the supermembrane in general backgrounds for which one expects a substantial degree of simplification, as may be seen already the level of the flat space action. We shall see that here the θ expansion terminates at order four.

Imposing the light cone condition $X^+ = p^+\,\tau$ and $\Gamma^+\Theta = 0$ by κ-symmetry the flat space action of the supermembrane reads[9]

$$\mathcal{L} = \tfrac{1}{2}(DX^a)^2 - \tfrac{1}{4}\{X^a, X^b\}^2 - i\theta\,D\theta - i\theta\,\gamma_a\{X^a,\theta\} \tag{2}$$

where $X^a = X^a(\tau,\sigma_1,\sigma_2)$ and $\theta^\alpha = \theta^\alpha(\tau,\sigma_1,\sigma_2)$ denote the transverse ($a = 1,\ldots 9$, $\alpha = 1,\ldots,16$) coordinates. We have used the covariant derivative $DX^a := \partial_\tau X^a - \{\omega, X^a\}$ along with the Poisson bracket $\{A,B\} := \epsilon^{ij}\,\partial_{\sigma_i}A\,\partial_{\sigma_j}B$. The gauge field ω entering the covariant derivative D stems from a residual symmetry of area preserving diffeomorphisms[10]

$$\delta X^a = \{\xi, X^a\}, \qquad \delta\theta = \{\xi,\theta\}, \qquad \delta\omega = \partial_\tau + \{\xi,\omega\} \tag{3}$$

[a] There exists an alternative and potentially more effective procedure via a normal coordinate expansion in superspace,[8] which has so far not been applied to the eleven dimensional case.

The action (2) is moreover invariant under the set of supersymmetry transformations

$$\begin{aligned} \delta X^a &= -2\,\epsilon\,\gamma^a\theta \qquad\qquad \delta\omega = -2\,\epsilon\,\theta \\ \delta\theta &= i\,DX^a\,\gamma_a\,\epsilon - \tfrac{i}{2}\{X^a, X^b\}\,\gamma_{ab}\,\epsilon + \eta \end{aligned} \tag{4}$$

with 32 components, 16 from η and 16 from ϵ, a remnant of the 11 dimensional origin of the model. Viewing the parameters (σ_1, σ_2) as internal degrees of freedom the light cone supermembrane may be understood as a supersymmetric quantum mechanical system equipped with an infinite dimensional gauge group of area preserving diffeomorphisms.

2. Vertex Operator Construction

The graviton, three form and gravitino emission operators that we seek to construct take the general form

$$\begin{aligned} V_h &= h_{ab} \int d\tau d^2\sigma\, \mathcal{O}^{ab}[X^a(\tau,\sigma_i), \theta(\tau,\sigma_i)]\, e^{ik\cdot X + ik_+ X^+} \\ V_C &= C_{abc} \int d\tau d^2\sigma\, \mathcal{O}^{abc}[X^a, \theta]\, e^{ik\cdot X + ik_+ X^+} \\ V_\psi &= \psi_a^\alpha \int d\tau d^2\sigma\, \mathcal{O}_\alpha^a[X^a, \theta]\, e^{ik\cdot X + ik_+ X^+} \end{aligned} \tag{5}$$

with the polarizations $(h_{ab}, C_{abc}, \psi_a^\alpha)$ and momenta k_a and k_+. The polarizations are subject to on-shell constraints, e.g. $k^a\, h_{ab} = h_{aa} = 0 = k^a\, C_{abc}$.[b] The strategy for the explicit construction of the operators $\mathcal{O}^{ab}$, $\mathcal{O}^{abc}$ and $\mathcal{O}_\alpha^a$ is rather simple. The (unknown) full background field action of the supermembrane transforms covariantly under supersymmetry as

$$\delta\mathcal{L}_{\text{full}}[X, \theta; h, C, \psi] = \mathcal{L}_{\text{full}}[X, \theta; \hat{\delta}h, \hat{\delta}C, \hat{\delta}\psi] \tag{6}$$

where δ denotes the supersymmetry variation of X^a, θ and ω of (4), whereas $\hat{\delta}$ are the induced light-cone supergravity variations of the background fields h_{ab}, C_{abc} and ψ_a^α whose precise expressions may be found in Refs. 11,12. Hence the vertex operators, being the linear background field couplings, must transform into each other under supersymmetry according to

$$\begin{aligned} \delta V_h &= V_{\hat{\delta}\psi[h]} \\ \delta V_C &= V_{\hat{\delta}\psi[C]} \\ \delta V_\psi &= V_{\hat{\delta}h} + V_{\hat{\delta}C} \end{aligned} \tag{7}$$

[b] A derivation of this may be found in Ref. 11. Note that in the above we are suppressing all terms coupling to k_-, thus effectively setting $k_- = 0$ in order to decouple X^- in the expressions, a complicated function in X^a and X^+. This is standard practice in light cone string theory, albeit questionable, as with it all transverse momenta become complex. A cleaner way to state this point is that one consistently ignores all terms coupling to k_- in the computations, assuming that they work out by themselves.

So for example under the linear part of the supersymmetry transformations of (4) $\delta X^a = 0$ and $\delta\theta = \eta$ the vertex operator V_ψ transforms into a sum of the graviton and three-form vertices, whose polarizations are then given by $\delta h_{ab} = -\tilde{\psi}_{(a}\gamma_{b)}\eta$ and $\delta C_{abc} = \frac{3}{2}\tilde{\psi}_{[a}\gamma_{bc]}\eta$ respectively, the linearized form of the supergravity transformations parametrized by the same η entering (4).

It turns out that this requirement of covariance under supersymmetry completely determines the form of (5). The results read[12]

$$\begin{aligned}
V_h &= h_{ab}\Big[DX^a\,DX^b - \{X^a,X^c\}\,\{X^b,X^c\} - i\theta\gamma^a\,\{X^b,\theta\} \\
&\quad -2DX^a\,R^{bc}\,k_c - 6\{X^a,X^c\}\,R^{bcd}\,k_d + 2R^{ac}\,R^{bd}\,k_c\,k_d\Big] \\
V_C &= -C_{abc}\,DX^a\,\{X^b,X^c\} + F_{abcd}\Big[(DX^a - \tfrac{2}{3}R^{ae}\,k_e)\,R^{bcd} \\
&\quad -\tfrac{1}{2}\{X^a,X^b\}\,R^{cd} - \tfrac{1}{96}\,\{X^e,X^f\}\,\theta\gamma^{abcdef}\theta\Big] \\
V_\psi &= \psi_a\Big[\Big(DX^a - 2R^{ab}\,k_b + \gamma_c\,\{X^c,X^a\}\Big)\theta\Big] \\
&\quad +\tilde{\psi}_a\Big[\gamma\cdot DX\,\Big(DX^a - 2R^{ab}\,k_b + \gamma_c\{X^c,X^a\}\Big)\,\theta \\
&\quad +\tfrac{1}{2}\gamma_{bc}\,\{X^b,X^c\}\,(DX^a - \{X^a,X^d\}\,\gamma^d\,)\theta \\
&\quad +8\gamma_b\theta\,\{X^b,X^c\}\,R^{cad}\,k_d + \tfrac{5}{3}\gamma_{bc}\theta\,\{X^b,X^c\}R^{ad}\,k_d \\
&\quad +\tfrac{4}{3}\gamma_{bc}\theta\,\Big(\{X^a,X^b\}\,R^{cd} + \{X^c,X^d\}R^{ab}\Big)k_d \\
&\quad +\tfrac{2}{3}i\,\Big(\gamma_b\theta\,\{X^a,\theta\}\gamma^b\theta - \theta\,\{X^a,\theta\}\theta\Big) \\
&\quad +\tfrac{8}{9}\gamma^b\theta\,R^{ac}\,R^{bd}\,k_c k_d\Big]
\end{aligned} \tag{8}$$

where $R^{ab} = \frac{1}{4}\theta\gamma^{ab}\theta$ and $R^{abc} = \frac{1}{12}\theta\gamma^{abc}\theta$, suppressing the overall $\exp(ik\cdot X)$ of (5). The vertices for $h_{+a}, h_{++}, C_{+ab}, \psi_+$ and $\tilde{\psi}_+$ are also known; ψ and $\tilde{\psi}$ are the light-cone decompositions of the gravitino.[11]

These results are subject to three stringent consistency checks. Firstly, the vertex operators are invariant under the following background field symmetries:

$$\begin{aligned}
\delta h_{ab} &= k_{(a}\,\xi_{b)} && \text{(coordinate transformations)} \\
\delta C_{abc} &= k_{[a}\,\xi_{bc]} && \text{(tensor gauge transformations)} \\
\delta\psi_a &= k_a\epsilon && \text{(field independent SUSY)}
\end{aligned}$$

Secondly, the point particle limit of the membrane vertices (8), which amounts to simply dropping the terms involving the Poisson brackets $\{.,.\}$ yields the d=11 superparticle vertex operators of Green, Gutperle and Kwon.[11]

Finally, a stronger check is to perform a double dimensional reduction[5] of the membrane vertices, which should reduce them to the type IIA superstring vertex operators. This reduction procedure is performed by wrapping the σ_2 coordinate of the membrane around a target space circle along X^9

$$X^a(\tau,\sigma_1,\sigma_2) \rightarrow \begin{pmatrix} X^i(\tau,\sigma_1) \\ X^9 = \sigma_2 \end{pmatrix} \qquad i = 1,\ldots,8 \tag{9}$$

along with

$$\theta_\alpha \to \begin{pmatrix} S_a(\tau,\sigma_1) \\ \tilde{S}_{\dot{a}}(\tau,\sigma_1) \end{pmatrix} \qquad a,\dot{a}=1,\ldots,8 \tag{10}$$

Quite remarkably under this description the vertex operators of (8) factorize into left and right moving contributions. Let us demonstrate this for the graviton vertex reduction. Under (9) and (10) we have

$$\{X^i,X^j\}=0 \qquad \{X^i,X^9\}=\partial_{\sigma_1}X^i \qquad \omega=0$$
$$\theta\gamma^{ij}\theta = S\Gamma^{ij}S+\tilde{S}\Gamma^{ij}\tilde{S} \qquad \theta\gamma^{ij9}\theta = S\Gamma^{ij}S-\tilde{S}\Gamma^{ij}\tilde{S} \tag{11}$$

where Γ^i are the standard SO(8) Γ-matrices. Then

$$\begin{aligned} V_h|_{\text{DDR}} \quad &\to \quad h_{ij}\Big[\partial_0X^i\partial_0X^j-\partial_1X^i\partial_1X^j-\tfrac{1}{2}\partial_0X^i(S\Gamma^{jm}S \\ &\qquad +\tilde{S}\Gamma^{jm}\tilde{S})k_m+\tfrac{1}{2}\partial_1X^i(S\Gamma^{jm}S-\tilde{S}\Gamma^{jm}\tilde{S})k_m\ +\tfrac{1}{4}S\Gamma^{im}S\,\tilde{S}\Gamma^{jn}\tilde{S}k_mk_n\Big] \\ &= \quad h_{ij}\Big(\partial_+X^i-\tfrac{1}{2}S\Gamma^{im}Sk_m\Big)\Big(\partial_-X^j-\tfrac{1}{2}\tilde{S}\Gamma^{jn}\tilde{S}k_n\Big) \end{aligned}$$

which is nothing but the IIA graviton vertex.

3. Matrix Theory in Background Fields

The obtained results may be directly translated to matrix theory, which emerges from a supersymmetry preserving "discretization" of the membrane spacesheet. For this one replaces the infinite dimensional gauge group of area preserving diffeomorphisms by the large N limit of SU(N).[10] This well known prescription here amounts to the replacements

$$X^a(\tau,\sigma^1,\sigma^2)\to \mathbf{X}^a_{mn}(\tau) \qquad \theta^\alpha(\tau,\sigma^1,\sigma^2)\to \mathbf{\Theta}^\alpha_{mn}(\tau) \qquad m,n=1,\ldots,N$$
$$\{.,.\}\to i\,[.,.] \qquad \tfrac{1}{4\pi}\int d^2\sigma(\ldots)\to\tfrac{1}{N}\,\mathrm{STr}[\ldots] \tag{12}$$

STr denotes the symmetrized trace whose introduction becomes necessary due to ordering ambiguities for the composite operators that we have been discussing. The STr prescription guarantees that all manipulations performed in section 2 for the continuous membrane model go through for the matrix model as well. In principle there may exist a less symmetric ordering that also works, but the differences to the STr prescription will be subleading in N.

Hence the weak (i.e. linear coupling) background field action of matrix theory is now known to *all* orders in $\mathbf{\Theta}$ and derivatives $\partial/\partial\mathbf{X}^a$

$$\mathcal{S}_{\text{MT}}=\int d\tau\Big(\mathcal{L}_0+V_{h(X)}+V_{C(X)}+V_{\Psi(X)}\Big) \tag{13}$$

where e.g. the graviton coupling takes the form

$$\begin{aligned} V_{h(X)} \quad = \quad & \mathrm{STr}\Big[\Big\{\dot{\mathbf{X}}^a\dot{\mathbf{X}}^b+[\mathbf{X}^a,\mathbf{X}^c]\,[\mathbf{X}^b,\mathbf{X}^c]++\mathbf{\Theta}\gamma^a\,[\mathbf{X}^b,\mathbf{\Theta}] \\ & -\tfrac{1}{2}\dot{\mathbf{X}}^a\,(\mathbf{\Theta}\gamma^{bc}\mathbf{\Theta})\,\tfrac{\partial}{\partial\mathbf{X}^c}-\tfrac{1}{2}i[\mathbf{X}^a,\mathbf{X}^c]\,(\mathbf{\Theta}\gamma^{bcd}\mathbf{\Theta})\,\tfrac{\partial}{\partial\mathbf{X}^d} \\ & +2(\mathbf{\Theta}\gamma^{ac}\mathbf{\Theta})\,(\mathbf{\Theta}\gamma^{bd}\mathbf{\Theta})\frac{\partial}{\partial\mathbf{X}^c}\,\frac{\partial}{\partial\mathbf{X}^d}\Big\}\,h_{ab}(\mathbf{X})\Big] \end{aligned} \tag{14}$$

Note that we have performed a Fourier transformation of (8) to target configuration space. These results agree with and go beyond the explicit matrix theory current calculations of Taylor and V.Raamsdonk,[13] which so far have been performed up to order $o(\theta^2)$ and linear order in $\partial/\partial \mathbf{X}^a$.

4. Scattering Amplitudes

We now turn to the discussion of three point tree level scattering amplitudes. For this it is advantageous to work in the framework of the finite N matrix theory. In order to define a tree level amplitude we first split off the center of mass degrees of freedom of the matrices by writing

$$\mathbf{X}^a = x^a \, \mathbb{1} + \hat{\mathbf{X}}^a \qquad \mathbf{\Theta}^a = \theta^a \, \mathbb{1} + \hat{\mathbf{\Theta}}^a \tag{15}$$

with traceless matrices $\hat{\mathbf{X}}^a$ and $\hat{\mathbf{\Theta}}$. An asymptotic 1-graviton state in matrix theory is then given by

$$|\mathrm{IN}\rangle\rangle = |k_1, h_1\rangle_{x,\theta} \otimes |\mathrm{GS}\rangle^{\mathrm{SU(N)}}_{\hat{\mathbf{X}},\hat{\Theta}} \tag{16}$$

where $|k_1, h_1\rangle$ is the graviton state of the superparticle[14,11] and $|\mathrm{GS}\rangle$ denotes the *exact* SU(N) normalized zero energy groundstate, whose explicit form is unknown but is known to exist.[15] The tree level three point amplitude is then defined by

$$\mathcal{A}_{\text{3-point}} = \langle\langle \, 1 \, | \, V_2 \, | \, 3 \, \rangle\rangle \tag{17}$$

where one inserts the graviton vertex operator

$$V_2 = h^{(2)}_{ab} \, \tfrac{1}{N} \, \mathrm{STr}\Big[(p^a \, p^b + 2 \, p^a \, \hat{\mathbf{P}}^b + \hat{\mathbf{P}}^a \, \hat{\mathbf{P}}^b + [\hat{\mathbf{X}}^a, \hat{\mathbf{X}}^c] \, [\hat{\mathbf{X}}^b, \hat{\mathbf{X}}^c] \,) \, e^{ik\cdot\hat{\mathbf{X}}}\Big] \, e^{ik\cdot x} + \text{fermions} \tag{18}$$

One may wonder how one could ever evaluate (17) upon inserting (18) without the knowledge of $|\mathrm{GS}\rangle$. The first contribution to (17) takes the form

$$\langle k_1, h_1 | p^a \, p^b e^{ik\cdot x} | k_3, h_3 \rangle \, h^{(2)}_{ab} \, \langle \mathrm{GS} | \mathrm{STr} \, e^{ik\cdot\hat{\mathbf{X}}} | \mathrm{GS} \rangle \tag{19}$$

Now by SO(9) covariance $\langle \mathrm{GS} | \mathrm{STr} \, e^{ik\cdot\hat{\mathbf{X}}} | \mathrm{GS} \rangle = N$, as the only SO(9) scalar it could depend on would be k^2 which vanishes on shell. It must then be a constant which is fixed to be N by considering the $k^a \to 0$ limit. Remarkably the remaining two terms in (17) upon inserting (18) vanish by a combination of SO(9) covariance and on-shell arguments:

$$h_{ab} \, \langle \mathrm{GS} | \mathrm{STr} \, \hat{\mathbf{P}^b} \, e^{ik\cdot\hat{\mathbf{X}}} | \mathrm{GS} \rangle \sim k^b \, h_{ab} = 0 \tag{20}$$

$$h_{ab} \, \langle \mathrm{GS} | \mathrm{STr} \Big[(\hat{\mathbf{P}^a} \, \hat{\mathbf{P}^b} + [\hat{\mathbf{X}}^a, \hat{\mathbf{X}}^c] \, [\hat{\mathbf{X}}^b, \hat{\mathbf{X}}^c] \,) \, e^{ik\cdot\hat{\mathbf{X}}} \Big] | \mathrm{GS} \rangle \sim (k^a \, k^b + c \, \delta^{ab}) \, h_{ab} = 0$$

But as the first correlator in (19) is nothing but the bosonic contribution to the 3-point d=11 *superparticle* amplitude[11] and as the fermionic terms work out in a similar fashion we see that our 3-point tree level amplitude

$$\langle\langle \, 1 \, | \, V_2 \, | \, 3 \, \rangle\rangle = \langle \, 1 \, | \, V_2 \, | \, 3 \, \rangle_{x,\theta} \, \langle \mathrm{GS} | \mathrm{GS} \rangle \tag{21}$$

agrees with the 3-point amplitude of d=11 supergravity!

Clearly the next step would be to study n-point tree level amplitudes which should be given by

$$\mathcal{A}_{\text{n-point}} = \langle\langle\, 1\,|\, V_2\, \Delta\, V_3\, \Delta \ldots \Delta\, V_{n-1}\,|\, n\,\rangle\rangle \tag{22}$$

where Δ denotes the propagator $1/(\frac{1}{2}\, p_0{}^2 + \hat{\mathbf{H}})$ built from the *interacting* membrane Hamiltonian $\hat{\mathbf{H}}$. However, now we expect the details of the groundstate $|\mathrm{GS}\rangle$ to enter the computation. Developing some perturbative scheme for calculating (22) would be highly desirable, but is conceivably very complicated as it must involve an expansion in both the propagator Δ and the groundstate $|\mathrm{GS}\rangle$.

Instead we shall briefly comment on ongoing attempts to compute loop amplitudes within this scenario. Here, led by the formalism in light cone superstring and superparticle theory, we propose to define a membrane n-point one-loop amplitude by the expression

$$\mathcal{A}_{\text{1-loop, n-point}} = \int d^{11}p_0\, \mathbf{Tr}(\Delta\, V_1\, \Delta\, V_2\, \Delta\, V_3\, \Delta\, V_4\, \ldots \Delta\, V_n) \tag{23}$$

where the trace is over the Hilbert space of $\hat{\mathbf{H}}$. Again this appears as a daunting task, however the zero mode sector of (23) already yields some amount of information. In particular the trace over the fermionic zero mode θ of (15) tells us that all 2 and 3 particle amplitudes vanish at one loop, as at least four vertex operators ($\leq 16\ \theta$'s) are needed to saturate the fermion zero mode trace

$$\mathbf{Tr}(\theta_{\alpha_1} \ldots \theta_{\alpha_N})_\theta = \delta_{N,16}\, \epsilon^{\alpha_1 \ldots \alpha_{16}} \tag{24}$$

In the pure graviton sector the first non-vanishing amplitude is then the 4-graviton amplitude whose leading momentum dependence is given by

$$\mathcal{A}_{4h} = \epsilon^{\alpha_1 \ldots \alpha_{16}}\, \gamma^{a_1\, a_2}_{\alpha_1 \alpha_2} \ldots \gamma^{a_{15}\, a_{16}}_{\alpha_{15} \alpha_{16}}\, R^{(1)}_{a_1 a_2 a_3 a_4} \ldots R^{(4)}_{a_{13} a_{14} a_{15} a_{16}} \int d^{11}\, p_0\, \mathbf{Tr}'\, \Delta^4\,. \tag{25}$$

We thus see the emergence of the expected $\mathcal{R}^4$ term[16] in the kinematical sector, but it remains to be seen what can be said about the remaining trace. Potentially BPS arguments could here come to ones aid.[17]

5. Outlook

In this talk we have constructed the supermembrane vertex operators in the light cone gauge, which hopefully provide us with a new tool in the study of quantum M theory. We have demonstrated their reduction to the corresponding vertices of the d=11 superparticle and d=10 type IIA strings. Moreover they yield complete weak background field matrix theory action. It would be interesting to clarify their relevance for the related IKKT matrix model[18] and matrix string theory.[19] N-point tree level and 1-loop amplitudes were defined for the model and it was argued that the resulting 3-point tree level amplitudes agree with d=11 supergravity.

Clearly there remains a host of open questions. On the conceptual side one may ask where the multiparticle interpretation of the membrane/matrix theory emerges in the outlined formalism. After all our construction proceeds in complete analogy with the first quantized particle and string theories. On the technical side progress is clearly necessary for the evaluation of higher loop amplitudes, both at tree and one loop level.

Acknowledgments

The material presented in this talk was obtained in collaboration with A. Dasgupta and H. Nicolai.[12] I wish to thank the organizers of the Strings 2000 conference for a stimulating meeting and for giving me the opportunity to present these results.

References

1. E. Bergshoeff, E. Sezgin and P.K. Townsend, *Phys. Lett.* **B189**, 75 (1987); *Ann. Phys.* **185**, 330 (1988).
2. E. Cremmer, B. Julia and J. Scherk, *Phys. Lett.* **76B**, 409 (1978).
3. M.B. Green and J.H. Schwarz, *Phys. Lett.* **B136**, 367 (1984); *Nucl. Phys.* **B127**, 285 (1984).
4. T. Banks, W. Fischler, S.H. Shenker and L. Susskind, *Phys. Rev.* **D55**, 5112 (1997) [hep-th/9610043].
5. M.J. Duff, P.S. Howe, T. Inami and K.S. Stelle, *Phys. Lett.* **191B**, 70 (1987).
6. B. de Wit, K. Peeters and J. Plefka, *Nucl. Phys.* **B532**, 99 (1998) [hep-th/9803209].
7. E. Cremmer and S. Ferrara, *Phys. Lett.* **B91**, 384 (1980).
8. M. Grisaru, M. Knutt-Wehlau and W. Siegel, *Nucl. Phys.* **B523**, 663 (1998) [hep-th/9711120].
9. B. de Wit, J. Hoppe and H. Nicolai, *Nucl. Phys.* **B305**, 545 [FS23] (1988).
10. J. Goldstone, unpublished;
    J. Hoppe, in Proc. Int. Workshop on Constraint's Theory and Relativistic Dynamics, eds. G. Longhi and L. Lusanna (World Scientific, 1987).
11. M.B. Green, H. Kwon and M. Gutperle, *JHEP* **9908**, 012 (1999) [hep-th/9907155].
12. A. Dasgupta, H. Nicolai and J. Plefka, *JHEP* **0005**, 007 (2000) [hep-th/0003280].
13. W. Taylor IV and M. Van Raamsdonk, *JHEP* **9904**, 013 (1999) [hep-th/9812239]; *Nucl. Phys.* **B558**, 63 (1999) [hep-th/9904095].
14. J. Plefka and A. Waldron, *Nucl. Phys.* **B512**, 460 (1998) [hep-th/9710104].
15. A.V. Smilga, *Nucl. Phys.* **B266**, 45 (1986);
    P. Yi, *Nucl. Phys.* **B505**, 307 (1997) [hep-th/9704098];
    S. Sethi and M. Stern, *Comm. Math. Phys.* **194**, 675 (1998) [hep-th/9705046];
    M. Porrati and A. Rozenberg, *Nucl. Phys.* **B515**, 184 (1998) [hep-th/9708119];
    J. Fröhlich and J. Hoppe, *Comm. Math. Phys.* **191**, 613 (1998);
    M.B. Halpern and C. Schwartz, *Int. J. Mod. Phys.* **A13**, 4367 (1998);
    J. Fröhlich, G.M. Graf, D. Hasler, J. Hoppe and S.T. Yau, *Nucl. Phys.* **B567**, 231 (2000) [hep-th/9904182].
16. M.B. Green, M. Gutperle and P. Vanhove, *Phys. Lett.* **B409**, 177 (1997);
    J.G. Russo and A.A. Tseytlin, *Nucl. Phys.* **B508**, 245 (1997).
17. B. de Wit and D. Lüst, *Phys. Lett.* **B477**, 299 (2000) [hep-th/9912225].
18. N. Ishibashi, H. Kawai, Y. Kitazawa and A. Tsuchiya, *Nucl. Phys.* **B498**, 467 (1997) [hep-th/9705046].

19. R. Dijkgraaf, E. Verlinde and H. Verlinde, *Nucl. Phys.* **B500**, 43 (1997) [hep-th/9703030];
R. Schiappa, *Matrix Strings in Weakly Curved Background Fields*, hep-th/0005145.

RS BRANEWORLDS IN TYPE IIB SUPERGRAVITY

K.S. STELLE

Theoretical Physics Group, Imperial College
Prince Consort Road, London SW7 2BW, U.K.

We show how the Randall-Sundrum geometry, which has been proposed as a scenario for the universe realized as a 3-brane embedded in a 5-dimensional spacetime, arises naturally as an S^5 dimensional reduction of a supersymmetric 3-brane of type IIB supergravity. In this construction, the S^5 scale-setting "breathing mode" plays a crucial rôle, permitting one to evade certain "no-go" claims for the possibility of such a supersymmetric realization.

The idea that our universe might be realized as a 3-brane embedded in a higher-dimensional spacetime has been considered at various times in recent years.[1,2,3,4,5] In the context of string duality, it was specifically the construction of Hořava and Witten[6,7] realizing heterotic to M-theory duality via an orbifold compactification that set a pattern for this scenario. In particular, one may obtain a 3-brane solution to M-theory reduced on a Calabi-Yau manifold down to five spacetime dimensions.[8,9,10] This solution has parallel 3-brane universes facing each other across a transverse fifth dimension, located at the fixed planes of the Hořava-Witten $S^1/\mathbb{Z}_2$ orbifold. The 3-branes are magnetically charged and saturate a BPS bound, so are supersymmetric. The solution is supported by a $D = 5$ scalar field which has a higher-dimensional interpretation as the volume modulus, or "breathing mode" of the compactifying space. This scalar field acquires a potential as a result of 4-form fluxes being turned on in the compactified dimensions. The dimensional reduction is thus an example of a generalized (aka Scherk-Schwarz) reduction with non-trivial field strengths turned on in the compactifying space.

Interest in such pictures became very much heightened when Randall and Sundrum showed[11,12] that in such a brane-world universe, gravity could behave as if it were effectively 4-dimensional even though the distance between the two 3-branes might be taken to infinity, provided the bulk geometry near the brane we live on is a $D = 5$ anti de Sitter space. Specifically, a model was considered that involved two segments of a pure AdS_5 spacetime patched together,

$$ds_5^2 = e^{\frac{-2|z|}{L}} dx^\mu dx_\mu + dz^2 \tag{1}$$

with a "kink" at $z = 0$ corresponding to a positive-tension δ-function stress-tensor source. In Ref. 12 it was shown that this gives rise to a "binding" of gravity to the $D = 5$ spacetime region near the (3+1) dimensional braneworld, with an effective

Newtonian gravitational potential plus eventual measurable corrections,

$$V_{\rm grav} \sim \frac{1}{r} + \frac{L^2}{r^3} \ . \qquad (2)$$

It was the potential measurability of these corrections to Newtonian gravity that attracted such strong attention within the scientific community.

No specific supergravity realization of such a construction was given, although clearly it seems natural to try to embed the RS braneworld into a $D = 5$ dimensional reduction of type IIB supergravity. Realizing the RS brane in a supergravity context ran into certain difficulties, however, primarily concerning the behavior of the scalar field that would need to be used to support the 3-brane solution. No known scalar field in any of the dimensionally-reduced versions of $D = 5$ supergravity has the properties needed to flow correctly to a fixed point at locations far from the RS brane, and this was encoded in a "no-go theorem.[13,14]" As is frequently the case with no-go theorems, however, the main result may be to direct attention towards the underlying assumptions that need to be relaxed. The key one in this case concerns the nature of the supporting scalar.

Even before the Randall-Sundrum work on our universe as a braneworld embedded in a $D = 5$ spacetime, a general study had been made[15] of the spherical dimensional reductions of various supergravity theories and of the branes and domain walls that exist in these reduced theories. For the specific case of the S^5 reduction of type IIB supergravity down to $D = 5$, it was shown that the familiar D3-brane geometry of $D = 10$ type IIB theory dimensionally reduces to a 3-brane in $D = 5$, supported by the "breathing mode" scalar modulus that determines the volume of the compactifying S^5. This works in a very similar way to the breathing-mode supported 3-brane in the Calabi-Yau reduction of M-theory.[8,9,10] It should be noted, however, that the breathing mode for an S^5 reduction does not itself belong to the massless supergravity multiplet. Instead, the breathing mode belongs to a massive spin-two multiplet, as is appropriate, since the dimensional reduction turns on a flux in the internal S^5 directions, and this gives the breathing mode φ a scalar potential that allows this mode to support a 3-brane solution. Without this potential, the breathing mode could not support a 3-brane solution. But the massive character of this mode places it outside the class of modes normally considered in $D = 5$ compactifications of supergravity theories. The importance of this mode for realizing the Randall-Sundrum braneworld as a supergravity construction was recognized in Refs. 16,17, although a main focus was still on the difficulty of realizing RS geometries as a fully "smoothed-out" solitonic solution.

To see how a construction analogous to the M-theory 3-brane solution can be made in S^5 reduced type IIB theory, consider a simplified theory just retaining the $D = 10$ metric and the self-dual 5-form field strength $H_{[5]}$,

$$\begin{aligned} R_{\mu\nu} &= \frac{1}{96}(H_{[5]})^2_{\mu\nu} \\ H^{[5]} &= {}^*H_{[5]} \qquad dH_{[5]} = 0 \ , \end{aligned} \qquad (3)$$

where the equations of motion for the five-form are implied by the Bianchi identity $dH \equiv 0$ taken together with the $H_{[5]} = {}^*H_{[5]}$ duality relation. Dimensionally reducing on S^5, one makes the Kaluza-Klein ansatz

$$\begin{aligned} ds_{10}^2 &= e^{2\alpha\varphi} ds_5^2 + e^{2\beta\varphi} ds^2(S^5) &(4)\\ H_{[5]} &= 4me^{8\alpha\varphi}\varepsilon_{[5]} + 4m\varepsilon_{[5]}(S^5) &(5)\\ &\alpha = \frac{1}{4}\sqrt{\frac{5}{3}} \qquad \beta = \frac{-3\alpha}{5}\ . \end{aligned}$$

This reduction yields the $D = 5$ bosonic theory

$$\mathcal{L}_5 = eR - \tfrac{1}{2}e\partial_\mu\varphi\partial^\mu\varphi - 8m^2 ee^{8\alpha\varphi} + R_5 ee^{\frac{16\alpha\varphi}{5}} + \text{more} \tag{6}$$

where the terms represented by "more" include bosonic fields that are not relevant for the 3-brane solution, plus all the fermions. Note that there are two potential terms in (6): the one with coefficient $-8m^2$ comes from the $H_{[5]}$ fluxes turned on in the reduction ansatz (5), while the one with the coefficient R_5 comes from the Einstein-Hilbert action in the five compactified directions, since S^5 is not Ricci-flat. The coefficient R_5 is equal to the constant Ricci scalar of the internal S^5.

The presence of two potential terms in (6) with opposite signs enables a particularly simple and maximally symmetric solution to the $D = 5$ reduced theory. In this case, one can find a solution with a constant breathing-mode scalar $\varphi = \varphi_*$, with

$$e^{\frac{24\alpha\varphi_*}{5}} = \frac{R_5}{20m^2} \qquad R_{\mu\nu} = -4m^2 e^{8\alpha\varphi_*} g_{\mu\nu}\ . \tag{7}$$

Solving this $D = 5$ Einstein equation with a cosmological term, one finds the $\mathrm{AdS}_5 \times S^5$ "vacuum" of the S^5 compactified theory. The existence of this vacuum makes this a simpler situation than the one obtained in M-theory reduced on a Calabi-Yau manifold, where only a single potential term is obtained, and where no maximally-symmetric solution in $D = 5$ is found.

In addition to the $\mathrm{AdS}_5 \times S^5$ solution (7), one can also search for brane solutions with less symmetry, but which tend asymptotically in appropriate regions to the above solution. Before pursuing this search, let us make a small change to the reduction ansatz (5) which is frequently made when considering domain-wall solutions (*i.e.* for codimension-one branes). The original type IIB theory in $D = 10$ has a $\mathbb{Z}_2$ symmetry (actually, it is just a discrete $D = 10$ proper Lorentz transformation) that couples an orientation-reversing transformation on the S^5 coordinates together with a sign flip on one of the lower $D = 5$ coordinates, say $y \to -y$. This symmetry is broken by the original ansatz (5), but will be restored if one generalizes the ansatz by inclusion of θ functions ($\theta(y) = 1$ for $y > 0$ and $\theta(y) = -1$ for $y < 0$):

$$H_{[5]} = 4m\theta(y)e^{8\alpha\varphi}\varepsilon_{[5]} + 4m\theta(y)\varepsilon_{[5]}(S^5)\ . \tag{8}$$

Note that both terms in (8) need to have θ functions in order to satisfy the $H_{[5]}$ self-duality condition in (3). With this modified ansatz, one has traded in translation invariance in the y coordinate for this preserved $\mathbb{Z}_2$ symmetry. Although the field strength $H_{[5]}$ is discontinuous in (8), the underlying four-form gauge potential $A_{[4]}$ can still be continuous. We shall adopt a basic boundary condition requirement of continuity for the metric and the gauge potentials at such "kink" locations.

Adopting the ansatz (8) and searching for a domain-wall solution, one finds the following:[15]

$$\begin{aligned} ds_5^2 &= e^{2A} dx^\mu dx^\nu \eta_{\mu\nu} + e^{2B} dy^2 \\ e^{4A} &= e^{-B} = \tilde{b}_1 H^{\frac{2}{7}} + \tilde{b}_2 H^{\frac{5}{7}} \\ e^{\frac{-7\varphi}{\sqrt{15}}} &= H = e^{\frac{-7\varphi_0}{\sqrt{15}}} + k|y| \\ \tilde{b}_1 &= \pm\frac{28m}{3k} \qquad \tilde{b}_2 = \pm\frac{14}{15k}\sqrt{5R_5}\ . \end{aligned} \tag{9}$$

Of the sign choices allowed in (9), we shall pick $\tilde{b}_2 > 0$, $\tilde{b}_1 < 0$ in order to ensure reality of the metric and to permit a $k \to 0$ limit so as to recover the pure AdS Randall-Sundrum bulk spacetime.[18] We shall also choose the integration constant φ_0 so that $H(0) > H_* = e^{\frac{-7\varphi_*}{\sqrt{15}}}$ and we shall take the slope parameter k to be negative. Then the "kink" at $y = 0$ faces downward, so that the function $H(y)$ reaches H_* at some finite value y_*. The solution (9) may then be considered to be a "semi-interpolating soliton" in the sense that, although the point $y = 0$ at which the domain-wall kink is located is not null, *i.e.* not a horizon, the solution evolves as one moves away from $y = 0$ through either positive or negative y values towards the $\mathrm{AdS}_5 \times S^5$ vacuum solution (7) at $y = y_*$. With the "kink-down" structure selected here, the surface at $y = 0$ corresponds to an extended object of positive tension.

The solution (9) is a fully supersymmetric solution, despite its kink singularity. The bulk geometry admits a 16-component Killing spinor, since it is none other than the regular D3-brane geometry of type IIB theory. Moreover, the $\mathbb{Z}_2$ invariant structure of (9) is precisely what is needed for the Killing spinor equation to be valid at all points, including at the kink location $y = 0$, with a continuous Killing spinor. The flip of sign in the 5-form flux value m as given in the modified Kaluza-Klein ansatz (8) is essential for the Killing spinor equations to be solved in this way.

The positive-tension nature of the $y = 0$ surface and the approach to the $\mathrm{AdS}_5 \times S^5$ vacuum solution at y_* suggests that one should be able to take a limit of the solution (9) and obtain the Randall-Sundrum spacetime.[18] This limit needs to be taken conjointly in both the integration constants φ_0 and k. We let $H_0 = e^{\frac{-7\varphi_0}{\sqrt{15}}} = H_* + \beta|k|$ and then take the limit $k \to 0_-$. In this conjoint limit, factors of k^{-1} cancel against factors of k, and the limiting metric becomes

$$ds^2 = \frac{2}{\sqrt{L}}(\beta - |y|)^{\frac{1}{2}} dx^\mu dx^\nu \eta_{\mu\nu} + \frac{L^2}{16}\frac{dy^2}{(\beta - |y|)^2}\ , \qquad L = m^{-1}\left(\frac{20m^2}{R_5}\right)^{\frac{5}{6}}\ . \tag{10}$$

This solution is a patched $D = 5$ anti de Sitter space; to recognize it in a more standard form, make a final coordinate change: $\beta - |y| = \beta e^{\frac{-4|\tilde{y}|}{L}}$, thus obtaining

AdS_5 spacetime in Poincaré coordinates:

$$ds^2 = e^{\frac{-2|\tilde{y}|}{L}} dx^\mu dx^\nu \eta_{\mu\nu} + d\tilde{y}^2 \ . \tag{11}$$

Let us now consider how this Randall-Sundrum metric has been successfully obtained as a solution of type IIB supergravity theory, despite the apparent implications of the various "no-go" theorems for the necessary scalar flows.[13,14] Consider a theory consisting of gravity coupled to a scalar field ϕ with a potential $V(\phi)$:

$$\mathcal{L} = e[R - \tfrac{1}{2}\nabla_\mu\phi\nabla^\mu\phi - V(\phi)] \ , \tag{12}$$

where the minimum of the potential is taken to occur at $\phi = \phi_0$. Then expand $V(\phi)$ near ϕ_0: $V(\phi) = -12g^2 + \frac{1}{2}\mu^2(\phi - \phi_0)^2 + \ldots$ (the constant g is chosen to make the AdS curvature equal to $-g^2(g_{MP}g_{NQ} - g_{MQ}g_{NP})$. Writing the metric as $ds^2 = e^{2A(y)}dx^\mu dx^\nu \eta_{\mu\nu} + e^{2B(y)}dy^2$ and solving the Einstein equations up to linear order in ϕ, one finds $A(y) = \pm gy$. Then one finds for static $\phi(y)$ near ϕ_0 the approximate field equation

$$\phi'' \pm 4g\phi' - \mu^2\phi \approx 0 \ . \tag{13}$$

This equation has two solutions:

$$\phi \approx \phi_0 + ce^{-E_0 A(y)} \tag{14}$$

$$\phi \approx \phi_0 + ce^{-(4-E_0)A(y)} \tag{15}$$

$$E_0 = 2 + \sqrt{\left(\frac{\mu}{g}\right)^2 + 4} \geq 2 \tag{16}$$

where E_0 is the AdS energy. Requiring a stable infrared flow to the vacuum value ϕ_0 as $A \to -\infty$ (*i.e.* $e^{2A} \to 0$, so one moves in to the horizon), one must take the second solution (15) and also impose a restriction that the scalar field's AdS energy be bounded below by 4: $E_0 > 4$. Now, the AdS energy is a fixed constant for a given field, determined by the Lagrangian. General fields in $D = 5$ AdS spacetime carry AdS representations $D(E_0, j_1, j_2)$, where j_1 and j_2 are spins. For "standard" supergravities in $D = 5$ (*i.e.* supergravities containing the massless graviton and vector multiplets, plus hypermultiplets and tensor multiplets), one finds scalars $D(E_0, 0, 0)$ with $E_0 = 2, 3, 4$ only, so an infrared stable flow of the above type is not possible. However, the solution (9) is supported by the breathing-mode scalar ϕ, obtained from the S^5 dimensional reduction down from $D = 10$. This mode belongs to a short massive multiplet of $D = 5$, $N = 4$ supergravity, which contains a massive spin-two mode, so it does not belong to one of the supermultiplets customarily considered in $D = 5$ massless supergravity models. Comparison of the breathing-mode potential $V(\phi) = 8m^2e^{8\alpha\varphi} - e^{\frac{16\alpha\varphi}{5}}R_5$, $\alpha = \frac{1}{4}\sqrt{\frac{5}{3}}$, with the formula (16) for E_0 gives $E_0 = 8$, clearly satisfying the required bound for a stable flow to $\varphi_0 = \varphi_*$.

Since the Kaluza-Klein ansatz (4,8) constitutes a consistent truncation of the $D = 10$ theory down to $D = 5$, one may automatically oxidize the solution (9) back up to $D = 10$ and consider its structure there. In this case, it becomes a patched set of domains of a standard type IIB D3-brane geometry. Each patch runs from a horizon at isotropic-coordinate radius $r = 0 \leftrightarrow y = y_*$ out to an outer radius $r = r_{\rm RS} \leftrightarrow y = 0$, where $r_{\rm RS} = \sqrt{\frac{20}{R_5}}[e^{-\sqrt{\frac{3}{5}}\varphi_0} - e^{-\sqrt{\frac{3}{5}}\varphi_*}]$. At this outer radius $r = r_{\rm RS}$, the solution is patched onto a $\mathbb{Z}_2$ mirror solution on another sheet of spacetime, corresponding to the $D = 5$ region with $y < 0$. The Randall-Sundrum limit $k \to 0_-$, $\varphi_0 \to \varphi_*$ corresponds to shrinking down to zero the radius $r_{\rm RS}$ at which the patch to the second sheet is made. Alternately, one could take a limit $m \to \infty$ for the flux parameter in the reduction ansatze (4,8). In either case, one obtains a spacetime that has a uniform AdS structure: in the first case, because one is restricting the spacetime ever more narrowly down to a solid annulus around the horizon, which is asymptotically $\text{AdS}_5 \times S^5$; in the second case because this asymptotic region spreads out to fill the whole spacetime. Regardless of the perspective one takes on this limit, the proper length running from a given radius $0 < r < r_{\rm RS}$ down to the horizon at $r = 0$ diverges. So, in this sense, the horizon is an infinite proper distance away along a radial (*i.e.* spacelike) geodesic. However, as is generally the case with extremal geometry horizons, one may also reach the horizon along a timelike or lightlike geodesic within a finite affine parameter interval. So the question of whether this Randall-Sundrum spacetime is really infinite or not requires careful interpretation.

At the horizon itself, one has a choice of interpretations for the structure of the solution (9) when oxidized back up to $D = 10$. The D3 brane geometry is actually non-singular and $\mathbb{Z}_2$ symmetric at the $r = 0$ horizon.[19] If one takes the horizons in the two sheets patched together at $r = r_{\rm RS}$ to be distinct, then one considers a patched-brane realization of RSII geometry,[12] which in $D = 5$ consists of a single kinked warp-factor AdS metric as in (11), extending out then to infinite proper distances in the $y > 0$ and $y < 0$ regions. On the other hand, if one decides to exploit the $\mathbb{Z}_2$ symmetry of the D3 brane solution at the horizon, one may alternatively make a second patch of the horizon at $y = y_*$ onto the second sheet horizon at $y = -y_*$. This produces a second, upwards-facing kink in the $D = 5$ geometry, corresponding to an extended object of negative tension, reproducing the RSI geometry[11] with two branes of opposite tension, facing each other across a compact dimension. This situation is clearly a type IIB analogue of the M-theory 3-brane solution obtained in a Calabi-Yau compactification.[8,9,10] The second patching surface can equally well be moved off from the horizon by moving the inner patching radius away from $r = 0$, corresponding to moving the second $D = 5$ brane in to a finite proper distance from the $y = 0$ surface.

Whatever the interpretation given to the horizon region, the kink surface at $y = 0 \leftrightarrow r = r_{\rm RS}$ possesses the essential properties of the Randall-Sundrum solution. This surface has a positive tension $\sigma_{\rm RS} > 0$, as can be verified using the Israel matching conditions

$$\Delta K_{\mu\nu} = K^+_{\mu\nu} - K^-_{\mu\nu} = -\frac{8\pi G}{3}\sigma_{\rm RS} g_{\mu\nu} \,, \tag{17}$$

for the discontinuity in the extrinsic curvature $K_{\mu\nu} = \frac{1}{2}n^\lambda \partial_\lambda g_{\mu\nu}$, where n^λ is the outward-pointing surface normal. Consequently, in accordance with the results of Ref. 12 this surface has the property of "binding" gravity to it: matter on this 3+1 dimensional surface gravitationally interacts as if the theory were in $D = 4$.

The above picture of the Randall-Sundrum spacetime as a patching of type IIB 3-brane geometries leaves some important questions unaddressed. The principal one of these is the nature of the singular sources that must be present as a result of the curvature delta-functions arising from the patching process. An immediate appreciation of this may be had by considering the signs of the source brane delta functions. The bulk geometry between the inner and outer patching radii in the $D = 10$ perspective is a normal D3-brane geometry with a positive energy. At the same time, if the outermost source is of positive tension, as it must be in order to agree with the Randall-Sundrum tension as obtained from (17) in $D = 5$, then the inner source would have to be of opposite, *i.e.* negative, tension. This is clearly inconsistent with the positive-energy D3-brane geometry in the solid annulus between the inner and outer sources. A related problem is that not only the sign, but also the magnitude of the tensions do not agree with D3-brane tensions: the D3-brane tension is only $\frac{2}{3}$ of the Randall-Sundrum value as determined by (17).[20]

Both of the above problems are resolved by a recognition that the sources at the inner and outer radii in $D = 10$ cannot simply be D3-brane sources alone.[21] Careful dimensional oxidation of the $D = 5$ source structure up to $D = 10$ shows that the needed delta functions do not have the characteristic form of brane sources. Brane-source stress tensors have delta-function contributions with indices spanning the brane worldvolume directions only. On the other hand, the dimensional oxidation of the source obtained from (9) definitely has components in the compactification S^5 directions. One may understand what is going on by turning off the 5-form flux parameter m, but keeping the patching structure, now just for flat space patched to another flat-space sheet at inner and outer radii. Computing the stress-tensor delta function for this flat-to-flat patch, one finds[21] that it is opposite in sign to the delta functions arising from the D3-brane sources, and has a magnitude $\frac{5}{3}$ that of the D3 source. Combining these observations, one sees that the d3-brane tension is $-\frac{2}{3}$ times the sum of D3-brane tension plus the flat-to-flat patching source. The sign and the magnitude of this result combine to explain the differences between the D3-brane and Randall-Sundrum tensions. So, the Randall-Sundrum delta-function sources are not pure D3-branes, but contain another component as well. Since this other component, the flat-to-flat patch, has Ricci-tensor singularities, it does not simply correspond to an orbifold, although there would seem to be a clear orbifold analogy in the compact and $\mathbb{Z}_2$ symmetric 5th dimension. Understanding this source structure in the context of string theory is obviously an important remaining problem.

References

1. K. Akama, *Pregeometry*, *Lect. Notes Phys.* **176**, 267 (1982) [hep-th/0001113].
2. V.A. Rubakov and M.E. Shaposhnikov, *Do We Live Inside A Domain Wall?*, *Phys. Lett.* **B125**, 136 (1983).
3. V.A. Rubakov and M.E. Shaposhnikov, *Extra Space-Time Dimensions: Towards A Solution To The Cosmological Constant Problem*, *Phys. Lett.* **B125**, 139 (1983).
4. G.W. Gibbons and K. Maeda, *Black Holes And Membranes In Higher Dimensional Theories With Dilaton Fields*, *Nucl. Phys.* **B298**, 741 (1988).
5. G.T. Horowitz and A. Strominger, *Black strings and P-branes*, *Nucl. Phys.* **B360**, 197 (1991).
6. P. Hořava and E. Witten, *Heterotic and type I string dynamics from eleven dimensions*, *Nucl. Phys.* **B460**, 506 (1996) [hep-th/9510209].
7. P. Horavǎ and E. Witten, *Eleven-Dimensional Supergravity on a Manifold with Boundary*, *Nucl. Phys.* **B475**, 94 (1996) [hep-th/9603142].
8. A. Lukas, B. Ovrut, K.S. Stelle and D. Waldram, *The Universe as a Domain Wall*, *Phys. Rev.* **D59**, 086001 (1999) [hep-th/9803235].
9. A. Lukas, B. Ovrut, K. S. Stelle and D. Waldram, *Heterotic M–theory in Five Dimensions*, *Nucl. Phys.* **B552**, 246 (1999) [hep-th/9806051].
10. A. Lukas and K.S. Stelle, *Heterotic anomaly cancellation in five dimensions*, *JHEP* **0001**, 010 (2000) [hep-th/9911156].
11. L. Randall and R. Sundrum, *A large mass hierarchy from a small extra dimension*, *Phys. Rev. Lett.* **83**, 3370 (1999) [hep-ph/9905221].
12. L. Randall and R. Sundrum, *An alternative to compactification*, *Phys. Rev. Lett.* **83**, 4690 (1999) [hep-th/9906064].
13. R. Kallosh and A. Linde, *Supersymmetry and the brane world*, *JHEP* **0002**, 005 (2000) [hep-th/0001071].
14. K. Behrndt and M. Cvetič, *Anti-de Sitter vacua of gauged supergravities with 8 supercharges*, *Phys. Rev.* **D61**, 101901 (2000) [hep-th/0001159].
15. M.S. Bremer, M.J. Duff, H. Lü, C.N. Pope and K.S. Stelle, *Instanton cosmology and domain walls from M-theory and string theory*, *Nucl. Phys.* **B543**, 321 (1999) [hep-th/9807051].
16. M. Cvetič, H. Lü and C.N. Pope, *Domain walls and massive gauged supergravity potentials*, *Class. Quant. Grav.* **17**, 4867 (2000) [hep-th/0001002].
17. M. Cvetič, H. Lü and C.N. Pope, *Localised gravity in the singular domain wall background?*, hep-th/0002054.
18. M.J. Duff, J.T. Liu and K.S. Stelle, *A supersymmetric type IIB Randall-Sundrum realization*, hep-th/0007120.
19. G.W. Gibbons, G.T. Horowitz and P.K. Townsend, *Class. Quant. Grav.* **12**, 297 (1995) [hep-th/9410073].
20. P. Kraus, *Dynamics of anti-de Sitter domain walls*, *JHEP* **9912**, 011 (1999) [hep-th/9910149].
21. M. Cvetič, M.J. Duff, J.T. Liu, C.N. Pope and K.S. Stelle, *Randall-Sundrum Brane Tensions*, hep-th/0011167.

STRINGS IN AdS_3 AND THE $SL(2,R)$ WZW MODEL

HIROSI OOGURI
California Institute of Technology, 453-48
Pasadena, CA 91125, USA
E-mail: ooguri@theory.caltech.edu

Strings in the three-dimensional anti-de Sitter space in the NS background is described by the $SL(2,R)$ WZW model. We present a complete description of the Hilbert space of the model. We then compute the one-loop partition function for Euclidean BTZ black hole backgrounds or equivalently thermal AdS_3 backgrounds which are quotients of Euclidean AdS_3. We see long strings and discrete states in agreement with the structure of the Hilbert space.

1. Introduction

In this talk, I will describe my work with Maldacena[1] and with Maldacena and Son[2] on strings in the three-dimensional anti-de Sitter space, AdS_3, in the NS background. The worldsheet of the string is described by $SL(2,R)$ WZW model with the action,

$$S = \frac{k}{8\pi}\int d^2z \mathrm{tr}\left(g^{-1}\partial g g^{-1}\bar{\partial}g\right) + k\Gamma_{WZ}, \tag{1}$$

where g is an element of the universal cover of the group $SL(2,R)$, Γ_{WZ} is the Wess-Zumino term, and k gives the level of the $SL(2,R)$ current algebra. Our motivation is to understand string theories in curved spacetimes where the metric component g_{00} is non-trivial, of which AdS_3 is the simplest example. Moreover, it is possible to construct black hole solutions as quotients of AdS_3,[3] so understanding string theory on AdS_3 would lead to an understanding of strings moving near black hole horizons.

There have been puzzles about the spectrum of the model, in particular about what types of representations of $SL(2,R)$ appear in the Hilbert space and whether the no ghost theorem holds after the Virasoro constraints are imposed.[a] These puzzles have been around for more than a decade, and they have made some of us worried that the model may not exist as a consistent quantum theory. Recently the AdS/CFT correspondence has motivated us to revisit these problems again, and I am happy to report to you today that we now have a coherent picture about the model.

In this talk, I will describe the complete structure of the Hilbert space of the model, corresponding to the propagation of string in AdS_3 with the Lorentzian signature metric. Since AdS_3 is globally static, one can perform the Euclidean rotation

[a] A long list of historical papers on this subject can be found in Ref. 2.

of the metric. It is possible to evaluate the Euclidean path-integral explicitly for one-loop free energy at finite temperature. I will show that the result of this computation agrees with expectations from the analysis of the model with the Lorentzian signature metric.

2. The Hilbert Space

Let us summarize the results of Ref. 1, where a concrete proposal for the spectrum of AdS_3 string theory was made. We consider a critical bosonic string theory on $AdS_3 \times \mathcal{M}$. The Hilbert space of the $SL(2,R)$ WZW model is generated by the action of the left-moving and right-moving current algebra $\widehat{SL(2,R)}_L \times \widehat{SL(2,R)}_R$, and all the states form representations of this algebra. The simplest representations are built by first choosing representations for the zero modes, then regarding them as the primary states annihilated by $J^{3,\pm}_{n>0}$. The raising operators $J^{3,\pm}_{n<0}$ are then used to generate the representations of the current algebra. From harmonic analysis, *i.e.* quantum mechanical limit, it is known that the left-right symmetric combinations $\mathcal{C}^\alpha_{j=1/2+is} \times \mathcal{C}^\alpha_{j=1/2+is}$ and $\mathcal{D}^\pm_{j>1/2} \times \mathcal{D}^\pm_{j>1/2}$ form a complete basis in $\mathcal{L}^2(AdS_3)$, where $\mathcal{C}^\alpha_{j=1/2+is}$ is the principal continuous representation and $\mathcal{D}^\pm_{j>1/2}$ the principal discrete representation of $SL(2,R)$. These representations are unitary, but the resulting current algebra representations $\widehat{\mathcal{C}}^\alpha_{j=1/2+is} \times \widehat{\mathcal{C}}^\alpha_{j=1/2+is}$ and $\widehat{\mathcal{D}}^\pm_{j>1/2} \times \widehat{\mathcal{D}}^\pm_{j>1/2}$, constructed as explained above, are not. This is not a surprise, for even in flat Minkowski space it is not until one imposes the Virasoro constraints

$$(L_n + \mathcal{L}_n - \delta_{n,0})|\text{physical}\rangle = 0, \quad n \geq 0 \tag{2}$$

that a unitary spectrum is obtained. Here $\mathcal{L}_n$ is the Virasoro generator for the internal conformal field theory corresponding to $\mathcal{M}$. The proposal of Ref. 1 is that one should consider not just these representations but also those obtained by the spectral flow

$$\begin{aligned} J^3_n &\to \tilde{J}^3_n = J^3_n - \frac{k}{2} w \delta_{n,0} \\ J^+_n &\to \tilde{J}^+_n = J^+_{n+w} \\ J^-_n &\to \tilde{J}^-_n = J^-_{n-w}. \end{aligned} \tag{3}$$

The Virasoro generators, given by the Sugawara form, then become $\tilde{L}_n = L_n + wJ^3_n - \frac{k}{4}w^2\delta_{n,0}$. Imposing on $\widehat{\mathcal{D}}^\pm_{j>1/2} \times \widehat{\mathcal{D}}^\pm_{j>1/2}$ the condition (2) with $\tilde{L}_n$ one finds that these states have a discrete energy spectrum

$$\begin{aligned} E &= J^3_0 + \bar{J}^3_0 = q + \bar{q} + kw + 2\tilde{j} \\ &= 1 + q + \bar{q} + 2w + \sqrt{1 + 4(k-2)\left(N_w + h - 1 - \frac{1}{2}w(w+1)\right)}, \end{aligned} \tag{4}$$

here N_w is defined to be the level of the current algebra after spectral flow by amount w, $N_w = \tilde{N} - wq$, and $\tilde{N}$ is the level before spectral flow. The state with energy

(4) is obtained from a lowest weight state by acting with the $SL(2,R)$ currents $\prod \tilde{J}^{\pm}_{n\leq 0}|\tilde{j},\tilde{j}\rangle$, with q the net number of $\pm$ signs in this expression. In other words, q is the number of spacetime energy raising operators J^{+}_{a} minus the number of spacetime energy lowering operators J^{-}_{a} that we have to apply to the lowest weight, lowest energy state $|\tilde{j}, m = \tilde{j}\rangle$ to get to the state whose spacetime energy is (4). $\bar{q}$ is the corresponding quantity for the generators $\bar{J}^{\pm}_{a}$. We also have a level matching condition of the form

$$N_w + h = \bar{N}_w + \bar{h} \tag{5}$$

which implies that the angular momentum in AdS_3, $\ell = J^3_0 - \bar{J}^3_0 = q - \bar{q}$, is an integer. We argued in Ref. 1 that $\tilde{j}$ is further restricted to the range

$$\frac{1}{2} < \tilde{j} < \frac{k-1}{2}, \tag{6}$$

which implies

$$\frac{k}{4}w^2 + \frac{1}{2}w < N_w + h - 1 + \frac{1}{4(k-2)} < \frac{k}{4}(w+1)^2 - \frac{1}{2}(w+1). \tag{7}$$

With this constraint, (4) gives the spectrum of <u>short string</u> whose size is small compared to the curvature radius of AdS_3.

A similar analysis on $\widehat{\mathcal{C}}^{\alpha}_{j=1/2+is} \times \widehat{\mathcal{C}}^{\alpha}_{\bar{j}=1/2+is}$ yields a continuous spectrum

$$E = \frac{k}{2}w + \frac{1}{w}\left(\frac{2s^2 + \frac{1}{2}}{k-2} + \tilde{N} + h + \bar{\tilde{N}} + \bar{h} - 2\right), \tag{8}$$

where s takes values over the real numbers. This gives the spectrum of <u>long string</u> which can grow indefinitely large toward the boundary of AdS_3. The parameter s is interpreted as the momentum in radial direction for the long strings. These states satisfy the level matching condition

$$\tilde{N} + h = \bar{\tilde{N}} + \bar{h} + w \times (\text{integer}). \tag{9}$$

It was proven in Ref. 1 that both the short string with the constraint (6) and the long string satisfy the no ghost theorem, *i.e.* the physical Hilbert space satisfying the Virasoro constraint (2) is unitary.

3. One-Loop Free Energy

We would like to verify these results by an explicit calculation of the one-loop partition function. As shown in Ref. 4, the Euclidean black hole background is equivalent to the thermal AdS_3 background. So we will consider string theory on AdS_3 at a finite temperature, which is described by strings moving on a Euclidean AdS_3 background with the Euclidean time identified. The calculation of the partition function for this geometry is a minor variation on the calculation of Gawedzki in Ref. 5. From this we can read off the spectrum of the theory in Lorentzian signature by interpreting the result as the free energy of a gas of strings.

To compute the free-energy at finite temperature T, we consider AdS_3 with the Euclidean signature metric and periodically identify the imaginary time t as $t \sim t+\beta$ with $\beta = 1/T$. Let us call this geometry as thermal AdS_3. The worldsheet action for the Euclidean AdS_3 is given by

$$S_{Euclidean} = \frac{k}{\pi} \int d^2 z \left(\partial\phi\bar{\partial}\phi + (\partial + \partial\phi)\bar{v}(\bar{\partial} + \bar{\partial}\phi)v \right) . \tag{10}$$

This model is exactly solvable in the sense that the path-integral can be performed by the iterative Gaussian integral.

To evaluate the one-loop amplitude, we consider a path-integral over maps from the worldsheet torus of moduli τ to the thermal AdS_3 with the period β. It turned out that the resulting amplitude $Z(\tau, \beta)$ has singularities at

$$\tau = \frac{i\beta}{2\pi w}, \ \frac{i\beta \pm 1}{2\pi w}, \ \frac{i\beta \pm 2}{2\pi w}, \cdots, \tag{11}$$

with w being an integer. These singularities are unusual in that they appear in the middle of the worldsheet moduli space. What happens at these special points is that the moduli of the worldsheet and the moduli of the boundary of thermal AdS_3 coincide, and therefore there is a holomorphic map from the worldsheet to the boundary of the target space. This allows the worldsheet to grow indefinitely large toward the boundary without costing an infinite amount of action. One can show that the singularity is associated to the infrared divergence due to the infinite volume of string configuration space.

The one-loop partition function Z of the string in thermal AdS_3 integrated over the moduli space of the worldsheet torus gives the free energy of a gas of free strings, and therefore it can be expressed as a sum over single string energy spectrum,

$$\int d^2\tau Z(\tau, \beta) = \sum_i \log \left(\frac{1}{1 - e^{-\beta E_i}} \right) . \tag{12}$$

The Euclidean functional integral for Z in the left-hand size of this equation was evaluated by Gawedzki in Ref. 5. On the other hand, in the right-hand size, we expect to see the spectra of the short string (4) and the long string (8) described in the previous paragraph.

As I mentioned, the integrand $Z(\tau, \beta)$ has poles at the special points (11). There is a well-defined procedure to separate a divergent part and a finite part of the integrand since they have different temperature dependence. This is described in detail in Ref. 4. It was found that the integral of the finite part reproduces the spectrum of the short string (4) including the constraint (6), and the correct multiplicity of each representation.

On the other hand, the poles in the integrand are due to the fact that the worldsheet can grow infinitely large and therefore they correspond to the presence of the long string states. By regularizing the infrared divergence by cutting of the size of the long string by L_{IR}, we find that the density of long string states derived from the Euclidean functional integral takes the following form.

$$\rho(E) \sim L_{IR} + \frac{d\delta_(E)}{dE}. \tag{13}$$

It turned out that $\delta(E)$ coincides with the phase shift of the long string scattering; In the Lorentzian picture, the long string starts in the infinite past at the boundary of AdS_3, shrinks to a point and then is scattered away toward the boundary in the infinite future. The phase shift for this process can be computed using the earlier result by Teschner and Zamolodchikov, and we found that it coincides with $\delta(E)$ that appears in (13). According to the scattering theory, the finite part of the density of scattering states is given by a derivative of the phase shift with respect to the energy, and we find that this completely agrees with our computation of the density of states obtained from the one-loop Euclidean functional integral.

4. Conclusions

We have demonstrated that the $SL(2,R)$ WZW model makes sense as a quantum theory. The Hilbert space of the model is obtained and it is consistent with the no-ghost theorem of string in propagating in AdS_3. We found that the Hilbert space contains new sectors; winding states obtained by spectral flow of the standard highest weight representations of the current algebras and states corresponding to long strings with continuous spectrum. The existence of these states fits well with expectations from the AdS_3/CFT_2 correspondence.

We have carried out Euclidean path integral for the one-loop free energy and found that the result is consistent with the Hilbert space structure of the model. String in Lorentzian AdS_3 is described by the $SL(2,R)$ WZW model, and the Euclidean case is described by the $SL(2,C)/SU(2)$ coset model. These two models are very different, for example their Hilbert spaces are completely different, and I do not know whether one can related computations in these models directly, for example by changing the contour of the functional integral of the one theory to bring it to another. On the other hand, since these models are related by the target space Euclidean rotation, we can expect that string theory observables, obtained after integrating over the worldsheet moduli, must agree. This is indeed what we found in the paper Ref. 2.

Acknowledgments

I would like thank the organizers of the conference *Strings 2000* for giving me the opportunity to present this work and for their hospitalities. I would like to thank Juan Maldacena and John Son for collaborations on the work. This work is supported in part by NSF grant PHY-95-14797, DOE grant DE-AC03-76SF00098, and the Caltech Discovery Fund.

References

1. J. Maldacena and H. Ooguri, *Strings in AdS_3 and the $SL(2,R)$ WZW model. Part I: Spectrum*, hep-th/0001053.
2. J. Maldacena, H. Ooguri, and J. Son, *Strings in AdS_3 and the $SL(2,R)$ WZW model. Part II: Euclidean black hole*, hep-th/0005183.
3. M. Banados, C. Teitelboim, and J. Zanelli, *The Black hole in three-dimensional spacetime*, *Phys. Rev. Lett.* **69**, 1849 (1992) [hep-th/9204099].
4. J. Maldacena and A. Strominger, *AdS_3 black holes and a stringy exclusion principle*, *JHEP* **9812**, 005 (1998) [hep-th/9804085].
5. K. Gawedzki, *Noncompact WZW conformal field theories*, hep-th/9110076.

SUPERSYMMETRY IN SINGULAR SPACES AND DOMAIN WALLS

RENATA KALLOSH
Department of Physics, Stanford University
Stanford, CA 94305, USA

We discuss some recent work on brane world construction and supersymmetry issues. We introduce a concept of bulk & brane supersymmetry in the framework of 5-dimensional supergravity which allows to formulate the rules for supersymmetry in singular spaces. The original fine-tuning of the Randall-Sundrum world brane which provides the flatness of the brane is shown to follow from bulk & brane supersymmetry on $\frac{S_1}{\mathbf{Z}_2}$ orbifold. Examples of RS-type domain walls are presented.

Recent ideas of the brane world have set up a basic question: do we live on the brane? A fascinating part of such idea is that if brane world is realized in Nature, one would be forced eventually to change the fundamentals. New approach to the old problem of cosmological constant may come out from this direction. Experimental tests of brane world ideas seem to be possible. Will it lead to a new cosmology? The worrisome part of this new development that it is an extremely speculative and controversial field with fine tuning, naked singularities, sometimes low standards.

The basic problem with cosmological constant at present time can be formulated as follows: why the cosmological constant is zero and why it is not? The natural value would be $\Lambda \sim M_{PL}^4 \sim 10^{94} g/cm^3$, or any other value which a particle physicists may suggest. Experimentally $\Lambda \sim \Omega_\Lambda \cdot 10^{-29} g/cm^3$ where $\Omega_\Lambda \sim 0.6$ [1] according to the analysis of the recent cosmological observations from Maxima-1, Boomerang and COBE/DMR CMB.

The required fine-tuning is of the order $\frac{\Lambda_{exp}}{\Lambda_{nat}} \sim 10^{-123}$. Thus the challenge is first to explain why the cosmological constant is almost zero and why there is such a tiny deviation from this zero, which is completely out of any scale one can imagine in particle physics. For example, one can try to use for the natural value of cosmological constant not M_{PL}^4 but, for example the grand unification scale M_{GUT}^4. This will not make an important difference and the puzzle remains.

The main purpose of this talk is to study the supersymmetric properties of brane world. Various aspects of the brane world scenario related to string theory, higher dimensions and supergravity are presented in the talks on this conference by S. de Alwis, M. Cvetič, J. Maldacena, B. Ovrut, L. Randall, E. Silverstein, K. Stelle and H. Verlinde.

The world brane scenario of Randall-Sundrum (RS)[2] suggests that in 5 dimensions there are 4 dimensional branes. The action is taken to be a combination of the Einstein action with cosmological constant in d=5 of the form $\int d^4x dy \sqrt{-G}\{2M^3 R-$

$\Lambda_5\}$ and two brane actions, S_{brane} at $y = 0$ and $S_{brane'}$ at $y = \pi r_c$. The branes are placed at the fixed points of the orbifold $\frac{S_1}{\mathbf{Z}_2}$ and $S_{br} = \int d^4x\sqrt{g_{brane}}\{V_{brane} + \mathcal{L}_{brane}\}$. The action requires a fine-tuning. $V_{brane} = -V_{brane'} = 24M^3k$ and $\Lambda_5 = -24M^3k^2 = -kV_{brane}$. With such fine-tuning the cosmological constant vanishes on the brane, $\Lambda_{brane} = 0$ and the metric is $ds^2 = a^2(y)\eta_{\mu\nu}dx^\mu dx^\nu + dy^2$ where the warp factor is decreasing away from the wall $a^2(y) = e^{-2k|y|}$. When the second wall is pushed to infinity out of the space, an alternative to Kaluza-Klein compactification scenario is suggested in Ref. 3 where gravity is trapped to the wall and on the brane the 4 dimensional Newton law takes place.

In RS scenario the vanishing of the cosmological constant on the brane follows from the fine-tuning conditions specified in above. The problem is to remove the fine-tuning, if possible. Also the *adS* nature of the metric, suggests that the supersymmetric embedding for such brane world may exists.

The first attempts to relax the fine tuning of brane world scenario was using supergravities with ADS vacua without introducing singular sources, i.e. only the bulk action was considered. A no-go theorem for smooth BPS domain walls with required properties of decreasing warp factor was established in certain class of supergravities.[4] This theorem is based on the study of *Supersymmetric Flow Equations*

$$(\phi^i)' = \pm 3g^{ij}\partial_j W \ , \qquad \frac{a'}{a} = \mp W \tag{1}$$

and the universality of the critical point.[5] The flow equations can be taken in the form $a\frac{\partial}{\partial a}\phi^i = -3g^{ij}\frac{\partial_j W}{W} \equiv \beta^i(\phi)$. At the critical points $(\phi^i_{cr})' = 0$ and $(\partial_i W)_{cr} = 0$.

In five-dimensional supergravities[6] one can also show that

$$(\partial_i\partial_j W)_{cr} = \frac{2}{3}g_{ij}W_{cr} \ . \tag{2}$$

which follows from the very special geometry in the moduli space.

It follows due to positivity of the metric g_{ij} of the moduli space that the stable critical point of moduli requires that the scale factor of the metric in the direction towards the fixed point grows. This means that far away from the wall there is an ADS boundary and not an ADS horizon, as required for RSII scenario.

The NO-GO theorem can be formulated as follows: *there are no smooth supersymmetric RS one-brane scenario as alternative to Kaluza-Klein compactification in a certain class of 5D supergravities.* These include the 5D $U(1)$ gauged supergravity interacting with vector and tensor multiplets. The details may be found in Ref. 4. It was also shown in Ref. 7 that the condition of smoothness is too restrictive for obtaining reasonable world brane pictures.

Later on a new class of bulk & brane supersymmetric theories were introduced[8] where the bulk and the brane actions are supersymmetric. The complete supersymmetry rules were formulated in singular spaces which has allowed the supersymmetric embedding[a] of the brane world.

[a]The relation to other approaches to supersymmetric brane world[9] is described in detail in Ref. 8.

Here we will proceed with explanations of bulk & brane supersymmetric theories developed in Ref. 8. We will explain the concept of supersymmetry in singular spaces, apply it in an example for 3-branes in $D = 5$ and comment on 8-branes in $D = 10$. The new construction has an interpretation that the brane is a sink for the flux and requires adding to the standard supergravity a $(D-1)$-form field and a supersymmetry singlet field. This allows a consistent definition of supersymmetry on a $S_1/\mathbf{Z}_2$ orbifold, the bulk and the brane actions being separately supersymmetric.

We will show that the Randall–Sundrum brane-worlds can be reproduced in this framework without fine tuning. For fixed scalars, the doubling of unbroken supersymmetries takes place and the negative tension brane can be pushed to infinity. In more general BPS domain walls with 1/2 of unbroken supersymmetries, the distance between branes in some cases may be restricted by the collapsing cycles of the Calabi–Yau manifold.

Finally, it turns out that the energy of any static x^5-dependent bosonic configuration vanishes, $E = 0$, in analogy with the vanishing of the Hamiltonian in a closed universe.

To set up a new general point of view on supersymmetry in singular spaces, and to clarify the issue of energy of brane worlds, we have introduced a set of rules defining *consistent supersymmetry on singular spaces.*

The most important new steps include: *i)* some constants (masses or gauge couplings) have to become "$\mathbf{Z}_2$-odd" to make supersymmetry commuting with $\mathbf{Z}_2$-symmetry. This can be achieved by promoting such a constant to the status of a supersymmetry singlet field; *ii)* moreover, one has to add to the theory the $(D-1)$-form potential to compensate the variation of the Lagrangian proportional to the derivative of the new field; *iii)* one can find afterwards the new bulk and the brane actions, which are separately invariant under supersymmetry.

The *unusual features of the new supersymmetry* are the presence of the supersymmetry singlet field and $(D-1)$-form field in the bulk, and the fact that the purely bosonic action on the brane is supersymmetric due to the fact that its fermionic partner is a $\mathbf{Z}_2$-odd fermion which vanishes on the brane.

In absence of brane actions, the new fields become irrelevant: on shell for the 4-form and supersymmetry singlet, the bulk action reduces to the standard supergravity action supersymmetric under the standard rules.

In general, when the brane actions are added, the new fields play an important role in understanding the energy issue. We have found that *the total energy of supersymmetric configurations vanishes locally at each brane. The positive (negative) energy of the brane tensions are compensated separately by the terms with the derivative of the supersymmetry singlet field.* The energy of any static y-dependent bosonic configuration vanishes locally, $E = 0$, in analogy with the vanishing of the Hamiltonian in a closed universe.

The strategy is applied in detail in the particular case of a 3-brane in $D = 5$ on the basis of $U(1)$ gauged $N = 2$ supergravity interacting with abelian vector multiplets. We expect that it will work in other cases as well: in $D = 5$ one can try to include more general gauging and tensor and hypermultiplets. The gauged $N = 4$ and $N = 8$ supergravities in $D = 5$ are also natural candidates for an analogous extension.

Our supersymmetric bulk & brane Lagrangian in a singular space is given by

$$S_{new}(x^5) = S_{bulk} + S_{brane}\,, \qquad S_{bulk} = S_{GST}(g \to G(x)) - \int \mathrm{d}^5x\, e\, F(x)G(x)\,. \quad (3)$$

Here F is a curl of the 4-form $F = \frac{1}{4!}\varepsilon^{\mu\nu\rho\sigma\tau}\partial_\tau A_{\mu\nu\rho\sigma}$ and $S_{GST}(g) = S_0 + gS_1 + g^2 S_2$ is the $U(1)$ gauged 5D supergravity of Günaydin, Sierra and Townsend.[6] The supergravity at $g = 0$ is called ungauged supergravity. We will use the notation $\mathcal{L}_0$ for its Lagrangian. Our new bulk theory has Lagrangian $\mathcal{L}_0$, there are terms linear in $G(x)$, which are proportional to the flux $\hat{F}$, and terms quadratic in the field $G(x)$,

$$\begin{aligned} \mathcal{L}_{bulk} &= \mathcal{L}_0 + 6eG^2(x)\left(W^2 - \frac{3}{4}W^2_{,x}\right) - e\,G(x)\hat{F} \\ &\quad + e\,\mathrm{i}G(x)Q_{ij}\overline{\lambda}^{ix}\lambda^{jy}\left(-\tfrac{1}{4}g_{xy}W + \sqrt{\frac{3}{2}}T_{xyz}W^{,z}\right). \end{aligned} \quad (4)$$

The brane actions are:

$$\mathcal{L}_{brane} = -2g\,\left(\delta(x^5) - \delta(x^5 - \tilde{x}^5)\right)\left(e_{(4)}3W + \tfrac{1}{4!}\varepsilon^{\underline{\mu\nu\rho\sigma}}A_{\underline{\mu\nu\rho\sigma}}\right), \quad (5)$$

The new supersymmetry rules are given by the following expressions

$$\begin{aligned} \delta(\epsilon)\{e_\mu^m, A_\mu^I, \varphi^x\} &= \delta_0\{e_\mu^m, A_\mu^I, \varphi^x\}\,, \\ \delta(\epsilon)\psi_{\mu i} &= \delta_0(\epsilon)\psi_{\mu i} + G(x)A_\mu^{(R)}Q_{ij}\epsilon^j + \mathrm{i}\tfrac{1}{2}G(x)\,\gamma_\mu\epsilon^j W Q_{ij}\,, \\ \delta(\epsilon)\lambda_i^x &= \delta_0(\epsilon)\lambda_i^x - \tfrac{3}{2}G(x)\,\epsilon^j W^{,x}Q_{ij}\,, \\ \tfrac{1}{4!}\delta(\epsilon)A_{\mu\nu\rho\sigma} &= \left[-6\overline{\epsilon}^i\gamma_{[\mu\nu\rho}\psi^j_{\sigma]}W + \mathrm{i}\overline{\epsilon}^i\gamma_{[\mu\nu}A_\rho^{(R)}\psi^j_{\sigma]} - \mathrm{i}\tfrac{3}{2}\overline{\epsilon}^i\gamma_{\mu\nu\rho\sigma}\lambda^j_x W^{,x}\right]Q_{ij}\,, \\ \delta(\epsilon)G &= 0\,. \end{aligned} \quad (6)$$

The action is defined by an integral over a product space $\mathbf{M}$ of the 4-dimensional manifold and an orbifold, $\mathbf{M} = \mathbf{M}_4 \times \frac{S^1}{\mathbf{Z}_2}$.

$$S^{new} = \int_{\mathbf{M}} d^4x dx^5 \mathcal{L}^{new}\,. \quad (7)$$

The 4d manifold is non-compact and the 5th dimension is compact but has no boundaries in $S^1/\mathbf{Z}_2$. Therefore, all surface terms in the variation of the action vanish assuming as usual that the parameters of supersymmetry decrease at infinities of the 4d space. The bulk and the brane actions are separately invariant. The supersymmetry transformations form an on-shell closed algebra.

Here we limit ourselves to configurations which depend only on x^5. We are using the warped metric in the form

$$ds^2 = a^2(x^5)dx^{\underline{\mu}}dx^{\underline{\nu}}\eta_{\underline{\mu\nu}} + (dx^5)^2\,. \quad (8)$$

Starting with the new Lagrangian (3), we may present the energy functional for static x^5-dependent bosonic configurations as follows

$$\begin{aligned} E(x^5) &= -6a^2a'^2 + \tfrac{1}{2}a^4(\varphi^{x\prime})^2 + a^4V - \tfrac{1}{4!}\varepsilon^{\mu\nu\rho\sigma 5}A_{\mu\nu\rho\sigma}G' \\ &\quad +2g\left(\delta(x^5) - \delta(x^5 - \tilde{x}^5)\right)\left(3a^4\alpha W + \tfrac{1}{4!}\varepsilon^{\underline{\mu\nu\rho\sigma}}A_{\underline{\mu\nu\rho\sigma}}\right). \end{aligned} \tag{9}$$

With account of the A and G equations and no boundary condition, which allows to ignore $+3\alpha[a^4GW]'$, the energy functional takes the form

$$E(x^5)_{BPS} = \frac{1}{2}a^4\left\{[\varphi^{x\prime} - 3\alpha GW^{,x}]^2 - 12[\frac{a'}{a} + \alpha GW]^2\right\}. \tag{10}$$

The squared terms in (10) suggest the BPS conditions:

$$g_{xy}\,(\varphi^y)' = +3\alpha G(x^5)W_{,x}\,, \qquad \frac{a'}{a} = -\alpha G(x^5)W\,, \tag{11}$$

where we can use the on shell value of $G(x^5)$, which is equal to $g\varepsilon(x^5)$, so that we obtain

$$g_{xy}\,(\varphi^y)' = +3\alpha g\varepsilon(x^5)W_{,x}\,, \qquad \frac{a'}{a} = -\alpha g\varepsilon(x^5)W\,. \tag{12}$$

From now on we will make the choice $\alpha = 1$, i.e. pick up a particular property of fermions under parity. Physics depends on the sign of αg. Therefore, the change in the sign of α can always be compensated by a change of the sign of g.

One can verify that the jump conditions on the branes, derived starting with the second order differential equations, are satisfied automatically due to the new supersymmetry (6), the on-shell condition and the presence of the 5-form flux, changing the sign when passing through the wall.

Let us also consider the Killing spinors. In the background with only non-vanishing scalars and the warped metric (8), which has as only non-zero components of the spin connection $\omega_{\underline{\mu}}^{m5} = a'\delta_{\underline{\mu}}^m$, the transformations of the spinors are

$$\begin{aligned} \delta(\epsilon)\lambda_i^x &= -\mathrm{i}\tfrac{1}{2}\gamma_5\varphi^{x\prime}\epsilon_i - \tfrac{3}{2}GW^xQ_{ij}\epsilon^j\,, \\ \delta(\epsilon)\psi_{\underline{\mu}i} &= \partial_{\underline{\mu}}\epsilon_i + \tfrac{1}{2}\delta_{\underline{\mu}}^m\gamma_m\left(a'\gamma_5\epsilon_i + \mathrm{i}aGWQ_{ij}\epsilon^j\right), \\ \delta(\epsilon)\psi_{5i} &= \epsilon_i' + \tfrac{1}{2}\mathrm{i}GW\gamma_5Q_{ij}\epsilon^j\,. \end{aligned} \tag{13}$$

To solve these, we split

$$\begin{aligned} \epsilon_i &= \epsilon_i^+ + \epsilon_i^-\,, \\ \epsilon_i^\pm &= \tfrac{1}{2}\left(\epsilon_i \pm \mathrm{i}\gamma_5Q_{ij}\epsilon^j\right) = \pm\mathrm{i}\gamma_5Q_{ij}\epsilon^{\pm j}\,. \end{aligned} \tag{14}$$

The solutions are

$$\epsilon_i = a^{1/2}\epsilon_i^{+(0)} + a^{-1/2}\left(1 - \frac{a'}{a}x^{\underline{\mu}}\gamma_{\underline{\mu}}\gamma_5\right)\epsilon_i^{-(0)}\,, \tag{15}$$

where $\epsilon_i^{\pm(0)}$ are constant spinors with each only 4 real components due to the projection (14). Remains the first Killing equation, which implies

$$\varphi^{x\prime}\epsilon_i^{-(0)} = 0\,. \tag{16}$$

There are thus two possibilities to solve the Killing equations.

- Maximal unbroken supersymmetry in the bulk ($N = 2$).

  Here we require that the scalars are strictly constant, $\varphi^{x\prime} = 0$, and the superpotential is independent of the scalars, $\frac{\partial W}{\partial \varphi^x} = 0$, at the solution. No constraints on Killing spinors arise from the gaugino.

  From the gravitino transformation, we have shown in (15) that in the warped geometry there is a doubling of supersymmetries: the constant spinors $\epsilon_i^{\pm(0)}$ give together 8 unbroken supersymmetries.

- 1/2 of the maximal unbroken supersymmetry in the bulk ($N = 1$).

  When the scalars are not constant, then there is the extra condition (16), leaving just one projected supersymmetry with 4 real components,

  $$\epsilon_i - \mathrm{i}\gamma_5 Q_{ij}\epsilon^j = 0\,, \qquad \epsilon_i = a^{1/2}\epsilon_i^{(0)}\,, \tag{17}$$

  where $\epsilon_i^{(0)}$ is constant. Thus, we remain in the bulk as well as on the brane with 1/2 of the original supersymmetries. Vice versa, imposing the projection (17) one derives from the vanishing of (13) that the conditions (11) should be satisfied.

Now we will study the fixed scalars, doubling of supersymmetries and an alternative to compactification world brane. The field equation for the 4-form and G-field Killing equations are solved if (11) are solved. In this section we look for the very particular solutions of these equations with maximal unbroken supersymmetry that have everywhere constant scalars.

The 'fixed scalar domain wall solution' is given by

$$(\varphi^y)' = 0\,, \qquad \left(\frac{\partial W}{\partial \varphi^x}\right)_{crit} = 0\,, \qquad \frac{a'}{a} = -g\varepsilon(x^5)W_{crit}\,. \tag{18}$$

The solution is given by the supersymmetric attractor equation[5] in the form

$$C_{IJK}\bar{h}^J\bar{h}^K = q_I\,, \quad \bar{h}^K \equiv \sqrt{W_{crit}}h^K\,. \tag{19}$$

Consistency implies

$$W_{crit}(C_{IJK}, q_I) = h^I_{crit}q_I = (\bar{h}^I q_I)^{2/3} = (C_{IJK}\bar{h}^I\bar{h}^J\bar{h}^K)^{2/3}\,. \tag{20}$$

In some cases the explicit solution of the attractor equation is known in the form $\bar{h}^I(C_{IJK}, q_I)$ The metric is

$$ds^2 = e^{-2gW_{crit}|x^5|}dx^{\underline{\mu}}dx^{\underline{\nu}}\eta_{\underline{\mu\nu}} + (dx^5)^2\,. \tag{21}$$

If we choose gW_{crit} to be positive (which means that at $x^5 = 0$ we have a positive tension brane), the metric is that introduced in Ref. 2, where two branes are present at some finite distance from each other. We will refer to this scenario as RSI. The second brane has a negative tension and can be sent to infinity, in principle, which leads to an alternative to compactification, discussed in Ref. 3. We will refer to this as RSII. Whether this limit is totally consistent is an independent issue, however the warp factor in the metric can be chosen to exponentially decrease away from the positive tension brane. This is not in a contradiction with the no-go theorem.[4] For constant scalars at the critical point, the metric behaves differently from the case when scalars are not constant but approaching the critical point. This can also be explained by the *doubling of unbroken supersymmetries in the bulk* for these solutions. Indeed, the gaugino transformations are vanishing without any constraint on the Killing spinors and the gravitino transformations also have an 8-dimensional zero mode as shown in (15). Note that the curvature is everywhere constant, except on the branes where the metric has a cusp. For example, the scalar curvature for this solution, is equal to

$$\tfrac{1}{4}R = 5g^2W_{crit}^2 - g(\delta(x^5) - \delta(x^5 - \tilde{x}^5))W_{crit} \,. \tag{22}$$

The simplest example of fixed scalar domain wall (related to double-extreme STU black holes) comes out from the M5-brane compactified on T^6 so that $W_{crit} = 3(q_S q_T q_U)^{1/3}$ (see sec. 4.3 in Ref. 10). Here q_S, q_T, q_U are FI terms.

In the context of our present work in the more general case with vector multiplets present and non-constant scalars, we will first change coordinates, write down the energy functional in the new coordinate system and proceed by solving the BPS conditions following from the energy functional. We take

$$ds^2 = a^2(y)dx^{\underline{\mu}}dx^{\underline{\nu}}\eta_{\underline{\mu\nu}} + a^{-4}(y)dy^2 \,, \tag{23}$$

so that $\frac{\partial}{\partial x^5} = a^2\frac{\partial}{\partial y}$ and we will use $\dot{}$ for $\frac{\partial}{\partial y}$. The stabilization equations for the energy are the BPS conditions:

$$a^2 g_{xy}\varphi^y = +3G(y)W_{,x} \,, \qquad a\dot{a} = -G(y)W \,, \tag{24}$$

where on shell for the 4-form field

$$G(y) = g\varepsilon(y) \,, \qquad \dot{G} = 2g\left(\delta(y) - \delta(y - \tilde{y})\right) \,. \tag{25}$$

Using the on-shell expression for $G(y)$ we may solve this and get that

$$C_{IJK}\tilde{h}^J\tilde{h}^K = H_I(y) \,, \tag{26}$$

where H_I is a harmonic function,

$$H_I = c_I - 2gq_I|y| \,, \tag{27}$$

satisfying the equation

$$\frac{d}{dy}\frac{d}{dy}H_I = -4gq_I[\delta(y) - \delta(y-\tilde{y})]\,. \tag{28}$$

This is an explicit answer for a given C_{IJK} and H_I under the condition that we know how to solve the algebraic attractor equation (19). If we know the solution of the algebraic attractor equation (19), we simply replace the q_I in this expression by the harmonic functions $H_I(y)$.

$$\tilde{h}^I(y) = \tilde{h}^I(C_{IJK}, H_I(y))\,, \tag{29}$$

and we can find the scalars and the metric. Contracting the attractor equation with $\tilde{h}^I$ we find that

$$C_{IJK}\tilde{h}^I\tilde{h}^J\tilde{h}^K = a^3(y)C_{IJK}h^Ih^Jh^K = a^3(y) \qquad \text{or} \qquad a^2(y) = h^I H_I\,. \tag{30}$$

Our solution for the domain wall metric is therefore given by (23) with $a(y)$ given by this expression.

The properties of the solutions with non-constant scalars depend on the choice of the model.

Here we briefly present a couple of examples.

The first one is an STU Wall. We consider again $C_{IJK}h^Ih^Jh^K = STU = 1$ and the harmonic functions are given by

$$H_S = c_S - 2gq_S|y|\,, \qquad H_T = c_T - 2gq_T|y|\,, \qquad H_U = c_U - 2gq_U|y|\,. \tag{31}$$

If all c are positive and all gq are negative, all harmonic functions are strictly positive. The metric and the moduli are

$$ds^2 = 3(H_SH_TH_U)^{1/3}(dx)^2 + 3^{-2}(H_SH_TH_U)^{-2/3}dy^2\,, \tag{32}$$

$$S = \left(\frac{H_TH_U}{H_S^2}\right)^{1/3}, \qquad T = \left(\frac{H_SH_U}{H_T^2}\right)^{1/3}, \qquad U = \left(\frac{H_TH_S}{H_U^2}\right)^{1/3}. \tag{33}$$

We use the normalization $3(c_Sc_Tc_U)^{1/3} = 1$ so that near $|y| = 0$ the domain wall metric (23) tends to a flat Minkowski metric.

$$ds^2_{y\to 0} \Rightarrow dx^{\underline{\mu}}dx^{\underline{\nu}}\eta_{\underline{\mu}\underline{\nu}} + dy^2\,. \tag{34}$$

If near the second wall at $|y| = |\tilde{y}|$ the values of the second terms in the harmonic functions are much larger than the first ones, the metric near this brane approaches the boundary of the adS space.

If in any of the harmonic functions the sign between the terms is opposite, and the harmonic function vanishes at $|y|_{sing}$, the second brane has to be at the position $|\tilde{y}| < |y|_{sing}$.

The second example is a Calabi-Yau Wall (Base $\mathbf{P}_2$ Vacuum). Here we take an example of a large complex structure limit of a particular Calabi-Yau threefold which defines a cubic form for 5D supergravity.

$$\mathcal{V}(h^1, h^2) = 9(h^1)^3 + 9(h^1)^2 h^2 + 3h^1 (h^2)^2 = 1 . \tag{35}$$

The attractor equation for this theory was solved in Ref. 10. We will use this solution to present one for the CY domain wall. We perform a coordinate redefinition to variables U and T, related to the basic cycles h^1 and h^2 by $U = 6^{\frac{1}{3}} h^1$, $T = 6^{\frac{1}{3}} (h^2 + \frac{3}{2} h^1)$. The relevant harmonic functions are given by $H_U = c_U - 2gq_U|y|$, $H_T = c_T - 2gq_T|y|$ and we choose $H_U > 0$ and $H_T > 0$. The domain wall metric takes the form (23), where

$$a^2(|y|) = \left[H_U \frac{2}{\sqrt{3}} \sqrt{H_U - \sqrt{H_U^2 - \frac{9}{4} H_T^2}} + H_T \sqrt{3} \sqrt{H_U + \sqrt{H_U^2 - \frac{9}{4} H_T^2}} \right]^{2/3} . \tag{36}$$

The range of the variables is determined by the positivity of the basic cycles h^1 and h^2, which translates for our solution into $2H_U - 3H_T > 0, H_U > 0, H_T > 0$. To avoid the singularity of the spacetime metric we may require that the second brane is at

$$|\tilde{y}| < |y|_{sing} = \frac{c_T}{2gq_T} . \tag{37}$$

We also find that if the condition $g(2q_U - 3q_T) > 0$ is satisfied, then there is also an upper limit for the distance between the branes:

$$|\tilde{y}| < \frac{2c_U - 3c_T}{2g(2q_U - 3q_T)} . \tag{38}$$

At this point $h^2 = 6^{-1/3}(T - \frac{3}{2}U) = 0$, i.e. the second cycle is collapsing. Thus to prevent this from happening, we have to put the second wall before this value of $|y|$.

As we already mentioned, a more careful and detailed analysis of domain walls in application to other CY spaces, as it was done for CY black holes, may lead to more interesting and diverse configurations.

We expect that the results of the present paper will clarify the issues of bulk and brane 10D supersymmetry and the construction of the 8-branes on the orbifold.

One would hope that the improved supersymmetric formulation of the 8-brane of string theory in the framework of 10D supergravity in a singular space will shed some light on the M-theory and its 9-brane. Since 11D supergravity does not admit a cosmological constant, one may in this way uplift from 10D to 11D some important information on extended objects of string theory, which is difficult to understand from the 11D massless supergravity point of view.

A related general problem to which the construction of this paper may be useful is to clarify the appearance of chiral fermions on the brane in the context of supersymmetric theory with extended supersymmetry in the bulk.

Acknowledgments

I am grateful to the organizers of Strings 2000 for the invitation to this conference and to the Theory Division of CERN where this work was performed. This work was supported by NSF grant PHY-9870115.

References

1. A.H. Jaffe *et al.*, astro-ph/0007333.
2. L. Randall and R. Sundrum, *Phys. Rev. Lett.* **83**, 3370 (1999).
3. L. Randall and R. Sundrum, *Phys. Rev. Lett.* **83**, 4690 (1999).
4. R. Kallosh and A. Linde, *JHEP* **0002**, 005 (2000); K. Behrndt and M. Cvetič, *Phys. Rev.* **D61**, 101901 (2000).
5. S. Ferrara, R. Kallosh and A. Strominger, *Phys. Rev.* **D52**, 5412 (1995); S. Ferrara and R. Kallosh, *Phys. Rev.* **D54**, 1514 (1996); S. Ferrara and R. Kallosh, *Phys. Rev.* **D54**, 1525 (1996).
6. M. Günaydin, G. Sierra and P.K. Townsend, *Nucl. Phys.* **B242**, 244 (1984); *Nucl. Phys.* **B253**, 573 (1985).
7. G.W. Gibbons and N.D. Lambert, *Phys. Lett.* **B488**, 90 (2000) [hep-th/0003197]; J. Maldacena and C. Nunez, hep-th/0007018.
8. E. Bergshoeff, R. Kallosh and A. Van Proeyen, *JHEP* **0010**, 033 (2000) [hep-th/0007044].
9. R. Altendorfer, J. Bagger and D. Nemeschansky, hep-th/0003117; T. Gherghetta and A. Pomarol, *Nucl. Phys.* **B586**, 141 (2000) [hep-ph/0003129]; N. Alonso-Alberca, P. Meessen and T. Ortin, *Phys. Lett.* **B482**, 400 (2000) [hep-th/0003248]; A. Falkowski, Z. Lalak and S. Pokorski, *Phys. Lett.* **B491**, 172 (2000) [hep-th/0004093].
10. M. Shmakova, *Phys. Rev.* **D56**, 540 (1997); A. Chou, R. Kallosh, J. Rahmfeld, S. Rey, M. Shmakova and W.K. Wong, *Nucl. Phys.* **B508**, 147 (1997).

OVERVIEW OF K-THEORY APPLIED TO STRINGS

EDWARD WITTEN
School of Natural Sciences, Institute for Advanced Study
Einstein Drive, Princeton, NJ 08540, USA

K-theory provides a framework for classifying Ramond-Ramond (RR) charges and fields. K-theory of manifolds has a natural extension to K-theory of noncommutative algebras, such as the algebras considered in noncommutative Yang-Mills theory or in open string field theory. In a number of concrete problems, the K-theory analysis proceeds most naturally if one starts out with an infinite set of D-branes, reduced by tachyon condensation to a finite set. This suggests that string field theory should be reconsidered for $N = \infty$.

1. Introduction And Definition Of $K(X)$

A D-brane wrapped on a submanifold S of spacetime may carry a nonzero Ramond-Ramond (RR) charge. RR fields are p-forms, and superficially it seems that the conserved charge should be measured by the cohomology class of the RR form (or of the cycle S itself). However, D-branes carry gauge fields; and gauge fields are not natural in (co)homology theory. They are natural in "K-theory." K-theory has been used to answer some questions about RR charges and fields; the aim of the present article[a] is to give an overview of this, along with some speculations (for which I have only very modest evidence) about how one might want to rethink open string field theory in the large N limit. These subjects fill sections 1-3. Some mathematical details have been postponed to section 4.

If X is spacetime and $\mathcal{A}(X)$ is the commutative, associative algebra of continuous complex-valued functions on X, then the K-theory of X can be defined in terms of representations of $\mathcal{A}(X)$. A representation of a ring is usually called a module. Here are some examples of $\mathcal{A}(X)$-modules.

The most obvious example of an $\mathcal{A}(X)$-module is $\mathcal{A}(X)$ itself. For $f \in \mathcal{A}(X)$ (regarded as a ring) and $g \in \mathcal{A}(X)$ (regarded as a module), we define

$$f(g) = fg, \tag{1}$$

where on the right hand side the multiplication occurs in $\mathcal{A}(X)$. This obviously obeys the defining condition of a module, which is that $(f_1 f_2)(g) = f_1(f_2(g))$.

More generally, consider a Dp-brane (or a collection of N Dp-branes for some $N > 0$) wrapped on a submanifold S of X, with any Chan-Paton gauge bundle W on the D-brane. Let $M(S)$ be the space of sections of W, that is, the space of

[a]This talk was based in part on lectures at the CIT-USC Center for Theoretical Physics. I thank both audiences for questions and comments.

one-particle states for a charged scalar coupled to the bundle W. Then $M(S)$ is an $\mathcal{A}(X)$-module; for $f \in \mathcal{A}(X)$, $g \in M(S)$, we simply set again $f(g) = fg$. On the right hand side, the multiplication is defined by restricting f (which is a function on X) to S and then multiplying f and g.

So in, say, Type IIB superstring theory, a collection of $D9$-branes defines a representation or module E of $\mathcal{A}(X)$. A collection of $\overline{D9}$-branes defines another module F. So any configuration of $D9$ and $\overline{D9}$-branes determines a pair (E, F).

To classify D-brane charge, we want to classify pairs (E, F) modulo physical processes. An important process[1] is brane-antibrane creation and annihilation – the creation or annihilation of a set of $D9$'s and $\overline{D9}$'s each bearing the same gauge bundle G. This amounts to

$$(E, F) \leftrightarrow (E \oplus G, F \oplus G). \tag{2}$$

The equivalence classes make up a group called $K(X)$ (or $K(\mathcal{A}(X))$ if we want to make the interpretation in terms of $\mathcal{A}(X)$-modules more explicit). The addition law in this group is just

$$(E, F) + (E'F') = (E \oplus E', F \oplus F'). \tag{3}$$

The inverse of (E, F) is (F, E); note that $(E, F) \oplus (F, E) = (E \oplus F, E \oplus F)$, and using the equivalence relation in Eq. 2, this is equivalent to zero. D-branes of Type IIB carry conserved charges that take values in $K(X)$.

In the above definition of $K(X)$, we used only ninebranes, even though, as we explained earlier, an $\mathcal{A}(X)$ module can be constructed using Dp-branes (or antibranes) for any p. In fact, we can classify D-brane charge just using the ninebranes, and then build the Dp-branes of $p < 9$ via pairs (E, F) with a suitable tachyon condensate. This construction (due to Atiyah, Bott, and Shapiro) was reviewed in Ref. 2, section 4, in the D-brane context. A more systematic explanation of the definition of $K(X)$ in terms of modules, clarifying the role of ninebranes, can be found in section 4.

1.1. *Advantages of K-theory description*

What do we gain by knowing that D-brane charge is classified by K-theory?

First of all, it is the right answer. Wherever one looks closely at topological properties of RR charges (or fields), one sees effects that reflect the K-theory structure. For example, there are stable D-brane states (like the nonsupersymmetric $D0$-branes of Type I) that would not exist if D-brane charge were classified by cohomology instead of K-theory. Conversely, it is possible to have a D-brane state that would be stable if D-brane charge were measured by cohomology, but which is in fact unstable (via a process that involves nucleation of $D9$-$\overline{D9}$ pairs in an intermediate state). This occurs[3] in Type II superstring theory, in which a D-brane wrapped on a homologically nontrivial cycle in spacetime is in fact in certain cases unstable. Finally, the K-theory interpretation of D-branes is needed[4] to make sense of a certain global worldvolume anomaly.

But I think that there is a deeper reason that it is good to know about the K-theory interpretation of D-branes: it may be naturally adapted for stringy generalizations. In fact (though some mathematical details have been postponed to section 4), we defined $K(X)$ in terms of representations of the algebra $\mathcal{A}(X)$ of functions on spacetime. We can similarly define $K(\mathcal{A})$ for any noncommutative algebra $\mathcal{A}$, in terms of pairs (E, F) of $\mathcal{A}$-modules. By contrast, we would not have an equally useful and convenient notion of "cohomology" if the algebra of functions on spacetime is replaced by a noncommutative ring.

For example, turning on a Neveu-Schwarz B-field, we can make $\mathcal{A}(X)$ noncommutative; the associated $K(\mathcal{A})$ was used by Connes, Douglas, and Schwarz in the original paper on noncommutative Yang-Mills theory applied to string theory.[5] This is an interesting example, even though it involves only the zero modes of the strings. One would much like to have a fully stringy version involving a noncommutative algebra constructed using all of the modes of the string, not just the zero modes.

What is the right noncommutative algebra that uses all of the modes? We do not know, of course. One concrete candidate is the $*$-algebra of open string field theory, defined in terms of gluing strings together. If I call this algebra $\mathcal{A}_{st}$, it seems plausible that D-brane charge is naturally labeled by $K(\mathcal{A}_{st})$. (For a manifold of very large volume compared to the string scale, I would conjecture that $K(\mathcal{A}_{st})$ is the same as the ordinary $K(X)$ of topological K-theory.) I will come back to $K(\mathcal{A}_{st})$ in section 3.3. First, I want to finish our survey of known applications of K-theory in string physics.

2. K-Theory and RR Fields

K-theory is relevant to understanding RR fields as well as charges.[6–8,3]

Naively speaking, an RR p-form field G_p obeys a Dirac quantization law according to which, for any p-cycle U in spacetime,

$$\int_U \frac{G_p}{2\pi} = \text{integer}. \tag{4}$$

If that were the right condition, then RR fields would be classified by cohomology.

But that is not the right answer, because the actual quantization condition on RR periods is much more subtle than Eq. 4. There are a variety of corrections to Eq. 4 that involve spacetime curvature and the gauge fields on the brane, as well as self-duality and global anomalies.[9–13,7,8,3]

The answer, for Type IIB superstrings, turns out to be that RR fields are classified by $K^1(X)$. For our purposes, $K^1(X)$ can be defined as the group of components of the group of continuous maps from X to $U(N)$, for any sufficiently large N.

This statement means that topological classes of RR fields on X are classified by a map $U : X \to U(N)$ for some large N. The relation of G_p to U is roughly $G_p \sim \text{Tr } (U^{-1}dU)^p$; here I have ignored corrections due to spacetime curvature and subtleties associated with self-duality of RR fields.

The physical meaning of U is not clear. For Type IIA, the analog is that RR fields are classified by a $U(N)$ gauge bundle (for some large N) with connection A and curvature F_A, the relation being $G_p \sim \mathrm{Tr}\ F_A^{p/2}$. The analog for M-theory involves E_8 gauge bundles with connection.[13,14,3] Again, the physical meaning of the $U(N)$ or E_8 gauge fields is not clear.

The value of using K^1 to classify RR fields of Type IIB is that this gives a concise way to summarize the otherwise rather complicated quantization conditions obeyed by the RR fields. In addition, this framework is useful in describing subtle phase factors that enter in the RR partition function. In hindsight, once it is known that RR charges are classified by K-theory, one should have suspected a similar classification for RR fields. After all, RR charges produce RR fields! So the math used to classify RR charges must be similar to the math used to classify RR fields.

Just like $K(X)$, $K^1(X)$ has an analog for any noncommutative algebra $\mathcal{A}$. Given $\mathcal{A}$, we let $\mathcal{A}_N$ denote the group of invertible $N \times N$ matrices whose matrix elements are elements of $\mathcal{A}$. Then $K^1(\mathcal{A})$ is the group of components of $\mathcal{A}_N$, for large N.

For example, for $\mathcal{A} = \mathcal{A}(X)$ the ring of complex-valued continuous functions on X, $\mathcal{A}_N$ is the group of maps of X to $GL(N, \mathbf{C})$. This is contractible to the group of maps of X to $U(N)$, so for large N the group of components of $\mathcal{A}_N$ is the same as $K^1(X)$, as we defined it initially.

The existence of a generalization of $K^1(X)$ for noncommutative rings means that the description of Type IIB RR fields by $K^1(X)$ in the long distance limit may be a useful starting point for stringy generalizations.

3. $N \to \infty$

In the last section, N was a sufficiently large but finite integer. Our next task will be to describe some things that depend on setting N equal to infinity.

Before doing so, let us recall the role of the $N \to \infty$ limit in physics. It is important in the conjectured link of gauge theory with strings; in the old matrix models that are used to give soluble examples of string theory; in the matrix model of M-theory; and in the correspondence between gravity in an asymptotically AdS spacetime and conformal field theory on the boundary.

My theme here will be to suggest that we should somehow study D-branes with $N = \infty$, with tachyon condensation to annihilate most of the branes and reduce to something more manageable. To motivate this, I will consider two concrete questions that seem to require taking the number of D-branes to be infinite. One question is the relation of Type IIA superstrings to K-theory; the other is the inclusion of a topologically non-trivial NS three-form field H in the K-theory classification of RR charges and fields.

After giving the talk, I became aware of a standard problem that has already been interpreted in terms of an infinite number of $D9$ and $\overline{D9}$-branes with tachyon condensation to something manageable (see Ref. 15, sections 2.2, 4, and 5). This is a problem involving a $D5$-brane probe of a $D5$-$D9$ system.

3.1. *Type IIA*

For Type IIB superstrings, we used $K(X)$ to classify RR charges, and $K^1(X)$ to classify RR fields.

The T-dual statement is that for Type IIA, $K^1(X)$ should classify RR charges, and $K(X)$ should classify RR fields. (By Bott periodicity, $K^{i+2}(X) = K^i(X)$, so the only K-groups of X are $K^0(X)$, which we have called simply $K(X)$, and $K^1(X)$.)

The most concrete and natural attempt to explain in general why $K^1(X)$ classifies RR charges for Type IIA is that of Horava.[16] The starting point here is to consider a system of N unstable $D9$-branes of Type IIA. The branes support a $U(N)$ gauge field and a tachyon field T in the adjoint representation of $U(N)$. There is a symmetry $T \to -T$.

The effective potential for the tachyon field is believed to have the general form

$$V(T) = \frac{1}{g_{st}} \mathrm{Tr}\ F(T), \tag{5}$$

where the function $F(T)$ is non-negative and, after scaling T correctly, vanishes precisely if $T = \pm 1$. Hence $V(T)$ is minimized if and only if every eigenvalue of T is ± 1.

It was argued in Ref. 16 that, in flat $\mathbf{R}^{10}$, one can make supersymmetric Dp-branes (for even p) as solitons of T. For example, to make a $D6$-brane, we set $N = 2$. Let $\vec{x}$ be the three coordinates normal to the $D6$-brane, and set

$$T = \frac{\vec{\sigma} \cdot \vec{x}}{|x|} f(|x|), \tag{6}$$

where $f(r) \sim r$ for small r, and $f(r) \to \infty$ for $r \to \infty$. So for $|x| \to \infty$, the eigenvalues of T are everywhere ± 1. Near $x = 0$, there is a topological knot that we interpret as the $D6$-brane.

In flat $\mathbf{R}^{10}$, one can similarly make Dp-branes for other even p. But on a general spacetime, this does not work for arbitrary Dp-branes unless we set $N = \infty$. The problem is most obvious if X, or at least its spatial part, is compact. The tachyon field T, being adjoint-valued, maps X to the Lie algebra of $U(N)$; since the Lie algebra is contractible, T carries no topology. So a map from X to the Lie algebra does not represent an element of $K^1(X)$; indeed, it does not carry topological information at all.

To define an element of $K^1(X)$, we need the group, not the Lie algebra; a map $U : X \to U(N)$ does the job, as I have stated before.

Amazingly, as Atiyah and Singer showed long ago, we get back the right topology from the Lie algebra if we set $N = \infty$! We have to interpret $U(\infty)$ to be the unitary group $\mathcal{U}$ of a Hilbert space $\mathcal{H}$ of countably infinite dimension. (Such a Hilbert space is also called a separable Hilbert space.) We interpret the $N = \infty$ analog of the space of hermitian $N \times N$ matrices to be the space of bounded self-adjoint operators T on $\mathcal{H}$ whose spectrum is as follows: there are infinitely many positive eigenvalues and infinitely many negative ones, and zero is not an accumulation point of the spectrum. (The last condition makes T a "Fredholm operator.") Physically, T should be required to obey these conditions, since they are needed to make the energy and the $D8$-brane charge finite. In fact, to make the energy finite, almost all the eigenvalues of T are very close to ± 1. Anyway, with these conditions imposed on T, it turns out that the space of T's has the same topology as that of $U(N)$ for large N.

So we can use tachyon condensation on a system of $D9$-branes to describe RR charges for Type IIA. But we have to start with infinitely many $D9$-branes, which then undergo tachyon condensation down to a configuration of finite energy.

3.2. *Obstruction to finite* N

Let us now explain in more concrete terms the obstruction to making D-branes in this way for finite N, and how it vanishes for $N = \infty$.

Let us go back to the example of a $D6$-brane constructed with 2 $D9$-branes. We took the transverse directions to be a copy of $\mathbf{R^3}$, and the tachyon field to be

$$T = \frac{\vec{\sigma} \cdot \vec{x}}{|x|} f(|x|). \tag{7}$$

If we try to compactify the transverse directions to $\mathbf{S^3}$, we run into trouble because T is not constant at infinity. The conjugacy class of T is constant at infinity – the eigenvalues of T are everywhere 1 and -1 – but T itself is not constant.

Moreover, T is not homotopic to a constant at infinity. If T were homotopic to a constant near infinity, we would deform it to be constant and then extend it over $\mathbf{S^3}$. But it is not homotopic to a constant.

The basic obstruction to making T constant at infinity is the "magnetic charge." Let $\mathbf{S^2}$ be a sphere at infinity in $\mathbf{R^3}$. Over $\mathbf{S^2}$, we can define a line bundle $\mathcal{L}_+$ whose fiber is the $+1$ eigenspace of T, and a line bundle $\mathcal{L}_-$ whose fiber is the -1 eigenspace of T. The line bundles L_+ and $\mathcal{L}_-$ are topologically nontrivial – their first Chern classes are respectively 1 and -1. As long as we try to deform T preserving the fact that its eigenvalues are 1 and -1, the line bundles $L_\pm$ are well-defined, and their first Chern classes are invariant. So the nontriviality of $\mathcal{L}_+$ (or $\mathcal{L}_-$) prevents us from making a homotopy to constant T.

Let us add some additional "spectator" $D9$-branes, and see if anything changes. Suppose there are $M = 2k$ additional branes, so that the total number of branes is $N = 2 + M = 2 + 2k$. Let the tachyon field be $T' = T \oplus U$, where T is as above and U is the sum of k copies of the matrix

$$\begin{pmatrix} 1 & 0 \\ 0 & -1 \end{pmatrix} \tag{8}$$

acting on the $2k$ additional branes.

Thus T' has near infinity $k+1$ eigenvalues $+1$ and $k+1$ eigenvalues -1. The $+1$ eigenspace of T' is a vector bundle V_+ of first Chern class 1 (since it is constructed by adding a trivial bundle to $\mathcal{L}_+$), and the -1 eigenspace of T' is similarly a vector bundle V_- of first Chern class -1. In particular, V_+ and V_- are nontrivial, so we have not gained anything by adding the spectator branes: T' is not homotopic to a constant, and cannot be extended over infinity. The nontriviality of V_+ is controlled by $\pi_1(U(k+1)) = \mathbf{Z}$, which is associated with the existence of a first Chern class.

Instead, what happens if we set $k = \infty$? To be more precise, we take the number of spectator $D9$-branes to be countably infinite, and assume $T' = T \oplus U$, where U is the direct sum of countably many copies of the matrix in Eq. 8. We can still define the bundles V_+ and V_-, their fibers are separable Hilbert spaces (that is, Hilbert spaces of countably infinite dimension). $U(k+1)$ is replaced by $\mathcal{U}$, the unitary group of a separable Hilbert space. Now we run into the fundamental fact (Kuiper's theorem) that $\mathcal{U}$ is contractible; its homotopy groups are all zero. Thus, any bundle of separable Hilbert spaces is trivial. In particular, V_+ and V_- are trivial, so T' is homotopic to a constant and can be extended over infinity.

So if the total number of unstable $D9$-branes is $N = \infty$, we can make a $D6$-brane localized at a point in $\mathbf{S^3}$. More generally, in view of the result of Atiyah and Singer, we can starting at $N = \infty$ build an arbitrary class in $K^1(X)$ via tachyon condensation.

In terms of applying this result to physics, there are a few issues that we should worry about. One question is simply whether it is physically sensible to start with infinitely many branes and rely on tachyon condensation to get us down to something of finite energy. We will have to leave this question to the future (but see Ref. 15 for an example).

Quite a different question is whether the answer that we have obtained by setting $N = \infty$ is the right one for physics. In the field of a $D6$-brane that is localized at a point on $\mathbf{S^3}$, the equation for the RR two-form field G_2 (of which the $D6$-brane is a magnetic source) has no solution, since "the flux has nowhere to go."

It seems that the situation is that $N = \infty$ corresponds to the correct answer in classical open string theory, where the effective action comes from worldsheets with the topology of a disc. The RR fields enter as a correction of relative order g_{st} (the closed string coupling constant) coming from worldsheets with cylinder topology, and should be ignored in the classical approximation.

The classification of D-branes by brane creation and annihilation holds at $g_{st} = 0$, and (as we recalled in section 1) leads for Type IIB to a classification of D-brane charge by $K(X)$. To get the analogous answer – namely $K^1(X)$ – for Type IIA via unstable $D9$-branes and tachyon condensation, we need to start at $N = \infty$.

Intuitively, in the absence of tachyon condensation, $N = \infty$ should correspond to $g_{st} = 0$, since the effective expansion parameter for open strings is $g_{st}N$. If N is infinite, then prior to tachyon condensation, g_{st} must be zero, or the quantum corrections diverge. If we want g_{st} to be nonzero, we need tachyon condensation to reduce to an effective finite value of N.

3.3. *Turning on an H-field*

A somewhat analogous problem is to consider D-branes when the Neveu-Schwarz three-form field H is topologically nontrivial. We will carry out this discussion in Type IIB (for Type IIA, we would have to combine what follows with what we said above in the absence of the H-field).

Just as at $H = 0$, we would like to classify D-brane states by pairs (E, F) (where E is a $D9$ state and F is a $\overline{D9}$ state) subject to the usual sort of equivalence relation. But there is a problem in having a $D9$ state in the presence of an H-field.

In fact, when H is topologically non-trivial, one cannot have a single $D9$-brane. On the $D9$-brane, there is a $U(1)$ gauge field with field strength F. The relation $dF = H$ shows, at the level of de Rham cohomology, that H must be topologically trivial if a single $D9$-brane is present. This conclusion actually holds precisely, not just in de Rham cohomology.

There is a special case in which there is a comparatively elementary cure for this difficulty (see Ref. 2, section 5.3 and Ref. 17). If H is torsion, that is if there is an integer $M > 0$ such that MH is topologically trivial, then it is possible to have a set of M $D9$-branes whose "gauge bundle" actually has structure group $U(M)/\mathbf{Z}_M$, rather than $U(M)$. (The obstruction to lifting the $U(M)/\mathbf{Z}_M$ bundle to a $U(M)$ bundle is determined by H.) We will call such a gauge bundle a twisted bundle. More generally, for any positive integer m, we can have $N = mM$ $D9$-branes with the structure group of the bundle being $U(mM)/\mathbf{Z}_M$. In such a situation, D-brane charge is classified, as one would guess, by pairs (E, F) of twisted bundles (or $D9$ and $\overline{D9}$ states) subject to the usual equivalence relation. The equivalence classes make a group $K_H(X)$.

If one wishes to interpret $K_H(X)$ as the K-theory of representations of an algebra, one must pick a particular twisted bundle W and consider a D-brane state with boundary conditions determined by W. The W-W open strings transform in the adjoint representation, so the gauge parameters of the zero mode sector of the open strings are sections of $W \otimes \overline{W}$. Notice that although W is a twisted bundle (with structure group $U(M)/\mathbf{Z}_M$ rather than $U(M)$), $W \otimes \overline{W}$ is an ordinary bundle, since the center acts trivially in the adjoint representation.

The sections of $W \otimes \overline{W}$ form an algebra, defined as follows: if $s^i{}_j$ and $t^k{}_l$ are sections of $W \otimes \overline{W}$ – where the upper and lower indices are respectively W- and $\overline{W}$-valued – then their product is $(st)^i{}_l = \sum_k s^i{}_k t^k{}_l$. This is the algebra $\mathcal{A}_W(X)$ of all endomorphisms or linear transformations of the bundle W. The algebra of open string field theory, for W-W open strings, reduces to $\mathcal{A}_W$ if one looks only at the zero modes of the strings. This is a sensible approximation at low energies in a limit in which X is very large compared to the string scale.

If H is zero and W is a trivial rank one complex bundle, then $\mathcal{A}_W(X)$ is our friend $\mathcal{A}(X)$. If H is zero and W is a trivial rank N complex bundle, then including W means simply that there are $N \times N$ Chan-Paton matrices everywhere. So in this case, $\mathcal{A}_W(X) = \mathcal{A}(X) \otimes M_N$, where M_N is the algebra of $N \times N$ complex-valued matrices. In general, whatever H is, W is always trivial locally, so locally $\mathcal{A}_W(X)$

is isomorphic to $\mathcal{A}(X) \otimes M_N$.

A twisted bundle is equivalent to an $\mathcal{A}_W$-module, and the group $K_H(X)$ of pairs (E, F) of twisted bundles (modulo the usual equivalence) coincides with $K(\mathcal{A}_W)$, the K-group of $\mathcal{A}_W$-modules. This will be explained in section 4. This assertion leads to an immediate puzzle; $K_H(X)$ as defined in terms of pairs (E, F) of twisted bundles is manifestly independent of W while $K(\mathcal{A}_W)$ appears to depend on W. Indeed, given any two distinct twisted bundles W and W', the corresponding algebras $\mathcal{A}_W$ and $\mathcal{A}_{W'}$ are distinct, but are "Morita-equivalent." This concept is explained in section 4, where we also show that the Morita equivalence implies that $K(\mathcal{A}_W) = K(\mathcal{A}_{W'})$.

So far, we have only considered the case that H is torsion. A typical example, important in the AdS/CFT correspondence, is the spacetime $X = \mathrm{AdS}_5 \times \mathrm{RP}^5$, where a torsion H-field on RP^5 is used to describe $Sp(n)$ rather than $SO(2n)$ gauge theory in the boundary CFT.

However, in most physical applications, H is not torsion. In that case, we must somehow take a large M limit of what has been said above. The right way to do this has been shown by Bouwknegt and Mathai[18] and Atiyah and Segal.[19] The suitable large M limit of $U(M)/\mathbf{Z}_M$ is $PU(\mathcal{H}) = U(\mathcal{H})/U(1)$. In other words, for $M = \infty$, one replaces $U(M)$ by the unitary group $U(\mathcal{H})$ of a separable Hilbert space $\mathcal{H}$; and one replaces $\mathbf{Z}_M$ by $U(1)$. This means, in particular, that when H is not torsion, one cannot have a finite set of $D9$- or $\overline{D9}$-branes, but one can have an infinite set, with a suitable infinite rank twisted gauge bundle E or F. Then D-brane charge is classified by the group K_H of pairs (E, F) modulo the usual equivalence relation. A detailed explanation can be found in Ref. 18. Kuiper's theorem – the contractibility of $\mathcal{U} = U(\mathcal{H})$ – plays an important role, as it did in section 3.1.

This construction, in the $M = \infty$ limit, has the beautiful property, explained in Ref. 18, that the noncommutative algebra whose K-group is K_H is *unique*, independent of any arbitrary choice of twisted bundle W or W'. This really depends on the number of $D9$ and $\overline{D9}$ branes being infinite.

3.4. *Stringy generalization?*

At the risk of going out on a somewhat shaky limb, I will now try to propose a stringy generalization of some of this.

We start with a closed string background and try to form an open string algebra. To do so, we must pick an open string boundary condition – call it α. Then, in open string field theory, the α-α open strings form an algebra $\mathcal{A}_\alpha$.

This algebra is not unique; we could instead pick another boundary condition β and define another algebra $\mathcal{A}_\beta$.

The example of strings in a background H-field that is torsion, and the associated algebras $\mathcal{A}_W$, $\mathcal{A}_{W'}$, suggests the nature of the relation between $\mathcal{A}_\alpha$ and $\mathcal{A}_\beta$: they are different algebras, but are Morita-equivalent and hence have the same K-theory.

The assertion that $\mathcal{A}_\alpha \neq \mathcal{A}_\beta$, if true, is quite troublesome. It is a sharp statement of the lack of manifest background independence of open string field theory. (Background independence in this context means independence of the open string background; of course, classical open string field theory depends on a closed string background in which the open strings propagate.) It means that the formalism depends on which open string background one uses in setting up the theory.

The example of open strings in an H-field suggests a cure: take the number of $D9$- and $\overline{D9}$-branes to be infinite.

Let $m\alpha$ and $m\beta$ denote m copies of α or β (that is, α or β supplemented with $U(m)$-valued Chan-Paton factors). Then the conjecture is that $\mathcal{A}_{m\alpha} \neq \mathcal{A}_{m\beta}$ for any finite m, but that they are equal for $m = \infty$. Actually, the statement about what happens for $m = \infty$ can be formulated much more precisely by analogy with statements in Ref. 18. Let $\mathcal{K}$ be the algebra of compact operators in a separable Hilbert space $\mathcal{H}$. Then the conjecture is that for any α and β,

$$\mathcal{A}_\alpha \otimes \mathcal{K} = \mathcal{A}_\beta \otimes \mathcal{K}. \tag{9}$$

The main evidence for the conjecture is that, according to Ref. 18, the corresponding statement ($\mathcal{A}_W \neq \mathcal{A}_{W'}$ but $\mathcal{A}_W \otimes \mathcal{K} = \mathcal{A}_{W'} \otimes \mathcal{K}$) holds for the zero mode algebra in the presence of a torsion H-field. In addition, if it is necessary to tensor with $\mathcal{K}$ before the algebras become isomorphic, this helps explain why background independence in open string field theory is so hard to understand; in other words it helps explain why the theory constructed with classical solution α and algebra $\mathcal{A}_\alpha$ looks different from the theory constructed with classical solution β and algebra $\mathcal{A}_\beta$. However, the evidence for the conjecture is quite limited; it may be that whenever α is continuously connected to β, $\mathcal{A}_\alpha = \mathcal{A}_\beta$, and that the difficulty in understanding background independence in open string field theory is "just" a technical problem.

If the conjecture is right, one would suspect that to get a greater degree of background independence in open string field theory, we should start with infinitely many $D9$-branes and rely on tachyon condensation to get us down to something reasonable.

For this to be useful, we would need a description of $\mathcal{A}_\alpha \otimes \mathcal{K}$ much simpler and more incisive than any description we have today for $\mathcal{A}_\alpha$ or $\mathcal{A}_\beta$. In such a hypothetical new description, the BRST operator Q might be harder to describe. If that happened, we would have to take our lumps!

4. Mathematical Details

This concluding section will be devoted to explaining a few of the mathematical points that we have skimmed over so far.

Let $\mathcal{A}$ be a ring – in fact, let us momentarily assume that $\mathcal{A}$ is commutative and associative. An $\mathcal{A}$-module M is called "free" if it is a direct sum of copies of $\mathcal{A}$: $M = \mathcal{A} \oplus \mathcal{A} \oplus \ldots \oplus \mathcal{A}$. An $\mathcal{A}$-module M is called "projective" if there is another $\mathcal{A}$-module M' such that $M \oplus M'$ is free.

Let us consider what these definitions mean in case $\mathcal{A} = \mathcal{A}(X)$ is the ring of continuous complex-valued functions on a manifold X. We can think of $\mathcal{A}$ as the space of sections of a trivial complex line bundle $\mathcal{O}$. Given a trivial rank n complex vector bundle $V = \mathcal{O} \oplus \mathcal{O} \oplus \ldots \oplus \mathcal{O}$, the space of sections of V is a free module M which is the sum of n copies of $\mathcal{A}$.

Now suppose we are given *any* complex vector bundle E over X. The space of sections of E is an $\mathcal{A}$-module $M(E)$ (given a function $a \in \mathcal{A}(X)$ and a section m of E, we simply define $a(m)$ to be the product am).[b] $M(E)$ is a free module if and only if E is a trivial vector bundle. But $M(E)$ is always projective. Indeed, there is always a "complementary" vector bundle F such that $E \oplus F$ is trivial, and hence $M(E) \oplus M(F) = M(E \oplus F)$ is free. Conversely (by a theorem of Serre and Swan), every projective $\mathcal{A}(X)$-module is $M(E)$ for some complex vector bundle E over X.

For any ring $\mathcal{A}$, the projective modules form a semigroup: if E and E' are projective modules, so is $E \oplus E'$. Given any semigroup (with an addition operation that we will write as $\oplus$), there is a canonical way to form a group. This is done the same way that one builds the group of integers, starting with the semigroup of positive integers. An element of the group is a pair (E, F), with E and F elements of the semigroup (so in our situation, E and F are projective modules) and subject to the equivalence relation $(E, F) \cong (E \oplus G, F \oplus G)$ for any element G in the semigroup. Pairs are added by $(E, F) + (E', F') = (E \oplus E', F \oplus F')$. The equivalence classes form a group (the zero element or additive identity is $0 = (G, G)$ for any G, and the additive inverse of (E, F) is (F, E), since $(E, F) + (F, E) = (E \oplus F, E \oplus F) = 0$). In case one starts with the semigroup of projective modules for a ring $\mathcal{A}$, the group formed this way is called $K(\mathcal{A})$.

For $\mathcal{A} = \mathcal{A}(X)$, we have in section 1 defined $K(\mathcal{A})$ just in terms of vector bundles or in other words $D9$- and $\overline{D9}$-brane configurations. Now we see how to describe this restriction in a way that has broader validity: the ninebranes correspond to projective modules, and the definition of $K(X)$ in terms of ninebranes is a special case of the definition of $K(\mathcal{A})$ for any ring $\mathcal{A}$ in terms of projective modules.

Now let us drop the assumption that $\mathcal{A}$ is commutative (but keep associativity). If $\mathcal{A}$ is a noncommutative ring, one can canonically associate with it a second ring $\mathcal{A}^{op}$ (called the opposite ring) whose elements are in one to one correspondence with elements of $\mathcal{A}$, but which are multiplied in the opposite order. Thus, for every element $a \in \mathcal{A}$, there is a corresponding $a^{op} \in \mathcal{A}^{op}$, with the multiplication law of $\mathcal{A}^{op}$ being $a^{op}b^{op} = (ba)^{op}$.

For a noncommutative ring, we must distinguish two types of modules, left-modules and right-modules. $\mathcal{A}$ acts on the left on a left-module M, the basic axiom being that for $a, a' \in \mathcal{A}$ and $m \in M$, one has $(aa')(m) = a(a'(m))$. $\mathcal{A}$ acts on the right on a right-module M, the basic axiom being that $m \cdot (aa') = (m \cdot a) \cdot a'$. A left-module of $\mathcal{A}$ is the same as a right-module of $\mathcal{A}^{op}$, and vice-versa.

The definition of projective module is the same as in the commutative case: a left or right $\mathcal{A}$-module M is projective if there is another left or right $\mathcal{A}$-module N such that $M \oplus N$ is a free left or right $\mathcal{A}$-module. From the semigroup of projective left

---

[b]In the informal spirit of our discussion, I have generally used the same name E for a bundle and the corresponding module. In this paragraph only, I distinguish E from $M(E)$ in the notation, to facilitate the statement of the Serre-Swan theorem.

$\mathcal{A}$-modules, we form a group called $K(\mathcal{A})$. We can also form a semigroup from the projective right $\mathcal{A}$-modules; it is the same as the semigroup of projective left $\mathcal{A}^{op}$-modules, so the associated abelian group is $K(\mathcal{A}^{op})$. For the rings we are interested in, there is generally an operation of complex (or hermitian) conjugation that maps $\mathcal{A}$ to $\mathcal{A}^{op}$. Consequently, the K-groups formed using left or right $\mathcal{A}$-modules are equivalent to each other under complex conjugation.

Given two rings $\mathcal{A}$ and $\mathcal{B}$, a bimodule (or $(\mathcal{A},\mathcal{B})$-bimodule) is a group M which is simultaneously a left $\mathcal{A}$-module and a right $\mathcal{B}$-module, with the left action of $\mathcal{A}$ and the right action of $\mathcal{B}$ commuting. For $a \in \mathcal{A}$, $b \in \mathcal{B}$, and $m \in M$, the left action of $\mathcal{A}$ is denoted $a(m) = am$, the right action of $\mathcal{B}$ is denoted $b(m) = mb$, and since we assume that $\mathcal{A}$ and $\mathcal{B}$ commute, we have $(am)b = a(mb)$; we abbreviate these expressions as amb.

We can now define a very important relationship between rings, called Morita equivalence. It was first exploited in connection with D-branes in Ref. 5; for further developments see Ref. 20. Let M be an $(\mathcal{A},\mathcal{B})$-bimodule. The choice of M enables us to define a map from left $\mathcal{B}$-modules to left $\mathcal{A}$-modules: given a left $\mathcal{B}$-module W, we map it to the left $\mathcal{A}$-module $W' = M \otimes_{\mathcal{B}} W$. Similarly, if we are given a $(\mathcal{B},\mathcal{A})$-bimodule N, we can map a left $\mathcal{A}$-module W' to a left $\mathcal{B}$-module W'' by $W'' = N \otimes_{\mathcal{A}} W'$. If these two operations are inverse to each other, in other words if one has $W'' = W$ for all W, we say that $\mathcal{A}$ and $\mathcal{B}$ are Morita-equivalent. In this case, the semigroup of left $\mathcal{A}$-modules is isomorphic to the semigroup of left $\mathcal{B}$-modules, and $K(\mathcal{A}) = K(\mathcal{B})$.

The meaning of this relationship in open string field theory was explained in Ref. 21. For any open string boundary condition α, one defines a ring $\mathcal{A}_\alpha$ consisting of the α-α open strings. For any other boundary condition γ, the α-γ open strings form a left $\mathcal{A}_\alpha$-module $M_{\alpha\gamma}$, and the γ-α open strings form a right $\mathcal{A}_\alpha$-module $M_{\gamma\alpha}$. Presumably these are in a suitable sense a basis of left and right $\mathcal{A}_\alpha$-modules. Now consider any other open string boundary condition β and the associated algebra $\mathcal{A}_\beta$. One forms left and right $\mathcal{A}_\beta$ modules $M_{\beta\gamma}$ and $M_{\gamma\beta}$ in the same way, using β-γ and γ-β open strings. Obviously, there is a natural association between left $\mathcal{A}_\alpha$ modules and left $\mathcal{A}_\beta$ modules, namely

$$M_{\alpha\gamma} \leftrightarrow M_{\beta\gamma}. \tag{10}$$

So $K(\mathcal{A}_\alpha) = K(\mathcal{A}_\beta)$ for all α and β. We established this last fact directly without talking about Morita equivalence, but in fact, it seems that the relation between $\mathcal{A}_\alpha$ and $\mathcal{A}_\beta$ is that they are Morita-equivalent. Indeed, one has natural $(\mathcal{A}_\alpha, \mathcal{A}_\beta)$ and $(\mathcal{A}_\beta, \mathcal{A}_\alpha)$ bimodules $M_{\alpha\beta}$ and $M_{\beta\alpha}$ constructed from the α-β and β-α open strings; it seems very likely that these bimodules have the right properties to establish a Morita equivalence between $\mathcal{A}_\alpha$ and $\mathcal{A}_\beta$.

Now let us consider a more elementary version of this, with the goal of adding some details to the discussion in section 3.2. Let M_n be the algebra of $n \times n$ complex matrices. The most obvious left M_n-module is the space W of n-component column vectors. This is a projective module because, as a left module, M_n consists of n

copies of W (M_n consists of $n \times n$ matrices; in the left action of M_n on itself, M_n acts separately on each of the n columns making up the $n \times n$ matrix, so these comprise n copies of W). So if W' is the sum of $n-1$ copies of W, then $W \oplus W'$ is free, and thus W is projective. It can be shown that every projective left M_n module is the sum of copies of W. So the semigroup of projective left M_n modules is isomorphic to the semigroup of non-negative integers, and $K(M_n) = \mathbf{Z}$.

Since $K(M_n)$ is thus independent of n, one might wonder if the M_n's of different n are Morita-equivalent. This is indeed so. For any positive integers n and k, the most obvious (M_n, M_k) bimodule is the space $C_{n,k}$ of $n \times k$ matrices, with M_n acting on the left and M_k on the right. Likewise, the most obvious (M_k, M_n) bimodule is $C_{k,n}$. It can be shown that these bimodules establish the Morita equivalence between M_n and M_k.

Now, we reconsider the torsion H-field that was discussed in section 3.2. For any twisted bundle W, there is an algebra $\mathcal{A}_W(X)$ generated by the ground states of the W-W open strings. An element of $\mathcal{A}_W(X)$ is a section of $W \otimes \overline{W}$. Given any two different twisted bundles W and W', we can try to make a Morita equivalence between $\mathcal{A}_W(X)$ and $\mathcal{A}_{W'}(X)$ by specializing to this situation the more abstract discussion of general open string algebras $\mathcal{A}_\alpha$ and $\mathcal{A}_\beta$ that was given above. This means that we should find $(\mathcal{A}_W(X), \mathcal{A}_{W'}(X))$ and $(\mathcal{A}_{W'}(X), \mathcal{A}_W(X))$ bimodules by looking at the ground states of the W-W' and W'-W open strings. The ground states of the W-W' open strings are sections of $W \otimes \overline{W'}$; the space of these sections is a $(\mathcal{A}_W(X), \mathcal{A}_{W'}(X))$ bimodule M. Likewise, the space of ground states of the W'-W open strings is a $(\mathcal{A}_{W'}(X), \mathcal{A}_W(X))$ bimodule N consisting of the sections of $W' \otimes \overline{W}$. The bimodules M and N do indeed establish a Morita equivalence between $\mathcal{A}_W(X)$ and $\mathcal{A}_{W'}(X)$. Indeed, in proving this one can work locally on X. Locally, W and W' are trivial, and (if n and k are the ranks of W and W') we have locally as noted in section 3.2, $\mathcal{A}_W(X) = \mathcal{A}(X) \otimes M_n$, $\mathcal{A}_{W'}(X) = \mathcal{A}(X) \otimes M_k$. The same computation that gives a Morita equivalence of M_n with M_k shows that $\mathcal{A}_W(X)$ is Morita equivalent to $\mathcal{A}_{W'}(X)$.

A projective right $\mathcal{A}_W(X)$ module is the space of sections of $E \otimes \overline{W}$, for any twisted bundle E. (If $s^\alpha{}_j$ is such a section – where α is an E index and j a $\overline{W}$ index – and $w^k{}_l$ is a section of $W \otimes \overline{W}$, then the product of s and w is defined by $(sw)^\alpha{}_j = \sum_k s^\alpha{}_k w^k{}_j$.) So projective right $\mathcal{A}_W(X)$ modules are in natural one-to-one correspondence with twisted bundles E, and the K-group of projective right $\mathcal{A}_W(X)$ modules is the group of pairs (E, F), subject to the usual equivalence relation. The Morita equivalence of $\mathcal{A}_W$ and $\mathcal{A}_{W'}$, for any two twisted bundles W and W' with the same H, is just a fancy way of describing the natural map between $\mathcal{A}_W$-modules and $\mathcal{A}_{W'}$-modules that comes from the correspondence

$$E \otimes \overline{W} \leftrightarrow E \otimes \overline{W'}. \tag{11}$$

This one-to-one map between right $\mathcal{A}_W$-modules and right $\mathcal{A}_{W'}$-modules (or the corresponding map for left modules) leads to the conclusion that $K(\mathcal{A}_W) = K(\mathcal{A}_{W'})$.

Acknowledgments

This work was supported in part by NSF Grant PHY-9513835 and the Caltech Discovery Fund. I am grateful to K. Hori and G. B. Segal for comments and explanations.

References

1. A. Sen, *Tachyon Condensation On The Brane-Antibrane System*, *JHEP* **9808**, 012 (1998) [hep-th/9805170].
2. E. Witten, *D-Branes And K-Theory*, *JHEP* **9812**, 019 (1998) [hep-th/9810188].
3. E. Diaconescu, G. Moore and E. Witten, *E_8 Gauge Theory, And A Derivation Of K-Theory From M-Theory*, hep-th/0005091.
4. D. Freed and E. Witten, *Anomalies In String Theory With D-Branes*, hep-th/9907189.
5. A. Connes, M. Douglas and A. Schwarz, *Noncommutative Geometry And Matrix Theory: Compactification On Tori*, *JHEP* **9802**, 003 (1998) [hep-th/9711162].
6. D.S. Freed and M.J. Hopkins, *On Ramond-Ramond Fields And K-Theory*, *JHEP* **0005**, 044 (2000) [hep-th/0002027].
7. E. Witten, *Duality Relations Among Topological Effects In String Theory*, *JHEP* **0005**, 031 (2000) [hep-th/9912086].
8. G. Moore and E. Witten, *Self-Duality, Ramond-Ramond Fields, And K-Theory*, *JHEP* **0005**, 032 (2000) [hep-th/9912279].
9. M. Bershadsky, C. Vafa and V. Sadov, *D-Branes And Topological Field Theories*, *Nucl. Phys.* **B463**, 420 (1996).
10. M.B. Green, J.A. Harvey and G. Moore, *I-Brane Inflow And Anomalous Couplings On D-Branes*, *Class. Quant. Grav.* **14**, 47 (1997) [hep-th/9605033].
11. E. Cheung and Z. Yin, *Anomalies, Branes, and Currents*, *Nucl. Phys.* **B517**, 69 (1998) [hep-th/9710206].
12. G. Moore and R. Minasian, *K-Theory And Ramond-Ramond Charge*, *JHEP* **9711**, 002 (1997) [hep-th/9710230].
13. E. Witten, *On Flux Quantization In M-Theory And The Effective Action*, *J. Geom. Phys.* **22**, 1 (1997) [hep-th/9609122].
14. M. Fabinger and P. Horava, *Casimir Effect Between World-Branes In Heterotic M-Theory*, *Nucl. Phys.* **B580**, 243 (2000) [hep-th/0002073].
15. K. Hori, *D-Branes, T-Duality, and Index Theory*, hep-th/9902102, sections 2.2, 4, and 5.
16. P. Horava, *Type IIA D-Branes, K Theory, and Matrix Theory*, *Adv. Theor. Math. Phys.* **2**, 1373 (1999) [hep-th/9812135].
17. A. Kapustin, *D-Branes In A Topologically Nontrivial B-Field*, hep-th/9909089.
18. P. Bouwknegt and V. Mathai, *D-Branes, B Fields, And Twisted K Theory*, *JHEP* **0003**, 007 (2000) [hep-th/0002023].
19. M.F. Atiyah and G.B. Segal, unpublished.
20. A. Schwarz, *Morita Equivalence And Duality*, *Nucl. Phys.* **534**, 720 (1998) [hep-th/9805034]; A. Schwarz and B. Pioline, *Morita Equivalence And T Duality (Or B Versus Theta)*, *JHEP* **9908**, 021 (1999) [hep-th/9908019].
21. N. Seiberg and E. Witten, *String Theory And Noncommutative Geometry*, *JHEP* **9909**, 032 (1999) [hep-th/9908142].

$\mathcal{N} = 2$ GAUGE-GRAVITY DUALS

JOSEPH POLCHINSKI
Institute for Theoretical Physics, University of California
Santa Barbara, CA 93106–4030, USA

We study the $\mathcal{N} = 2$ analog of the Klebanov-Strassler system. We first review the resolution of singularities by the enhançon mechanism, and the physics of fractional branes on an orbifold. We then describe the exact $\mathcal{N} = 2$ solution. This exhibits a duality cascade as in the $\mathcal{N} = 1$ case, but the singularity resolution is the characteristic $\mathcal{N} = 2$ enhançon. We discuss some related solutions and open issues.

1. Introduction

Maldacena's duality[1] between IIB string theory on $AdS_5 \times S^5$ and $\mathcal{N} = 4$ supersymmetric Yang-Mills theory is a remarkable relation between string theory and quantum field theory, and it is important to extend it to situations of less supersymmetry. As with all dualities, information flows in both directions. In one direction, the less supersymmetric gauge/gravity duals give quantitative solutions to strongly coupled gauge theories with confinement and chiral symmetry breaking. In the other, they give information about string compactifications with realistic amounts of supersymmetry as well as large warp factors.

In this talk I will focus on $D = 4$, $\mathcal{N} = 2$ duals. Systems with $\mathcal{N} = 2$ supersymmetry are often studied as a step towards $\mathcal{N} = 1$, taking advantage of the extra supersymmetry. Happily, recent progress in $\mathcal{N} = 1$ gauge/gravity duals has been quite rapid, so this motivation is no longer as great. Nevertheless, it is interesting to study the $\mathcal{N} = 2$ systems, both for the perspective that they give on $\mathcal{N} = 1$ and because there are some interesting open questions.

The supersymmetry and conformal invariance can be broken in various ways. I will focus on breaking by the combination of orbifolding and fractional brane fluxes, but will also discuss briefly breaking by small perturbations (the latter is based on work with Buchel and Peet[2]). A central issue is that the breaking of the conformal invariance of the AdS space often leads to the horizon being replaced by a naked singularity. It has now been seen in many examples[3,4,5,6] that the singularity is not actually present, though the mechanism that eliminates it varies from case to case. There is a particular phenomenon, the enhançon, that is characteristic of $\mathcal{N} = 2$ systems.[3]

In section 2 I review singularity resolution by the enhancçn. In section 3 I review gauge/gravity duality with orbifolding and fractional branes, and describe an $\mathcal{N} = 2$ analog of Seiberg duality. In section 4 I study the $\mathcal{N} = 2$ version of the Klebanov-Strassler solution,[5] showing that there is a duality cascade in the UV as in the $\mathcal{N} = 1$ case, but that the infrared behavior is the characteristic $\mathcal{N} = 2$ enhançon. Finally, I discuss dualities relating various solutions, and some open issues.[a]

2. The Enchançon

For convenience let us focus on pure $\mathcal{N} = 2$ $SU(N)$ Yang-Mills theory. The Coulomb branch is N complex dimensional. Correspondingly the brane system is made up of N constituents, each moving in a two-dimensional transverse space: these are 3-branes in six dimensions, with four dimensions having been compactified or reduced.

When the constituents are well-separated, the metric g_{ij} on the moduli space of their positions is flat, but at sufficiently small separations it becomes negative. This is unphysical: the condition

$$g_{ij} > 0 \tag{1}$$

thus excludes a region in the interior of the moduli space. On the boundary of this region the metric has a zero, so at least one of the 3-branes has vanishing tension. A 3-brane of vanishing tension is of course a rather special object, and so it must be at a spacetime point of nontrivial infrared physics. In the $2+1$ dimensional case it is a point of enhanced gauge symmetry, hence the name *enhançon*[3] for such point; in the $3+1$ dimensional case there are tensionless strings. We will also use the term enhançon for the codimension one surface in moduli space where the metric has a zero.

Generically the metric has only one zero, but there are points in the N-dimensional moduli space where multiple 3-branes are tensionless. If we try to bring the 3-branes as close together as possible, they will all be tensionless, and their positions will map out some closed curve in spacetime — the enhançon points fill out, in the large-N limit, a closed curve.

In Refs. 3 and 9 the supersymmetry was broken to $\mathcal{N} = 2$ by different means, but the IR physics in both is governed by the enhançon. In Ref. 3 the constituents were D7-branes wrapped on K3, with $\mathcal{N} = 4$ broken to $\mathcal{N} = 2$ by curvature. For simplicity it was assumed that these were distributed on a ring; as its radius was reduced they all became tensionless simultaneously. In Ref. 9 the breaking to $\mathcal{N} = 2$ was by a perturbation of the $AdS_5 \times S^5$ background, corresponding[10] to a mass perturbation of the $\mathcal{N} = 4$ theory. A family of solutions were obtained by the lift of five-dimensional solutions, correspond to a one-parameter ($\gamma < 0$) curve in moduli space. At the limiting value $\gamma = 0$ the constituent D3-branes all lay on enhanọn points, which in this case formed a segment of the real axis.[2]

---

[a] I should point out that Igor Klebanov, with a series of collaborators, has pioneered the study of this fascinating system. Much of my talk is a review of his work, with a few new ingredients — the exact $\mathcal{N} = 2$ solution, the analysis of the $\mathcal{N} = 2$ duality cascade and singularity resolution, and some of the final remarks. See also the talks by Klebanov and Strassler.
After this was written I learned of the recent Ref. 7, with which it has substantial overlap. Ref. 8, which has since appeared, presents a new $\mathcal{N} = 2$ supergravity dual; how it is related to the solutions described in section 5.2 is not yet clear.

The enhançon has a simple gauge theory interpretation.[3,2,11] In the $\mathcal{N} = 2$ theory the metric on moduli space receives corrections only at tree level, one loop, and nonperturbatively. The whole is positive definite, but the perturbative part alone goes negative in parts of moduli space.[12] The instanton corrections are of order $e^{-O(g_{\rm YM}^{-2})}$. Since $g_{\rm YM}^2 = O(N^{-1})$, the perturbative part is accurate until one reaches a point on moduli space where $g_{\rm YM}^{-2} \to 0$, at which point the corrections must suddenly become important. Noting that the metric on moduli space is essentially $g_{\rm YM}^{-2}$, the enhançon surface in moduli space is precisely where the instanton effects become important. Of course, the moduli space does not end there. By holomorphy it must continue into the interior, but as argued in Ref. 3 it can no longer be interpreted as the moduli space of the positions of pointlike constituents: naive continuation of the supergravity solution into the interior would be singular. Rather, the moduli represent the internal state of the theory at the enhançon; it would be interesting to make this more precise.

In other words, the enhançon is the large-N manifestation of the Seiberg-Witten curve,[12] which takes the form[13]

$$y^2 = x^{2N} - \phi^{2N} + \Lambda^{2N} \tag{2}$$

for a $\mathbf{Z}_N$-invariant Higgs v.e.v. of magnitude ϕ. For $\phi > \Lambda$, the Λ^{2N} term is negligible and the physics is perturbative; for $\phi < \Lambda$ the instanton correction Λ^{2N} is dominant. Note that for $\phi = \Lambda$ the curve degenerates and one has an Argyres-Douglas point; it would be interesting to see whether this has any special geometric interpretation in the enhançon context.[b]

3. Orbifolds and Fractional Branes

3.1. *D3-Branes on the* $\mathbf{Z}_2$ *orbifold*

The symmetries of $AdS_5 \times S^5$ can be broken by a perturbation which becomes linear at the boundary, corresponding to a relevant perturbation of the gauge theory Hamiltonian,[10] or by a large deformation at the boundary. Generally the latter corresponds to placing the D3-branes at a singular point of the transverse space.[14,15,16,17,18] I will consider the simplest case, of a $\mathbf{Z}_2$ orbifold

$$R: \quad x^{4,5} \to x^{4,5} \ , \quad x^{6,7,8,9} \to -x^{6,7,8,9} \ . \tag{3}$$

This breaks half of the supersymmetry, leaving $\mathcal{N} = 2$. In this section I will review the gauge/gravity duality for this orbifold.[14,19]

On the string side one restricts as usual to $\mathbf{Z}_2$-invariant configurations and adds in twisted sectors, states localized on the fixed plane $x^{6,7,8,9} = 0$. On the gauge side, the massless open string spectrum is obtained by standard technology.[20,21] To describe N D3-branes on the orbifolded space we also need their N images (fig. 1a), giving a total of $2N$ Chan-Paton degrees of freedom labeled $i \in 1, \ldots, 2N$. The orbifolding acts both on the oscillator state ψ of an open string and on its

[b] This remark is due to Cumrun Vafa.

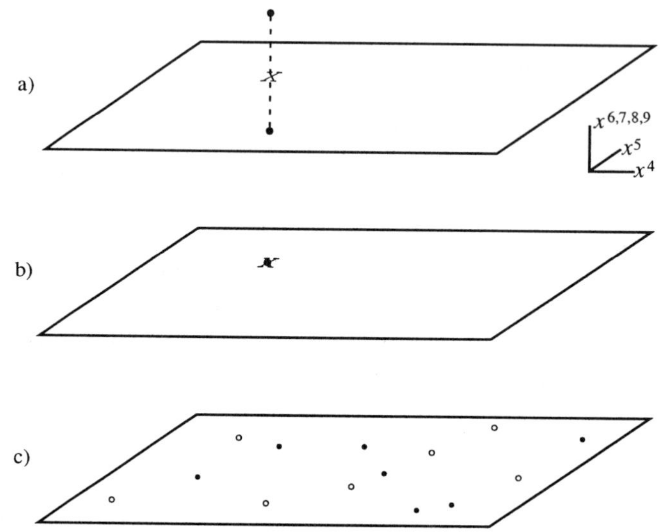

Fig. 1. a) N D3-branes above the fixed plane, and their images below. b) All $2N$ branes+images coincident on the plane. c) N D5-branes (closed circles) and N anti-D5-branes (open circles) separated on the fixed plane.

Chan-Paton degrees of freedom:

$$R|i,j,\psi\rangle = \gamma_{ii'}\gamma_{jj'}|i',j',\hat{R}\psi\rangle\ , \tag{4}$$

where the matrix γ interchanges each D3-brane with its image,

$$\gamma = \begin{bmatrix} 0 & I_N \\ I_N & 0 \end{bmatrix} . \tag{5}$$

The orbifold projection retains states even under R.

We are interested in the case that the D3-branes (and so their images) lie in the fixed plane, as in fig. 1b. To analyze this case it is useful to go to a different basis for the Chan-Paton factors,

$$\gamma = \begin{bmatrix} I_N & 0 \\ 0 & -I_N \end{bmatrix} . \tag{6}$$

In this basis, states in the diagonal blocks survive the projection if their oscillator state is even under $\hat{R}$d and states in the off-diagonal blocks survive if it is odd. This leaves the massless states

$$\begin{bmatrix} A^\mu,\, X^{4,5} & X^{6,7,8,9} \\ X^{6,7,8,9} & A^\mu,\, X^{4,5} \end{bmatrix} \to \begin{bmatrix} A^\mu,\, \phi & A_{1,2} \\ B_{1,2} & \tilde{A}^\mu,\, \tilde{\phi} \end{bmatrix} . \tag{7}$$

In the second form we have collected the scalars into complex pairs. The gauge group is $U(N) \times U(N)$, with an adjoint $\mathcal{N} = 2$ vector multiplet and two $(\mathbf{N}, \overline{\mathbf{N}}) + (\overline{\mathbf{N}}, \mathbf{N})$ hypermultiplets. The notation (7) corresponds to $\mathcal{N} = 1$ multiplets.

The overall $U(1)$ decouples from the bulk dynamics, so the orbifolded string theory is the dual to the $\mathcal{N} = 2$ $SU(N) \times SU(N) \times U(1)$ gauge theory with two bifundamental hypermultiplets. Not surprisingly, since there is still an AdS_5 factor, this is a conformally invariant theory.[14,22,c]

3.2. *Fractional branes*

The system becomes more interesting when generalized as follows. In the basis (6), we can generalize to blocks of different size,

$$\gamma = \begin{bmatrix} I_{N+M} & 0 \\ 0 & -I_N \end{bmatrix} . \tag{8}$$

This corresponds to a $U(N+M) \times U(N)$ gauge theory with two bifundamental hypermultiplets. This γ can no longer be brought to the form (5). A $2N$-dimensional submatrix can be, and so N D3-branes can be moved off the fixed plane, but this leaves M 'half-D3-branes.' The 'half' is because these do not have images, and so for example couple with half the strength of a D3-brane to closed string states.

We can also form these half-D3-branes directly in the $U(N)\times U(N)$ system. The vevs of ϕ and $\tilde{\phi}$ give $2N$ independent coordinates in the fixed plane, corresponding to the N D3-branes separating into $2N$ half-D3-branes. These are of two types (fig. 1c), according to whether γ acts as $+1$ or -1 on the given Chan-Paton index.

These half-branes have a simple interpretation.[24] There are five massless scalar twisted states in the IIB theory. Three correspond to blowups of the fixed point $x^{6,7,8,9} = 0$. Hidden in the fixed point is a zero-size S^2, which acquires a finite size when the fixed point is blown up into a smooth space. The other two scalars are θ-parameters from the NS-NS and R-R two-form potentials integrated over the S^2,

$$\theta_{\rm B} = \frac{1}{2\pi\alpha'} \int_{S^2} B_2 \ , \quad \theta_{\rm C} = \frac{1}{2\pi\alpha'} \int_{S^2} C_2 \ . \tag{9}$$

c The two types of half-brane correspond to D5-branes or anti-D5-branes wrapped on this S^2. Being D-branes, these should have a (magnetic) coupling to $\theta_{\rm C}$. This coupling is given by a disk amplitude with a single vertex operator. The vertex operator has a branch cut extending to the boundary, and includes a factor of γ acting on the boundary state. Thus the coupling is proportional to the trace of γ, and so is zero for a full D3-brane, but ± 1 for the two types of half-brane.

The twisted state coupling measures the D5 charge. The half unit of D3-brane charge can be understood as follows. A D5-brane couples to the Ramond background as

$$\int (C_6 + B_2 \wedge C_4) \ . \tag{10}$$

The first term is its magnetic coupling to C_2. The second is a D3 charge equal to $\theta_{\rm B}/2\pi$, and the free orbifold CFT is known[25] to corresponds to $\theta_B = \pi$; indeed, this

[c] To be precise, the $U(1)$ is not conformal but IR free.[23] The β-function is proportional to the 't Hooft parameter, so the coupling will flow quickly to zero in the supergravity approximation.

is probably the most physical way to derive that fact. Thus, fig. 1c corresponds to N D3-branes separating into N [D5 + $\frac{1}{2}$D3] plus N [$\overline{\text{D5}}$ + $\frac{1}{2}$D3]. The more general matrix γ corresponds to

$$(N+M)\left[\mathrm{D5}+\tfrac{1}{2}\mathrm{D3}\right]\ \oplus\ N\left[\overline{\mathrm{D5}}+\tfrac{1}{2}\mathrm{D3}\right] \tag{11}$$

Thus, the $U(N+M)$ gauge factor is associated with the wrapped D5-branes, and the $U(N)$ factor with the wrapped anti-D5-branes.

3.3. *An* $\mathcal{N}=2$ *Seiberg duality*

The final issue to understand is the effect of varying θ_{B}, particularly on the couplings of the two gauge groups. First let us stay in the range $0 \leq \theta_{\mathrm{B}} \leq 2\pi$. The system (11) evolves to

$$(N+M)\left[\mathrm{D5}+\tfrac{1}{2\pi}\theta_{\mathrm{B}}\mathrm{D3}\right]\ \oplus\ N\left[\overline{\mathrm{D5}}+(1-\tfrac{1}{2\pi}\theta_{\mathrm{B}})\mathrm{D3}\right]. \tag{12}$$

For these objects, the tension is simply proportional to the magnitude of the D3-brane charge,

$$\tau=\tau_3|Q_3|\ ,\quad \tau_3=\frac{1}{(2\pi)^3\alpha'^2 g_{\mathrm{s}}}\ , \tag{13}$$

and so is respectively $\frac{1}{2\pi}\theta_{\mathrm{B}}\tau_3$ and $(1-\frac{1}{2\pi}\theta_{\mathrm{B}})\tau_3$ for the constituents (12). The D5-brane tension makes no contribution because the sphere on which the D5-brane is wrapped has zero size. Note that — as long as all constituents have the same sign of Q_3 — the full system satisfies the BPS property (13) and so is a BPS state, in spite of having both D5-branes and anti-D5-branes. Of course, if the S^2 is blown up to finite size, the BPS mass will have term involving $|Q_5|$ and the fully separated system of fig. 1c will no longer be supersymmetric — the fractional branes of opposite Q_5 will bind into full D3-branes.

The gauge field action comes from expanding out the Born-Infeld Lagrangian

$$-\tau\sqrt{-\det(G_{\mu\nu}+2\pi\alpha' F_{\mu\nu})}\ . \tag{14}$$

It follows that the gauge coupling is just proportional to the tension, and so for the constituents (12)[22]

$$\frac{4\pi}{g^2_{SU(N+M)}}=\frac{\theta_{\mathrm{B}}}{2\pi g_{\mathrm{s}}}\ ,\quad \frac{4\pi}{g^2_{SU(N)}}=\frac{2\pi-\theta_{\mathrm{B}}}{2\pi g_{\mathrm{s}}}\ . \tag{15}$$

The angle θ_{B} is a periodic variable, but the system that we are studying undergoes an interesting spectral flow. Consider increasing θ_{B} past 2π. The D3 charge of the anti-D-branes becomes negative, so the system (12) is no longer BPS. However, it can remain BPS by rearranging into new constituents as θ_{B} goes through 2π,

$$(N+2M)\left[\mathrm{D5}+(\tfrac{1}{2\pi}\theta_{\mathrm{B}}-1)\mathrm{D3}\right]\ \oplus\ (N+M)\left[\overline{\mathrm{D5}}+(2-\tfrac{1}{2\pi}\theta_{\mathrm{B}})\mathrm{D3}\right]. \tag{16}$$

Thus the gauge group changes as well, from $U(N+M) \times U(N)$ to $U(N+2M) \times U(N+M)$. This is the same Seiberg duality[26] that appears in $\mathcal{N}=1$.[5] Note that we are treating $\theta_{\rm B}$ as a constant, but the gauge couplings will run with energy; we will take account of this in the next section.

For $M=0$ there is no spectral flow and $\theta_{\rm B}$ is simply a periodic variable; it is remarkable that the inverse gauge coupling is thus also a periodic variable,[27] which can be smoothly taken to 0 and then back to positive values. For $M \neq 0$ it would be interesting to understand better the nature of the flow. Note that there is a $U(N+M)$ factor on both sides, whose gauge coupling is finite and continuous through the transition. Note also that we can think of $U(N+2M)$ as related to $U(N)$ through Higgsing by the scalar field ϕ in the vector multiplet, with the $U(N+M)$ group as a spectator.[d] However, this simple picture is probably not correct: note that the $U(N+M)$ factor initially acts on the D5-brane Chan-Paton factors, but after the transition it acts on the anti-D5-brane Chan-Paton factors; simple Higgsing would not have this effect. Also, we might expect some form of electric-magnetic duality to relate the two sides;[26] it is not clear how to see this.

4. The Supergravity Dual

We now consider the supergravity solution corresponding to a distribution of D5- and D3-branes. We first consider the case that the D5-branes can be treated as a perturbation, as in Ref. 19. With the three-form flux having components only transverse components, its equation of motion takes the simple form

$$d[Z^{-1}(*_6 G_3 - iG_3)] = 0 \ , \tag{17}$$

where Z is the harmonic (with sources) function appearing in the D3-brane background and $G_3 = F_3 - \tau H_3$. In the present case, $G_3 = dA_2$ with

$$A_2 = \theta(x^4, x^5)\omega_2 \ , \tag{18}$$

where $\theta = \theta_{\rm C} - \tau\theta_{\rm B}$ and ω_2 is the two-form localized at the fixed point, with unit integral over the small two-sphere. In the orbifold limit ω_2 has delta function support, but expressions can be regularized by slightly blowing up the fixed point. The equation of motion then takes the form

$$\partial_z(Z^{-1}\partial_{\bar z}\theta) = 0 \tag{19}$$

in terms of $z = x^4 + ix^5$.

A simple class of solutions is that θ be analytic, and this moreover is the condition for $\mathcal{N}=2$ supersymmetry. The wrapped D5-branes are magnetic sources for

[d] The $U(N+2M)\times U(N+M)$ can also be broken to $U(N+M)\times U(N)\times U(1)$'s by a combination of ϕ and $\tilde\phi$, or by the hypermultiplets A_i and B_i. The former most closely reflects the actual distribution of D-brane charges, but it is unlikely that the rearrangement between constituents (12) and (16) can be understood in terms of classical Higgsing. I would like to thank Ofer Aharony for extensive discussions of this point.

$\theta_{\rm C}$, so[e]

$$\theta = -2i \sum_i q_i \ln(z - z_i) , \tag{20}$$

where $q_i = \pm 1$ and z_i are the D5 charge and position of each wrapped 5-brane. The branch cut in the logarithm corresponds to the equivalence $\theta_{\rm C} \cong \theta_{\rm C} + 2\pi$, with a factor of 2 arising because this is the self-intersection number of the S^2. This is all as in Ref. 19. The one new ingredient I have to add is the exact solution with backreaction. With θ analytic, the solution is in the class found in Ref. 28: the three-form flux acts precisely as a D3-brane source in the other field equations, so there backreaction appears in a modified harmonic function Z'. That is, in addition to a source from any explicit D3-branes, there is a source on the fixed plane, proportional to $|\partial_z \theta|^2$. For example, the source for dF_5 is proportional to $H_3 \wedge F_3$.

Consider now M D5-branes at the origin, so that

$$\theta = -2iM \ln z . \tag{21}$$

The first point to notice is that $\theta_{\rm B}$ is a function of radius, $gM \ln r$. Through Eq. (15) this translates into a running of the couplings, which agrees with the $\mathcal{N} = 2$ beta function.[19] Thus there is a duality cascade, with the successive duality steps separated in scale by a factor $e^{2\pi/gM}$; note that the exponent is small in the supergravity regime.

Now consider the backreaction. The effective D3 charge density is proportional to $|\partial_z \theta|^2 = M^2 |z|^{-2}$ and so logarithmically divergent both in the UV and IR. The normalization is such that the total D3 charge between two duality cascades is exactly M. This is as expected: the size of the gauge group matches the D3 charge, but as in the Klebanov-Strassler (KS) solution[5] these D3-branes are realized as flux, not mobile D-objects (more on this in the next section).

The UV divergence is as in the KS solution, reflecting the unbounded growth of the gauge group. One difference is that in the $\mathcal{N} = 1$ KS case, there is no known way to terminate the cascade in a four-dimensional field theory, while in the $\mathcal{N} = 2$ case it is possible: the $U(N+M) \times U(N)$ theory can be obtained as the low energy limit of a spontaneously broken $U(N+M) \times U(N+M)$ theory, which is conformal in the UV. In terms of branes, beginning with $N + M$ D3-branes we can pull M anti-D5-branes out to long distance.

The IR divergence, however, is a problem. It can be 'renormalized,' leaving a finite backreaction, by adding a negative infinite D3 charge at the origin. However, this solution is singular. For a sufficiently small r the total D3 charge within is negative, and so the warp factor Z' changes sign. Where it goes to zero, the metric coefficients $g_{\mu\nu}$ vanish as well. This is the characteristic 'repulson' singularity,[29] whose resolution is the enhançon:[3] the configuration with all the D5-branes at the origin is unphysical.

---

[e] I would like to thank the authors of Ref. 7 for correcting a factor of 2 in this and the subsequent equations.

Let us see this as in Ref. 3, by starting with the constituent branes dispersed and trying to contract the distribution. For convenience we will regulate the UV also, as described above: we start with a ring of $N + M$ D3-branes at radius r_0, in the orbifold limit $\theta_{\rm B} = \frac{1}{2}$. Leave M anti-D5-branes at r_0 and contract the rest. When the inner ring is at radius r' we have

$$\theta_{\rm B} = \begin{cases} \frac{1}{2} \, , & r > r_0 \\ \frac{1}{2} - 2gM \ln(r_0/r) \, , & r_0 > r > r' \\ \frac{1}{2} - 2gM \ln(r_0/r') \, , & r' > r \end{cases} . \tag{22}$$

The inner ring is initially composed of $N + M$ D5-branes and N anti-D5-branes, but as it moves inward and $\theta_{\rm B}(r')$ passes through integer values, the constituents reorganize as described in the previous section and the system cascades downwards, the number of branes of each type decreasing by M at each step.

For N a multiple of M, after $K = N/M$ steps there remain M D5-branes. When these reach the enhançon radius

$$r_{\rm e} = r_0 \exp[-\pi(K + \tfrac{1}{2})/gM] \tag{23}$$

they become massless and the contraction must stop. At this point, all of the D3-brane charge of the inner ring has been stored in the three-form flux background. The density is nonnegative everywhere and there is no repulson singularity. If N is not a multiple of M, there remain at the end some whole D3-branes.

In summary, the $\mathcal{N} = 2$ fractional brane system has several features in common with the $\mathcal{N} = 1$ system, but the singularity resolution and IR physics is the same as in the $\mathcal{N} = 2$ system considered in Ref. 3. In the conclusions we will discuss some relations among these systems.

5. Open Questions

5.1. *Exotic Physics*

An important open question in this and related systems is the nature of the low energy physics, the physics of the enhançon ring.[f] The final D5-branes sit at a radius where $\theta_{\rm B}$ is an integer and so there are tensionless strings. Thus the effective low energy theory may be rather exotic. Note that in the present case there are additional low energy degrees of freedom on every ring where $\theta_{\rm B}$ passes through an integer. As the constituents move inward and pass through a radius of integer $\theta_{\rm B}$, they rearrange, for example from the form (16) to (12), and the number of moduli drops by $2M$. The dimension of the $\mathcal{N} = 2$ moduli space cannot change, so these moduli must remain as degrees of freedom of the low energy theory on the ring. Thus the system exhibits an interesting analog of 'spin-charge separation': the D3-brane charge is spread throughout space in the form $F_3 \wedge H_3$, while the D3 moduli are localized at discrete radii.

[f] See also the talk by Clifford Johnson.

A related question concerns the dual at very large g. In the $\mathcal{N} = 4$ system, the effective description at $g \ll 1/N$ is a perturbative field theory, and at $g \gg N$ it is the dual field theory; in between is the supergravity description. In the present case, the description at $g \gg N$ is the field theory on wrapped NS5-branes, and it is not clear what this should be — there seems to be no simple orbifold interpretation. Note that in this limit the rings of integer $\theta_{\rm B}$ are very closely spaced and so there are many exotic low energy degrees of freedom.

5.2. *Related Solutions*

It is interesting to take the T-dual on an 'angular' $U(1)$ at the $\mathbf{Z}_2$ singularity.[30,3] This gives a pair of NS5-branes, extended in the 12345-directions and separated in the periodic 6-direction, with M D4-branes stretched between. The NS5-branes bend logarithmically away from one another and repeatedly intersect due to the periodicity — these intersections are at the radii of exotic low energy physics.

Now imagine the same brane configuration, rotated so that the separation and bending are in the 7-direction, while the 6-direction remains periodic. The shape of the branes is the same, though they no longer intersect. This T-dualizes back to a different solution: the 6-separation is T-dual to $\theta_{\rm B}$, while the 7-separation is T-dual to the volume of the S^2. Thus one obtains a solution in which the S^2 does not remain at zero volume but expands logarithmically. This is the original context in which the enhancon was discussed,[3] though the focus there was a yet-different T-dual form in which the volume of an entire K3 varies logarithmically.

Rotating one of the NS5-branes into the 12389 directions gives a T-dual to the $\mathcal{N} = 1$ KS solution.[31,5] Thus the $\mathcal{N} = 1$ and 2 singularity resolutions can be continuously connected, even though one involves supergravity fields only and the other involves branes. Further, if this entire brane configuration is now rotated from the 6- into the 7-direction, one obtains a solution with vanishing NS-NS flux but a logarithmically growing S^3 — presumably the Maldacena-Nuñez (MN) solution.[6] The MN and KS solutions are different in the UV, the former being six-dimensional and the latter four-dimensional but with an unbounded gauge group; evidently the same number of degrees of freedom can be arranged in these two different ways.

The solutions in the 6-direction have constant dilaton and imaginary-self-dual three-form flux; the solutions in the 7-direction have real three-form flux with a related dilaton gradient. By rotating to an intermediate angle and taking the T-dual, one will obtain a solution of more general form.

Acknowledgments

I would like to thank Ofer Aharony, Alex Buchel, Ehud Fuchs, Mariana Grana, Steve Gubser, Clifford Johnson, Andreas Karch, Igor Klebanov, Nikita Nekrasov, Amanda Peet, Matt Strassler, and Cumrun Vafa for discussions and collaboration. This work was supported by National Science Foundation grants PHY99-07949 and PHY97-22022.

References

1. J. Maldacena, *Adv. Theor. Math. Phys.* **2**, 231 (1998) [hep-th/9711200].
2. A. Buchel, A.W. Peet and J. Polchinski, hep-th/0008076.
3. C.V. Johnson, A.W. Peet and J. Polchinski, *Phys. Rev.* **D61**, 086001 (2000) [hep-th/9911161].
4. J. Polchinski and M. Strassler, hep-th/0003136.
5. I.R. Klebanov and M.J. Strassler, *JHEP* **0008**, 052 (2000) [hep-th/0007191].
6. J.M. Maldacena and C. Nunez, hep-th/0008001.
7. M. Bertolini, P. Di Vecchia, M. Frau, A. Lerda, R. Marotta and I. Pesando, hep-th/0011077.
8. A. Rajaraman, hep-th/0011279.
9. K. Pilch and N.P. Warner, hep-th/0004063;
   A. Brandhuber and K. Sfetsos, *Phys. Lett.* **B488**, 373 (2000) [hep-th/0004148].
10. E. Witten, *Adv. Theor. Math. Phys.* **2**, 253 (1998) [hep-th/9802150];
    S.S. Gubser, I.R. Klebanov and A.M. Polyakov, *Phys. Lett.* **B428**, 105 (1998) [hep-th/9802109].
11. N. Evans, C.V. Johnson and M. Petrini, *JHEP* **0010**, 022 (2000) [hep-th/0008081].
12. N. Seiberg and E. Witten, *Nucl. Phys.* **B426**, 19 (1994) [hep-th/9407087].
13. P.C. Argyres and A.E. Faraggi, *Phys. Rev. Lett.* **74**, 3931 (1995) [hep-th/9411057];
    A. Klemm, W. Lerche, S. Yankielowicz and S. Theisen, *Phys. Lett.* **B344**, 169 (1995) [hep-th/9411048];
    M.R. Douglas and S.H. Shenker, *Nucl. Phys.* **B447**, 271 (1995) [hep-th/9503163].
14. S. Kachru and E. Silverstein, *Phys. Rev. Lett.* **80**, 4855 (1998) [hep-th/9802183].
15. A. Kehagias, *Phys. Lett.* **B435**, 337 (1998) [hep-th/9805131].
16. I.R. Klebanov and E. Witten, *Nucl. Phys.* **B536**, 199 (1998) [hep-th/9807080].
17. B.S. Acharya, J.M. Figueroa-O'Farrill, C.M. Hull and B. Spence, *Adv. Theor. Math. Phys.* **2**, 1249 (1999) [hep-th/9808014];
    D.R. Morrison and M.R. Plesser, *Adv. Theor. Math. Phys.* **3**, 1 (1999) [hep-th/9810201].
18. O. Aharony, A. Fayyazuddin and J. Maldacena, *JHEP* **9807**, 013 (1998) [hep-th/9806159].
19. S.S. Gubser and I.R. Klebanov, *Phys. Rev.* **D58**, 125025 (1998) [hep-th/9808075];
    I.R. Klebanov and N.A. Nekrasov, *Nucl. Phys.* **B574**, 263 (2000) [hep-th/9911096].
20. E.G. Gimon and J. Polchinski, *Phys. Rev.* **D54**, 1667 (1996) [hep-th/9601038].
21. M.R. Douglas and G. Moore, hep-th/9603167.
22. A. Lawrence, N. Nekrasov and C. Vafa, *Nucl. Phys.* **B533**, 199 (1998) [hep-th/9803015].
23. E. Fuchs, *JHEP* **0010**, 028 (2000) [hep-th/0003235].
24. J. Polchinski, *Phys. Rev.* **D55**, 6423 (1997) [hep-th/9606165];
    C.V. Johnson and R.C. Myers, *Phys. Rev.* **D55**, 6382 (1997) [hep-th/9610140];
    M.R. Douglas, *JHEP* **9707**, 004 (1997) [hep-th/9612126];
    D. Diaconescu, M.R. Douglas and J. Gomis, *JHEP* **9802**, 013 (1998) [hep-th/9712230].
25. P.S. Aspinwall, *Phys. Lett.* **B357**, 329 (1995) [hep-th/9507012];
26. N. Seiberg, *Nucl. Phys.* **B435**, 129 (1995) [hep-th/9411149].
27. E. Witten, *Nucl. Phys.* **B500**, 3 (1997) [hep-th/9703166];
    S. Katz, P. Mayr and C. Vafa, *Adv. Theor. Math. Phys.* **1**, 53 (1998) [hep-th/9706110].
28. M. Grana and J. Polchinski, hep-th/0009211;
    S.S. Gubser, hep-th/0010010.
29. R. Kallosh and A. Linde, *Phys. Rev.* **D52**, 7137 (1995) [hep-th/9507022].
30. H. Ooguri and C. Vafa, *Nucl. Phys.* **B463**, 55 (1996) [hep-th/9511164];
    R. Gregory, J.A. Harvey and G. Moore, *Adv. Theor. Math. Phys.* **1**, 283 (1997) [hep-th/9708086].

C.V. Johnson, *Nucl. Phys.* **B507**, 227 (1997) [hep-th/9706155];
A. Karch, D. Lust and D. Smith, *Nucl. Phys.* **B533**, 348 (1998) [hep-th/9803232];
K. Dasgupta and S. Mukhi, *JHEP* **9907**, 008 (1999) [hep-th/9904131].

31. A.M. Uranga, *JHEP* **9901**, 022 (1999) [hep-th/9811004];
K. Dasgupta and S. Mukhi, *Nucl. Phys.* **B551**, 204 (1999) [hep-th/9811139].

THE SUPERGRAVITY BRANE-WORLD

JAMES T. LIU
Michigan Center for Theoretical Physics
Randall Laboratory, Department of Physics, University of Michigan
Ann Arbor, MI 48109–1120, USA
E-mail: jimliu@umich.edu

We examine the Randall-Sundrum brane-world and demonstrate how it may be realized in terms of a doubled D3-brane configuration of type IIB theory. We further investigate the supersymmetry of the brane-world and show that, despite claims to the contrary, it is indeed compatible with supersymmetry. This result depends on having a higher dimensional origin for the model, which the type IIB realization provides.

1. Introduction

We show that the type IIB domain wall solution of Bremer *et al.*[1] provides a supersymmetric realization of the Randall-Sundrum brane-world.[2,3] In particular, we pay close attention to the constraints on supersymmetric flows in the brane-world geometry and demonstrate the criteria for stability of such flows. While the connections between 3-brane solutions of the type IIB theory and the brane-world were previously investigated in Refs. 1,4,5,6,7,8,9, the prior focus was not specifically on supersymmetric flows. Here we report on the work of Refs. 10,11,12, which demonstrates that from a five-dimensional point of view the "breathing mode" realization of the brane-world evades the no-go theorems of Refs. 13,14,15.[a] This occurs because the breathing mode falls in a massive representation of AdS, thus yielding a stable flow to the IR away from the brane.

While there are many ways of representing a metric on anti-de Sitter space, we start by noting that it may be given in terms of Poincaré coordinates as

$$ds^2 = e^{-2gy}\eta_{\mu\nu}dx^\mu dx^\nu + dy^2. \tag{1}$$

Written in this manner, the Minkowski signature boundary of AdS is reached when $y \to -\infty$, while the point $y \to \infty$ is instead a null surface, the AdS Killing horizon. In the AdS/CFT correspondence, this metric is viewed as the near-horizon geometry of N coincident D3-branes, which is described by $\mathcal{N} = 4$ super Yang-Mills living on the boundary. Furthermore, the distance to the boundary is regarded as an energy; from the bulk point of view $y \to -\infty$ is a flow to the UV, while $y \to \infty$ is a flow to the IR.

The Randall-Sundrum brane-world is obtained by taking two Poincaré patches of AdS, both given by (1), and joining them at the brane location $y = 0$. The

[a]Note that for the present we are not particularly concerned about smoothness of the brane-world. In the case of smooth flows, see Ref. 16, also Ref. 17.

resulting Randall-Sundrum metric has the form

$$ds^2 = e^{-2g|y|}\eta_{\mu\nu}dx^\mu dx^\nu + dy^2, \tag{2}$$

and its geometry gives rise to a localized graviton on the 'Planck' brane. From the form of (2), it is apparent that the Planck brane only lives in a tiny portion of AdS, and that movement away from the brane flows towards the Killing horizon and not towards the Minkowski boundary of AdS. Had one instead chosen to join together the $y < 0$ regions of (1), the resulting geometry would preserve the vast majority of the original space, including all of the portion of AdS near the boundary. This then would yield a divergent 'localization' volume and give rise to a brane of opposite character to the Randall-Sundrum brane, namely one that does not bind gravity.

The five-dimensional brane-world metric, (2), may be lifted ten dimensions, where it is clearly the near-horizon limit of a doubled D3-brane configuration. Going away from the near-horizon limit then corresponds to turning on the breathing mode in the compactification.[1,10] To be explicit, we start with a single stack of N D3-branes in ten dimensions

$$ds_{10}^2 = H_{D3}^{-1/2}\eta_{\mu\nu}dx^\mu dx^\nu + H_{D3}^{1/2}(dr^2 + r^2 d\Omega_5^2), \tag{3}$$

with a harmonic function $H_{D3} = 1 + \tilde{k}/r^4$. The doubled D3-brane metric is then obtained by defining a new radial coordinate $\xi \in [-r_0, r_0]$ such that $r = r_0 - |\xi|$. The metric then has the form

$$ds_{10}^2 = \left(1 + \tfrac{\tilde{k}}{(r_0 - |\xi|)^4}\right)^{-\frac{1}{2}}\eta_{\mu\nu}dx^\mu dx^\nu + \left(1 + \tfrac{\tilde{k}}{(r_0 - |\xi|)^4}\right)^{\frac{1}{2}}(d\xi^2 + (r_0 - |\xi|)^2 d\Omega_5^2). \tag{4}$$

The Randall-Sundrum brane is located at $\xi = 0$, while the AdS horizons are at $\xi = \pm r_0$. This act of patching together two stacks of D3-branes essentially compactifies the six-dimensional space transverse to the branes, and also introduces a curvature discontinuity at $\xi = 0$.

Furthermore, this compactification introduces a charge conservation condition, implying that the net D3 charge must vanish. Thus the resulting kink at $\xi = 0$ must include a stack of $2N$ negative tension D3-branes, with $-2N$ units of charge soaking up the $N + N$ units of charge from the two stacks of positive tension D3-branes.

2. Stability of the Brane-World and Flows to the IR

For the Randall-Sundrum picture to be realistic, where the Planck brane is a dynamical object, it would have to be supported by bulk scalar fields. Thus, in the language of bulk renormalization group flow, we seek a brane solution with stable flows to AdS critical points in the IR on both sides of it. This approach has been studied extensively in both the AdS/CFT[18,19] and brane-world[20,21,13,14] pictures, with considerable overlap. Nevertheless, the distinction between flows of massless and massive scalars has not always been made clear, so we wish to do so here.

To be specific, we consider the case of a brane supported by a single scalar coupled to gravity, where the Lagrangian takes the form

$$e^{-1}\mathcal{L} = R - \tfrac{1}{2}\partial\varphi^2 - V(\varphi). \tag{5}$$

While one may generalize by including more scalars, this single scalar example is sufficient to bring out our conclusion. Since we demand that the flow away from the brane is towards an AdS background, the scalar must reach some fixed value, φ_*, corresponding to a critical point in the potential. Then, independent of any specific model, we may expand φ about this fixed value. To quadratic order, the potential thus has the form

$$V = -12g^2 + \tfrac{1}{2}m^2(\varphi - \varphi_*)^2 + \cdots, \tag{6}$$

where the constant factor is chosen to give the conventional normalization of the AdS curvature,

$$R_{MNPQ} = -g^2(g_{MP}g_{NQ} - g_{MQ}g_{NP}). \tag{7}$$

While in some cases g may coincide with the coupling constant of gauged supergravity, here we only take it to parameterize the AdS background at the specific fixed point we are interested in.

To study the scalar behavior, we choose a domain-wall ansatz

$$ds^2 = e^{2A(y)}\eta_{\mu\nu}dx^\mu dx^\nu + dy^2. \tag{8}$$

The resulting equations of motion are then

$$\begin{aligned} A'' &= -\tfrac{1}{6}\varphi'^2, \\ A'^2 &= \tfrac{1}{24}\varphi'^2 - \tfrac{1}{12}V, \\ \varphi'' + 4A'\varphi' &= \partial_\varphi V, \end{aligned} \tag{9}$$

where primes denote y derivatives. The first two equations were obtained by combining components of the Einstein equation. Note that the three equations are not all independent, and we find it convenient to focus only on the first two.

As emphasized in Refs. 19,18, the first of these equations indicates that $A'' \leq 0$, with saturation of the inequality corresponding to sitting in the pure AdS vacuum. For the present case, this has the consequence that the function $A(y)$ must be concave-down, which is in fact exactly what is needed to support a 'kink-down' (*i.e.* positive tension) Randall-Sundrum brane of the form (2) with a continuous metric function.

We now insert (6) into the second equation of (9) to find that $A(y) \approx \pm gy$, at least up to linear order in φ. Thus we recover the expected linear behavior giving rise to an AdS background. Continuing with the φ equation of motion, and again working to linear order in φ, we find

$$\varphi'' \pm 4g\varphi' - m^2(\varphi - \varphi_*) \approx 0, \tag{10}$$

which has in general two solutions,

$$(\varphi - \varphi_*) \sim e^{-E_0 A(y)}, \quad \text{and} \quad (\varphi - \varphi_*) \sim e^{-(4-E_0)A(y)}, \tag{11}$$

where $E_0 = 2 + \sqrt{(m/g)^2 + 4} \geq 2$ is given by the mass-E_0 relation for a scalar field in AdS space. For either flow, the metric function behaves like

$$A \approx \pm gy - \tfrac{1}{24}(\varphi - \varphi_*)^2. \tag{12}$$

Finally, this allows us to examine the IR flow, corresponding to the behavior in the direction $A \to -\infty$ (*i.e.* the AdS horizon). We see that IR stability is ensured for $E_0 > 4$ by taking the second solution of (11), while the flow is always unstable for $2 \leq E_0 < 4$, and the massless case, $E_0 = 4$, is marginal.

3. Supersymmetry and IR Stable Flows

The analysis of the previous section indicates that $E_0 > 4$ is a necessary condition for IR stability, and hence for the construction of a Randall-Sundrum brane. This result was in fact already present in the discussion of Refs. 13,14. However Refs. 13,14 only consider scalars residing in massless vector multiplets of $N = 2$ gauged supergravity (*i.e.* the $\mathcal{D}(2,0,0,0)$ representation, where the last value denotes the $U(1)_r$ charge). In particular, Ref. 13 relied on the $N = 2$ very special geometry relation $(\partial_i\partial_j W)_{cr} = \frac{1}{3}g_{ij}W_{cr}$ (in our normalization), where the superpotential is related to the potential by

$$V = (\partial_\varphi W)^2 - \tfrac{2}{3}W^2. \tag{13}$$

By expanding the superpotential with the above constraint in mind

$$W = \pm 3\sqrt{2}g(1 + \tfrac{1}{6}(\varphi - \varphi_*)^2 + \cdots), \tag{14}$$

we find a resulting potential

$$V = -12g^2 - 2g^2(\varphi - \varphi_*)^2 + \cdots. \tag{15}$$

Thus such scalars always have $m^2 = -4g^2$ (*i.e.* $E_0 = 2$), yielding the negative reported result.

Furthermore, while scalars in the decomposition of an $N = 8$ gauged supergravity multiplet reside in $N = 2$ tensor and hypermatter multiplets as well as vector multiplets, such $N = 8$ scalars all have $E_0 = 2$, 3 or 4 and belong in the representations

$$D(2,0,0,\mathbf{20'}), \qquad D(3,0,0,\mathbf{10}+\overline{\mathbf{10}}), \qquad D(4,0,0,\mathbf{1}+\mathbf{1}), \tag{16}$$

where the last value indicates the $SO(6)$ representation. Thus they also do not lead to IR stable branes, and hence the brane-world must be rescued by scalars outside of the ordinary massless supergravity sector.

3.1. *The breathing and squashing modes*

As mentioned above, the Randall-Sundrum brane-world is intimately connected to type IIB string theory compactified on S^5. This sphere reduction of the ten-dimensional theory, with the inclusion of a single squashing mode along with the breathing mode, was investigated in Ref. 1. Focusing only on the scalar modes, the resulting five dimensional theory has a scalar potential of the form[10]

$$V(\varphi, f) = 8m^2 e^{\frac{10}{\sqrt{15}}\varphi} + e^{\frac{4}{\sqrt{15}}\varphi}(4\mu^2 e^{\frac{6}{\sqrt{10}}f} - R_4 e^{\frac{1}{\sqrt{10}}f}), \tag{17}$$

where the constants (m, μ, R_4) are parameters of the compactification.[1] Here φ denotes the breathing mode while f denotes the squashing mode.

The supersymmetry of this S^5 compactification with breathing and squashing modes was considered in Ref. 12. The supersymmetry transformations were found to be

$$\begin{aligned} \delta\psi_\mu &= [\nabla_\mu - \tfrac{1}{6\sqrt{2}} W \gamma_\mu]\epsilon, \\ \delta\lambda &= \tfrac{1}{2}[\gamma \cdot \partial\varphi + \sqrt{2}\partial_\varphi W]\epsilon, \\ \delta\chi &= \tfrac{1}{2}[\gamma \cdot \partial f + \sqrt{2}\partial_f W]\epsilon, \end{aligned} \tag{18}$$

where the superpotential, satisfying (13), is given by[12]

$$W = 2\sqrt{2}m e^{\frac{5}{\sqrt{15}}\varphi} \mp e^{\frac{2}{\sqrt{15}}\varphi}(2\sqrt{2}\mu e^{\frac{3}{\sqrt{10}}f} + \frac{R_4}{4\sqrt{2}\mu} e^{-\frac{2}{\sqrt{10}}f}). \tag{19}$$

The choice of sign corresponds to the orientation of S^5, which is an important consideration, as we will see below. We take the Freund-Rubin flux m to be $m > 0$ ($m < 0$) for the top (bottom) sign in (19). Expanding V to quadratic order then demonstrates explicitly that the breathing mode φ has $E_0 = 8$ while the squashing mode f has $E_0 = 6$.

From the $N = 2$ point of view, the breathing and squashing modes belong to a massive vector representation of the AdS supergroup $SU(2,2|1)$. The content of this shortened multiplet is given by

$$\begin{aligned} \mathcal{D}(6,0,0;0) = {} & D(7,\tfrac{1}{2},\tfrac{1}{2};0) \\ & + D(6\tfrac{1}{2},\tfrac{1}{2},0;-1) + D(6\tfrac{1}{2},0,\tfrac{1}{2};1) + D(7\tfrac{1}{2},0,\tfrac{1}{2};-1) + D(7\tfrac{1}{2},\tfrac{1}{2},0;1) \\ & + D(6,0,0;0) + D(7,0,0;-2) + D(7,0,0;2) + D(8,0,0;0). \end{aligned} \tag{20}$$

The breathing mode and squashing mode scalars can be identified with $D(8,0,0;0)$ and $D(6,0,0;0)$, respectively. Note that this is the only surviving multiplet at the second Kaluza-Klein mass level in the $N = 2$ truncation of the $N = 8$ Freund-Rubin compactification. The evasion of the no-go theorems of Refs. 13,14 then follows since the massive breathing mode clearly satisfies the $E_0 > 4$ stability criteria given above.

4. Supersymmetry and Kinks in the Brane-world

While the E_0 condition was derived without any requirement of supersymmetry, it is nevertheless important to see if supersymmetry imposes any further conditions on the scalar domain wall solution. In particular, for the thin-brane solution, a question arises whether the kink in the solution destroys supersymmetry or not.

To examine this issue, it is sufficient to consider the case when the squashing mode is truncated out. This is simply accomplished by setting $f = 0$ and $R_4 = 24\mu^2 (\equiv \frac{6}{5}R_5)$. The resulting breathing mode superpotential now has the form

$$W = \sqrt{2}m\Big[2e^{\frac{5}{\sqrt{15}}\varphi} - 5\sqrt{\frac{R_5}{20m^2}}e^{\frac{2}{\sqrt{15}}\varphi}\Big], \tag{21}$$

valid for either sign of m. Regardless of the specific form of this superpotential, we see from (18) that the Killing spinor conditions yield the first order equations

$$\begin{aligned} A' &= \pm\frac{1}{3\sqrt{2}}W, \\ \varphi' &= \mp\sqrt{2}\partial_\varphi W, \end{aligned} \tag{22}$$

for a solution preserving exactly half of the supersymmetries. The resulting Killing spinors are given by

$$\epsilon = e^{A/2}(1 \pm \gamma^{\overline{y}})\epsilon_0, \tag{23}$$

with the choice of sign corresponding to that in (22).

4.1. *Construction of brane-world Killing spinors*

For a thin brane supported by the breathing mode, we note that the parameter m in (21) is essentially the 5-form flux of the Freund-Rubin compactification. Thus the two parameters m and R_5 of the five dimensional superpotential, (21), have their origin in the Kaluza-Klein compactification from ten dimensions. From a strictly five-dimensional point of view, the Freund-Rubin parameter m ought to be a constant parameter. However, because of its Kaluza-Klein origin, it is easy to see that m must be odd under transformations $y \to -y$ (corresponding to a simultaneous orientation reversal of S^5 along with the flipping of the transverse y coordinate). This orientation reversal has important consequences for the supersymmetry transformations (18), since the superpotential W also flips, $W \to -W$, under these transformations.

In fact this result is exactly what is needed to prevent the complete breaking of supersymmetry by the domain wall. To see this, we may consider the discussion of Ref. 22, which takes the approach of realizing the Randall-Sundrum brane-world in the context of pure $D = 5$, $N = 2$ Maxwell-Einstein supergravity. This corresponds, in the present case, to retaining only the gravitino transformation in (18) and to taking a constant $W = 6\sqrt{2}g$ that does not flip sign. Then, since the metric factor $A(y) = -2g|y|$ changes sign at the brane, the Killing spinors resulting from (22) must likewise change form on opposite sides of the brane:

$$\begin{aligned} &y > 0: \qquad A' = \tfrac{1}{3\sqrt{2}}W \qquad \epsilon_> = e^{A/2}(1+\gamma^{\overline{y}})\epsilon_0, \\ &y < 0: \qquad A' = -\tfrac{1}{3\sqrt{2}}W \qquad \epsilon_< = e^{A/2}(1-\gamma^{\overline{y}})\epsilon_0. \end{aligned} \tag{24}$$

It is apparent that $\epsilon_>$ and $\epsilon_<$ cannot be continuously joined at the brane.[b] Thus the assumption that W does not flip a sign runs into difficulty when attempting to patch together separate Killing spinors on both sides of $y = 0$.

Of course, the orientation reversal of S^5 demands that the superpotential *does* change sign when passing through the brane. Thus the first order BPS condition $A' = \frac{1}{3\sqrt{2}}W$ may be satisfied with the same sign on both sides of the brane, resulting in global Killing spinors of the form $\epsilon = e^{A/2}(1+\gamma^{\overline{y}})\epsilon_0$. [23,24,25,10,11] This demonstrates that the brane-world configuration is indeed BPS, and preserves half of the supersymmetries.

4.2. *Killing spinor equations and the equations of motion*

Unlike the approach taken in Ref. 22, in Ref. 10 we have seen that the kink at the brane location has different consequences for supersymmetry and for the equations of motion. Since the supersymmetry variations and Killing spinor equations are of first order, the kinks give rise to possibly discontinuous quantities, but no δ-function singularities. On the other hand, the equations of motion are indeed sensitive to the additional δ-function source. Although one may view the equations of motion as a composition of two supersymmetries, there is no contradiction in the presence and absence of the δ-function terms since the Killing spinor equations only give rise to a subset of the full equations of motion. To see this, we may start from the first order equations (22), and take the top sign to be definite. By taking an additional derivative of the first equation in (22), we obtain

$$A'' = \frac{1}{3\sqrt{2}}W'. \tag{25}$$

For a continuous W, one simply uses the chain rule, $W' = \partial_\varphi W \varphi'$, and substitutes in the second equation of (22) to arrive at the A'' equation of motion given in (9).

This would then suggest that 'source-free' Killing spinor equations yield similarly source-free second order equations of motion. However, the assumption of a continuous W is actually too strong. Since W changes sign at the $y = 0$ brane, one would pick up a source term upon differentiating, namely:

$$W' = \partial_\varphi W \varphi' + 2W_+\delta(y). \tag{26}$$

This results in the equation of motion

$$A'' = -\tfrac{1}{6}\varphi'^2 + \tfrac{\sqrt{2}}{3}W\delta(y). \tag{27}$$

The δ-function source is in fact what is needed to generate the kink in the brane-world scenario.[10] Thus, while supersymmetry implies most of the equations of motion, it does not in fact determine all of them. The harmonic function condition,

---

[b]Note that Ref. 22 allows for a discontinuous Killing spinor by introducing δ-function 'sources' into the supersymmetry transformation itself. However it is unclear whether this resolution is complete, especially in light of the higher dimensional interpretation given here.

of primary importance in brane constructions, is generally a consequence of the equations of motion, and *not* that of supersymmetry itself.[26,27]

5. Conclusions

We have demonstrated that the Randall-Sundrum brane-world may be realized in terms of a doubled D3-brane configuration. While focusing only on thin branes, we have investigated the supersymmetry of the brane-world and shown that the no-go theorems of Refs. 13,14,15 are overcome by use of the S^5 breathing mode scalar. From the five-dimensional point of view, this breathing mode has energy $E_0 = 8$, thus allowing a stable domain-wall configuration with the breathing mode flowing to its fixed value away from the brane.

Additionally, we have seen that the five-dimensional superpotential must change sign on opposite sides of the brane. While unnatural from a purely five-dimensional point of view, this is actually to be expected from the ten-dimensional type IIB origin. In the latter case, the sign of the superpotential is determined from the Freund-Rubin parameter m, which necessarily has to change sign as the D3-brane itself is a source of five-form charge. This flip in sign may be viewed as an orientation reversal of S^5, which accompanies the flip $y \to -y$ for the coordinate transverse to the brane.

Finally, while the breathing mode brane-world may be lifted to ten dimensions according to (4), one must still address the nature of the kink itself. While $2N$ negative tension D3-branes contribute to the kink as pointed out above, it turns out that that in itself is insufficient to 'fold up' the space transverse to the brane. This was already noted in Ref. 28. Thus there must be a second contribution to the tension of the Randall-Sundrum brane. In work completed after this conference,[29] it was shown that this second contribution arises from the Z_2 identification of flat space itself, and may hint at a seven-brane, or F-theory, picture of the brane-world.[30]

Acknowledgments

I would like to thank Mirjam Cvetič, Renata Kallosh and Chris Pope for stimulating discussions. This work was done in collaboration with Mike Duff, Ruben Minasian, Herve Partouche, Hisham Sati and Kelly Stelle. This work was supported in part by DOE Grant DE-FG02-95ER40899 Task G.

References

1. M.S. Bremer, M.J. Duff, H. Lü, C.N. Pope and K.S. Stelle, *Instanton cosmology and domain walls from M-theory and string theory*, *Nucl. Phys.* **B543**, 321 (1999) [hep-th/9807051].
2. L. Randall and R. Sundrum, *A large mass hierarchy from a small extra dimension*, *Phys. Rev. Lett.* **83**, 3370 (1999) [hep-ph/9905221].
3. L. Randall and R. Sundrum, *An alternative to compactification*, *Phys. Rev. Lett.* **83**, 4690 (1999) [hep-th/9906064].
4. D. Youm, *Solitons in brane worlds*, *Nucl. Phys.* **B576**, 106 (2000) [hep-th/9911218].

5. D. Youm, *Probing solitons in brane worlds*, *Nucl. Phys.* **B576**, 123 (2000) [hep-th/9912175].
6. M. Cvetič, H. Lü and C.N. Pope, *Domain walls and massive gauged supergravity potentials*, *Class. Quant. Grav.* **17**, 4867 (2000) [hep-th/0001002].
7. M. Cvetič, H. Lü and C.N. Pope, *Localised gravity in the singular domain wall background?*, hep-th/0002054.
8. S.P. de Alwis, *Brane world scenarios and the cosmological constant*, hep-th/0002174.
9. S.P. de Alwis, A.T. Flournoy and N. Irges, *Brane worlds, the cosmological constant and string theory*, hep-th/0004125.
10. M.J. Duff, J.T. Liu and K.S. Stelle, *A supersymmetric type IIB Randall-Sundrum realization*, hep-th/0007120.
11. J.T. Liu, R. Minasian and H. Partouche, unpublished.
12. J.T. Liu and H. Sati, *Breathing mode compactifications and supersymmetry of the brane-world*, hep-th/0009184.
13. R. Kallosh and A. Linde, *Supersymmetry and the brane world*, *JHEP* **0002**, 005 (2000) [hep-th/0001071].
14. K. Behrndt and M. Cvetič, *Anti-de Sitter vacua of gauged supergravities with 8 supercharges*, *Phys. Rev.* **D61**, 101901 (2000) [hep-th/0001159].
15. A. Ceresole and G. Dall'Agata, *General matter coupled $\mathcal{N} = 2$, $D = 5$ gauged supergravity*, *Nucl. Phys.* **B585**, 143 (2000) [hep-th/0004111].
16. J. Maldacena and C. Nuñez, *Supergravity description of field theories on curved manifolds and a no-go theorem*, hep-th/0007018.
17. M. Wijnholt and S. Zhukov, *On the uniqueness of black hole attractors*, hep-th/9912002.
18. K. Skenderis and P.K. Townsend, *Gravitational stability and renormalization-group flow*, *Phys. Lett.* **B468**, 46 (1999) [hep-th/9909070].
19. D.Z. Freedman, S.S. Gubser, K. Pilch and N.P. Warner, *Renormalization group flows from holography—supersymmetry and a c-theorem*, hep-th/9904017.
20. K. Behrndt and M. Cvetič, *Supersymmetric domain wall world from $D = 5$ simple gauged supergravity*, *Phys. Lett.* **B475**, 253 (2000) [hep-th/9909058].
21. O. DeWolfe, D.Z. Freedman, S.S. Gubser and A. Karch, *Modeling the fifth dimension with scalars and gravity*, *Phys. Rev.* **D62**, 046008 (2000) [hep-th/9909134].
22. R. Altendorfer, J. Bagger and D. Nemeschansky, *Supersymmetric Randall-Sundrum scenario*, hep-th/0003117.
23. A. Falkowski, Z. Lalak and S. Pokorski, *Supersymmetrizing branes with bulk in five-dimensional supergravity*, *Phys. Lett.* **B491**, 172 (2000) [hep-th/0004093].
24. N. Alonso-Alberca, P. Meessen and T. Ortin, *Supersymmetric brane-worlds*, *Phys. Lett.* **B482**, 400 (2000) [hep-th/0003248].
25. E. Bergshoeff, R. Kallosh and A. Van Proeyen, *Supersymmetry in singular spaces*, *JHEP* **0010**, 033 (2000) [hep-th/0007044].
26. M.J. Duff, R.R. Khuri and J.X. Lu, *String solitons*, *Phys. Rept.* **259**, 213 (1995) [hep-th/9412184].
27. M.J. Duff, J.T. Liu, H. Lü and C.N. Pope, *Gauge dyonic strings and their global limit*, *Nucl. Phys.* **B529**, 137 (1998) [hep-th/9711089].
28. P. Kraus, *Dynamics of anti-de Sitter domain walls*, *JHEP* **9912**, 011 (1999) [hep-th/9910149].
29. M. Cvetič, M.J. Duff, J.T. Liu, H. Lü, C.N. Pope and K.S. Stelle, *Randall-Sundrum brane tensions*, hep-th/0011167.
30. C.S. Chan, P.L. Paul and H. Verlinde, *A note on warped string compactification*, *Nucl. Phys.* **B581**, 156 (2000) [hep-th/0003236].

THE COSMOLOGICAL CONSTANT IN BRANE WORLD SCENARIOS AND STRING THEORY*

S.P. DE ALWIS
Physics Department Box 390, University of Colorado
Boulder, CO 80303, USA
E-mail: dealwis@pizero.colorado.edu

The possibility of getting a zero (or small) cosmological constant within the brane world scenario is discussed in some detail. We find that the best one can do is to argue that within a certain class of compactifications (on squashed spheres) there may be a way of obtaining flat (or nearly flat) effective four dimensional theories by adjusting integration constants/parameters.

1. Introduction

The natural cutoff for an effective field theory description of our observable world is the Planck scale M_P, so that in general the cosmological constant (CC) is of order M_P^4 which is too large by 120 orders of magnitude. Even in a theory with broken supersymmetry (with the breaking scale at the weak scale) the effective CC is 70 orders of magnitude too large. The problem of the cosmological constant is to understand how to account for this discrepancy and to obtain a dynamical mechanism which would give a CC that is within observational limits.[a]

In an effective field theory one can always fine tune the CC to zero simply by adding a constant to the effective potential (or superpotential in a supersymmetric theory) to cancel that which is generated by renormalization effects[b] and phase transitions (eg electroweak, QCD etc.). This is of course extremely unnatural involving the adjustment of this constant to 120 (or 70 in a broken SUSY theory) significant figures. Nevertheless in the absence of some microscopic theory one would usually do that, and that is the sense in which we make practical use of the standard model. When there is a microscopic theory however (such as string theory) from which this effective field theory emerges at low energies even this freedom is in general absent.

For instance take the standard discussion of low energy phenomenology in supergravity models coming from string theory compactified on a Calabi-Yau manifold (or an orbifold) to get $\mathcal{N}=1$ supergravity in flat four dimensions. $M_4 \to M_4 \times CY_3$. The four D theory is expected to break supersymmetry (say by gaugino condensation) and generate a potential for the moduli (represented by the four dimensional field S here) of the form $W \sim ae^{\alpha S} + be^{\beta S} + ce^{\gamma S} + \ldots$. Here we have assumed at least three

*This is an expanded version of an invited talk given at Strings 2000. It is partially based on work done with Alex Flournoy and Nicos Irges.

[a] For a critical review of attempts to solve this problem in field theory, see Ref. 1.

[b] In Wilsonian language the CC is the most relevant operator in the theory!

different gauge groups with beta function coefficients α, β, γ. Now if a, b, c are adjustable parameters one could fine tune them to get (hidden sector) supersymmetry breaking at some intermediate scale and a vanishing cosmological constant. However these numbers are not really adjustable and typically one gets a cosmological constant Λ at the scale of supersymmetry breaking $m^2_{graviton} M_P^2 > 10^{90} \Lambda_{observed}$. Recent attempts to break supersymmetry by constructing models with anti-D-branes as well as D-branes suffer similar problems. They will have dilaton tadpoles and will thus effectively give rise to theories with cosmological constant at the string scale.

In this talk we will discuss brane world scenarios in which the CC might set to zero by adjusting integration constants.[c] In particular we will argue that there may exist within a string theoretic brane-world context, that it is difficult to find a Calabi-Yau (or orbifold) compactification of the Horava-Witten theory such that the effective four dimensional theory has a parameter (integration constant) that can be adjusted to set the long distance CC to zero. On the other hand for Ricci non-flat compactifications such as on a squashed sphere, this might be possible, though a full string theoretic understanding is not yet available.

2. Branes in a Five Dimensional Effective Theory

Let us first discuss some generalities of brane world scenarios with a five dimensional bulk theory with gravity and a scalar field.[3,4] Consider the action

$$S[G, \phi] = \frac{1}{2\kappa^2} \int d^5x \sqrt{-G}(R - \frac{1}{2}(\nabla\phi)^2 - V(\phi) + \ldots) \tag{1}$$

This could be an action that comes from a low energy effective action of string theory compactified to five dimensions. It is taken to be an effective action valid at some scale with all RG effects down to that scale taken into account. Let us consider in addition two 3 branes sitting in this space. In the static gauge (where we identify the world volume coordinates of the brane with the bulk coordinates) we add the following brane actions to the above bulk action.

$$-\sum_i \int_{u=u_i} d^4x T_i(\phi, \epsilon)\sqrt{-G_{4(i)}}, \tag{2}$$

where we have called the transverse dimension u. Note that since we are looking for a flat four dimensional brane world in its ground state we have ignored the derivative terms in this action. Now we look for solutions to the equations of motion using the following metric ansatz.

$$\begin{aligned} ds^2 &= e^{2\omega(u)}\eta_{\mu\nu}dx^\mu dx^\nu + du^2 \\ \phi &= \phi(u). \end{aligned} \tag{3}$$

There are two equations coming from the metric variations and one equation from the scalar field variation. General covariance implies that only two of these

[c] These ideas originated from the work of Randall and Sundrum.[2]

equations are independent. Furthermore one of the metric equations (the δG_{uu} equation) is a constraint equation.

Writing $\frac{d}{du} \equiv'$ we have the equations of motion.

$$\begin{aligned} 6\omega'^2 &= \frac{1}{4}\phi'^2 - V(\phi) \\ 3\omega'' &= -\frac{1}{2}\phi'^2 - \frac{1}{2}\sum_i 2\kappa^2 T_i(\phi,\epsilon)\delta(u-u_i) \\ \phi'' + 4\omega'\phi' &= \frac{1}{2}\frac{dV}{d\phi} + \kappa^2 \sum_i \frac{dT_i}{d\phi}\delta(u-u_i). \end{aligned} \tag{4}$$

We now take the u coordinate to describe a orbifold S^1/Z_2 where the Z_2 symmetry that we impose is under the reflection $u \to -u$, i.e. we take $\omega(u) = \omega(-u)$, $\phi(u) = \phi(-u)$ The branes will sit at the fixed points and the above equations imply that we would have to satisfy the following matching conditions at $u_0 = 0$ and $u_1 = \pi\rho$ (say).

$$\begin{aligned} 3\omega'|_{u=u_i} &= \mp\frac{1}{2}\kappa^2 T_0(\phi)|_{u=u_i} \\ \phi'|_{u=u_i} &= \pm k^2 \frac{dT_0}{d\phi}(|_{u=u_i} \end{aligned} \tag{5}$$

where the upper (lower) sign corresponds to u_1 (u_2).

Now let us count the integration constants. There are 2 second order equations for ω and ϕ implying four integration constants. However $\omega(0)$ can be absorbed into a rescaling of x^μ and there is a constraint equation so we have only two independent integration constants, say $\phi(0)$ and $\omega'(0)$.[4]

First of all this means that one can generically get one flat brane solution (say when the orbifold size goes to infinity) without fine tuning of parameters of the Lagrangian. Indeed from the constraint equation in (4) and the matching conditions (5) at $u = u_0 = 0$ we have

$$\left(\frac{\kappa^2}{2}\frac{dT_0}{d\phi}\right)^2 |_{u=0+} - \frac{2}{3}\left(\frac{\kappa^2}{2}T_0\right)^2 |_{u=0+} = V|_{u=0+} \tag{6}$$

In general this equation will be solved for a discrete set of values of $\phi(0)$. By plugging these into the first matching condition determines $\omega'(0)$. It should be stressed here that such one brane solutions typically either fail to confine gravity or have naked singularities somewhere in the bulk[5,6] thus requiring some additional physical input to resolve them. We believe this means that one should really look at two brane solutions.

There is however one degenerate case that we should mention here. First as has been noticed in Refs. 7,4 one should be able to express the potential as

$$V = \frac{1}{2}\left(\frac{\partial W}{\partial \phi}\right)^2 - \frac{1}{3}W^2. \tag{7}$$

where W is a sort of superpotential. In terms of W the equations of motion become first order and the extra integration constant comes in solving the above first order equation for W in terms of V. In the supersymmetric brane world scenario the tension is proportional to W and then the equation (6) is automatically satisfied.

Let us now discuss the two brane situation. Now the orbifold size is finite and there is a brane (or collection of branes) at each fixed point. There is then a second pair of matching conditions to satisfy. We have however used up the two independent integration constants in satisfying the first pair of matching conditions. We have however one adjustable modulus namely the size of the orbifold. Nevertheless we need one more parameter. From the five dimensional point of view we thus require a fine tuning either of the bulk potential or of the tension. Thus the question is whether we can make such a parameter dynamical. Later we will see an example where this fine tuning becomes the adjustment of a parameter of compactification from ten dimensions.

3. Horava-Witten on CY_3

Let us now apply these general considerations to the case of the compactification of the Horava-Witten theory[8] on CY_3. In the supersymmetric case this has been discussed extensively by Lucas *et al.*[9]

After reducing on a Calabi-Yau manifold the Horava-Witten theory gives a brane world scenario with two three branes (containing the gauge degrees of freedom) localized on the boundary walls (fixed planes of the orbifold). The relevant metric ansatz is

$$ds^{(11)^2} = ds_{str}^{(5)^2} + ds_{CY}^{(6)^2}. \tag{8}$$

The subscript "*str*" indicates string frame and superscripts indicate the dimensionality. As pointed out in Refs. 8,9 one also requires (for the so-called standard embedding)[d] non-trivial $\mathcal{G}$-flux of the form

$$\mathcal{G} = \frac{i}{2\mathcal{V}} n_i G^{ij} (*\omega_j), \tag{9}$$

where the n_i are integers. The low energy spectrum includes the five dimensional graviton multiplet, $h^{1,1}-1$ vector multiplets containing one shape modulus scalar (b^i) each ($i = 1 \dots h^{1,1}$ but there is one constraint, see below), and the universal hypermultiplet that includes the volume modulus (ϕ), the scalar dual to the five dimensional three-form (σ) and a pair of complex scalars (ξ and $\bar{\xi}$). The metric on the vector moduli space is

$$G_{ij}(a) = \frac{i}{2\mathcal{V}} \int_{CY} \omega_i \wedge *\omega_j = -\frac{1}{2} \partial_i \partial_j \ln \mathcal{V}(a), \tag{10}$$

The Kähler form is $J = a^i \omega_i$, with the ω_i a basis of $h^{1,1}$ 2-forms and a^i are the $h^{1,1}$

[d] The form of the $\mathcal{G}$-flux for non-standard embeddings and additional five branes in the bulk is similar; the only modification is that the n_i in the above is replaced by a sum over fluxes.[10]

Kähler moduli of the Calabi-Yau. The Calabi-Yau volume $\mathcal{V}$ is

$$\mathcal{V}(a) = \frac{1}{3!}\int_{CY} J\wedge J\wedge J = \frac{1}{6}c_{ijk}a^i a^j a^k, \tag{11}$$

with c_{ijk} the intersection numbers of the Calabi-Yau.

For our purposes, it is sufficient to consider the part of the effective action which is 5D gravity coupled to the scalars of the vector multiplets and the volume modulus (breathing mode) of the universal hypermultiplet. The corresponding five dimensional string frame bulk effective action is[9]

$$S = -\frac{1}{2{\kappa^{(5)}}^2}\int d^5x\sqrt{-g_{str}^{(5)}}\Big[\mathcal{V}\mathcal{R}^{(5)} + (\mathcal{V}G_{ij}(a) + \partial_i\partial_j\mathcal{V})\partial_I a^i\partial^I a^j - \frac{1}{4\mathcal{V}}G^{ij}(a)n_i n_j\Big]. \tag{12}$$

The five dimensional index is $I = \{\mu, r\}$. After the Weyl rescaling ${ds_E^{(5)}}^2 = \mathcal{V}^{\frac{2}{3}}{ds_{str}^{(5)}}^2$ and separation of the volume modulus from the shape moduli which can be done by defining $b^i = a^i\mathcal{V}^{-1/3}$, we arrive at a bulk action of the form

$$\begin{aligned} S_{bulk} &= -\frac{1}{2{\kappa^{(5)}}^2}\int d^5x\sqrt{-g_E^{(5)}}\Big[\mathcal{R}^{(5)} - G_{ij}(b)\partial_I b^i\partial^I b^j - \frac{1}{2}\partial_I\phi\partial^I\phi \\ &\quad -\frac{1}{4}e^{2\phi}G^{ij}(b)n_i n_j + \lambda(c_{ijk}b^i b^j b^k - 6)\Big], \end{aligned} \tag{13}$$

where we have defined $\mathcal{V} = e^{-\phi}$ and λ is a Lagrange multiplier. This is not the whole relevant Horava-Witten 5D action, because we have not taken into account yet the Horava-Witten Wall/M5-brane action sitting at the fixed points of the orbifold. The additional brane terms are

$$S_{branes} = \frac{1}{2{\kappa^{(5)}}^2}\int d^4x\sqrt{-g_E^{(4)}}T_1(\phi)\delta(r) + \frac{1}{2{\kappa^{(5)}}^2}\int d^4x\sqrt{-g_E^{(4)}}T_2(\phi)\delta(r-\pi\rho), \tag{14}$$

with $g_E^{(4)}$ the induced metric on the brane (we will assume static gauge and ignore fluctuations) and $T_{1,2}$ are the tensions of the branes.[e]

Let us first consider solutions to the equations that keep the shape moduli constant. It turns out that they are determined by

$$\zeta b_k = \frac{3}{2}n_k, \tag{15}$$

where $\zeta \equiv n_l b^l$. By solving the above system of equations, we obtain the values that the squashing modes take in the vacuum.

Now we turn to the breathing mode ϕ. The bulk action, in the Einstein frame, with all the moduli besides the breathing mode ϕ stabilized, is

$$S_{bulk} = -\frac{1}{2{\kappa^{(5)}}^2}\int d^5x\sqrt{-g_E^{(5)}}\Big[\mathcal{R}^{(5)} - \frac{1}{2}(\partial\phi)^2 - \frac{1}{6}\zeta^2 e^{2\phi}\Big]. \tag{16}$$

[e] Note the change in normalization of the tensions compared to the previous section.

In the above, the constant ζ^2 in front of the potential is determined in terms of the integer n_i (therefore it is not a continuous quantity) and it corresponds to contributions from fluxes associated with wrapping M5 branes around the non trivial $(1,1)$ cycles of the Calabi Yau.

Now we use the ansatz (3) to get equations for the warp factor ω and the bulk modulus φ. In terms of a 'superpotential' W these equations become first order:

$$\phi' = \frac{\partial W}{\partial \phi} \tag{17}$$

$$\omega' = -\frac{1}{6} W \tag{18}$$

The superpotential is the solution of the equation

$$V(\phi) = \frac{1}{2}\left(\frac{\partial W}{\partial \phi}\right)^2 - \frac{1}{3} W^2. \tag{19}$$

In our case the potential is of course $V(\phi) = e^{2\phi}\frac{1}{6}\zeta^2$. The obvious particular solution for W i.e. $W = \zeta e^{\phi}$, corresponds to the unbroken supersymmetry case. In order to discuss supersymmetry breaking we need the general solution. Indeed to have any hope of satisfying the matching conditions (as discussed at the end of the last section we need the integration constant coming from the differential equation for W.

This solution turns out to be[11]

$$W = \sqrt{3V} \sinh f(\phi) \tag{20}$$

where f is determined implicitly by

$$c_1 e^{-\phi} = (\tanh f - \sqrt{\frac{2}{3}})^3 \frac{(\tanh f + 1)^{\sqrt{\frac{3}{2}}(1-\sqrt{\frac{2}{3}})}}{(\tanh f - 1)^{\sqrt{\frac{3}{2}}(1+\sqrt{\frac{2}{3}})}}, \tag{21}$$

with c_1 an integration constant. One may now integrate the first order equations for ϕ and ω (17,18) to get

$$r + c_2 = \frac{\sqrt{3}}{\zeta} \int d\phi e^{-\phi} \sqrt{1 - \tanh^2 f(\phi)} \tag{22}$$

and

$$\omega = -\frac{1}{6}\sqrt{\frac{3}{2}} \int d\phi \tanh f(\phi). \tag{23}$$

Thus we have solutions for the warp factor and the breathing mode that depend explicitly on two integration constants c_1, c_2. The matching conditions become,

$$2\phi'(r) = +\frac{\partial T_1}{\partial \phi}(\phi(r)) \mid_{r=0} \tag{24}$$

$$2\omega'(r) = -\frac{1}{6} T_1(\phi(r)) \mid_{r=0} \tag{25}$$

$$2\phi'(r) = -\frac{\partial T_2}{\partial \phi}(\phi(r))\,|_{r=\pi\rho} \tag{26}$$

$$2\omega'(r) = +\frac{1}{6}T_2(\phi(r))\,|_{r=\pi\rho}\,. \tag{27}$$

In the supersymmetric limit these are satisfied with $\tanh f(\phi(0)) = \sqrt{\frac{2}{3}}$, $\tanh f(\phi(\pi\rho)) = -\sqrt{\frac{2}{3}}$ and with the solution $W = \zeta e^{\phi}$ we get $\cosh f(\phi(r)) = \pm\sqrt{3}$ and $\sinh f(\phi(r)) = \pm\sqrt{2}$. The equations of motion (22) and (23) can be solved easily, yielding

$$\phi(r) = -\ln(\zeta|r| + c) \quad \text{and} \quad A(r) = \frac{1}{6}\ln(\zeta|r| + c)\,, \tag{28}$$

with $c \equiv \zeta c_2$ the (only) integration constant. This is a trivial solution, trivial in the sense that the jump conditions are satisfied identically, without restriction on ζ and the integration constant c. It follows that after supersymmetry breaking, the simple ansatz $W = \zeta e^{\phi}$ can not be used anymore to satisfy (24)-(27). But this is expected, since in general it is not possible to satisfy the four jump equations with the three parameters ζ, c and R (even if ζ were continuous). Thus, for a consistent model with zero cosmological constant away from the supersymmetric point, we have to look for more general solutions to (19).

After supersymmetry breaking (on the brane), the brane tensions get renormalized as

$$T_1(\phi) \to 2\zeta e^{\phi}(1 + \epsilon\psi_1(\zeta e^{\phi})), \qquad T_2(\phi) \to 2\zeta e^{\phi}(1 + \epsilon\psi_2(\zeta\epsilon^{\phi})) \tag{29}$$

with ϵ being a small parameter characterizing the size of supersymmetry breaking.

The jump conditions therefore, after supersymmetry breaking become

$$\cosh f(\phi(0)) = \sqrt{3}\Big(1 + \epsilon\frac{d}{dx}(x\psi_1(x))|_{x=x(0)}\Big) \tag{30}$$

$$\sinh f(\phi(0)) = \sqrt{2}\Big(1 + \epsilon\psi_1(x)|_{x=x(0)}\Big) \tag{31}$$

$$\cosh f(\phi(\pi R)) = -\sqrt{3}\Big(1 + \epsilon\frac{d}{dx}(x\psi_2(x))|_{x=x(\pi R)}\Big) \tag{32}$$

$$\sinh f(\phi(\pi R)) = -\sqrt{2}\Big(1 + \epsilon\psi_2(x)|_{x=x(\pi R)}\Big), \tag{33}$$

where $x = \zeta e^{\phi}$. Now let us recall that the difference of $\cosh f$ ($\sinh f$) from $\sqrt{3}$ ($\sqrt{2}$) vanishes like c_1 which is of $O(\epsilon)$. Thus we may write the above equations as relations between $O(1)$ functions. Now in accordance with the general argument of Ref. 4 (see the previous section) the integration constants c_1, c_2 may be adjusted to satisfy say the first two equations above. In addition the distance πR may be freely adjusted to satisfy one more matching condition leaving us with one more condition to satisfy. To satisfy the fourth condition then requires a fine tuning of the parameters in the bulk potential. In our case this means a fine tuning of the quantity ζ. However, as noted earlier, the latter is completely determined in terms of integers n_i governing the fluxes and the integers d_{ijk} and h_{11} of the Calabi-Yau manifold. There is no way these can be chosen to satisfy this last equation since any change of these integers results in a O(1) change of the functions $\psi_{1,2}$ and it would be a miracle if for any choice of the integers the condition could be satisfied.[f] In other words there is no adjustable continuous parameter that could be tuned to satisfy the equation.

Let us now discuss possible generalizations. One possible non-vanishing supersymmetric correction to the scalar potential comes from the instanton gas obtained from wrapping M5-branes around the Calabi-Yau manifold or M2-branes wrapping around 3-cycles. Clearly, the original problem associated with the discreteness of the fourth adjustable parameter still remains since we are still talking about contributions that are completely determined by a set of integers.

Another possible generalization would be to look for solutions to the 5D equations of motion with r-dependent b^i. Such a solution, for the supersymmetric case, was found in Ref. 9. After supersymmetry breaking, we would have to satisfy two additional jump conditions for each b^i, which is possible, since we get two additional integration constants for each b^i from their equations of motion. However, as before, we still have one fine tuning to do and the previous argument, that since there is no continuous parameter there is no possibility of doing this, still applies.

4. Ricci Non-Flat Compactifications

In contrast to the above situation where the discreteness of ζ makes it impossible to find a flat or nearly flat solutions we will show in this section that a compactification on a Ricci non-flat manifold will lead to a continuous parameter that can be adjusted to get a flat brane solution.[12] Consider type IIB compactified to five dimensions on S^5 as in Ref. 13 (section 2.4). Actually in order to get a more realistic scenario we would have to compactify on a squashed sphere so that the supersymmetry on the brane is reduced to $\mathcal{N}=1$ but we will focus on the simplest compactification since this highlights the main issue that we wish to focus on.

We assume that the ten dimensional dilaton φ is fixed by string scale dynamics, but we keep the breathing mode that arises in the compactification, in the action. The relevant part of the low energy effective action is

$$S=\frac{1}{2\kappa_5^2}\int d^5x\sqrt{-G}(R-(\nabla\phi)^2)-8m^2e^{8\alpha\phi}+e^{\frac{16\alpha}{5}\phi}R_5) \tag{34}$$

In getting the above from the 10 D action (or equations of motion) of IIB supergravity, the ansatz

$$ds_{10}^2=e^{2\alpha\phi}ds_5^2+e^{2\beta\phi}ds^2(S^5)$$

with $\alpha=\frac{1}{4}\sqrt{\frac{5}{3}}$, $\beta=-\frac{3}{5}\alpha$ has been made. R_5 is the Ricci scalar of the compact 5-manifold (in this case S^5) and is an unconstrained positive constant. The ansatz

---

[f] Of course one does not require exact satisfaction of the matching conditions. Strictly speaking all we need is that the cosmological constant on the brane be of $O(10^{-120}M_P^4)$. This would still imply an adjustment of parameters to this accuracy as explained in Ref. 4 and is clearly ruled out in the present situation.

for the self-dual 5-form, is

$$F_5 = 4me^{8\alpha\phi}\epsilon_{(5)} + 4m\epsilon_{(5)}(S^5) \tag{35}$$

where m is an integration constant.

At this point let us show out how the quantization of m mentioned earlier comes about if this scenario emerges from string theory. In this case the sources of the F_5 are D_3 branes (and/or orientifolds). The consistency of the the coupling of these objects leads to the standard Dirac quantization rule,

$$\tau_3 \int_{S_5} F_5 = 2\pi n$$

where n is an integer. Using the ansatz (35) for F_5 then gives $\tau_3 4m\pi^3 r^5 = 2\pi n$ where r is the radius of S_5. This then gives us (after using the D3-brane tension formula $\tau_3 = 2\pi M_s^4$)

$$m = \frac{n}{4\pi^3 M_s^4 r^5} = \frac{nM_s}{4\pi^3 \hat{r}^5} \tag{36}$$

where we have introduced the dimensionless parameter $\hat{r} \equiv rM_s$. Also the Ricci scalar of S_5 may then be written as $R_5 = \frac{20M_s^2}{\hat{r}^2}$. Then the potential in (34) may be rewritten

$$\begin{aligned} V &= M_s^2 \left[\frac{n^2}{2\pi^6 \hat{r}^{10}} e^{8\alpha\phi} - \frac{20}{\hat{r}^2} e^{\frac{16}{5}\alpha\phi} \right] \\ &= M_s^2 \hat{r}^{\frac{10}{3}} \left[\frac{n^2}{2\pi^6} e^{8\alpha\tilde{\phi}} - 20 e^{\frac{16}{5}\alpha\tilde{\phi}} \right] \end{aligned} \tag{37}$$

where in the last line of the above equation we put $\alpha\phi = \alpha\tilde{\phi} + \frac{5}{3}\ln\hat{r}$.

The minimum of this potential is given by $e^{\frac{24}{5}\alpha\tilde{\phi}_0} = \frac{16\pi^6}{n^2}$. As discussed in Ref. 13 the 5D action allows a AdS_5 solution with the cosmological constant being an integration constant, with the breathing mode being fixed at $(\tilde{\phi}_0)$. The potential at this point gives a cosmological constant $\Lambda = V(\tilde{\phi}_0) = -(2\pi)^4 \frac{16}{n^2} M_s^2 \hat{r}^{\frac{10}{3}}$. It should be noted that this five dimensional cosmological constant is dependent on the compactification parameter $\hat{r}$ and can be adjusted even though there are no adjustable constants in the 10 dimensional action.

Let us now compactify one of the spatial dimensions on S^1/Z_2 as before. In the original string theory scenario we would have $D3$ branes and orientifolds distributed over the five sphere but we will just consider the effective theory in five dimensions with two branes sitting at the ends of the compact fifth dimension giving us the scenario we had earlier.

Now the fifth dimension is an interval S_1/Z_2 with 16 orientifold fixed planes at the fixed points $u = 0$ and $u = R$. One also needs to introduce D-branes in order to cancel tadpoles.[g] In the presence of D-branes and orientifold planes that are charged under the F_5 field we have as in Ref. 14 a discontinuity in m by an

[g]Indeed such a model is T-dual to the type IA theory discussed in Ref. 14.

amount equal to the brane charge/tension at the position of the brane. Imposing also the Z_2 symmetry the constant m in (34) would be fixed in terms of the brane charge as in Ref. 14. This is of course just another way of verifying the quantization discussed earlier. When supersymmetry is broken however the brane tension would get renormalized so that the supersymmetric relation between tension and charge will be lost. Nevertheless the constant $\hat{r}$ can adjust itself now to track the brane tension. Thus according to the discussion in section 2 one may have solutions with two flat branes.

Let us give some more details of this scenario. The effective five dimensional theory in the presence of branes is given by (after putting in also the value of the five D Newton constant)

$$\begin{aligned} S &= 2\pi^4 M_s^3 \hat{r}^5 \int d^5x\sqrt{g_5}\left(R - \frac{1}{2}(\partial\tilde{\phi})^2 + M_s^2 \hat{r}^{\frac{10}{3}} U(\tilde{\phi})\right) \\ &+ \frac{2\pi M_s^4}{\pi^3}\left\{\int_{u=0} d^4x\sqrt{g_4}T_0(\tilde{\phi}) + \int_{u=R} d^4x\sqrt{g_4}T_1(\tilde{\phi})\right\}. \end{aligned} \tag{38}$$

In the above we have written the potential $V = M_s^2 R^{\frac{10}{3}} U(\tilde{\phi})$ where U is independent of $\hat{r}$ (see (37)). Let us first look at what might happen in the supersymmetric case. Here in analogy with the case discussed by Lucas *et al.*[9] for the Supersymmetry preserving compactification of M theory on $\frac{S^1}{Z_2} \times CY_3$ we expect the BPS equations

$$T_0(\tilde{\phi}) = -T_1(\tilde{\phi}) = \pi^3 \hat{r}^{5/3} W(\tilde{\phi})$$

where

$$U(\tilde{\phi}) = \left(\frac{\partial W}{\partial\tilde{\phi}}\right)^2 - \frac{2}{3}W^2.$$

(Note that apart from normalization this is the same W as before). In fact these BPS conditions have been justified for this system very recently in Refs. 15,16. In this case it is easily seen that the parameter $\hat{r}$ drops out of the matching conditions which however can be satisfied for arbitrary inter-brane distance R and integration constants. This is exactly like what happens in the M theory case investigated by Lukas *et al.*[9]

In the broken supersymmetric case however the situation is quite different. We may imagine that the compactification is now on a squashed sphere so that the supersymmetry is $N = 1$ which is then broken for instance by gaugino condensation effects on one or other brane. Now the tension will be renormalized from its BPS value so that one would expect

$$T_i(\tilde{\phi}) \to \pm\pi^3 \hat{r}^{\frac{5}{3}} W(\tilde{\phi}) + \epsilon\psi_i(\hat{r}, \tilde{\phi})$$

Here the function ψ is dependent on the details of the low energy field theories on the branes. Thus now $\hat{r}$ will not drop out of the matching conditions and indeed we need to adjust the two integration constants $\tilde{\phi}(0), \omega'(0)$ the inter-brane distance

R as well as the compactification parameter $\hat{r}$ in order to get a flat four-dimensional brane world.

5. Conclusions

There are three main conclusions of our investigations.

- It appears that in a string/M theory context to get a zero (or nearly zero) cosmological constant one needs a brane world scenario.
- A phenomenologically relevant brane world scenario coming from M theory is the Horava-Witten theory compactified on a Calabi-Yau three fold (or an orbifold). However in this case it is hard to see how one can preserve a zero cosmological constant after supersymmetry breaking.
- In Ricci flat compactifications such as type IIB orientifolds on (squashed) five-spheres one may have enough freedom to set the cosmological constant to zero.

The last is by no means a solution to the cosmological constant problem. All it indicates is that there may be string theoretic scenarios in which flat four dimensional spaces may be obtained by adjusting integration constants and compactification parameters. A justification of such an adjustment (i.e. a dynamical mechanism) is still beyond our reach.

Clearly to do meaningful string phenomenology we need to look for solutions of string theory with enough freedom to be able to set $\Lambda = 0$ after SUSY breaking. To this end one should perhaps investigate string theory compactifications on squashed spheres. Based on our considerations it appears that in such compactifications the moduli superpotential of the four dimensional theory may take the form $W = \sum \alpha_i e^{\alpha_i S} + \ldots + c$ where S represents a typical modulus and c is an adjustable integration constant that may be set to zero in order to have a vanishing cosmological constant. Unlike much of the string theoretic phenomenology that is found in the literature a phenomenology based on this[17] would be meaningful.

Acknowledgments

I wish to acknowledge discussions with Nicos Irges and Alex Flournoy. This work is partially supported by the Department of Energy contract No. DE-FG02-91-ER-40672.

References

1. S. Weinberg, *The Cosmological Constant Problem*, *Rev. Mod. Phys.* **61**, 1 (1989).
2. L. Randall and R. Sundrum, *A large mass hierarchy from a small extra dimension*, *Phys. Rev. Lett.* **83**, 3370 (1999) [hep-ph/9905221]; *An alternative to compactification*, *Phys. Rev. Lett.* **83**, 4690 (1999) [hep-th/9906064].
3. W. D. Goldberger and M. B. Wise, *Phenomenology of a stabilized modulus*, *Phys. Lett.* **B475**, 275 (2000) [hep-ph/9911457].

4. O. DeWolfe, D. Z. Freedman, S. S. Gubser and A. Karch, *Modeling the fifth dimension with scalars and gravity*, *Phys. Rev.* **D62**, 046008 (2000) [hep-th/9909134].
5. N. Arkani-Hamed, S. Dimopoulos, N. Kaloper and R. Sundrum, *A small cosmological constant from a large extra dimension*, *Phys. Lett.* **B480**, 193 (2000) [hep-th/0001197].
6. S. Kachru, M. Schulz and E. Silverstein, *Self-tuning flat domain walls in 5d gravity and string theory*, *Phys. Rev.* **D62**, 045021 (2000) [hep-th/0001206].
7. K. Skenderis and P. K. Townsend, *Gravitational stability and renormalization-group flow*, *Phys. Lett.* **B468**, 46 (1999) [hep-th/9909070].
8. P. Horava and E. Witten, *Heterotic and type I string dynamics from eleven dimensions*, *Nucl. Phys.* **B460**, 506 (1996) [hep-th/9510209]; *Eleven-Dimensional Supergravity on a Manifold with Boundary*, *Nucl. Phys.* **B475**, 94 (1996) [hep-th/9603142]; E. Witten, *Strong Coupling Expansion Of Calabi-Yau Compactification*, *Nucl. Phys.* **B471**, 135 (1996) [hep-th/9602070].
9. A. Lukas, B. A. Ovrut, K. S. Stelle and D. Waldram, *Heterotic M-theory in five-dimensions*, *Nucl. Phys.* **B552**, 246 (1999) [hep-th/9806051].
10. A. Lukas, B. A. Ovrut, K. S. Stelle and D. Waldram, *Non-standard embedding and five-branes in heterotic M-theory*, *Phys. Rev.* **D59**, 106005 (1999) [hep-th/9808101].
11. S. P. de Alwis and N. Irges, *The cosmological constant in Calabi-Yau threefold compactifications of the Horava-Witten theory*, *Phys. Lett.* **B492**, 171 (2000) [hep-th/0007223].
12. S. P. de Alwis, *Brane world scenarios and the cosmological constant*, hep-th/0002174; S. P. de Alwis, A. T. Flournoy and N. Irges, *Brane worlds, the cosmological constant and string theory*, hep-th/0004125.
13. M. S. Bremer, M. J. Duff, H. Lu, C. N. Pope and K. S. Stelle, *Instanton cosmology and domain walls from M-theory and string theory*, *Nucl. Phys.* **B543**, 321 (1999) [hep-th/9807051].
14. J. Polchinski and E. Witten, *Evidence for Heterotic - Type I String Duality*, *Nucl. Phys.* **B460**, 525 (1996) [hep-th/9510169].
15. E. Bergshoeff, R. Kallosh and A. Van Proeyen, *Supersymmetry in singular spaces*, *JHEP* **0010**, 033 (2000) [hep-th/0007044].
16. M. J. Duff, J. T. Liu and K. S. Stelle, *A supersymmetric type IIB Randall-Sundrum realization*, hep-th/0007120.
17. R. Brustein and S.P. de Alwis, Work in progress.

ASYMPTOTICALLY ANTI-DE SITTER SPACETIMES AND THEIR STRESS ENERGY TENSOR

KOSTAS SKENDERIS
Physics Department, Princeton University
Princeton, NJ 08544, USA
E-mail: kostas@feynman.princeton.edu

We consider asymptotically anti-de Sitter spacetimes in general dimensions. We review the origin of infrared divergences in the on-shell gravitational action, and the construction of the renormalized on-shell action by the addition of boundary counterterms. In odd dimensions, the renormalized on-shell action is not invariant under bulk diffeomorphisms that yield conformal transformations in the boundary (holographic Weyl anomaly). We obtain formulae for the gravitational stress energy tensor, defined as the metric variation of the renormalized on-shell action, in terms of coefficients in the asymptotic expansion of the metric near infinity. The stress energy tensor transforms anomalously under bulk diffeomorphisms broken by infrared divergences.

1. Introduction

Anti-de Sitter (AdS) spacetimes were studied in the eighties because they appear as ground states of many gauged supergravities. In recent times they have attracted a lot of attention due to the AdS/CFT correspondence. The construction of conserved charges for asymptotically AdS spaces was addressed in the eighties, see for instance Ref. 1. Recent work appeared in Ref. 2. The AdS/CFT duality has provided us with new insights and results about AdS gravity. It is the purpose of this contribution to extract these results and put them in a purely gravitational context.

Our considerations will be at the classical level. We would however like to view our results as the lowest order approximation to the quantum theory. The (logarithm of the) partition function of gravity is given to lowest order by the on-shell value of the gravitational action. It is thus essential that the on-shell action is finite. For asymptotically AdS solutions the on-shell action is proportional to the volume of spacetime and thus diverges since the spacetime is non-compact. One may deal with this divergence by regulating the on-shell action, computing all infinities, adding counterterms to cancel them and then removing the regulator. The resulting renormalized on-shell action is manifestly finite. This procedure was carried out in Ref. 3.

The renormalized on-shell action does not preserve all symmetries that the unrenormalized (and thus infinite) on-shell action formally does. This is so because some of the infrared divergences do not preserve all symmetries. Adding counterterms to the action to cancel the infinities results in a renormalized on-shell action that transforms anomalously under some of the original symmetries. In particular, it was found in Ref. 3 that in odd dimensions the counterterms break the bulk diffeomorphisms that generate Weyl transformations on the boundary. This means that the renormalized on-shell action depends on the chosen metric on the boundary, not just on its conformal class. This anomaly corresponds, via the AdS/CFT correspondence, to the conformal anomaly of the dual CFT. We would like to emphasize that the anomaly does not depend on the specific way one chooses to do the computation. Any scheme that results in a finite on-shell action for arbitrary solutions of Einstein's equations with negative cosmological constant will have this property.

Motivated by the AdS/CFT correspondence, it was proposed in Ref. 4 to define the stress energy tensor associated with asymptotically AdS spacetimes as the metric variation of the renormalized on-shell action. Explicit formulae for the stress energy tensor in terms of coefficients in the asymptotic expansion of the metric near infinity was obtained in Ref. 5. The stress energy tensor so defined does not transform covariantly under all bulk diffeomorphisms because the on-shell renormalized action transforms anomalously. Notice that the energy momentum tensor may transform anomalously even if the anomaly in the action vanishes for a given background. In other words, the value of the anomaly may vanish, but not the value of its metric variation. This case is encountered for global AdS. Only if the anomaly vanishes identically will the stress energy tensor transform covariantly. This is true for even dimensional asymptotically AdS spaces. In Ref. 2 it was argued that the counterterm subtraction method is not covariant (in odd dimensions). Indeed, this is the case. The non-covariance, however, is forced upon us by infrared divergences in the on-shell action, and these divergences cannot be ignored.

This paper is organized as follows. In the next section we recall the definition of asymptotically AdS spacetimes. In section 3, we review the computation of the infrared divergences and the construction of the on-shell renormalized action. Section 4 contains the computation of the stress energy tensor.

2. Asymptotically AdS Spacetimes

Anti-de Sitter (AdS) spacetime is a maximally symmetric solution of Einstein's equations with negative cosmological constant,

$$R_{\mu\nu} - \frac{1}{2} R G_{\mu\nu} = \Lambda G_{\mu\nu}, \tag{1}$$

where Λ is the negative cosmological constant. AdS spacetime is homogeneous and isotropic and its isometry group is $SO(2,d)$. These conditions imply that the AdS solution is conformally flat and satisfy the stronger equation,

$$R_{\kappa\lambda\mu\nu} = l^2 (G_{\kappa\mu} G_{\lambda\nu} - G_{\kappa\nu} G_{\lambda\mu}), \tag{2}$$

where l is related to the cosmological constant as $\Lambda = -d(d-1)/2l^2$.

The metric for AdS_{d+1} is given by

$$ds^2 = l^2 \left[-(1+r^2) dt^2 + \frac{dr^2}{(1+r^2)} + r^2 d\Omega_{d-1} \right]. \tag{3}$$

At infinity the space has a boundary with topology $R \times S^{d-1}$ (we actually consider the covering space of AdS). Let us introduce a new coordinate $\tan\theta = r$. The metric becomes

$$ds^2 = \frac{l^2}{\cos^2\theta}\left[-dt^2 + d\theta^2 + \sin^2\theta d\Omega_{p-1}\right]. \tag{4}$$

The boundary is now located at $\theta = \pi/2$. The metric has a second order pole at the boundary, so it does not induce a metric there. To obtain a metric one picks a positive function r with a single zero at the boundary, and then evaluates $r^2 G$ at the boundary,

$$g_{(0)} = r^2 G|_{R\times S^{d-1}}. \tag{5}$$

In our case, one can pick $r = \cos\theta$. Then $g_{(0)}$ is the standard metric on a Lorentzian cylinder. The metric $g_{(0)}$, however, depends on the choice of defining function. Different defining functions are related by $r' = re^w$. It follows that $g_{(0)}$ is well-defined up to conformal transformation. Thus the AdS metric induces a conformal structure (i.e. a metric up to conformal transformations) at the boundary.

To define asymptotically AdS spacetimes it is useful to recall the notion of conformal infinity introduced by Penrose. Let X be the interior of a manifold-with-boundary $\bar{X}$, M the (compact) boundary of $\bar{X}$, and G a metric on X. The metric G is said to be conformally compact if there exists a defining function r (i.e. $r > 0$ in X and $r = 0, dr \neq 0$ on M) such that $r^2 G$ can be smoothly extended as a metric on $\bar{X}$. This procedure defines a conformal structure at the boundary in the way described in the previous paragraph. If in addition G satisfies (1) we will say that X is an "asymptotically *AdS* spacetime". Notice that that this definition is more general than the one employed in Ref. 2 since we do not impose any restriction on the topology of the boundary.

Given a conformal structure, can one obtain an asymptotically AdS spacetime with this conformal structure at infinity? One may view this question as a Dirichlet problem for AdS gravity. This question has been investigated in the mathematics literature. It was shown by Fefferman and Graham in Ref. 6 that one can obtain an asymptotic solution of Einstein's equations given a representative of the conformal structure. The solution is of the form

$$\begin{aligned} ds^2 &= G_{\mu\nu}dx^\mu dx^\nu = l^2\left(\frac{d\rho^2}{4\rho^2} + \frac{1}{\rho}g_{ij}(x,\rho)dx^i dx^j\right), \\ g(x,\rho) &= g_{(0)} + \cdots + \rho^{d/2}g_{(d)} + h_{(d)}\rho^{d/2}\log\rho + \ldots \end{aligned} \tag{6}$$

Several comments are in order here:

(i) Any asymptotically AdS metric can be brought into this form near the boundary. The boundary is located at $\rho = 0$.

(ii) Einstein's equations can be solved order by order in the r variable. Given a metric $g_{(0)}$ one can uniquely determine the coefficients $g_{(2)}, \ldots, g_{(d-2)}$ and $h_{(d)}$ in terms of $g_{(0)}$.

(iii) The coefficient $h_{(d)}$ is present only when d is even. It is equal to the metric variation of the coefficient of the infrared logarithmic divergence in the on-shell gravitational action (discussed below). It is traceless and covariantly conserved.

(iv) The asymptotic analysis determines only the trace and covariant divergence of $g_{(d)}$.

The explicit expression for $g_{(2)}, ..., g_{(d-2)}, h_{(d)}$ and the equations for the trace and covariant divergence of $g_{(d)}$ can be found in appendix A of Ref. 5.

A case where the Dirichlet problem can be solved exactly is when the bulk metric is conformally flat.[7] In this case one obtains an exact solution given by the metric (6) with

$$\begin{aligned} g(x,\rho) &= g_{(0)}(x) + g_{(2)}(x)\rho + g_{(4)}(x)\rho^2 \ , \\ g_{(2)ij} &= \frac{1}{d-2}(R_{ij} - \frac{1}{2(d-1)} R g_{(0)ij}) \ , \ g_{(4)} = \frac{1}{4}(g_{(2)})^2 \ , \end{aligned} \tag{7}$$

where the curvatures are of the metric $g_{(0)}$.[a]

So far we have discussed the construction of a bulk metric given a representative of a conformal structure. One may ask how these results change if one picks a different representative of the given conformal class, i.e. how the coefficients $g_{(i)}$ transform if we let $g_{(0)} \to e^{2\sigma(x)} g_{(0)}$. One may use the explicit formulae of $g_{(i)}$ in terms of $g_{(0)}$ in order to obtain these transformation rules. An alternative way to determine them is to note that there are bulk diffeomorphism that preserve the form of the metric (6) and induce the transformation $g_{(0)} \to e^{2\sigma(x)} g_{(0)}$. The infinitesimal form of these bulk diffeomorphisms has been worked out in Ref. 8. Here we discuss the corresponding finite transformations. Consider the coordinate transformation

$$\rho = \rho' e^{-2\sigma(x')} + \sum_{k=2} a_{(k)}(x')\rho'^k, \qquad x^i = x'^i + \sum_{k=1} a^i_{(k)}(x')\rho'^k. \tag{8}$$

The requirement that the transformation leaves the form of the metric invariant uniquely fixes the coefficients $a_{(i)}$ and $a^i_{(i)}$. The first few are given by

$$\begin{aligned} a_{(2)} &= -\frac{1}{2}(\partial\sigma)^2 e^{-4\sigma}, \qquad a_{(3)} = \frac{1}{4} e^{-6\sigma}\left(\frac{3}{4}(\partial\sigma)^2 + \partial^i\sigma\partial^j\sigma g_{(2)ij}\right), \\ a^i_{(1)} &= \frac{1}{2}\partial^i\sigma e^{-2\sigma}, \qquad a^i_{(2)} = -\frac{1}{4}e^{-4\sigma}\left(\partial_k\sigma g^{ik}_{(2)} + \frac{1}{2}\partial^i\sigma(\partial\sigma)^2 + \frac{1}{2}\Gamma^i_{kl}\partial^k\sigma\partial^l\sigma\right), \end{aligned} \tag{9}$$

where indices are raised and lowered with $g_{(0)}$. With these results one can calculate the transformation rules of $g_{(i)}$:

$$\begin{aligned} g'_{(0)ij} &= e^{2\sigma} g_{(0)ij} \\ g'_{(2)ij} &= g_{(2)ij} + \nabla_i\nabla_j\sigma - \nabla_i\sigma\nabla_j\sigma + \frac{1}{2}(\nabla\sigma)^2 g_{(0)ij} \end{aligned}$$

[a]The formula for $g_{(2)}$ is valid for $d \neq 2$. The $d = 2$ case is also covered in Ref. 7.

$$
\begin{aligned}
g'_{(4)ij} &= e^{-2\sigma}\Big[g_{(4)ij} - 2\sigma h_{(4)ij} - \frac{1}{4}\nabla^k\sigma(\nabla_i g_{(2)jk} + \nabla_j g_{(2)ik} - 2\nabla_k g_{(2)ij}) \\
&+ \frac{1}{4}(\nabla_i\nabla^k\sigma g_{(2)kj} + \nabla_j\nabla^k\sigma g_{(2)ki}) + \frac{1}{4}R_{kilj}\nabla^k\sigma\nabla^l\sigma + \frac{1}{4}\nabla_i\nabla_j\sigma(\nabla\sigma)^2 \\
&+ (\frac{1}{16}(\nabla\sigma)^4 - \frac{1}{4}\nabla^k\sigma\nabla^l\sigma g_{(2)kl})g_{(0)ij} + \frac{1}{4}\nabla_i\nabla_k\sigma\nabla_j\nabla^k\sigma \\
&- \frac{1}{8}(\nabla_i(\nabla\sigma)^2\nabla_j\sigma + \nabla_j(\nabla\sigma)^2\nabla_i\sigma)\Big]. \\
g'_{(d)ij} &= e^{-(d-2)\sigma}g_{(d)ij}, \qquad d = 2k+1.
\end{aligned} \tag{10}
$$

The infinitesimal version of (9) and (10) agree with the infinitesimal ones derived in Ref. 8.

3. Infrared Divergences

We have argued in the previous section that asymptotically AdS spacetimes come equipped with a bulk Einstein metric and a corresponding boundary conformal structure. We will now show that infrared divergences in the on-shell value of the action force a dependence on the boundary metric, not just on its conformal class.

The gravitational equations (1) can be derived from the action

$$
S[G] = \frac{1}{16\pi G_{\rm N}}[\int_X d^{d+1}x\,\sqrt{G}\,(R[G]+2\Lambda) - \int_{\partial X} d^dx\,\sqrt{\gamma}\,2K], \tag{11}
$$

where K is the trace of the second fundamental form and γ is the induced metric on the boundary.

The on-shell gravitational action diverges because of the infinite bulk volume and the fact that the induced metric diverges at the boundary. To regulate the theory we cutoff the radial integration, $\rho \geq \epsilon$, and evaluate the boundary term at $\rho = \epsilon$, where ϵ is the regulator. This regularization procedure was proposed in Ref. 9 and implemented in Ref. 3.

Using the asymptotic solution discussed in the previous section one can evaluate the on-shell action. The result is[3]

$$
\begin{aligned}
S_{\rm reg} &= \frac{1}{16\pi G_{\rm N}}\left[\int_{\rho\geq\epsilon} d^{d+1}x\,\sqrt{G}\,(R[G]+2\Lambda) - \int_{\rho=\epsilon} d^dx\sqrt{\gamma}\,2K\right] \\
&= \frac{1}{16\pi G_{\rm N}}\int d^dx\sqrt{\det g_{(0)}}\,\Big(\epsilon^{-d/2}a_{(0)} + \epsilon^{-d/2+1}a_{(2)} + \ldots + \epsilon^{-1}a_{(d-2)} \\
&\qquad - \log\epsilon\, a_{(d)}\Big) + \mathcal{O}(\epsilon^0),
\end{aligned} \tag{12}
$$

where the coefficients $a_{(n)}$ are local covariant expressions of the metric $g_{(0)}$ and its curvature tensor. The explicit expressions can be found in appendix B of Ref. 5. Here we only give the coefficients of the logarithmic divergences in $d = 2$ and $d = 4$: $a_{(2)} = \frac{1}{2}R$ and $a_{(4)} = -R^{ij}R_{ij}/8 + R^2/24$.

This computation is universal and applies to any asymptotically AdS spacetime. It is simple to show that the infrared divergences depend only on the coefficients

$g_{(2)}, \ldots, g_{(d-2)}$ in the asymptotic expansion of the metric. As we have seen in the previous section, these coefficients are universal in the sense that they uniquely determined in terms of $g_{(0)}$. The result in (12) does depend on the chosen regularization. The power law divergences are regularization dependent. For instance, a manifestly supersymmetric regularization will yield $a_{(0)} = 0$. The logarithmic divergence, however, is regularization independent.

We now proceed to renormalize the on-shell gravitational action by adding counterterms to cancel the infrared divergences.[3] Using minimal subtraction, we obtain for the renormalized action

$$\begin{aligned} S_{\text{ren}}[g_{(0)}] &= \lim_{\epsilon \to 0} \left(S_{\text{reg}} - \frac{1}{16\pi G_{\text{N}}} \int_{\rho=\epsilon} \sqrt{\gamma}[2(1-d) + \frac{1}{d-2} R \right. \\ &\quad \left. - \frac{1}{(d-4)(d-2)^2} (R_{ij} R^{ij} - \frac{d}{4(d-1)} R^2) - \log \epsilon \, a_{(d)}] \right), \end{aligned} \tag{13}$$

where we have written the counterterms in terms of fields on the regulating hypersurface $\rho = \epsilon$, as in Ref. 4. This formula should be understood as containing only divergent counterterms in each dimension. This means that in even dimension $d = 2k$ one should include only the first k counterterms and the logarithmic one. In odd $d = 2k + 1$, only the first $k + 1$ counterterms should be included. The logarithmic counterterms appear only for d even. The renormalized action (13) is finite up to $d = 6$. It is straightforward but tedious to compute the necessary counterterms for $d > 6$. As we have remarked above, only the logarithmic term has an invariant meaning. The other counterterms are regularization and scheme dependent. In particular one can add further boundary terms provided they do not diverge. We will make use of this freedom in the next section. The logarithmic term was incorrectly (and surprisingly) omitted in a large part of the literature on the subject. Because of this counterterm the on-shell action depends on the chosen representative of the boundary conformal structure, i.e.

$$S_{\text{ren}}[e^{2\sigma} g_{(0)}] = S_{\text{ren}}[g_{(0)}] + \mathcal{A}[g_{(0)}, \sigma]. \tag{14}$$

The anomalous term has been computed for infinitesimal σ in Ref. 3. It is proportional to $a_{(d)}$. This anomalous transformation has been called the holographic Weyl anomaly because it corresponds to the Weyl anomaly of the corresponding CFT in the AdS/CFT correspondence. Notice that in even dimensions $a_{(2k+1)}$ vanishes identically, so in these cases the on-shell action depends only on the conformal class of the boundary metric. This is in agreement with the fact that conformal field theories in odd dimensions do not have conformal anomalies.

As we have discussed in the previous section, boundary conformal transformations are induced by a specific class of bulk diffeomorphisms. It follows that the finite on-shell action in odd dimensions is *not* invariant under these diffeomorphisms. In other words, infrared divergences break part of bulk diffeomorphisms. Only the bulk diffeomorphisms that do not yield a boundary Weyl transformation are true symmetries. The anomaly itself is a conformal invariant, so its vanishing depends only on the conformal class of the boundary metric. It may vanish in one background, but its metric variation may not. This means that in backgrounds where the anomaly vanishes non-trivially, so the on-shell action is invariant under all diffeomorphisms, the stress energy tensor may still have an anomalous variation. This is exactly what happens for (globally) AdS spacetimes.

4. Stress Energy Tensor

In the previous section we have obtained the on-shell value of the gravitational action as a functional of the boundary metric. In Ref. 10, Brown and York proposed defining the gravitational energy momentum tensor through a Hamilton-Jacobi analysis as the functional derivative of the on-shell action with respect to the boundary metric. In Ref. 4 Balasubramanian and Kraus considered the Brown-York energy momentum tensor for asymptotically AdS spaces. Including the counterterms of Ref. 3 (except for the logarithmic one), they showed that the resulting energy momentum tensor was finite in the cases they considered. Their analysis was extended in Ref. 5, where completely explicit expressions for the energy momentum tensor in any even dimension and up seven dimensions were obtained, as we now review.

The stress energy tensor is defined to be

$$T_{ij} = \frac{2}{\sqrt{\det g_{(0)}}} \frac{\delta S_{\text{ren}}}{\delta g^{ij}_{(0)}}. \tag{15}$$

This can be evaluated by first computing the energy momentum tensor in the regulated theory and then removing the regulator,

$$T_{ij} = \lim_{\epsilon\to 0} \frac{2}{\sqrt{\det g(x,\epsilon)}} \frac{\delta S_{\text{ren}}}{\delta g^{ij}(x,\epsilon)} = \lim_{\epsilon\to 0} \left(\frac{1}{\epsilon^{d/2-1}} T_{ij}[\gamma] \right), \tag{16}$$

where $T_{ij}[\gamma]$ is the stress energy tensor in terms of the induced metric γ of the hypersurface $\rho = \epsilon$. It receives a contribution $T^{reg}[\gamma]$ from the original (regulated) action (12), and a contribution $T^{ct}[\gamma]$ from the counterterms. The former is equal to $T^{reg}[\gamma] = -\frac{1}{8\pi G_{\text{N}}}(K_{ij} - K\gamma_{ij})$; the latter can be easily calculated from the explicit form of the counterterms.

Equipped with the explicit asymptotic solutions one can now evaluate (16). This is a rather tedious exercise. The details can be found in Ref. 5. Here we only give the final result,

$$T_{ij} = \frac{dl}{16\pi G_{\text{N}}} (g_{(d)ij} + X^{(d)}_{ij}), \tag{17}$$

where $X^{(d)}_{ij}$ depends on the dimension. The result for all odd d, and even d up to six are:

$$\begin{aligned} X^{(2k+1)}_{ij} &= 0, \qquad X^{(2)}_{ij} = -g_{(0)ij}\,\text{Tr}\; g_{(2)} \\ X^{(4)}_{ij} &= -\frac{1}{8} g_{(0)ij}[(\text{Tr}\; g_{(2)})^2 - \text{Tr}\; g^2_{(2)}] - \frac{1}{2}(g^2_{(2)})_{ij} + \frac{1}{4} g_{(2)ij}\text{Tr}\; g_{(2)} \\ X^{(6)}_{ij} &= -A_{(6)ij} + S_{ij} \end{aligned} \tag{18}$$

where

$$A_{(6)ij} = \frac{1}{3}\Big(2(g_{(2)}g_{(4)})_{ij} + (g_{(4)}g_{(2)})_{ij} - (g^3_{(2)})_{ij} + \frac{1}{8}\left[\text{Tr } g^2_{(2)} - (\text{Tr } g_{(2)})^2\right] g_{(2)ij}$$
$$-\text{Tr } g_{(2)}\left[g_{(4)ij} - \frac{1}{2}(g^2_{(2)})_{ij}\right] - \left[\frac{1}{8}\text{Tr } g^2_{(2)}\text{Tr } g_{(2)} - \frac{1}{24}(\text{Tr } g_{(2)})^3\right.$$
$$\left.-\frac{1}{6}\text{Tr } g^3_{(2)} + \frac{1}{2}\text{Tr } (g_{(2)}g_{(4)})\right] g_{(0)ij}\Big) \tag{19}$$
$$S_{ij} = \frac{1}{24}\Big(\nabla^2 C_{ij} - 2R^k{}_i{}^l{}_j C_{kl} + 4(g_{(2)}g_{(4)} - g_{(4)}g_{(2)})_{ij} + \frac{1}{10}(\nabla_i\nabla_j B - g_{(0)ij}\nabla^2 B)$$
$$+\frac{2}{5}g_{(2)ij}B + g_{(0)ij}(-\frac{2}{3}\text{Tr } g^3_{(2)} - \frac{4}{15}(\text{Tr } g_{(2)})^3 + \frac{3}{5}\text{Tr } g_{(2)}\text{Tr } g^2_{(2)})\Big),$$
$$C_{ij} = (g_{(4)} - \frac{1}{2}g^2_{(2)} + \frac{1}{4}g_{(2)}\text{Tr } g_{(2)})_{ij} + \frac{1}{8}g_{(0)ij}B \ , \quad B = \text{Tr } g_2^2 - (\text{Tr } g_2)^2 \ .$$

The vanishing of $X^{(2k+1)}_{ij}$ reflects the fact that the holographic Weyl anomaly identically vanishes in even spacetimes. In the computation we used the freedom to add finite counterterms in order to to remove terms proportional to $h_{(d)}$. We remind the reader that $h_{(d)}$ is proportional to the metric variation of $a_{(d)}$ so by adding a finite counterterm proportional to $a_{(d)}$ one can remove all $h_{(d)}$ dependence from the energy momentum tensor.

One can check by explicit computation that the stress energy tensors so obtained are covariantly conserved and their trace correctly reproduces the holographic Weyl anomaly.

The stress energy transforms anomalously under bulk diffeomorphisms that induce Weyl transformations in the boundary. Indeed, using the definition (15) and the transformation of the on-shell action one gets,

$$T_{ij}[e^{2\sigma}g_{(0)}] = e^{-(d-2)\sigma}\left(T_{ij}[g_{(0)}] + \frac{1}{\sqrt{g_{(0)}}}\frac{\delta \mathcal{A}[g_{(0)}]}{\delta g^{ij}_{(0)}}\right). \tag{20}$$

From here it follows that the stress energy tensor may transform anomalously even if $\mathcal{A}$ evaluated on a particular background is equal to zero. An alternative way to compute the dependence of the stress energy tensor on a given representative of the boundary conformal structure is to use the transformation rules of $g_{(i)}$ in (10) and formula (17)-(19). The infinitesimal transformations can be found in Ref. 5.

To summarize: given a solution of Einstein's equation with negative cosmological constant one can obtain the corresponding energy momentum tensor by first reaching the coordinate system (6) near the boundary, and then plugging in the coefficients $g_{(i)}$ in (17).

We finish this section with a few comments on the addition of matter fields. As in the case of pure gravity one first needs to obtain asymptotic solutions of the coupled gravity-matter system. It can happen that the leading behavior of the matter stress energy tensor is more singular than the leading behavior of the

Einstein tensor. In this case the methods discussed here apply only if the fields parameterizing the matter boundary conditions are considered infinitesimal. In the other cases one can proceed as in the case of pure gravity to regularize, renormalize the theory and obtain the stress energy tensor as the metric variation of the on-shell action. A more detailed discussion can be found in Ref. 5.

4.1. *Example: AdS$_5$*

As an example we discuss in some detail the case of AdS$_5$ spacetime. The coordinate transformation that brings the metric (3) to the coordinate system in (6) is

$$r^2 = \frac{1}{\rho}(1 - \frac{\rho}{4})^2. \tag{21}$$

The metric coefficients are given by $g_{(0)} = \mathrm{diag}(-1, 1, \sin^2\theta, \sin^2\theta \sin^2\phi)$, $g_{(2)} = -\frac{1}{2}\mathrm{diag}(1, 1, \sin^2\theta, \sin^2\theta\sin^2\phi)$, and $g_{(4)} = \frac{1}{16}g_{(0)}$. The expansion terminates at ρ^2 because AdS$_5$ is conformally flat.[7] The expressions for $g_{(2)}$ and $g_{(4)}$ agree with formula (7). Using (17) we now get

$$T_{ij} = \frac{l^3}{64\pi G_{\mathrm{N}}}(4\delta_{i,0}\delta_{j,0} + g_{(0)ij}) \tag{22}$$

One can explicitly check that this stress energy tensor is conserved and traceless. It is traceless because the conformal anomaly evaluated for global AdS vanishes.

The boundary metric is conformally flat. The coordinate transformation

$$\tau \pm r = \tan\frac{1}{2}(t \pm \theta) \tag{23}$$

brings the metric to the form

$$ds^2_{(0)} = 4\cos^2\frac{1}{2}(t+\theta)\cos^2\frac{1}{2}(t-\theta)\left(-d\tau^2 + dr^2 + r^2 d\Omega_2\right) \tag{24}$$

We can implement this Weyl transformation using the bulk diffeomorphism (8) with $e^{-2\sigma} = 4\cos^2\frac{1}{2}(t+\theta)\cos^2\frac{1}{2}(t-\theta)$. One can obtain the new metric by working out (10). After some algebra one obtains $g'_{(2)} = g'_{(4)} = 0$. Alternatively, one can notice that the Dirichlet boundary problem for conformally flat bulk metrics has a unique answer. In our case, after the transformation $g'_{(0)}$ is flat, and from (7) we get $g'_{(2)} = g'_{(4)} = 0$. It follows that T_{ij} vanishes identically.

We thus see explicitly that the energy momentum tensor, defined as the metric variation of the renormalized on-shell action, is *not* covariant with respect to diffeomorphisms that yield a Weyl transformation in the boundary. The anomalous transformation has its origin in the infrared divergences in the computation of the on-shell action.

Notice that these results are in exact agreement with expectations from the AdS/CFT correspondence. The gravitational energy momentum tensor is identified with the expectation value of the CFT stress energy tensor. This expectation value is zero for a CFT on flat space. When we consider the CFT on $R \times S^3$, however, the expectation value is non-zero due to the Casimir energy. Using the AdS/CFT dictionary for the gravity/gauge theory parameters one finds exact agreement.[4] One could infer this agreement from the fact that the Casimir energy follows from the trace anomaly, and the latter was shown to be reproduced exactly by a gravity computation in Ref. 3. These quantities had to agree, even though the gravity computation is at strong coupling and the CFT one in weak coupling, because there is a non-renormalization theorem that protects them.

To shed some more light on the results for the stress energy tensor, one can work out the relation between the generator $H = \partial/\partial t$ of global time translations to generators of isometries of AdS with flat boundary. Using the coordinate transformation (23), one obtains

$$H = \frac{1}{2}(P_\tau + K_\tau), \tag{25}$$

where $P_\tau = \partial/\partial\tau$ is the generator of τ-time translations and $K_i = x^2\partial_i - 2x_i x^i \partial_j$ is the generator of special conformal transformations. In a theory with conformal anomalies, the dilatations and special conformal transformations are broken. This allows for a non-zero eigenvalue of the generator of global time translations acting on the ground state. Notice that in the case of Euclidean AdS global time translations are mapped to dilatations, which are also broken.

Acknowledgments

This research is supported in part by the NSF grant PHY-9802484.

References

1. L.F. Abbott and S. Deser, *Nucl. Phys.* **B195**, 76 (1982); A. Ashtekar and A. Magnon, *Class. Quant. Grav.* **1**, L39 (1984); M. Henneaux and C. Teitelboim, *Phys. Lett.* **B142**, 355 (1984); *Comm. Math. Phys.* **98**, 391 (1985).
2. A. Ashtekar and S. Das, *Class. Quant. Grav.* **17**, L17 (2000) [hep-th/9911230].
3. M. Henningson and K. Skenderis, *JHEP* **9807**, 023 (1998) [hep-th/9806087]; *Fortsch. Phys.* **48**, 125 (2000) [hep-th/9812032].
4. V. Balasubramanian and P. Kraus, *Comm. Math. Phys.* **208**, 413 (1999) [hep-th/9902121].
5. S. de Haro, K. Skenderis and S.N. Solodukhin, hep-th/0002230.
6. C. Fefferman and C. Robin Graham, in *Elie Cartan et les Mathématiques d'aujourd'hui* (Astérisque, 1985), p. 95.
7. K. Skenderis and S.N. Solodukhin, *Phys. Lett.* **B472**, 316 (2000) [hep-th/9910023].
8. C. Imbimbo, A. Schwimmer, S. Theisen and S. Yankielowicz, *Class. Quant. Grav.* **17**, 1129 (2000) [hep-th/9910267].
9. E. Witten, *Adv. Theor. Math. Phys.* **2**, 253 (1998) [hep-th/9802150].
10. J.D. Brown and J.W. York, *Phys. Rev.* **D47**, 1407 (1993).

TOWARDS A SUPERSYMMETRIC NON-ABELIAN BORN-INFELD THEORY

E.A. BERGSHOEFF, M. DE ROO
Institute for Theoretical Physics,
Nijenborgh 4, 9747 AG Groningen,
The Netherlands
E-mail: e.a.bergshoeff@phys.rug.nl, m.de.roo@phys.rug.nl

A. SEVRIN
Theoretische Natuurkunde,
Vrije Universiteit Brussel,
Pleinlaan 2, B–1050 Brussels, Belgium
E-mail: asevrin@tena4.vub.ac.be

We define an iterative procedure to obtain a non-abelian generalization of the Born-Infeld action. This construction is made possible by the use of the severe restrictions imposed by kappa-symmetry. We have calculated all bosonic terms in the action up to terms quartic in the Yang-Mills field strength and all fermion bilinear terms up to terms cubic in the field strength. Already at this order the fermionic terms do not satisfy the symmetric trace-prescription.

1. Abelian Born-Infeld

One of the most intriguing features of D-branes is their close connection to gauge theories. Indeed, the effective theory describing the world-volume dynamics of a Dp-brane is a $p+1$–dimensional field theory with, in the static gauge, as bosonic degrees of freedom the transversal coordinates X^m $(m = 1, \cdots, 9-p)$ of the brane, and the massless states of the open strings ending on the brane which appear as a $U(1)$ gauge field describing $p-1$ degrees of freedom. When these fields vary slowly, the effective action governing their dynamics is known to all orders in α'. It is the ten-dimensional Born-Infeld action,[1] dimensionally reduced to $p+1$ dimensions. In the supersymmetric case there are additional fermionic degrees of freedom χ_α $(\alpha = 1, \cdots 16)$ describing 8 fermionic degrees of freedom so that we have a total of $8+8$ degrees of freedom.

Clearly, the $9-p$ transversal scalars break the 10-dimensional Lorentz covariance. The fully covariant worldvolume theory of a single D-brane in a type II theory can be formulated in terms of the following worldvolume fields:[a] the embedding coordinates $X^\mu(\sigma)$ (of which only the transverse coordinates represent physical degrees of freedom), the Born-Infeld vector field $V_i(\sigma)$, and $N=2$ spacetime fermionic fields $\theta(\sigma)$. The presence of a local fermionic symmetry (κ-symmetry) makes it possible

[a] The indices $\mu, \nu = 0, \ldots, 9$ are spacetime indices whereas the indices $i, j = 0, \ldots, p$ label the worldvolume coordinates σ^i.

to gauge away half the fermionic degrees of freedom. The field content then corresponds, in a static gauge, to that of a supersymmetric Yang-Mills theory in $p+1$ dimensions describing $8+8$ degrees of freedom.

In this talk our main emphasis will be on the IIB D9–brane. This case is special in the sense that, for $p = 9$, there are no transverse scalars and the worldvolume theory is given by a D=10 supersymmetric Maxwell multiplet (V_μ, χ). The bosonic part of the action is given by the ten-dimensional Born-Infeld Lagrangian[2,3]

$$\mathcal{L}(\text{bosonic}) = -\sqrt{-\det\left(\eta_{\mu\nu} + 2\pi\alpha' F_{\mu\nu}\right)} \ . \tag{1}$$

Expanding the square root this action gives rise to an infinite number of terms with even powers of F. The term quadratic in F is the usual Maxwell kinetic term.

In the supersymmetric case every F^n term gets a supersymmetric partner and the complete action has the schematic form

$$\begin{aligned}\mathcal{L}(\text{supersymmetric}) \;&=\; \text{tr}F^2 + (2\pi\alpha')^2\left[\text{tr}F^4 - \tfrac{1}{4}\left(\text{tr}F^2\right)^2\right] + \cdots \\ &+\; \bar{\chi}\partial\chi \;+ (2\pi\alpha')^2\, F^2\bar{\chi}\partial\chi + \cdots\end{aligned}$$

The fermionic $F^2\bar{\chi}\partial\chi$ terms were found in Refs. 4,5 from the requirement that the action must possess a linear supersymmetry. Complete results were obtained in D=4 dimensions via superspace techniques.[6]

Recently a new development took place which led to a construction of the complete supersymmetric Born-Infeld action. A crucial role in this development was played by kappa-symmetry. The kappa-symmetric covariant D-brane actions were constructed in flat,[7] as well as in curved backgrounds.[8,9]

Schematically, the kappa-symmetric Lagrangian is given by[b]

$$\mathcal{L}_{\text{BI}} = -e^{-\phi}\sqrt{-\det\left(g + \mathcal{F}\right)} \;+\; Ce^{\mathcal{F}}\,, \tag{2}$$

with $\mathcal{F} = 2dV + B$. The first term is often called the kinetic term whereas the second one is the Wess-Zumino term. It is understood that all NS-NS background fields g, ϕ, B and all R-R background fields C are superfields defined over a superspace with coordinates (X^μ, θ). The local κ-symmetry acts on the fermions as

$$\delta\bar{\theta}(\sigma) = \bar{\eta}(\sigma) \equiv \bar{\kappa}(\sigma)\,(1+\Gamma)\,, \tag{3}$$

where Γ, which depends on worldvolume as well as background fields, satisfies

$$\Gamma^2 = \mathbb{1}. \tag{4}$$

The projection (3) makes it possible to gauge away half the fermionic degrees of freedom. The variation of the D-brane action takes the form

$$\delta\mathcal{L} = -\bar{\eta}\,(1-\Gamma)\mathcal{T}\,, \tag{5}$$

[b]From now on we set $2\pi\alpha' = 1$.

where $\mathcal{T}$ is some expression in terms of the worldvolume and background fields. The variation (5) indeed vanishes if $\bar{\eta}$ is given by (3). This variation has the following source: the first term within the round brackets comes from the kinetic term in (2), the term with Γ arises from the Wess-Zumino contribution.

From the above it is clear that a crucial role is played by the projection operator $1+\Gamma(X,\mathcal{F})$. In the case of the IIB D9–brane the $\mathcal{F}$-independent part $\Gamma^{(0)}(X)$ of $\Gamma(X,\mathcal{F})$ is given by

$$\Gamma^{(0)}(X) = \frac{1}{10!\sqrt{-\det g}}\epsilon^{i_1\dots i_{10}}\gamma_{i_1\dots i_{10}}\,, \qquad \left(\Gamma^{(0)}(X)\right)^2 = \mathbb{1}. \tag{6}$$

The worldvolume metric reads

$$g_{ij} = \eta_{ij} + \bar{\theta}\gamma_{(i}\partial_{j)}\theta\,, \qquad \eta_{ij} \equiv \partial_i X^\mu \partial_j X_\mu\,, \tag{7}$$

and

$$\gamma_i \equiv \Gamma^\mu \partial_i X^\mu\,. \tag{8}$$

The complete expression for $\Gamma(X,\mathcal{F})$ is given by[9]

$$\Gamma = \frac{\sqrt{-\det g}}{\sqrt{-\det(g+\mathcal{F})}}\Gamma^{(0)}\sum_{k=0}^{5}\frac{1}{2^k k!}\mathcal{P}_{(k)}\gamma^{i_1\dots i_{2k}}\mathcal{F}^k_{i_1\dots i_{2k}}\,, \tag{9}$$

where

$$\mathcal{P}_{(k)} = \sigma_1 \text{ (for } k=1,3,5)\,, \quad \mathcal{P}_{(k)} = i\sigma_2 \text{ (for } k=2,4)\,. \tag{10}$$

The Pauli-matrices refer to the fact that we are using an N=2 superspace notation.

It was realized in Ref. 7 that for the special case of a IIB D9–brane in a flat background, the results take on a particularly simple form. We consider a flat background, i.e. we take $e_\mu{}^a = \delta_\mu{}^a$ and all other fields zero. Furthermore, we fix the kappa-transformations and worldvolume parametrizations (we choose the static gauge) by setting

$$X^\mu = \delta^{i\mu}\sigma^i\,, \qquad \theta = \begin{pmatrix}\theta_1 \equiv \chi \\ \theta_2 = 0\end{pmatrix}. \tag{11}$$

The result of this gauge-fixing is that the complete Wess-Zumino term vanishes and the kinetic term is given by the simple expression[7]

$$\mathcal{L}_{\rm BI} = -\sqrt{-\det\left(\eta_{\mu\nu} + F_{\mu\nu} + \bar{\chi}\Gamma_\mu\partial_\nu\chi + \tfrac{1}{4}\bar{\chi}\Gamma^a\partial_\mu\chi\bar{\chi}\Gamma_a\partial_\nu\chi\right)}\,. \tag{12}$$

This action has 16 linear supersymmetries as well as 16 nonlinear supersymmetries. The explicit form of these supersymmetry rules can be obtained from the original 32 supersymmetries of the N=2 superspace after taking care of the fact that they get deformed with a field-dependent kappa-transformation upon fixing the kappa-gauge.

Our goals, which we will discuss in this talk, are two-fold:

(1) We want to find the non-abelian version of the projection operator $1+\Gamma$. One application we have in mind is that knowing the explicit expression of Γ enables us to investigate non-abelian BPS configurations. The condition of supersymmetry for such configurations is given by $(1-\Gamma)\epsilon = 0$.[10]

(2) A second, and more ambitious goal is to construct the non-abelian supersymmetric Born-Infeld action by the use of the severe restrictions of kappa-symmetry as suggested in Ref. 11. We restrict ourselves to a flat background for reasons to be discussed later (see the comments section). Non-abelian N=2, D=4 supersymmetric Born-Infeld actions (with a symmetrized trace condition) have recently been constructed in Refs. 12,13.

It should be stressed that we only discuss the IIB D9–brane. We expect that at least some of the structure of the lower-dimensional branes can be obtained by applying T-duality.[14,15] These lower-dimensional cases have the additional complication that they contain non-trivial transverse scalars. It is an open question what kind of "D-geometry" these scalars are supposed to describe.[16,17]

2. Non-Abelian Born-Infeld

In the case of a single D–brane the worldvolume theory, in lowest order in α' is given by a supersymmetric Maxwell theory. The result to all orders in α' is given by the supersymmetric BI action.

Once several D-branes are present, the situation changes. The mass of the strings stretching between two branes is proportional to the shortest distance between the two branes. Starting off with n well separated D-branes, we end up with a $U(1)^n$ theory. However, once the n branes coincide, additional massless states appear which complete the gauge multiplet to a non-abelian $U(n)$-theory.[18] Contrary to the abelian case, the effective action is not known to all orders in α'. The first term, quadratic in the field strength, is nothing but a dimensionally reduced $U(n)$ Yang-Mills theory. The next order, which is quartic in the field strength, was obtained from the four-gluon scattering amplitude in open superstring theory[19] and from a three-loop β-function calculation.[20]

The notion of an effective action for slowly varying fields is subtle in the non-abelian case.[1] In the effective action higher derivative terms are dropped. However because of

$$D_iD_jF_{kl} = \frac{1}{2}\{D_i, D_j\}F_{kl} - \frac{i}{2}[F_{ij}, F_{kl}], \tag{13}$$

this is ambiguous. The analysis of the mass spectrum seems to indicate that the symmetrized product of derivatives acting on a fieldstrength should be viewed as an acceleration term which can safely be neglected, while the anti-symmetrized products should be kept.

Different order prescriptions of the non-abelian matrices occurring in the non-abelian Born-Infeld action have been given in the literature. In Ref. 21 a symmetrized trace prescription was suggested: the non-abelian Born-Infeld action assumes essentially the same form as the abelian one, however, all Lie algebra valued objects have to be symmetrized first before taking the trace. Other trace prescriptions, involving commutators, have been given as well.[22]

Recently, it was found that the symmetrized trace prescription could not be correct as it did not reproduce the mass spectrum of certain D-brane configurations.[23,11] It was shown in Ref. 24 that by adding commutator terms to the action the problem might be cured.

For n overlapping D9-branes the completely gauge-fixed result should be the supersymmetric version of the non-abelian Born-Infeld theory. Since the vector fields $V_i^A(\sigma)$, $A = 1, \ldots, n^2$, are in the adjoint representation of $U(n)$, we have to make the same choice for the fermion fields θ. Therefore we start out with fields $\theta^A(\sigma)$, which form a doublet ($N = 2$) of Majorana-Weyl spinors for each A, satisfying $\Gamma_{11}\theta^A = \theta^A$. After κ-gauge-fixing only half of each doublet will remain, and we have the correct number of degrees of freedom for the supersymmetric Yang-Mills theory.

This requires, that there are as many κ-symmetries as θ's, so that also the parameter of the κ transformations will have to be in the adjoint of $U(n)$. Thus the θ^A transform as follows under ordinary supersymmetry (ϵ), κ-symmetry (κ), Yang-Mills transformations (Λ^A), and worldvolume reparametrisations (ξ^i):

$$\delta\bar{\theta}^A(\sigma) = -\bar{\epsilon}^A + \bar{\eta}^A(\sigma) + f^A{}_{BC}\Lambda^B(\sigma)\bar{\theta}^C(\sigma) + \xi^i(\sigma)\partial_i\bar{\theta}^A(\sigma)\,, \tag{14}$$

with $\bar{\eta}^A \equiv \bar{\kappa}^B(\mathbb{1}\delta^{BA} + \Gamma^{BA})$. Here ϵ^A are constant, Γ^{AB} depends on the worldvolume fields, and therefore on σ. It must satisfy

$$\Gamma^{AB}\Gamma^{BC} = \delta^{AC}\mathbb{1}. \tag{15}$$

Useful information is obtained by considering commutators of these transformations. Because ϵ^A is constant we find from the commutator of Yang-Mills and supersymmetry transformations that

$$f^A{}_{BC}\Lambda^B\epsilon^C = 0 \rightarrow f_{ABC}\epsilon^C = 0\,. \tag{16}$$

Therefore $\epsilon = \epsilon^A T_A$, where T_A are the $U(n)$ generators, must be proportional to the unit matrix, i.e. we can choose a basis in which there is only one nonvanishing ϵ parameter. So only a subset of the θ^A transform under supersymmetry, and there is only one independent supersymmetry parameter. The θ's which are presently inert under supersymmetry will obtain their supersymmetry transformations after κ-gauge fixing.

The only spacetime scalars we have are the embedding coordinates $X^\mu(\sigma)$ for worldvolume directions. There are several options that one could consider for the X^μ:

1. We could assume that we are in the static gauge, i.e.

$$X^\mu(\sigma) = \delta_i^\mu\,\sigma^i\,, \tag{17}$$

from the beginning, so that the X^μ are absent. In this case there are no worldvolume reparametrisations, i.e. $\xi^i = 0$ in (14).

2. We could decide that the X^μ are in the singlet representation of the Yang-Mills group. The idea is that the n branes overlap, there is only one set of worldvolume coordinates, and the corresponding reparametrization group would be sufficient to gaugefix a singlet set of embedding coordinates.

3. We could choose the X^μ in the adjoint representation of Yang-Mills in analogy with transverse coordinates for $p < 9$. Here one thinks of starting with n separate branes where each has its own worldvolume and embedding coordinates. When the branes overlap the embedding coordinates "fill up" to form elements of the adjoint representation. Clearly this requires a different approach towards the world-volume reparametrisation invariance, which must then correspond to a sufficiently large symmetry group to gaugefix all these embedding functions.

We have investigated the first two possibilities in the non-abelian case, and we have found that only the first approach is consistent with the iterative procedure that we employ.

Our strategy will be to construct a kappa-symmetric action via an iterative procedure. We found that the flat background expressions of the abelian NS-NS and R-R fields cannot be generalized to the non-abelian case while remaining YM singlets. We therefore decided to make the most general Ansatz for the action and to require kappa-symmetry, order by order in F and up to terms quartic in the fermions. The iteration is obtained by expanding Γ and $\mathcal{T}$ in the variation (5) order by order in F:

$$\begin{aligned} \delta\mathcal{L} &= -\bar{\eta}\,(1-(\Gamma_0+\Gamma_1+\ldots))(\mathcal{T}_0+\mathcal{T}_1+\ldots) \\ &= -\bar{\eta}\,(\mathcal{T}_0-\Gamma_0\mathcal{T}_0+\mathcal{T}_1-\Gamma_1\mathcal{T}_0-\Gamma_0\mathcal{T}_1+\ldots)\,, \end{aligned} \tag{18}$$

where the subindices reflect the order in F. We completed this procedure up to variations of the action quadratic in F. Due to lack of space we give the result, both in its kappa-symmetric as well as in its gauge-fixed form, not here, but in a forth-coming publication.[25]

3. Comments

We have been able to embed a D=10 supersymmetric $U(n)$ YM theory (up to terms in the action quartic in F) into a kappa-symmetric system. In particular, we find fermionic terms of the form $[F^2,\bar{\chi}]\,\partial\chi$ which violate the symmetrized trace prescription.

We hope that our partial results will contribute to a better understanding of the non-abelian Born-Infeld action and perhaps may lead to the complete answer like in the abelian case. In this respect it is of interest to remember that the abelian expression (9) for $\Gamma(X,F)$ can be written as[10]

$$\Gamma(X,F) = e^{-a/2}\Gamma^{(0)}(X)e^{a/2}\,, \tag{19}$$

with

$$a = \tfrac{1}{2}Y_{ij}\Gamma^{ij}\sigma_3\,. \tag{20}$$

The matrix Y is related to the matrix F by a so-called "tan" relation defined in Ref. 10: F = "tan" Y. These results were obtained by looking to branes at angles.[26,27] Via T-duality these systems are related to branes with $F \neq 0$ as we discuss in this talk. It would be interesting to see whether a similar simple expression can be found for the nonabelian case

Another approach to find the complete answer could be to use the superembedding techniques developed in Refs. 28,29. Finally, yet another way to get the complete answer could be to extend to the non-abelian case the analysis of Ref. [30] where it was shown how the superworldvolume dynamics of superbranes can be obtained form nonlinear realizations.

We would like to end by mentioning a few important open issues.

(A) Can the results we obtained be generalized to a supersymmetric curved background? On the one hand the flat background expressions we find do not fit into general supergravity background field expressions that are YM singlets. We do not allow the supergravity fields to be in a nontrivial representations of the YM group. One possible scenario could be that the D-brane only couples to the U(1) part of the N=1 supergravity background fields and that the off-diagonal terms only involve worldvolume fields. The price to pay is that not all worldvolume fields are on the same footing whereas, at least after fixing the kappa-symmetries, they should be.[c]

(B) A somewhat related issue is: is there a natural (non-abelian) superspace geometry that describes the kappa-symmetric result? It is well-known that, to describe the abelian kappa-symmetric action, an important role is played by ordinary superspace geometry. In the non-abelian case we are dealing with many fermionic coordinates on which just a single N=2 supersymmetry is realized. Usually, introducing more fermionic coordinates, means going to an extended superspace with extended supersymmetries but this is not the case in the present situation.

(C) It would be interesting to see what happens when we T-dualize our results and obtain the worldvolume theory of non-abelian Dp–branes with $p < 9$. Our hope is that this will teach us something about the D-geometry which is involved in describing such systems.

(D) Finally, our results can be used to search for non-Abelian BPS configurations, i.e. configurations with nontrivial F-terms that take values outside the Cartan subalgebra.

---

[c] We thank A. Tseytlin for a discussion on this issue.

Acknowledgments

One of us (E.B.) would like to thank the organizers of Strings 2000 for providing such a stimulating environment. We like to thank I. Bandos, M. Cederwall, S. Ferrara, S.F. Hassan, E. Ivanov, R. Kallosh, U. Lindström, C. Nappi, A. Peet, V. Periwal, D. Sorokin, J. Troost and A. Tseytlin. for useful discussions. This work is supported by the European Commission RTN programme HPRN-CT-2000-00131, in which E.B. and M.d.R. are associated to the university of Utrecht and A.S. is associated to the university of Leuven.

References

1. A recent review of Born-infeld theory can be found in hep-th/9908105, to appear in the Yuri Golfand memorial volume.
2. E.S. Fradkin and A.A. Tseytlin, *Phys. Lett.* **B163**, 123 (1985).
3. A. Abouelsaood, C.G. Callan, C.R. Nappi and S.A. Yost, *Nucl. Phys.* **B280**, 599 (1987).
4. E. Bergshoeff, M. Rakowski and E. Sezgin, *Phys. Lett.* **B185**, 371 (1987).
5. R.R. Metsaev and M.A. Rakhmanov, *Phys. Lett.* **B193**, 202 (1987).
6. S. Cecotti and S. Ferrara, *Phys. Lett.* **B187**, 335 (1987).
7. M. Aganagic, C. Popescu and J.H. Schwarz, *Nucl. Phys.* **B495**, 99 (1997) [hep-th/9612080].
8. M. Cederwall, A. von Gussich, B.E.W. Nilsson, P. Sundell and A. Westerberg, *Nucl. Phys.* **B490**, 179 (1997) [hep-th/96100148].
9. E. Bergshoeff and P.K. Townsend, *Nucl. Phys.* **B490**, 145 (1997) [hep-th/9611173].
10. E. Bergshoeff, R. Kallosh, T. Ortin and G. Papadopoulos, *Nucl. Phys.* **B502**, 149 (1997) [hep-th/9701127].
11. F. Denef, A. Sevrin and J. Troost, *Nucl. Phys.* **B581**, 135 (2000) [hep-th/0002180].
12. S.V. Ketov, *Phys. Lett.* **B491**, 207 (2000) [hep-th/0005265]; S.V. Ketov, *Class. Quant. Grav.* **17**, L91 (2000) [hep-th/0005126].
13. A. Refolli, N. Terzi and D. Zanon, *Phys. Lett.* **B486**, 337 (2000) [hep-th/0006067].
14. W.I. Taylor and M. Van Raamsdonk, *Nucl. Phys.* **B573**, 703 (2000) [hep-th/9910052].
15. R.C. Myers, *JHEP* **9912**, 022 (1999) [hep-th/9910053].
16. M.R. Douglas, *Class. Quant. Grav.* **17**, 1057 (2000) [hep-th/9910170].
17. J. de Boer, talk presented at the Euresco conference *Particle and Physics, Euroconference on Quantum Fields and Strings*, September 9-15 2000, Kolymbari, Crete, Greece.
18. E. Witten, *Nucl. Phys.* **B460**, 35 (1996) [hep-th/9510135].
19. D.J. Gross and E. Witten, *Nucl. Phys.* **B277**, 1 (1986); A.A. Tseytlin, *Nucl. Phys.* **B276**, 391 (1986); **B291**, 876 (1987).
20. D. Brecher and M.J. Perry, *Nucl. Phys.* **B527**, 121 (1998) [hep-th/9801127];
K. Behrndt, proceedings of the XXIII Int. Symp. Ahrenshoop 1989, 174, Akademie der Wissenschaften der DDR; *Untersuchung der Weyl-Invarianz im Verallgemeinter σ-Modell für Offene Strings*, PhD thesis, Humboldt-Universität zu Berlin, 1990.
21. A.A. Tseytlin, *Nucl. Phys.* **B501**, 41 (1997) [hep-th/9701125].
22. P.C. Argyres and C.R. Nappi, *Nucl. Phys.* **B330**, 151 (1990).
23. A. Hashimoto and W. Taylor, *Nucl. Phys.* **B503**, 193 (1997) [hep-th/9703217].
24. P. Bain, proceedings of the Cargèse '99 Summer School, hep-th/9909154.
25. E. Bergshoeff, M. de Roo and A. Sevrin, in preparation.
26. M. Berkooz, M.R. Douglas and R.G. Leigh, *Nucl. Phys.* **B480**, 265 (1996) [hep-th/9606139].
27. V. Balasubramanian and R.G. Leigh, *Phys. Rev.* **D55**, 6415 (1997) [hep-th/9611165].

28. I.A. Bandos, D. Sorokin, M. Tonin, P. Pasti and D.V. Volkov, *Nucl. Phys.* **B446**, 79 (1995) [hep-th/9501113].
29. P.S. Howe, E. Sezgin and P.C. West, proceedings of the Volkov memorial conference, hep-th/9705093.
30. S. Bellucci, E. Ivanov and S. Krivonos, *Phys. Lett.* **B482**, 233 (2000) [hep-th/0003273].

COMPLEX GRAVITY AND NONCOMMUTATIVE GEOMETRY

ALI H. CHAMSEDDINE
Center for Advanced Mathematical Sciences (CAMS)
and Physics Department
American University of Beirut
College Hall, Room 425
P.O. Box 11–0236, Beirut, Lebanon

The presence of a constant background antisymmetric tensor for open strings or D-branes forces the space-time coordinates to be noncommutative. An immediate consequence of this is that all fields get complexified. By applying this idea to gravity one discovers that the metric becomes complex. Complex gravity is constructed by gauging the symmetry $U(1, D-1)$. The resulting action gives one specific form of nonsymmetric gravity. In contrast to other theories of nonsymmetric gravity the action is both unique and gauge invariant. It is argued that for this theory to be consistent one must prove the existence of generalized diffeomorphism invariance. The results are easily generalized to noncommutative spaces.

At planckian energies, the manifold structure of space-time will not hold, and a new geometrical setting is needed. At present there are two possible candidates to describe space-time at high energies, one is string theory and the other is noncommutative geometry.[1] Recently these two approaches came together when it was realized that the presence of constant background B-field for open strings or D-branes implies that the coordinates of space-time become noncommuting.[2,3,4,5,6,7,8] This result is expected to generalize to the case of a non-constant B-field. The resulting geometrical space is expected to be noncommuting and curved. The question I will address in this talk is how to describe the dynamics of the gravitational field in such spaces.

One possibility is to use the tools of noncommutative geometry of Alain Connes as specified by the spectral triple $(\mathcal{A}, \mathcal{H}, \mathcal{D})$ where $\mathcal{A}$ is an associative algebra with a $*$ product and identity, $\mathcal{H}$ a Hilbert space and D a self-adjoint operator on $\mathcal{H}$ such that $[D, a]$, $a \in \mathcal{A}$ defines a bounded operator on $\mathcal{H}$.[9] In this setting it is possible to develop the noncommutative analogue of Riemannian geometry. A good example of the realization of noncommutative geometry is the data encoded in superconformal field theory.[1] The operator D encodes the metric, differential calculus, integration and dynamics. For simple noncommutative spaces such as the noncommutative space defined by the standard model all information about the bosonic and fermionic action is encoded in the spectrum of the Dirac operator. This is known as the spectral action principle.[10] The difficulty in this approach is that in order to make progress one must know the Dirac operator. Enough information must be available about D to define geometrical quantities. In the problem at hand it is not easy to guess what D should one start with. The strategy I will adopt is to first gather information about noncommutative spaces with constant background B-fields.

Open strings or D-branes in presence of constant background B-field can be realized by deforming the algebra of functions on the classical world volume. The operator product expansion for vertex operators is identified with the star (Moyal) product of functions on noncommutative spaces.[11,12] In this respect it was shown that noncommutative U(N) Yang-Mills theory does arise in string theory.

The star product is defined by

$$f(x) * g(x) = e^{\frac{i}{2}\theta^{\mu\nu}\frac{\partial}{\partial\zeta^{\mu}}\frac{\partial}{\partial\eta^{\nu}}} f(x+\zeta)\, g(x+\eta)\,|_{\zeta=\eta=0} \tag{1}$$

This definition forces the gauge fields to become complex. Indeed the noncommutative Yang-Mills action is invariant under the gauge transformations

$$A_{\mu}^{g} = g * A_{\mu} * g_{*}^{-1} - \partial_{\mu} g * g_{*}^{-1} \tag{2}$$

where $g_{*}^{-1}$is the inverse of g with respect to the star product:

$$g * g_{*}^{-1} = g_{*}^{-1} * g = 1 \tag{3}$$

The contributions of the terms $i\theta^{\mu\nu}$ in the star product forces the gauge fields to be complex. Only conditions such as $A_{\mu}^{\dagger} = -A_{\mu}$ could be preserved under gauge transformations provided that g is unitary: $g^{\dagger} * g = g * g^{\dagger} = 1$. It is not possible to restrict A_{μ} to be real or imaginary to get the orthogonal or symplectic gauge groups as these properties are not preserved by the star product.[8,13] For open strings in constant background B-field the effective metric is[14,8]

$$\begin{aligned} g^{\mu\nu} &= \left(G_{\mu\nu} + 2\pi\alpha' B_{\mu\nu}\right)_{S}^{-1} \\ \theta^{\mu\nu} &= \left(G_{\mu\nu} + 2\pi\alpha' B_{\mu\nu}\right)_{A}^{-1} \\ g_{\mu\nu} &= G_{\mu\nu} - \left(2\pi\alpha'\right)^{2}\left(BG^{-1}B\right)_{\mu\nu} \end{aligned} \tag{4}$$

One can imagine a general setting where the closed string theory metric arise as an effective metric coming from open strings, or where the D-branes become dynamical. Under such circumstances one can get an effective metric of the form

$$g_{\mu\nu} = e_{\mu}^{a} * e_{\nu a} \tag{5}$$

Because of θ contributions the metric must become complex. This also seems inevitable as the star product appears in the operator product expansion of the string vertex operators. We are therefore led to investigate whether the metric can become complex.

Assume that we start with the $U(1, D-1)$ gauge fields $\omega_{\mu\ b}^{\ a}$. The $U(1, D-1)$ group of transformations is defined as the set of matrix transformations leaving the quadratic form

$$\left(Z^{a}\right)^{\dagger} \eta_{b}^{a} Z^{b} \tag{6}$$

invariant, where Z^a are D complex fields and

$$\eta^a_b = diag\,(-1, 1, \cdots, 1) \tag{7}$$

with $D-1$ positive entries. The gauge fields $\omega_{\mu\ b}^{\ a}$ must then satisfy the condition

$$\left(\omega_{\mu\ b}^{\ a}\right)^\dagger = -\eta^b_c \omega_{\mu\ d}^{\ c} \eta^d_a \tag{8}$$

The curvature associated with this gauge field is

$$R_{\mu\nu\ b}^{\ \ a} = \partial_\mu \omega_{\nu\ b}^{\ a} - \partial_\nu \omega_{\mu\ b}^{\ a} + \omega_{\mu\ c}^{\ a} \omega_{\nu\ b}^{\ c} - \omega_{\nu\ c}^{\ a} \omega_{\mu\ b}^{\ c} \tag{9}$$

Under gauge transformations we have

$$\widetilde{\omega}_{\mu\ b}^{\ a} = M^a_c \omega_{\mu\ d}^{\ c} M_b^{-1d} - M^a_c \partial_\mu M_b^{-1c} \tag{10}$$

where the matrices M are subject to the condition:

$$\left(M^a_c\right)^\dagger \eta^a_b M^b_d = \eta^c_d \tag{11}$$

The curvature then transforms as

$$\widetilde{R}_{\mu\nu\ b}^{\ \ a} = M^a_c R_{\mu\nu\ d}^{\ \ c} M_b^{-1d} \tag{12}$$

Next we introduce the complex vielbein e^a_μ and its inverse e^μ_a defined by

$$e^\nu_a e^a_\mu = \delta^\nu_\mu, \quad e^a_\nu e^\nu_b = \delta^a_b \tag{13}$$

which transform as

$$\widetilde{e}^a_\mu = M^a_b e^b_\mu, \quad \widetilde{e}^\mu_a = \widetilde{e}^\mu_b M_a^{-1b} \tag{14}$$

It is also useful to define the complex conjugates

$$e_{\mu a} \equiv \left(e^a_\mu\right)^\dagger, \quad e^{\mu a} \equiv \left(e^\mu_a\right)^\dagger \tag{15}$$

With this, it is not difficult to see that

$$e^\mu_a R_{\mu\nu\ b}^{\ \ a} \eta^b_c e^{\nu c} \tag{16}$$

is hermitian and $U(1, D-1)$ invariant. The metric is defined by

$$g_{\mu\nu} = \left(e^a_\mu\right)^\dagger \eta^a_b e^b_\nu \tag{17}$$

satisfy the property $g^\dagger_{\mu\nu} = g_{\nu\mu}$. When the metric is decomposed into its real and imaginary parts:

$$g_{\mu\nu} = G_{\mu\nu} + iB_{\mu\nu} \tag{18}$$

the hermiticity property then implies the symmetries

$$G_{\mu\nu} = G_{\nu\mu}, \quad B_{\mu\nu} = -B_{\nu\mu} \tag{19}$$

The gauge invariant Hermitian action is given by

$$I = \int d^D x \sqrt{e}^{\dagger} e^{\mu}_{a} R_{\mu\nu}{}^{a}{}_{b} \eta^{b}_{c} e^{\nu c} \sqrt{e} \tag{20}$$

where $e = \det\left(e^a_\mu\right)$. One goes to the second order formalism by integrating out the spin connection and substituting for it its value in terms of the vielbein. The resulting action depends only on the fields $g_{\mu\nu}$. It is worthwhile to stress that the above action, unlike others proposed to describe nonsymmetric gravity[15] is unique, except for the measure, and unambiguous. Similar ideas have been proposed in the past based on gauging the groups $O(D,D)$[16] and $GL(D)$,[17] in relation to string duality, but the results obtained there are different from what is presented here. The ordering of the terms in writing the action is done in a way that generalizes to the noncommutative case. The idea of a hermitian metric was first forwarded by Einstein and Strauss,[18] which resulted in a nonsymmetric action for gravity, with two possible contractions of the Riemann tensor.

The infinitesimal gauge transformations for e^a_μ is $\delta e^a_\mu = \Lambda^a_b e^b_\mu$, which can be decomposed into real and imaginary parts by writing $e^a_\mu = e^a_{0\mu} + i e^a_{1\mu}$, and $\Lambda^a_b = \Lambda^a_{0b} + i\Lambda^a_{1b}$.

From the gauge transformations of $e^a_{0\mu}$ and $e^a_{1\mu}$ one can easily show that the gauge parameters Λ^a_{0b} and Λ^a_{1b} can be chosen to make $e_{0\mu a}$ symmetric in μ and a and $e_{1\mu a}$ antisymmetric in μ and a. This is equivalent to the statement that the Lagrangian should be completely expressible in terms of $G_{\mu\nu}$ and $B_{\mu\nu}$ only, after eliminating $\omega_{\mu}{}^{a}{}_{b}$ through its equations of motion. In reality we have

$$\begin{aligned} G_{\mu\nu} &= e^a_{0\mu} e^b_{0\nu} \eta_{ab} + e^a_{1\mu} e^b_{1\nu} \eta_{ab} \\ B_{\mu\nu} &= e^a_{0\mu} e^b_{1\nu} \eta_{ab} - e^a_{1\mu} e^b_{0\nu} \eta_{ab} \end{aligned} \tag{21}$$

In this special gauge, where we define $g_{0\mu\nu} = e^a_{0\mu} e^b_{0\nu} \eta_{ab}$, $g_{0\mu\nu} g_0^{\nu\lambda} = \delta^\lambda_\mu$, and use $e^a_{0\mu}$ to raise and lower indices we get

$$\begin{aligned} B_{\mu\nu} &= -2 e_{1\mu\nu} \\ G_{\mu\nu} &= g_{0\mu\nu} - \frac{1}{4} B_{\mu\kappa} B_{\lambda\nu} g_0^{\kappa\lambda} \end{aligned} \tag{22}$$

The last formula appears in the metric of the effective action in open string theory.[14]

We can express the Lagrangian in terms of e^a_μ only by solving the $\omega_{\mu}{}^{a}{}_{b}$ equations of motion

$$\begin{aligned} e^\mu_a e^{\nu b} \omega_\nu{}^c{}_b + e^\nu_b e^{\mu c} \omega_\nu{}^b{}_a - e^{\mu b} e^\nu_a \omega_\nu{}^c{}_b - e^\mu_b e^{\nu c} \omega_\nu{}^b{}_a &= \\ \frac{1}{\sqrt{G}} \partial_\nu \left(\sqrt{G} \left(e^\nu_a e^{\mu c} - e^\mu_a e^{\nu c} \right) \right) &\equiv X^{\mu c}{}_a \end{aligned} \tag{23}$$

where $X^{\mu c}{}_a$ satisfy $\left(X^{\mu c}{}_a\right)^\dagger = -X^{\mu a}{}_c$. One has to be very careful in working with a nonsymmetric metric

$$g_{\mu\nu} = e^a_\mu e_{\nu a}, \quad g^{\mu\nu} = e^{\mu a} e_{\nu a}, \quad g_{\mu\nu} g^{\nu\rho} = \delta^\rho_\mu \tag{24}$$

but $g_{\mu\nu}g^{\mu\rho} \neq \delta^{\rho}_{\mu}$. Care also should be taken when raising and lowering indices with the metric.

Before solving the ω equations, we point out that the trace part of $\omega_{\mu\ b}^{\ a}$ (corresponding to the $U(1)$ part in $U(D)$) must decouple from the other gauge fields. It is thus undetermined and decouples from the Lagrangian after substituting its equation of motion. It imposes a condition on the e_{μ}^{a}

$$\frac{1}{\sqrt{G}}\partial_{\nu}\left(\sqrt{G}\left(e_{a}^{\nu}e^{\mu a} - e_{a}^{\mu}e^{\nu a}\right)\right) \equiv X^{\mu a}{}_{a} = 0 \tag{25}$$

We can therefore assume, without any loss in generality, that $\omega_{\mu\ b}^{\ a}$ is traceless $\left(\omega_{\mu\ a}^{\ a} = 0\right)$.

The $\omega-$equation gives

$$\omega_{\kappa\rho}{}^{\mu} + \omega_{\rho\ \kappa}^{\ \mu} = \frac{1}{8}\delta_{\kappa}^{\mu}\left(3X^{\mu}_{\ \rho\mu} - X^{\mu}_{\ \mu\rho}\right) + \frac{1}{8}\delta_{\rho}^{\mu}\left(-X^{\mu}_{\ \kappa\mu} + 3X^{\mu}_{\ \mu\kappa}\right) - X^{\mu}_{\ \rho\kappa} \equiv Y^{\mu}_{\ \rho\kappa} \tag{26}$$

We can rewrite this equation after contracting with $e_{\mu c}e_{\sigma}^{c}$ to get

$$\omega_{\kappa\rho\sigma} + e_{a}^{\mu}e_{\mu c}e_{\sigma}^{c}\omega_{\rho\ \kappa}^{\ a} = g_{\sigma\mu}Y^{\mu}_{\ \rho\kappa} \equiv Y_{\sigma\rho\kappa} \tag{27}$$

By writing $\omega_{\rho\ \kappa}^{\ a} = \omega_{\rho\nu\kappa}e^{\nu a}$ we get

$$\left(\delta_{\kappa}^{\alpha}\delta_{\rho}^{\beta}\delta_{\sigma}^{\gamma} + g^{\beta\mu}g_{\sigma\mu}\delta_{\rho}^{\alpha}\delta_{\kappa}^{\gamma}\right)\omega_{\alpha\beta\gamma} = Y_{\sigma\rho\kappa} \tag{28}$$

To solve this equation we have to invert the tensor

$$M_{\kappa\rho\sigma}^{\alpha\beta\gamma} = \delta_{\kappa}^{\alpha}\delta_{\rho}^{\beta}\delta_{\sigma}^{\gamma} + g^{\beta\mu}g_{\sigma\mu}\delta_{\rho}^{\alpha}\delta_{\kappa}^{\gamma} \tag{29}$$

In the conventional case when all fields are real, the metric $g_{\mu\nu}$ is symmetric and $g^{\beta\mu}g_{\sigma\mu} = \delta_{\sigma}^{\beta}$ so that the inverse of $M_{\kappa\rho\sigma}^{\alpha\beta\gamma}$ is simple. In the present case, because of the nonsymmetry of $g_{\mu\nu}$ this is fairly complicated and could only be solved by a perturbative expansion. Writing $g_{\mu\nu} = G_{\mu\nu} + iB_{\mu\nu}$, and defining $G^{\mu\nu}G_{\nu\rho} = \delta_{\rho}^{\mu}$ implies that

$$\begin{aligned} g^{\mu\alpha}g_{\nu\alpha} &\equiv \delta_{\nu}^{\mu} + L_{\nu}^{\mu} \\ L_{\nu}^{\mu} &= iG^{\mu\rho}B_{\rho\nu} - 2G^{\mu\rho}B_{\rho\sigma}G^{\sigma\alpha}B_{\alpha\nu} + O(B^{3}) \end{aligned} \tag{30}$$

The inverse of $M_{\kappa\rho\sigma}^{\alpha\beta\gamma}$ defined by

$$N_{\alpha\beta\gamma}^{\sigma\rho\kappa}M_{\sigma\rho\kappa}^{\alpha'\beta'\gamma'} = \delta_{\alpha}^{\alpha'}\delta_{\beta}^{\beta'}\delta_{\gamma}^{\gamma'} \tag{31}$$

is evaluated to give

$$\begin{aligned} N_{\alpha\beta\gamma}^{\sigma\rho\kappa} = &\frac{1}{2}\left(\delta_{\gamma}^{\sigma}\delta_{\beta}^{\rho}\delta_{\alpha}^{\kappa} + \delta_{\beta}^{\sigma}\delta_{\alpha}^{\rho}\delta_{\gamma}^{\kappa} - \delta_{\alpha}^{\sigma}\delta_{\gamma}^{\rho}\delta_{\beta}^{\kappa}\right) \\ &-\frac{1}{4}\left(\delta_{\beta}^{\kappa}\delta_{\alpha}^{\sigma}L_{\gamma}^{\rho} + \delta_{\alpha}^{\kappa}\delta_{\gamma}^{\sigma}L_{\beta}^{\rho} - \delta_{\gamma}^{\kappa}\delta_{\beta}^{\sigma}L_{\alpha}^{\rho}\right) \\ &+\frac{1}{4}\left(L_{\gamma}^{\kappa}\delta_{\beta}^{\sigma}\delta_{\alpha}^{\rho} + L_{\beta}^{\kappa}\delta_{\alpha}^{\sigma}\delta_{\gamma}^{\rho} - L_{\alpha}^{\kappa}\delta_{\gamma}^{\sigma}\delta_{\beta}^{\rho}\right) \\ &-\frac{1}{4}\left(\delta_{\alpha}^{\kappa}L_{\gamma}^{\sigma}\delta_{\beta}^{\rho} + \delta_{\gamma}^{\kappa}L_{\beta}^{\sigma}\delta_{\alpha}^{\rho} - \delta_{\beta}^{\kappa}L_{\alpha}^{\sigma}\delta_{\gamma}^{\rho}\right) + O(L^{2}) \end{aligned} \tag{32}$$

This enables us to write

$$\omega_{\alpha\beta\gamma} = N_{\alpha\beta\gamma}^{\sigma\rho\kappa} Y_{\rho\sigma\kappa} \tag{33}$$

It is clear that the leading term reproduces the Einstein-Hilbert action plus contributions proportional to $B_{\mu\nu}$ and higher order terms. We can check that in the flat approximation for gravity with $G_{\mu\nu}$ taken to be $\delta_{\mu\nu}$, the $B_{\mu\nu}$ field gets the correct kinetic terms. First we write

$$e_\mu^a = \delta_\mu^a - \frac{i}{2} B_{\mu a}, \quad e_{\mu a} = \delta_\mu^a + \frac{i}{2} B_{\mu a} \tag{34}$$

The $\omega_{\mu\ a}^{\ a}$ equation implies the constraint

$$X^{\mu a}{}_a = \partial_\nu \left(e_a^\mu e^{\nu a} - e_a^\nu e^{\mu a} \right) = 0 \tag{35}$$

This gives the gauge fixing condition $\partial^\nu B_{\mu\nu} = 0$. We then evaluate

$$\omega_{\mu\nu\rho} = -\frac{i}{2} \left(\partial_\mu B_{\nu\rho} + \partial_\nu B_{\mu\rho} \right) \tag{36}$$

When the $\omega_{\mu\nu\rho}$ is substituted back into the Lagrangian, and after integration by parts one gets

$$L = \omega_{\mu\nu\rho}\omega^{\nu\rho\mu} - \omega_\mu{}^{\mu\rho}\omega_{\nu\rho}{}^\nu = -\frac{1}{4} B_{\mu\nu}\partial^2 B^{\mu\nu} \tag{37}$$

This is identical to the usual expression $\frac{1}{12} H_{\mu\nu\rho} H^{\mu\nu\rho}$, where $H_{\mu\nu\rho} = \partial_\mu B_{\nu\rho} + \partial_\nu B_{\rho\mu} + \partial_\rho B_{\mu\nu}$. The later developments of nonsymmetric gravity showed that the occurrence of the trace part of the spin-connection in a linear form would result in the propagation of ghosts in the field $B_{\mu\nu}$.[19] This can be traced to the fact that there is no gauge symmetry associated with the field $B_{\mu\nu}$. For the theory to become consistent one must show that the action above has an additional gauge symmetry, which generalizes diffeomorphism invariance to complex diffeomorphism. This would protect the field $B_{\mu\nu}$ from having non-physical degrees of freedom. It is therefore essential to identify whether there are additional symmetries present in the above proposed action. This is presently under investigation.

Having shown that it is possible to formulate a theory of gravity with nonsymmetric complex metric, based on the idea of gauge invariance of the group $U(1, D-1)$ it is not difficult to generalize the steps that led us to the action for complex gravity to spaces where coordinates do not commute, or equivalently, where the usual products are replaced with star products.

First the gauge fields are subject to the gauge transformations

$$\widetilde{\omega}_{\mu\ b}^{\ a} = M_c^a * \omega_{\mu\ d}^{\ c} * M_{*b}^{-1d} - M_c^a * \partial_\mu M_{*b}^{-1c} \tag{38}$$

where M_{*a}^{-1b} is the inverse of M_b^a with respect to the star product. The curvature is now

$$R_{\mu\nu}{}^a{}_b = \partial_\mu \omega_{\nu\ b}^{\ a} - \partial_\nu \omega_{\mu\ b}^{\ a} + \omega_{\mu\ c}^{\ a} * \omega_{\nu\ b}^{\ c} - \omega_{\nu\ c}^{\ a} * \omega_{\mu\ b}^{\ c} \tag{39}$$

which transforms according to

$$\widetilde{R}_{\mu\nu}{}^a{}_b = M_c^a * R_{\mu\nu}{}^c{}_d * M_{*b}^{-1d} \tag{40}$$

Next we introduce the vielbeins e_μ^a and their inverse defined by

$$e_{*a}^{\nu} * e_\mu^a = \delta_\mu^\nu, \quad e_\nu^a * e_{*b}^{\nu} = \delta_b^a \tag{41}$$

which transform to

$$\widetilde{e}_\mu^a = M_b^a * e_\mu^b, \quad \widetilde{e}_{*a}^{\mu} = \widetilde{e}_b^{\mu} * M_{*a}^{-1b} \tag{42}$$

The complex conjugates for the vielbeins are defined by

$$e_{\mu a} \equiv \left(e_\mu^a\right)^\dagger, \quad e_*^{\mu a} = \left(e_{*a}^{\mu}\right)^\dagger \tag{43}$$

Finally we define the metric $g_{\mu\nu} = \left(e_\mu^a\right)^\dagger \eta_b^a * e_\nu^b$. The $U(1, D-1)$ gauge invariant Hermitian action is

$$I = \int d^D x \left(\sqrt{e}^\dagger * e_{*a}^{\mu} * R_{\mu\nu}{}^a{}_b \eta_c^b * e_*^{\nu c} * \sqrt{e}\right) \tag{44}$$

where $e = \det\left(e_\mu^a\right)$. This action differs from the one considered in the commutative case by higher derivatives terms proportional to $\theta^{\mu\nu}$. It would be very interesting to see whether these terms could be reabsorbed by redefining the field $B_{\mu\nu}$, or whether the Lagrangian reduces to a function of $G_{\mu\nu}$ and $B_{\mu\nu}$ and their derivatives only.

The connection of this action to the gravity action derived for noncommutative spaces based on spectral triples[20,21,22] remains to be made. In order to do this one must understand the structure of Dirac operators for spaces with deformed star products.

References

1. A.H. Chamseddine and J. Fröhlich, *Some Elements of Noncommutative and Space-Time Geometry*, Yang-Festschrift, eds. C.S. Liu and S.-T. Yau (International Press, 1995) 10.
2. A. Connes, M.R. Douglas and A. Schwarz, *JHEP* **9802**, 003 (1998).
3. M.R. Douglas and C. Hull, *JHEP* **9802**, 008 (1998).
4. Y.K.E. Cheung and M. Krogh, *Nucl. Phys.* **B528**, 185 (1998).
5. C.-S.Chu and P.-M. Ho, *Nucl. Phys.* **B528**, 151 (1999).
6. V. Schomerus, *JHEP* **9906**, 030 (1999).
7. F. Ardalan, H. Arfaei and M.M. Sheikh-Jabbari, *JHEP* **9902**, 016 (1999).
8. N. Seiberg and E. Witten, *JHEP* **9909**, 032 (1999).
9. A. Connes, *Noncommutative Geometry* (Academic Press, 1994).
10. A.H. Chamseddine and A. Connes, *Comm. Math. Phys.* **186**, 731 (1997).
11. J. Hoppe, *Phys. Lett.* **B250**, 44 (1990).
12. D.B. Fairlie, P. Fletcher and C.K. Zachos, *Phys. Lett.* **B218**, 203 (1989).
13. J. Madore, S. Schraml, P. Schupp and J. Wess, *Eur. Phys. J.* **C16**, 161 (2000) [hep-th/0001203].
14. C.G. Callan, C. Lovelace, C.R. Nappi and S.A. Yost, *Nucl. Phys.* **B288**, 525 (1987).
15. J. Moffat, *J. Math. Phys.* **36**, 3722 (1995) and references therein.
16. J. Maharana and J.H. Schwarz, *Nucl. Phys.* **B390**, 3 (1993).
17. W. Siegel, *Phys. Rev.* **D47**, 5453 (1993).
18. A. Einstein and E. Strauss, *Ann. Math.* **47**, 731 (1946).

19. T. Damour, S. Deser and J. McCarthy, *Phys. Rev.* **D47**, 1541 (1993).
20. A.H. Chamseddine, G. Felder and J. Fröhlich, *Comm. Math. Phys.* **155**, 109 (1993).
21. A.H. Chamseddine, O. Grandjean and J. Fröhlich, *J. Math. Phys.* **36**, 6255 (1995).
22. A. Connes, *J. Math. Phys.* **36**, 6194 (1995).

ASPECTS OF COLLAPSING CYCLES*

BRIAN R. GREENE
Departments of Physics and Mathematics, Columbia University
New York, NY 10027, USA

Much has been learned about string theory over the last few years by studying properties of cycles and branes in a given background geometry. Here we discuss three situations (quantum volume, attractor flows/D-brane stability, and dynamical topology change) in which cycles in a Calabi-Yau background evolve and/or degenerate in some manner, yielding various insights into aspects of quantum geometry.

1. Introduction

Over the last few years we have learned much by pushing string theory into extreme geometrical realms. Here we would like to discuss three recent works that continue this type of exploration. In the first, we will discuss aspects of quantum volume, focusing on some of the unusual features of volume as probed by wrapped branes.[1,2] In the second, we will discuss aspects of attractor flows, seeking to find qualitative differences between the spectra of large and small volume Calabi-Yau compactifications.[3] In the third, we will discuss collapsing cycles in the context of topology changing transitions, seeking to find situations in which fairly generic boundary conditions force the background Calabi-Yau space to undergo topology change as a function of the eleventh dimension in M-theory.[4]

2. Quantum Volume

We will work in the context of type II string theory compactified on a Calabi-Yau manifold M. If the real Kähler form on M is denoted by J, the classical area and volume of the nontrivial 2, 4, and 6-cycles of M are determined by the usual expression: $\int_{C_r} J^r$ where r is the complex dimensions of the cycle C_r. The question that naturally arises, though, is: How do these expressions change when quantum effects are taken into account? Now, in any situation in which a classical concept is extended into the quantum domain, without further information, some ambiguity creeps in as many quantum expressions have the same classical limit. To sharpen our choice of what quantum volume means, therefore, we should introduce some physical guide. In perturbative string theory, the contributions of world sheet instantons provide one such guide. Namely, consider a correlation function between three states corresponding to elements of $H^{1,1}(M)$. As is well known, the lowest order α' contribution involves the classical intersection form on M, but there are

*Invited talk presented at Strings 2000, Ann Arbor, Michigan.

nonperturbative corrections in α' coming from string world sheets that wrap around holomorphic curves in M. These contributions are sensitive to the volume of such curves as they are weighted by the exponential of the worldsheet action, which takes the form $\exp(2\pi i \int_{C_2} B + iJ)$. If one uses mirror symmetry to determine the *exact* value of such correlation functions[5] (by doing a variation of Hodge structure calculation on the mirror manifold W), one can then adjust the value of $\int_{C_2} B + iJ$ to ensure agreement with the calculation on M, thereby giving one definition of the quantum volume of C_2. As found in Ref. 6, this approach yields

$$\int_{C_2} B + iJ = \frac{\int_{\gamma} \Omega}{\int_{\gamma_0} \Omega} \tag{1}$$

where Ω is the holomorphic 3-form on W and γ, γ_0 are suitably chosen 3-cycles on W.

It is worth noting a few features of this definition of the quantum volume of 2-cycles on M. First, we are naturally led to a notion of complex volumes. Second, in the large radius limit on M, if one sets the B-field to zero, we recover the usual classical values for the volumes of 2-cycles. Third, in the small radius regime, there are departures from classical expectations. For instance, as in Refs. 6,7, on the quintic three-fold (and other one-parameter Calabi-Yau examples), there is a *lower* bound on the quantum volume of 2-cycles—explicitly, on the quintic, 2-cycle volumes satisfy

$$\mathrm{Vol}_{\mathrm{quintic}}(2 - \mathrm{cycles}) \geq .6881. \tag{2}$$

while classically the only constraint on such volumes is that they are non-negative.

The fourth point is simply that this approach, since it relies on worldsheet instantons in perturbative string theory, is only applicable to 2-cycles. To go beyond this limitation it is natural to turn to developments in nonperturbative string theory.

In particular, if we are interested in the quantum volume of a p-cycle, we can wrap a D-p-brane around it in a BPS manner, and then note that the mass of this state is a physical probe of the volume of the cycle. Hence, we can declare that the quantum volume of the p-cycle *is* the mass of the state obtained by such a BPS wrapping of a p-brane, suitably normalized. Of course, to make this definition useful, one needs to be able to reliably calculate these masses. For 3-cycles this is immediate: in Type IIB string theory, we wrap a 3-brane on a supersymmetric representative γ of given 3-cycle homology class, and the resulting mass is

$$\mathrm{Mass}(3 - \mathrm{brane}) = \frac{|\int_{\gamma} \Omega|}{|\int_M \Omega \wedge \overline{\Omega}|^{1/2}}. \tag{3}$$

For even dimensional cycles on M, we can write down a similar exact formula by invoking mirror symmetry. Namely, mirror symmetry maps $\oplus H^{2p}(M, \mathbb{Z}) \rightarrow \mathbb{H}^{3}(\mathbb{W}, \mathbb{Z})$ and hence any even cycle C on M is mapped to some 3-cycle γ on W. The mass of the corresponding BPS wrapped brane is then given by the formula above evaluated for γ on W. In turn, this gives us the corresponding quantum volume of C on M. Notice that in the case that C is a 2-cycle, the magnitude of the resulting quantum volume agrees with that of perturbative string theory, up to an overall normalization factor.

We see, therefore, that everything boils down to (a) being able to calculate the period integrals of Ω over arbitrary 3-cycles at arbitrary locations in the moduli space of a given Calabi-Yau manifold (and its mirror) and (b) explicitly realizing the map $\oplus H^{2p}(M, \mathbb{Z}) \to \mathbb{H}^{\not{k}}(\mathbb{W}, \mathbb{Z})$. The subtlety here is that the calculations in (a) quickly become involved and the map in (b) is not known in general. In practice, we deal with this in the following way.

First, we are often interested in understanding where in moduli space the quantum volume of a given cycle vanishes. If this cycle is an odd-dimensional cycle, then there is no need for the map in (b). If the cycle is even-dimensional (which we shall henceforth assume), we can get by with a weaker form of the map in (b). Namely, if we know the map in (b) up to an overall scale factor, then this is enough to determine ratios of quantum volumes, as well as the precise locations of the zeroes of the quantum volumes. So, if we find a basis γ_j of $H^3(W)$ which are *proportional* to integral cycles (with the same, generally unknown, overall proportionality factor) then by this reasoning we are free to use these cycles in the range of the map in (b). Such cycles, it turns out, are much easier to identify than are integral cycles. As for the map itself, we make use of the general insights of Refs. 6,8 which relates 3-cycles with $log^j(z)$ monodromy about a large complex structure point $z = 0$ on W (assuming for simplicity a one-dimensional moduli space) to cycles of complex dimension j on M. Now the subtlety here is that this identification is true up to admixtures of cycles of lower dimension, and it is generally a challenge to work out these lower dimensional contributions explicitly. Hence, when we describe cycles of complex dimension j this implicitly means up to such undetermined admixtures. Finally, as for (a), the difficulty in carrying out the period calculations at arbitrary points in the moduli space, we make use of the classical theory of Meijer functions which naturally encode all the required analytic continuations of solutions of Picard-Fuchs equations which these computations require (for details see Ref. 2).

Let us turn to some examples. For simplicity, let's begin with the quintic hypersurface whose Kähler form we parameterize as $J = se$ where s is a real number and e is an integral generator of H^2(Quintic). As the quintic hypersurface has $h^{1,1} = 1$, we can think of s as parameterizing a real slice through the one-complex dimensional Kähler moduli space. Classically, the volumes of the single homologically nontrivially 2, 4, and 6 cycles are just: 2-cycle volume: s, 4-cycle volume: $5s^2/2$; 6-cycle volume: $5s^3/6$. In particular, note that only at $s = 0$ does any cycle vanish, and at that point *every* cycle vanishes.

Quantum mechanically the story is different. By carrying out the procedure as above, we find that the quantum volumes of the 2, 4, and 6 cycles as a function of s are as given in Figure 2 (where in Figure 1, for comparison, we plot the classical result). Notice that for the quantum volumes there is only one location where a cycle vanishes, and it is not at $s = 0$. Moreover, the cycle that vanishes is the 6-

cycle,[1,9] not the 2-cycle as one would have naturally speculated. This is a bit odd, since although the entire Calabi-Yau has zero quantum volume at this point, the 2 and 4 cycle have nonzero volumes, as they are bounded below by positive numbers throughout the moduli space. In Refs. 2,11 this phenomenon has been found to hold in a range of other one parameter examples, so appears to be generic.

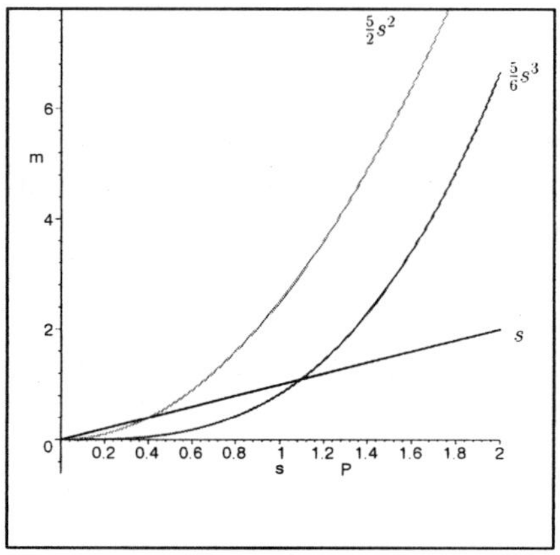

Fig. 1. Classical 2, 4, 6 cycle volumes on the Quintic.

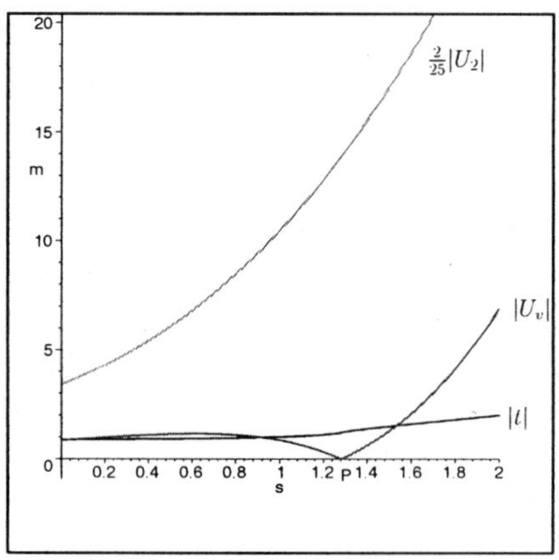

Fig. 2. Quantum 2, 4, 6 cycle volumes on the Quintic.

As a second example, let's consider two-dimensional orbifolds. In the case of a $\mathbb{Z}_{\not{p}}$ orbifold, it was shown in Ref. 7 — using the perturbative string theory approach to quantum volume— that there is a point in the moduli space where a 2-cycle collapses to zero quantum volume (the orbifold point itself). But for the case of

$\mathbb{Z}_{\not{k}}$ orbifolds, there was a bit of a puzzle in Ref. 7 because calculations showed that the 2-cycle does not vanish anywhere. Yet by virtue of there being a singularity in the moduli space of the associated physical model, we expect that some state has gone to zero mass at that point. But which state? Well, with our nonperturbative extension of quantum volume and our experience with the quintic above, we can answer that question and, moreover, not be too surprised by the result. As shown in Figure 3, we see that that the 4-cycle in this example goes to zero volume at a point in the moduli space, even though the 2-cycle never does.

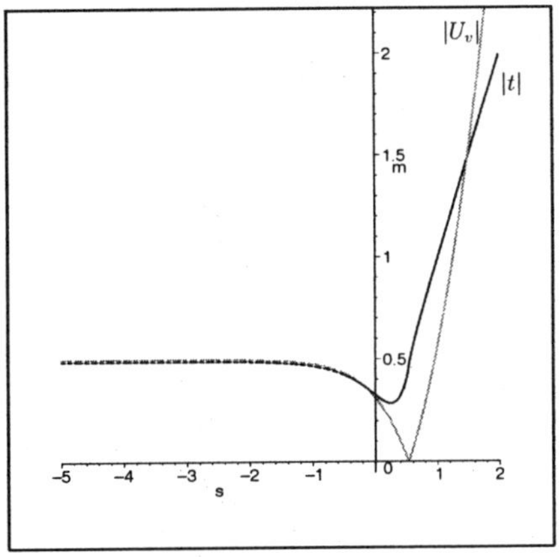

Fig. 3. Quantum 2 and 4 cycles volumes for a $\mathbb{Z}_{\not{k}}$ orbifold.

Finally, in Ref. 2 these calculations have been extended to one dimensional loci in two-dimensional moduli spaces, with a range of similar violations of classical geometric intuition making themselves apparent, and in Ref. 11 the whole series of one parameter examples has been studied with special attention paid to an interesting manifestation of T-duality in this context.

3. Attractor Flows

In the discussion of the last section, we implicitly assumed that in any given homology class in $H_3(M)$ or $H_3(W)$ there is a supersymmetric representative. As a matter of fact, however, the existence of supersymmetric cycles is a difficult and as yet unsettled question. So, an interesting an important question to ask is (a) under what conditions are we assured that the relevant supersymmetric cycles exist thereby implying that the associated BPS state exists, and (b) are there examples in which the answer to (a) can be used to qualitatively distinguish large radius and small radius compactifications? That is, might it be that the existence of a supersymmetric cycle in a given homology class might have a Kähler form dependent answer, existing, say, when the radius is large but not when it is small?

A full answer to these questions is beyond current understanding, but, nevertheless, much can be gleaned through a variety of techniques. One such method makes use of *attractor flows.*

Recall that in Ref. 12, spherically symmetry black hole solutions to the equations of $\mathcal{N} = 2$ supergravity were studied and shown to obey an attractor mechanism: the value of the fields at the horizon of the hole are largely insensitive to the boundary values of the fields at spatial infinity. Explicitly, these equations take the form:

$$\partial_\tau U = -e^U |Z| \tag{4}$$

$$\partial_\tau z^a = -2e^U g^{a\bar{b}} \partial_{\bar{b}} |Z| \tag{5}$$

where $\tau = 1/r$, r being the radial direction in space, U arises from the spacetime metric in the form $ds^2 = -e^{2U}dt^2 + e^{-2U}dx^i dx^i$, z^a are coordinates on the Kähler moduli space, $g^{a\bar{b}}$ is the inverse metric on the Kähler moduli space in these coordinates, and $Z = Z(\gamma) = \frac{\int_\gamma \Omega}{\int_M (\Omega \wedge \bar{\Omega})^{\frac{1}{2}}}$. As one can see, these equations relate flows in moduli space to the spatial profile of the fields in a black hole background. As such, they provide an interesting and novel link between moduli space and spacetime physics.

One expects that these BPS solutions are the supergravity description of the BPS D-brane states in the full string theory. In the beautiful work of Ref. 13, these equations were studied for both their physical and mathematical content and one observation was this. If we study an attractor flow solution to these equations which has the property that $|Z|$ vanishes at a *regular* point in the moduli space, we don't expect the corresponding BPS state to exist. The reason is that $|Z|$ corresponds to the mass of a BPS brane wrapped on γ. If it vanishes along a flow, at that point we expect to have a new massless degree of freedom, and this is generally accompanied by a singularity in the moduli space. If there is no such singularity, we expect that the corresponding BPS state simply does not exist, i.e. the corresponding supersymmetric cycle in the homology class γ does not exist. Roughly, we associate $|Z| = 0$ with a collapsing cycle, and the latter give rise to singularities in the moduli space. If there is no singularity, the existence of a supersymmetric cycle in the given class is thrown into question.

This is a natural conjecture, but in Refs. 14,15 it was suggested that there are circumstances in which the physics is more rich. Namely, there might be a would-be attractor solution in which there is a zero of $|Z|$ at a regular point (a "regular zero") but for which the flow crosses a curve of marginal stability before that point is encountered. At the crossing point, the state decays and the attractor flow splits. As for the physical realization of this state, as shown in Refs. 14,15, if γ splits into $\gamma_1 + \gamma_2$, the spacetime configuration can be realized by a spherical shell carrying the charge associated with γ_2 which surrounds the charge of γ_1 sitting at the origin.

An interesting aspect of these solutions, then, is that they provide a means of identifying candidate states that exist in one region of the moduli space but not at another. For instance, in Refs. 14,15 it was shown how the state identified by Ref. 16 — which exists at the Gepner point on the quintic (which is at small radius)

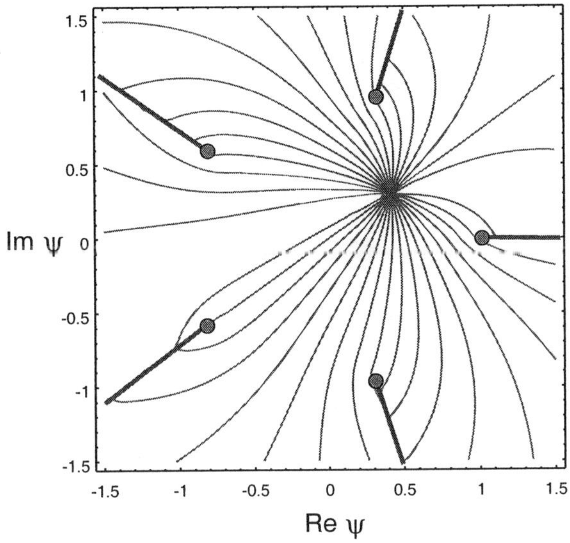

Fig. 4. Attractor flows in the quintic moduli space.

but apparently decays before getting to large radius — can be realized as a splitted attractor flow starting from the Gepner point. Naturally, then, one wonders about examples in which the reverse would be true: Start at large radius where classical geometry is a good guide to physics and identify states in the string spectrum. Then, go to ever smaller radius where non-classical geometrical effects become increasingly pronounced. Are there states for which these effects cause the state to decay, giving us a qualitative impact of short distance geometry on large distance expectations. Indeed in Ref. 3 such states have been found, and in Figures 4, 5, and 6 we give an example.

Namely, in Figure 4 we show a family of attractor flows on the Kähler moduli space of the quintic hypersurface, all associated with the charge vector $(Q_6, Q_4, Q_2.Q_0)$ $(0, 1, 3, -5)$. The flows terminate at the attractor point $\psi = .4146 + .3009i$. As we see in Figure 5, at this point the state has $|Z| = 0$, but this is a regular point (since the only singular point is the conifold point, indicated by the red circle(s) in Figure 4). In Figure 6, we show how this regular zero is physically interpreted in terms of split flow, crossing a curve of marginal stability. This is but one of many such examples, as discussed in greater detail in Ref. 3, thereby giving a nice qualitative distinction between long and short distance geometry.

4. Dynamical Topology Change

The final arena of collapsing cycles we will discuss has to do with topology change. Namely, over the years, several works[17,18] have established definitively that there

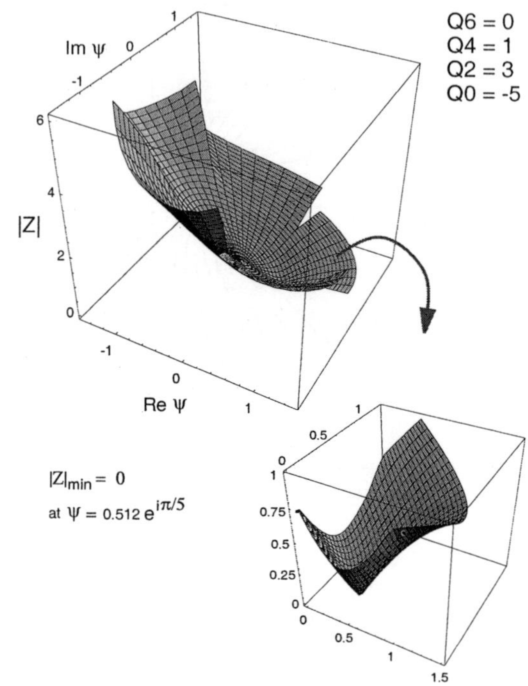

Fig. 5. Moduli dependence of the mass of a BPS wrapped brane.

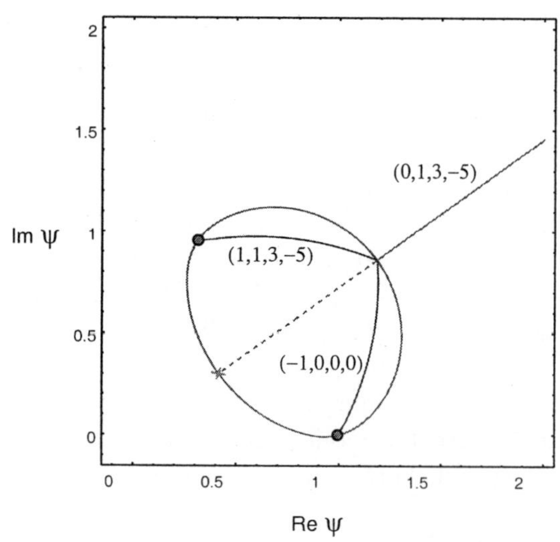

Fig. 6. A splitted attractor flow crossing marginal stability line.

are physically smooth processes in string theory which result in a change in the topology of spacetime. In these studies, as well as studies of topology change in M theory,[19] one considers a one parameter family of vacuum solutions — a one parameter family of spacetimes — that passes from one Calabi-Yau manifold to another which is topologically distinct. The referenced works succeeded in showing that there is no obstruction to such topology change, but no dynamics was ascribed to motion through the family. In this section, we discuss[4] a variation on this theme of topology change in which dynamics does drive the evolution from one topology to another. Specifically, (a) the topology change occurs within a single (not a family of) spacetime background and (b) for generic choices of initial conditions, the dynamics (*i.e.*, the field equations) drive us through a topology change.

To be concrete, we focus our attention on Calabi-Yau compactifications of M theory to five dimensions in the presence of G flux (the four form field strength). As discussed in Refs. 20,21,22,23, the effective five dimensional theory does not admit a flat space vacuum solution. Rather, the spacetime metric is warped and the solution is of the domain wall type with one of the five dimensions singled out as the transverse direction. In addition to the effective five dimensional spacetime metric, the moduli of the Calabi-Yau will generically vary along the transverse direction. In Ref. 4, we show that there exist Calabi-Yau compactifications in which the field equations force the Kähler moduli to pass from one Kähler cone into an adjacent cone, while the overall volume of the Calabi-Yau manifold remains large. This implies that the Calabi-Yau manifold undergoes a flop transition and continues on to a topologically distinct Calabi-Yau manifold as we move along the transverse dimension.

One may think of this work as being complementary to that of Refs. 24,25,26 in which it was shown that in the presence of certain dyonic black holes, a Calabi-Yau with particular moduli at spatial infinity can be driven by the attractor equations through a flop transition on the way to the black hole's horizon. Here we briefly describe vacuum solutions whose structure requires topology change, referring the reader to Ref. 4 for details.

Following the pioneering work of Ref. 21, the supersymmetric domain-wall or three-brane solution to the effective five dimensional field equations is given by

$$\begin{aligned} ds_5^2 &= e^{2A}dx_4^2 + e^{8A}dy^2 \\ V &= \left(\frac{1}{3!}d_{ijk}f^i f^j f^k\right)^2 , \\ e^A &= V^{1/6} , \\ b^i &= V^{-1/6} f^i , \\ F^i_{11,\mu\nu\rho\sigma} &= i\sqrt{2}\epsilon_{\mu\nu\rho\sigma}\partial_{11}V^{-1/2}f^i \end{aligned} \tag{6}$$

where the f^i's are defined in terms of one-dimensional harmonic functions

$$d_{ijk}f^j f^k = H_i , \qquad H_i = \sum_n \alpha_i^{(n)}|y - y_n| + c_i \tag{7}$$

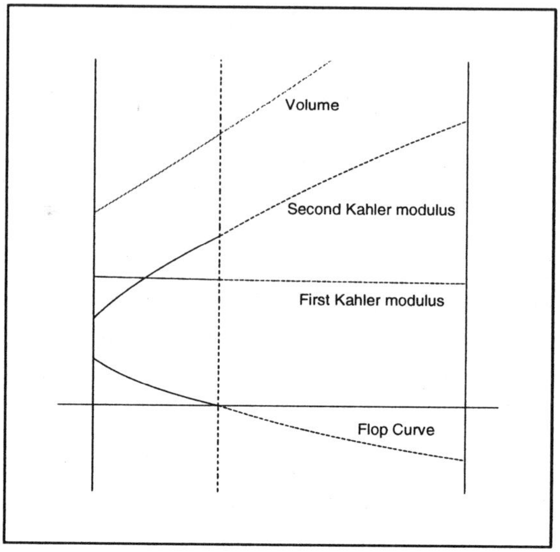

Fig. 7. Profile of the areas of the two cycles and the Calabi-Yau volume along y. Solid lines belong to the validity region $y < y_*$ of the Calabi-Yau $\widetilde{\mathcal{M}}$. Dashed lines belong to the Calabi-Yau $\mathcal{M}$ and hold for $y > y_*$.

$$= \sum_{n=0}^{k} 2\alpha_i^{(n)} y + k_i \ , \qquad y_k < y < y_{k+1} \tag{8}$$

and the k_i are arbitrary constants of integration. In these expressions, V is the volume of the Calabi-Yau manifold, the b^i are the Kähler moduli, and y is the coordinate in the "eleventh" dimension (the coordinate transverse to the end of the world branes in strongly coupled heterotic string theory).

We see explicitly that the moduli are y dependent, giving rise to the possibility that in a given example they might pass from one Kähler cone to another as y varies from one wall in spacetime to another. Indeed this possibility is borne out by studying a simple example of a pair of Calabi-Yau manifolds connected by a flop transition—namely, the well studied $(h^{1,1}, h^{2,1}) = (3, 243)$ Calabi-Yau manifolds considered in Refs. 27,28,24,25,29. Figure 7 shows a topology changing solution to these equations for this example, giving rise to the configuration schematically illustrated in Figure 8.

One particularly interesting issue regarding these solutions arises from examining the Bianchi identity for the G-field. Namely, in the eleven-dimensional theory on $S_1/\mathbb{Z}_{\not 2}$, the Bianchi identity for G is modified by boundary sources[30] to the form

$$\begin{aligned} dG &= \left(\left[\mathrm{tr} F_{(1)} \wedge F_{(1)} - \frac{1}{2} \mathrm{tr} R_{(1)} \wedge R_{(1)} \right] \delta(y) \right. \\ &\quad \left. + \left[\mathrm{tr} F_{(2)} \wedge F_{(2)} - \frac{1}{2} \mathrm{tr} R_{(2)} \wedge R_{(2)} \right] \delta(y - \pi R_{11}) \right) \wedge dy \ , \end{aligned} \tag{9}$$

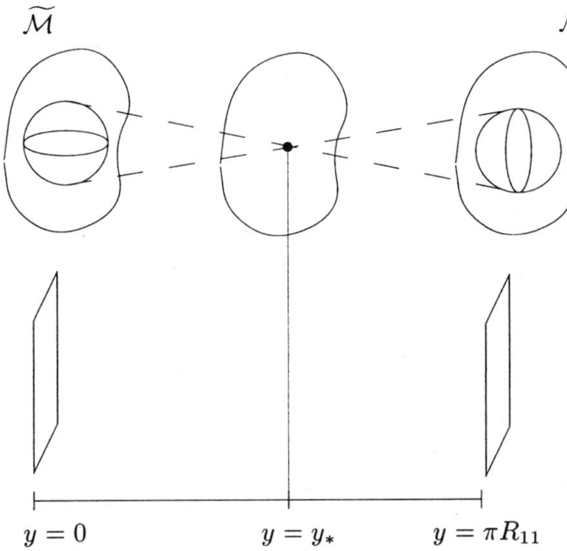

Fig. 8. Calabi-Yau configuration in strongly coupled heterotic theory which undergoes a flop.

where we are now taking care to distinguish the curvatures of the different Calabi-Yau spaces at each end of the interval. In the usual case, where $\mathrm{tr}R_{(1)} \wedge R_{(1)} = \mathrm{tr}R_{(2)} \wedge R_{(2)}$, we have the familiar standard embedding solution

$$\mathrm{tr}F_{(1)} \wedge F_{(1)} = \mathrm{tr}R \wedge R \quad , \quad \mathrm{tr}F_{(2)} \wedge F_{(2)} = 0 \ , \tag{10}$$

to the global consistency constraint that

$$\int_{S_1/\mathbb{Z}_{\not{k}}\times\mathbb{D}} dG = \int_{S_1/\mathbb{Z}_{\not{k}}\times\mathbb{D}} dy \, \left(\frac{1}{2}\mathrm{tr}R \wedge R \; \delta(y) - \frac{1}{2}\mathrm{tr}R \wedge R \; \delta(y - \pi R_{11})\right) = 0. \tag{11}$$

But if $\mathrm{tr}R_{(1)} \wedge R_{(1)} \neq \mathrm{tr}R_{(2)} \wedge R_{(2)}$ (cohomologically) then the mismatch implies that solely embedding either spin connection into the gauge group is no longer a solution.

A natural suggestion, then, is to seek out different holomorphic stable bundles to place at $y = 0$ and $y = \pi R_{11}$ with different second Chern classes, so as to find new consistent solutions to the Bianchi identity. In Ref. 4, though, we give evidence that such an approach will overlook an essential contribution. Namely, when $\mathrm{tr}R_{(1)} \wedge R_{(1)} \neq \mathrm{tr}R_{(2)} \wedge R_{(2)}$ because the Calabi-Yau has flopped somewhere along y, there is a new contribution to the Bianchi identity associated with the collapsed flop curves.

In particular, a theorem of Tian and Yau[31] states that starting from a Calabi-Yau manifold $\widetilde{\mathcal{M}}$ and a collection of holomorphic curves $\{C^\beta\}$ on $\widetilde{\mathcal{M}}$, the second

Chern numbers of $\tilde{M}$ and its flopped cousin M are related by

$$c_2(\mathcal{M}) = c_2(\widetilde{\mathcal{M}}) + 2\sum_\beta \int_D [C^\beta] \,, \tag{12}$$

with D an arbitrary divisor and $[C^\beta] \in H^4(\widetilde{\mathcal{M}})$, the Poincare dual of C^β. This implies that

$$-\frac{1}{2}\mathrm{tr}_{\mathcal{M}} R_{(2)} \wedge R_{(2)} + \sum_\beta \delta_{C^\beta_{(1)}} = -\frac{1}{2}\mathrm{tr}_{\widetilde{\mathcal{M}}} R_{(1)} \wedge R_{(1)} \,. \tag{13}$$

As suggested in Ref. 4 this jump should be compensated by associating a magnetic charge with the collapsed flop curve, (a charge that might well be interpretable — in a manner described precisely in Ref. 4 — as that of a 5-brane wrapped on the curve, but no definitive conclusion on this interpretation was reached).

This work suggests that the study of G_2 manifolds with Calabi-Yau boundaries (a study that has been initiated in Ref. 32) would be a fruitful way of learning more about these kind of string solutions.

Acknowledgments

I would like to thank Calin Lazaroiu, Frederik Denef, Mark Raugas, Koenraad Schalm, and Gary Shiu with whom I collaborated on the works presented here. I would also like to thank Sergei Gukov, Shamit Kachru, Renata Kallosh, Chien-Hao Liu, David Morrison, Burt Ovrut, and Edward Witten for useful discussions. This research was partially supported by the DOE grant DE-FG02-92ER40699B.

References

1. B.R. Greene and Y. Kanter, *Small Volumes in Compactified String Theory*, *Nucl. Phys.* **B497**, 127 (1997) [hep-th/9612181].
2. C.I. Lazaroiu and B.R. Greene, *Collapsing D-Branes in Calabi-Yau Moduli Space: I*, hep-th/0001025.
3. F. Denef, B. Greene and M. Raugas, in preparation.
4. B.R. Greene, K. Schalm and G. Shiu, *Dynamical Topology Change in M Theory*, hep-th/0010207.
5. B.R. Greene and M.R. Plesser, *Duality in Calabi-Yau Moduli Space*, *Nucl. Phys.* **B338**, 15 (1990).
6. P. Candelas, X.C. De La Ossa, P.S. Green and P. Parkes, *A Pair of Calabi-Yau Manifolds as an Exactly Soluble Superconformal Theory*, *Nucl. Phys.* **B359**, 21 (1991).
7. P.S. Aspinwall, B.R. Greene and D.R. Morrison, *Measuring small distances in N=2 sigma models*, *Nucl. Phys.* **B420**, 184 (1994) [hep-th/9311042].
8. D.R. Morrison, *Picard-Fuchs equations and mirror maps for hypersurfaces*, in *Essays on mirror manifolds* (Internat. Press, Hong Kong, 1992), pp. 241–264 [hep-th/9111025].
9. J. Polchinski and A. Strominger, *New Vacua for Type II String Theory*, *Phys. Lett.* **B388**, 736 (1996) [hep-th/9510227].
10. A. Strominger, *Massless Black Holes and Conifolds in String Theory*, *Nucl. Phys.* **B451**, 96 (1995) [hep-th/9504090].

11. C.I. Lazaroiu, *Collapsing D-branes in one-parameter models and small/large radius duality*, hep-th/0002004.
12. S. Ferrara, R. Kallosh and A. Strominger, *N=2 Extremal Black Holes*, *Phys. Rev.* **D52**, 5412 (1995).
13. G. Moore, *Attractors and Arithmetic*, hep-th/9807056.
14. F. Denef, *Supergravity flows and D-brane stability*, *JHEP* **0008**, 050 (2000).
15. F. Denef, *On the correspondence between D-branes and stationary supergravity solutions of type II Calabi-Yau compactifications*, hep-th/0010222.
16. I. Brunner, M.R. Douglas, A. Lawrence and C. Romelsberger, *D-branes on the Quintic*, *JHEP* **0008**, 015 (2000).
17. E. Witten, *Nucl. Phys.* **B403**, 159 (1993);
P.S. Aspinwall, B.R. Greene and D.R. Morrison, *Phys. Lett.* **B303**, 249 (1993); *Nucl. Phys.* **B416**, 414 (1994).
18. B.R. Greene, D.R. Morrison and A. Strominger, *Nucl. Phys.* **B451**, 109 (1995).
19. E. Witten, *Nucl. Phys.* **B471**, 195 (1996).
20. E. Witten, *Nucl. Phys.* **B471**, 135 (1996).
21. A. Lukas, B.A. Ovrut, K.S. Stelle and D. Waldram, *Phys. Rev.* **D59**, 086001 (1999); See, for example, B.A. Ovrut, hep-th/9905115 and references therein.
22. K. Behrndt and S. Gukov, *Nucl. Phys.* **B580**, 225 (2000).
23. B. R. Greene, K. Schalm and G. Shiu, *Nucl. Phys.* **B584**, 480 (2000).
24. A. Chou, R. Kallosh, J. Rahmfeld, S.-J. Rey, M. Shmakova and W.K. Wong, *Nucl. Phys.* **B508**, 147 (1997).
25. I. Gaida, S. Mahapatra, T. Mohaupt and W.A. Sabra, *Class. Quant. Grav.* **16**, 419 (1999).
26. K. Behrndt, D. Lust and W.A. Sabra, *Phys. Lett.* **B418**, 303 (1998).
27. J. Louis, J. Sonnenschein, S. Theisen and S. Yankielowicz, *Nucl. Phys.* **B480**, 185 (1996).
28. D.R. Morrison and C. Vafa, *Nucl. Phys.* **B473**, 74 (1996); *Nucl. Phys.* **B476**, 437 (1996).
29. P. Candelas, A. Font, S. Katz and D.R. Morrison, *Nucl. Phys.* **B429**, 626 (1994).
30. P. Horava and E. Witten, *Nucl. Phys.* **B460**, 506 (1996);
E. Witten, *Nucl. Phys.* **B471**, 135 (1996);
P. Horava and E. Witten, *Nucl. Phys.* **B475**, 94 (1996).
31. G. Tian and S.T. Yau, in *San Diego 1986, Proc. Mathematical aspects of String Theory*, pp. 543-559.
32. V. Braun and C-H. Liu, *On extremal transitions of Calabi-Yau threefolds and the singularity of the associated 7-space*, hep-th/9801175.

LOCALLY LOCALIZED GRAVITY

ANDREAS KARCH and LISA RANDALL
Center for Theoretical Physics, Massachusetts Institute of Technology
Cambridge, MA 02139–4307, USA
E-mail: karch@mit.edu, randall@.mit.edu

We study the fluctuation spectrum of linearized gravity around non-fine-tuned branes. We focus on the case of an AdS_4 brane in AdS_5. In this case, for small cosmological constant, the warp factor near the brane is essentially that of a Minkowski brane. However, far from the brane, the metric differs substantially. The space includes the AdS_5 boundary, so it has infinite volume. Nonetheless, for sufficiently small AdS_4 cosmological constant, there is a bound state graviton in the theory, and four-dimensional gravity is reproduced. However, it is a massive bound state that plays the role of the four-dimensional graviton.

1. Introduction

The realization that even spin 2 excitations can be localized by a gravitating brane[1] suggests an alternative to compactification, since one has an infinite extra dimension but nonetheless sees four-dimensional gravity at low scales. However, it seemed a critical aspect of this proposal was the finite "volume" of the extra dimensional space since that is what permitted a normalizable zero-mode and a finite value of the four-dimensional Planck scale. However, it would be perplexing if this finite volume is truly essential, since we should expect physics to be local, so it would be strange if there were strong dependence on the warp factor far from the brane on which gravity potentially localizes.

For this reason, it is very interesting to study an AdS_4 brane in AdS_5, in which the tuning of cosmological terms required for a flat brane is modified. The metric is known in this case, and it is a warped geometry in which the warp factor blows up at large distance from the brane, rather than asymptoting to zero as it does in the flat brane case. Therefore, the geometry includes the boundary of AdS_5 and the volume of the extra-dimensional space is infinite. Clearly, the zero mode, corresponding to a perturbation whose wavefunction in the 5th dimension is proportional to the warp factor, is nonnormalizable.

Nonetheless, since such a theory can be obtained from the flat brane scenario through an arbitrarily small reduction of the cosmological term on the brane, it is difficult to believe that there is no bound state graviton to generate four-dimensional physics. Indeed, we find that four-dimensional gravity is reproduced, due to a very light, but massive bound state Kaluza-Klein mode of the graviton. Of course the gravity we reproduce is AdS gravity, so the cosmological constant would have to be sufficiently small so that physics looks Minkowskian within the horizon size.

This is an important result, as it demonstrates that localization is a local phenomenon, as indeed it must be. A warped geometry in a local region can be sufficient to generate four-dimensional gravity independently of the behavior of the higher dimensional geometry far away. Although we explicitly analyze the AdS_4 brane in AdS_5, we expect this result is far more general.

For those familiar with the c-theorem in five dimensions, it might be surprising that we can have a warp factor that localizes gravity near the brane, but then rises again. However, we will show this is consistent in AdS space.

This theory probably has interesting implications from the AdS/CFT vantage point, since it includes both a Planck brane and the AdS_5 boundary, which we briefly comment on here.

Furthermore, this theory gives rise to four-dimensional gravitational physics on a brane embedded into an AdS_5 bulk that asymptotes to the boundary of AdS and not the interior as its Minkowski or dS cousin, thereby avoiding the no-go theorems about smooth RS branes and supergravity and possibly facilitating a realization of localized gravity in that context.

In Section 2, we review the background geometry for a single brane with a non-finetuned tension, that is with an effective cosmological constant. In Section 3 we will discuss the fluctuation spectrum around such a background. While in the dS case one still has a massless bound graviton, we will show that in the AdS case, the massless bound state is lost. However one does find an almost massless graviton bound on the brane, which we refer to as the almost zero mode. In Section 4, we argue that four-dimensional gravity is correctly reproduced. We conclude in the final section.

2. The Background

2.1. *Action and background solution*

As in the original setup,[1] the action we consider is just that of 5d gravity with a negative cosmological constant $\Lambda^{5d} = -3/L^2$ coupled to a brane of tension λ:

$$S = \int d^5x \sqrt{g} \left[-\frac{1}{4} R - \Lambda^{5d} \right] - \lambda \int d^4x \, dr \sqrt{|\det g_{ij}|} \, \delta(r), \tag{1}$$

where g_{ij} is the metric induced on the brane by the ambient metric $g_{\mu\nu}$.

We use the ansatz for the solution to be a warped product with warp factor $A(r)$,

$$ds^2 = e^{2A(r)} \bar{g}_{ij} dx^i dx^j - dr^2 \ , \tag{2}$$

allowing for the 4d metric to be Minkowski, de Sitter or anti-de Sitter with the 4d cosmological constant Λ being zero, positive or negative respectively following the conventions of Ref. 2.

The solutions to Einstein's equations are:[2,3,4,5]

$$
\begin{aligned}
dS_4 :& A = \log(\sqrt{\Lambda} L \sinh \frac{c-|r|}{L}) , \quad \lambda = \frac{3}{L} \coth \frac{c}{L} \\
M_4 :& A = \frac{c-|r|}{L} , \quad \lambda = \frac{3}{L} \\
AdS_4 :& A = \log(\sqrt{-\Lambda} L \cosh \frac{c-|r|}{L}) , \quad \lambda = \frac{3}{L} \tanh \frac{c}{L} .
\end{aligned} \tag{3}
$$

The geometry of the curved solutions is characterized by a parameter c, which is determined in terms of the brane tension. In the dS case c is the distance between the brane and the horizon, whereas in the AdS case c is the distance to the turn around point in the warp factor. As is well known, in order to have a Minkowski solution, one has to fine-tune λ relative to L; in our conventions $\lambda = \frac{3}{L}$. Since $\coth(x) > 1$ and $\tanh(x) < 1$ for any x, we see that, $\lambda > \frac{3}{L}$ implies that the effective 4d cosmological constant is positive and $\lambda < \frac{3}{L}$ implies a negative 4d cosmological constant. The 4d cosmological constant in 4d Planck units is given by

$$4\pi \int dr\, e^{2\left(A(r)-\frac{1}{2}\ln|\Lambda|\right)} \tag{4}$$

and is hence determined by c alone, that is by the mismatch of brane tension and bulk cosmological constant. Since only the combination Λe^{2A} appears in the equations of motion, not both, Λ and A, will be determined independently. One can use this freedom to set $A(0) = 0$, as in Ref. 1. This way the cosmological constant becomes

$$\Lambda_{dS} = \frac{1}{L^2 \sinh^2 \frac{c}{L}}, \quad \Lambda_{AdS} = \frac{-1}{L^2 \cosh^2 \frac{c}{L}}. \tag{5}$$

Using the value of c determined by the jump equations one finds that the 4d cosmological constant is indeed only given by the detuning

$$M = \frac{\lambda L}{3} \tag{6}$$

of the brane tension without any exponential suppression:

$$\Lambda_{dS} = \frac{1}{L^2}(M^2 - 1), \quad \Lambda_{AdS} = \frac{1}{L^2}(1 - M^2). \tag{7}$$

The general behavior of the three solutions is sketched in figure 1. In the fine-tuned Minkowski solution, the warp factor is just $A = -|r|$ and the graviton is marginally bound on the brane; that is, it is a bound state at threshold. The onset of the continuum is not separated by a mass gap. Increasing the tension the backreaction on the warp factor is even stronger and even though $A(r)$ starts out linear in r, as in the Minkowski case, it goes to $-\infty$ at a finite value of r, at $r = c$. $A(r)$ diverges as $\log(c - |r|)$, but all components of the curvature tensor stay finite. Since the bulk solution is nothing but AdS_5, the coordinate singularity we see in this particular coordinate system is just a horizon. Gravity is still trapped in the

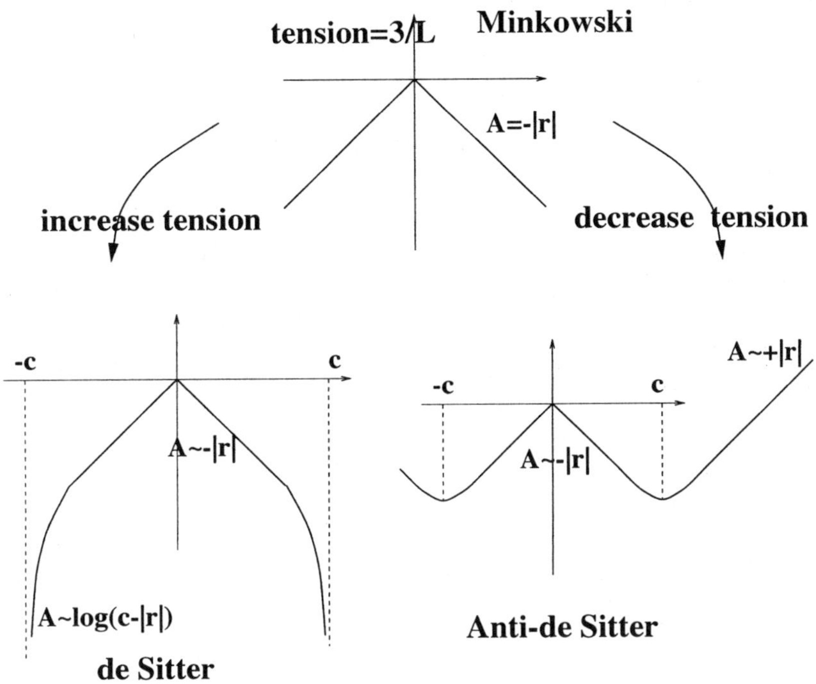

Fig. 1. The behavior of the warp-factor for $\Lambda = -1$, 0 and 1.

usual way. Since we increased the tension of the brane, one should expect that the marginal bound state becomes a true bound state. Formally one can even take the limit where the bulk cosmological constant vanishes altogether, $L \to \infty$, in which case one still traps gravity. The curvature vanishes identically in the bulk. The coordinate singularity reduces to the Rindler horizon of an accelerated observer in flat Minkowski space; see Ref. 3.

If we however decrease the tension of the brane as compared to the fine-tuned value, the backreaction on the warp factor becomes "weaker". Even though it still starts out as $A \sim -|r|$ close to the brane it reaches a minimum at a $r = c$ and then turns around and runs toward $+\infty$ as $A \sim +|r|$, approaching the boundary of AdS_5 and not the interior. So even though close to the brane it still looks like we trap gravity in the standard way, much of the amplitude for the zero mode state is concentrated near the boundary of the bulk space. However, since the cause of losing the bound graviton arises only far from the brane, at what should correspond to low energies, one should expect localization. In the next section, we will show that four-dimensional gravity is reproduced, first by a propagator analysis, and subsequently, by showing the interesting mode analysis that gives the same result. The authors of Ref. 2 gave the warp factor but did not study this physics, presumably due to the absence of phenomenologists.

2.2. *The modified c-theorem*

In the language of AdS flows the positive energy condition has been rephrased as the so called "c-theorem[6,7]" which asserts that $A'' \leq 0$. This is one of the reasons that it was believed that localization of gravity on a positive tension brane goes together with a warpfactor that tends asymptotically to the AdS horizon, since it can not turn around. In the solution we are considering A'' is obviously positive even though our brane has positive tension, so it does not violate positivity of stress energy. Let us see how this is consistent with the c-theorem.

The derivation of the c-theorem required Lorentz invariance and hence is only valid for Minkowski solutions. Even the AdS_4 brane violates $A'' \leq 0$, since

$$A''(r) = \frac{1}{L^2} \frac{1}{\cosh^2 \frac{c \pm |r|}{L}} \tag{8}$$

is positive everywhere in the bulk. Using the fact that the energy momentum tensor supporting our geometry has to be of the form

$$T^{\alpha}_{\beta} = \mathrm{diag}\{\rho, -p_1, -p_2, -p_3, -p_4\} \tag{9}$$

what is called the "weaker energy condition" in Ref. 6

$$T_{\alpha\beta}\xi^{\alpha}\xi^{\beta} \geq 0 \tag{10}$$

where ξ^{α} is an arbitrary null vector translates into the condition

$$\rho + p_i \geq 0. \tag{11}$$

It is easy to see that energy momentum tensors supporting solutions of the general form of a warped product based on 4d Minkowski, de Sitter or anti-de Sitter space will automatically satisfy the conditions for $i = 1, 2, 3$. Since

$$-3A'' - 3\Lambda e^{-2A} = R^0_0 - R^5_5 = G^0_0 - G^5_5 = (T^0_0 - T^5_5) \geq 0 \tag{12}$$

the constraint on A'' from the weaker energy condition reads

$$A'' \leq -\Lambda e^{-2A}. \tag{13}$$

For positive Λ this implies an even stronger version $A'' < 0$ of the standard c-theorem, $A'' \leq 0$. However for negative Λ in general positive A'' is allowed. The solutions written down above actually saturate the bound.

3. Spectrum of Linearized Fluctuations

3.1. *Gravity on the AdS brane*

Despite the absence of a massless graviton, we expect physics to be continuous and to react smoothly to small changes in the input parameters, like the brane tension.

If we were to lose localized gravity by going from zero to tiny negative cosmological constant, this would mean that by the very fact that four-dimensional gravity exists, we could rule out an AdS_4 brane world. Similarly, one can argue that low energy physics on the brane should only probe the warp factor close to the brane, which is basically indistinguishable for the Minkowski and AdS cases, and hence should be insensitive to the eventual turn around in the warp factor. From a holographic perspective, if we probe sufficiently short distances on the brane, physics should not depend on the warp factor far away. This fact becomes even more striking when we heat up the brane universe to finite temperature. In this case a black hole will form in the bulk, and space time gets cut off at the black hole horizon, manifestly localizing gravity and hiding all the differences between the zero and negative cosmological constant scenario behind an event horizon.

Indeed, we argue that four-dimensional gravity is localized to the brane. We first argue that the propagator between two points that are sufficiently close relative to the scale set by the AdS_4 curvature will reduce to the Minkowski brane propagator.

In Minkowski space, one finds[8] that in the conformal "z-coordinates" in which the metric reads $ds^2 = e^{2A(z)}(g_{ij}dx^i dx^j - dz^2)$ the propagator can be written as

$$\Delta_{d+1}(x, z; x', z') = \int \frac{d^4p}{(2\pi)^4} e^{ip(x-x')} \Delta_p(z, z') \tag{14}$$

where

$$e^{-2A}(\partial_z^2 + 3A'\partial_z - p^2)\Delta_p(z, z') = e^{-5A}\delta(z - z'). \tag{15}$$

In the AdS case, the integral in the Fourier transform will be replaced by a sum over discrete modes, whose spacing is of order $|\Lambda|$. So unless we want to study cosmological distance scales, the sum can still be approximated by an integral and then (15) tells us that the propagator for processes on the brane only depends on the warp factor in the vicinity of the brane.

Performing the coordinate change to the "z-coordinate", as we will work out later, the warp factor reads

$$e^A = \frac{L\sqrt{|\Lambda|}}{\sin\left((|z| + z_0)\sqrt{|\Lambda|}\right)}. \tag{16}$$

For small $|z|$, the sin can be expanded and we obtain to leading order in $|z|$

$$e^A = \frac{1}{1 + \cos\left(z_0\sqrt{|\Lambda|}\right)\ |z|/L}, \tag{17}$$

which is identical to the flat space form $e^A = \frac{1}{|z|/L+1}$ up to a rescaling in L, so that the form of the differential equation stays unchanged. The higher order corrections are accompanied with powers of $|\Lambda| \sim e^{-2c}$ and hence are very small for near critical tension.

To complete this analysis, and demonstrate the correct polarization structure of the propagator, one would have to incorporate the effect of brane bending. This will be discussed in the mode analysis below. However, one can use the propagator analysis, as performed in Ref. 9, to see that the correct gravitational potential is reproduced.

In Ref. 9, one explicitly solves for the trace and the longitudinal components of the metric perturbation in terms of the sources (rather than changing coordinates so that the brane is bent). These solutions then set up an "effective source term" for the TT-components, which includes contributions from the trace and longitudinal components of the metric. In terms of the trace reversed metric perturbation $\bar{h}_{\mu\nu} = h_{\mu\nu} - \frac{1}{2}h\eta_{\mu\nu}$ this effective source with stress energy only on the brane leads in flat space leads to a linearized response

$$\bar{h}_{\mu\nu}(x) = -\frac{1}{16}\int d^4x' \left\{ \Delta_5(x,0;x',0)T_{\mu\nu}(x') - \eta_{\mu\nu}\left[\Delta_5(x,0;x',0) - \frac{2}{L}\Delta_4(x,x')\right]\frac{T^\lambda_\lambda(x')}{4} \right\} \tag{18}$$

that contains an extra piece corresponding to the contribution of the extra scalar, that cancels the extra scalar contribution from the propagator of a massive mode. A similar analysis works in our AdS case with the scalar picking up a mass 4Λ, so that o ordinary gravity plus small corrections is reproduced.

It is important to see explicitly how the above propagator, which was treated approximately in the above, is reproduced in a mode analysis. In the next section, we solve for the modes of linearized fluctuations around the brane, so that one can explicitly construct the propagator.

3.2. *Analog quantum mechanics for transverse traceless modes*

In this section, we study the spectrum of linearized gravity fluctuations around the dS_4 and AdS_4 solutions. We first consider the transverse-traceless (TT) modes, defined by

$$D^j h_{ij} = g^{ij}\, h_{ij} = 0\,, \tag{19}$$

where the perturbed metric is

$$ds^2 = e^{2A(r)}(g_{ij} + h_{ij})\, dx^i dx^j - dr^2\,, \tag{20}$$

and we have chosen axial gauge where

$$h_{\mu 5} = 0\,, \tag{21}$$

with $\mu = 0,1,2,3,5$. We will consider the non-TT modes in the next section, and in more detail in Ref. 10. As suggested in Ref. 1, the TT modes are best analyzed by transforming Einstein's equations for the transverse traceless modes of the linearized fluctuations into the form of an analog quantum mechanics, where they satisfy a standard Schrödinger equation with a "volcano potential", which can be derived for arbitrary warp factor, see e.g. Refs. 2,11.

Their linearized equation of motion reads, see e.g. Refs. 12,13

$$\left(\partial_r^2 + 4A'\partial_r - e^{-2A}(\Box_{4d} + 2\Lambda)\right) h_{ij} = 0\,. \tag{22}$$

We are looking for a mode corresponding to a 4d graviton, or more generally, a spin 2 excitation of mass m; that is

$$(\Box_{4d} + 2\Lambda)h_{ij} = m^2 h_{ij} \tag{23}$$

where $\Box_{4d}$ is the 4-dimensional covariant d'Alembertian and m^2 the mass of the excitation. In the curved case, the constant shift by 2Λ is precisely the one needed in order to recover the equation of motion for a massless graviton (defined via the reduced number of polarization states) for $m^2 = 0$. Following the by now standard procedure, going to the conformal "z-coordinate" in terms of which the metric reads

$$ds^2 = e^{2A(z)} \left((g_{ij} + h_{ij})\, dx^i dx^j - dz^2 \right) . \tag{24}$$

and rescaling $H_{ij}(z) = e^{3A(z)/2}\, h_{ij}$, we obtain the Schrödinger equation of the analog quantum mechanics

$$\left(-\partial_z^2 + \frac{9}{4} A'(z)^2 + \frac{3}{2} A''(z) \right) H_{ij}(z) = m^2 H_{ij}(z) . \tag{25}$$

with the volcano potential

$$V(z) = \tfrac{9}{4} A'(z)^2 + \tfrac{3}{2} A''(z). \tag{26}$$

3.3. *The volcano potentials for dS and AdS*

The change of coordinates to the conformally flat metric is given by

$$\begin{aligned} dS_4 : \quad z(r) &= \frac{1}{\sqrt{\Lambda}} \operatorname{sgn}(r) \left\{ \operatorname{arcsinh} \left(\frac{1}{\sinh\left(\frac{c-|r|}{L}\right)} \right) - z_0 \sqrt{\Lambda} \right\} \\ AdS_4 : \quad z(r) &= \frac{1}{\sqrt{-\Lambda}} \operatorname{sgn}(r) \left\{ \arcsin \left(\frac{1}{\cosh\left(\frac{c-|r|}{L}\right)} \right) - z_0 \sqrt{-\Lambda} \right\} \end{aligned} \tag{27}$$

where the parameter z_0 is defined as

$$\begin{aligned} dS_4 : \quad z_0 &= \frac{1}{\sqrt{\Lambda}} \operatorname{arcsinh} \left(\frac{1}{\sinh\left(\frac{c}{L}\right)} \right) \\ AdS_4 : \quad z_0 &= \frac{1}{\sqrt{-\Lambda}} \arcsin \left(\frac{1}{\cosh\left(\frac{c}{L}\right)} \right) . \end{aligned} \tag{28}$$

Note that while in the dS case the z coordinate actually runs from $-\infty$ to $+\infty$ in the AdS case it only runs from $-(\frac{\pi}{\sqrt{|\Lambda|}} - z_0)$ to $(\frac{\pi}{\sqrt{|\Lambda|}} - z_0)$.

The resulting volcano potentials are

$$\begin{aligned} dS_4 : \quad V(z) &= \frac{9\Lambda}{4} + \frac{15}{4} \frac{\Lambda}{\sinh^2\left(\sqrt{\Lambda}(|z| + z_0)\right)} - 3\, \coth(\sqrt{\Lambda} z_0)\, \sqrt{\Lambda}\, \delta(z) \\ AdS_4 : \quad V(z) &= \frac{-9(-\Lambda)}{4} + \frac{15}{4} \frac{(-\Lambda)}{\sin^2\left(\sqrt{-\Lambda}(|z| + z_0)\right)} - 3\, \cot(\sqrt{-\Lambda} z_0)\, \sqrt{-\Lambda}\, \delta(z). \end{aligned} \tag{29}$$

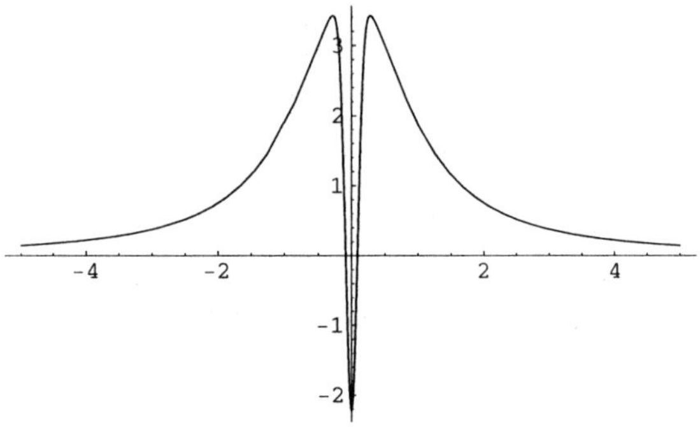

Fig. 2. The Minkowski volcano potential.

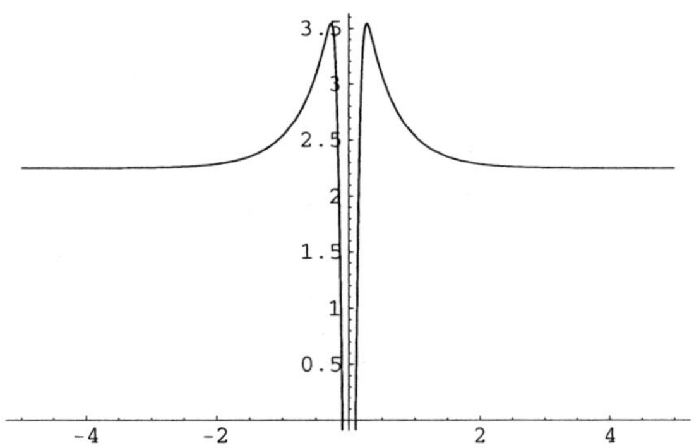

Fig. 3. The deSitter volcano potential.

3.4. *The de Sitter brane spectrum*

The volcano potential for the deSitter case is plotted in Figure 3. The general features of this potential are very similar to those of the flat space volcano of Figure 2. Both potentials have a normalizable zero mode trapped in the delta potential, whose wave function is given by the warp factor $e^{\frac{3A(z)}{2}}$ and decays like $1/z^3$. In the Minkowski case this massless graviton is a marginal bound state and the continuum of KK modes starts at $m = 0$. In the dS case, since at infinity the potential approaches a constant value $\frac{9}{4}\Lambda$, the onset of the continuum is only at $m^2 = \frac{9}{4}\Lambda$ and the graviton is separated by a finite mass gap from the continuum.[13]

3.5. *The embedding of AdS_4*

The Anti-deSitter volcano potential, as shown in Figure 4, looks quite different. As remarked above, the z-coordinate only stretches over the finite range from $-(\frac{\pi}{\sqrt{|\Lambda|}} -$

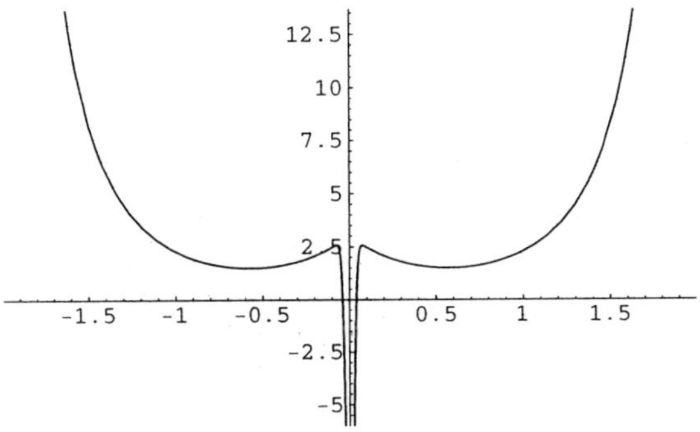

Fig. 4. The Anti-deSitter volcano potential.

z_0) to $(\frac{\pi}{\sqrt{|\Lambda|}} - z_0)$. The potential diverges at the boundaries, so that we effectively create a box. As we tune the effective 4D cosmological constant from positive to negative values, the flanks of the volcano move down from $\frac{9}{4}\Lambda$ to zero and then the potential "curls" in and we are left with a discrete spectrum of modes without a zero mode. The would be zero mode wavefunction diverges together with the warp factor $e^{A(z)}$ at the boundaries of the box.

Before we solve for the discrete spectrum of the quantum mechanics explicitly let us discuss why we see this behavior. The crucial point is that while the region of space bounded by the worldvolume of a single positive tension dS or Minkowski brane does not contain any part of the boundary of AdS_5, for the AdS brane this is not the case.

Figures 5, 6 summarize the difference between the flat and deSitter universes on the one hand and the AdS universe on the other hand. For the former the RS cutting and pasting procedure leads to a space that ends in a horizon on both sides. In the AdS_4 case even though n the space-time we keep it looks close to the brane like we are moving towards the horizon (the warp factor decreases) eventually we have to go all the way out to the boundary. While for the flat brane, gravitons sent out from the brane never come back but fall through the horizon, for the AdS_4 brane some eventually bounce back from the boundary and make it back to the brane. Effectively the 5d spacetime is a box and the gravity spectrum is discrete, at least when considered with respect to the Hamiltonian generating global time translations.

In Ref. 2 a coordinate transformation was given for the Poincare patch of AdS_5 from written as a warped product with an AdS_4 Poincare patch to the standard form. One can extend this coordinate transformation to the full global solution:

$$ds^2 = L^2 \cosh^2(\frac{c-r}{L})\, ds_4^2 \; - \; dr^2 \tag{30}$$

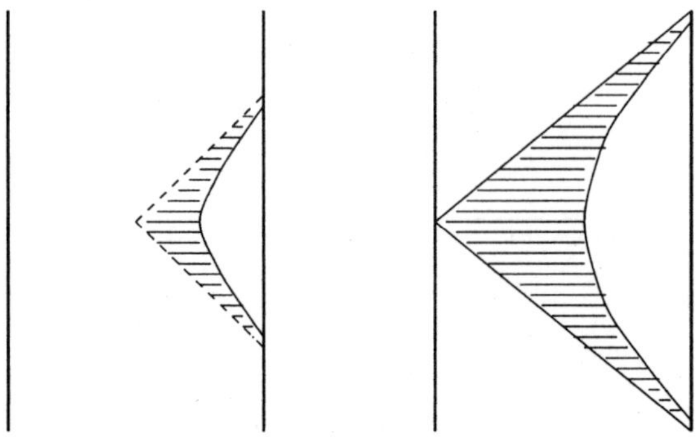

Fig. 5. Schematics of the Penrose diagrams of the dS and Minkowski. The spacetime one is instructed to keep is shaded. Since the branes are accelerated, they have their own horizon, only for the Minkowski brane does this coincide with the Poincare patch horizon. The spacetime one wants to keep is between the brane and the horizon. The dS brane really is a full hyperboloid, the part drawn corresponding to the static slice of dS. In the case of the AdS brane the brane falls through the horizon of the Poincare patch. It is more useful to study the embedding in terms of an constant time slice through global AdS.

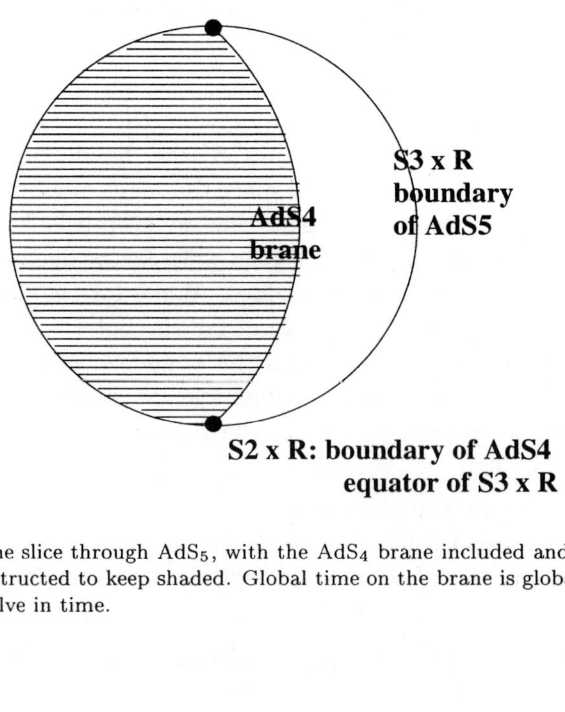

Fig. 6. Constant time slice through AdS_5, with the AdS_4 brane included and again the region of spacetime we are instructed to keep shaded. Global time on the brane is global time in AdS_5, the picture does not evolve in time.

where

$$ds_4^2 = \frac{1}{\cos^2(\rho)} \left(dt^2 - d\rho^2 - \sin^2(\rho)\, d\Omega_2^2\right). \tag{31}$$

Applying the following change of variables

$$\tan(\tilde{\rho}) = \coth(\frac{c-r}{L})\, \tan(\rho)\ , \quad \cosh(\tilde{r}) = \frac{\cosh(\frac{c-r}{L})}{\cos(\rho)}\ , \qquad \tilde{t} = t\ , \quad \tilde{\Omega}_2 = \Omega_2 \tag{32}$$

we end up with

$$ds^2 = L^2 \left(\cosh^2(\tilde{r})\, dt^2 - d\tilde{r}^2 - \sinh^2(\tilde{r}) d\tilde{\Omega}_3^2\right) \tag{33}$$

where $\tilde{\Omega}_3$ is the 3-sphere metric on $\tilde{\rho}$ and $\tilde{\Omega}_2$, $d\tilde{\Omega}_3^2 = d\tilde{\rho}^2 + \sin^2(\tilde{\rho})\tilde{\Omega}_2^2$. This can be recognized as global AdS_5, see for example Ref. 14.[a] Using this change of variables, it is easy to see, that an $r = const.$ surface is embedded in global AdS_5 as indicated in Figure 6.

Note that this implies an interesting realization of holography. Gravity in the bulk of our space should be dual to a theory living on the full boundary, which is composed of two pieces, the AdS_4 brane and the S^3/Z_2 half-boundary of the original AdS_5, with two different field theories living on them. They touch along their common boundary. According to AdS/CFT lore, any theory in AdS can be completely encoded in its boundary data, so that in the end all AdS_4 bulk physics should be captured by the CFT on the disk with the brane physics encoded in the boundary data of the disk. Instead of a CFT with momentum somehow cutoff, we get a CFT with spatial cutoff (that is formulated on a disk). The reduction of the symmetry from $SO(4,2)$ to $SO(3,2)$ corresponds to the well known fact that this is the subgroup of conformal transformations that leaves a boundary invariant.

3.6. *The discrete spectrum in the volcano potential*

We now solve the quantum mechanics problem with the box-like volcano potential (29) . In order to solve the spectrum with a delta function at the bottom of a box, we first find the eigenfunctions in the box, and then look for the right linear combinations to satisfy the jump imposed on the first derivative of the wavefunction due to the delta function.

We first consider the zero tension case, where formally the delta contribution vanishes.[b] The turn around point c in the warp factor (3) goes to zero, the solution becomes just $A = \log\left(L\sqrt{\Lambda}\cosh\frac{r}{L}\right)$, and we recover pure AdS_5. The parameter z_0

[a]Note that while ρ runs from 0 to $\frac{\pi}{2}$ and r from $-\infty$ to ∞, $\tilde{\rho}$ runs from 0 to π and $\tilde{r}$ from 0 to $+\infty$. When we take r to zero, $\tilde{\rho}$ has to go to $\frac{\pi}{2}$. As r goes negative, on the other side $\tilde{\rho}$ now explores the region between $\frac{\pi}{2}$ and π

[b]If one actually wants to analyse the spectrum of an almost tensionless brane, the spectrum will be the same, but every other mode is projected out by the boundary condition at the brane. By just omitting the delta function we are really looking at 'no brane' instead of a zero tension brane.

in (29) becomes $\sqrt{|\Lambda|}z_0 = \frac{\pi}{2}$ and the resulting potential is just

$$V(z) = \frac{-9(-\Lambda)}{4} + \frac{15}{4} \frac{(-\Lambda)}{\cos^2\left(\sqrt{-\Lambda}z\right)} \tag{34}$$

with z running from $-\frac{\pi}{2}$ to $\frac{\pi}{2}$. Rescaling $z \to \sqrt{-\Lambda}z$, Schrödinger's equation with energy eigenvalue $\frac{E}{-\Lambda}$ in the potential (34) can be solved explicitly and the resulting wavefunction is

$$\begin{aligned}\psi(z) = \quad & c_1 \cdot {}_2F_1(-\frac{3}{4} - \frac{\sqrt{9+4E}}{4}, \frac{5}{4} - \frac{\sqrt{9+4E}}{4}; 1 - \frac{\sqrt{9+4E}}{2}; \frac{1}{\cos^2 z}) \cos^{\frac{\sqrt{9+4E}}{2}} z + \\ & c_2 \cdot {}_2F_1(-\frac{3}{4} + \frac{\sqrt{9+4E}}{4}, \frac{5}{4} + \frac{\sqrt{9+4E}}{4}; 1 + \frac{\sqrt{9+4E}}{2}; \frac{1}{\cos^2 z}) \cos^{\frac{-\sqrt{9+4E}}{2}} z.\end{aligned} \tag{35}$$

In order to not diverge on either side of the box we need $c_2 = 0$ and

$$E = n\,(n+3) \quad \text{where} \quad n = 1, 2, 3, \ldots. \tag{36}$$

This spectrum indeed agrees with the spectrum of gravity fluctuations of pure AdS_5, as can be seen to be the case by group-theoretic considerations. Unitary representations of the AdS_5 group are classified (for a recent discussion see e.g. Ref. 15) by the eigenvalue of the global time Hamiltonian E_0, which is the generator of the $SO(2)$ subgroup of $SO(4) \times SO(2) \subset SO(2,4)$, and $SO(4)$ spins j_1 and j_2 for the groundstate from which the infinite dimensional representation is built by raising operators. The massless graviton (where masslessness is defined via the smaller number of polarization states) has $E_0 = 4$ and $j_1 = j_2 = 1$. The analysis of the analog quantum mechanics amounts to decomposing the $SO(4,2)$ representation under the $SO(3,2)$ subgroup that is manifest in the warped product based on AdS_4. As in 5d, these are classified by E_0 and s, where E_0 in 4d is identified with E in 5d (the $SO(2)$ subgroup corresponding to global time translations of $SO(4,2)$ is the same as in $SO(3,2)$) and s is the $SO(3)$ spin obtained by tensoring together j_1 and j_2. The Casimir now reads

$$C_2(E_0, s) = E_0(E_0 - 3) + s(s+1) \tag{37}$$

and a massless representation obeys

$$E_0 = s + 1. \tag{38}$$

So the appropriate definition of mass for a spin 2 representation is

$$m^2 = C_2(E_0, 2) - C_2(3, 2) = E_0(E_0 - 3) \tag{39}$$

The single massless graviton decomposes into a tower of spin 0,1 and 2 representations in four dimensions. The masses of the spin 2 excitations are

$$m^2 = E_{0,n}(E_{0,n} - 3) \quad \text{with} \quad E_{0,n} = 4, 5, 6, \ldots \tag{40}$$

which obviously agrees with the spectrum (36) found via the Schrödinger equation.

Upon including the brane the solution in terms of the linear combination of hypergeometric functions becomes quite complicated and we chose to study it numerically.[c] A more detailed analysis of the spectrum will appear elsewhere. Let us briefly exhibit the crucial features of the spectrum as obtained via a numerical solution of Schrödinger's equation using the shooting method.

In order to find the eigenvalues and wavefunctions for non-zero tension, we use

$$\frac{\psi'}{\psi} = -3\tan(z_0) \tag{41}$$

as required by the delta function in the potential as the start condition for wavefunction on the brane and then numerically solve Schrödinger's equation outwards. For generic energy values, the would-be wavefunction will diverge at the boundary. Eigenvalues are the values of the energy for which this expression goes to zero at the boundary. Of course, this is never really zero in any numerical approximation. The shooting method relies on the fact that (for any reasonable potential) the would be wavefunction diverges to (say) +infinity if you are below an eigenvalue and then will diverge to -infinity above the eigenvalue (and then vice versa for the next eigenvalue). If one plots the wavefunction at the boundary as a function of energy, the zeroes of this function are the eigenvalues.

Figures 7, 8, 9 show the spectrum of low lying modes as we increase the tension. For small tension, we keep every other mode of the pure AdS_5 analysis. As we increase the tension, one mode comes down to mass[d] $m^2 << \Lambda$, while all the other modes stay at $m^2 = \mathcal{O}(\Lambda)$. As we approach the critical tension, the potential becomes "two copies" of the pure AdS_5 potential pasted together along a delta function. Indeed we have all the heavy modes of the AdS_5 analysis at $m^2 = (n-3)n\Lambda$ and in addition a very light mode trapped in the delta function. In the critical limit the trapped very light mode becomes the trapped massless graviton, while the densely spaced excited modes become the continuum of KK modes. Figures 10, 11 show the wavefunctions of the almost massless mode and of a highly excited mode. Comparison with the shape of the zero mode and the KK-modes in the RS scenario show indeed that in the flat space limit, the former approaches the zero mode localized on the brane, while the latter approaches a highly excited KK mode, oscillating without feeling the brane at all. The low KK modes also look as expected: oscillating in the bulk but suppressed on the brane.

As we decrease the tension, the wavefunction of the almost zero mode gets broader and loses its amplitude enhancement over the KK modes, as is studied in detail in Ref. 17. Therefore, we would never actually be able to see four-dimensional AdS gravity and at the same time a significant deviation due to the graviton mass. For small Λ, the mass is tiny, and by the time we reach distance scales of order $1/m$ where the mass of the graviton should become significant, all excitations have

[c] For an analytic treatment see Ref. 16.
[d] A more detailed numerical analysis[17] or an analytic treatment relying on the factorization of the hamiltonian[16] both show that the actual behavior for the almost zero mode is $m^2 \sim \frac{3}{2}\Lambda^2$.

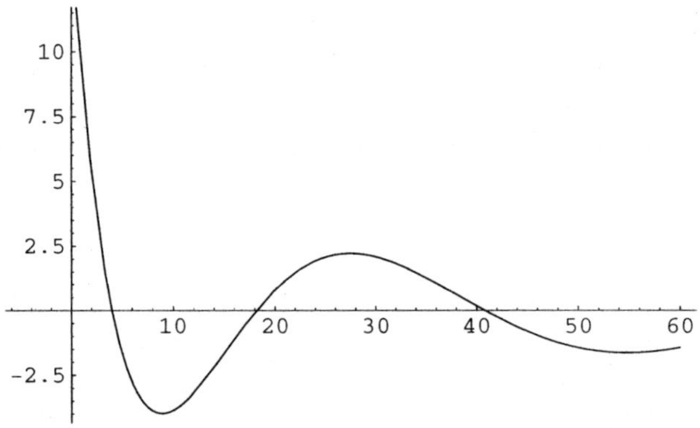

Fig. 7. Spectrum at zero tension, $\sqrt{|\Lambda|}z_0 = \frac{\pi}{2}$.

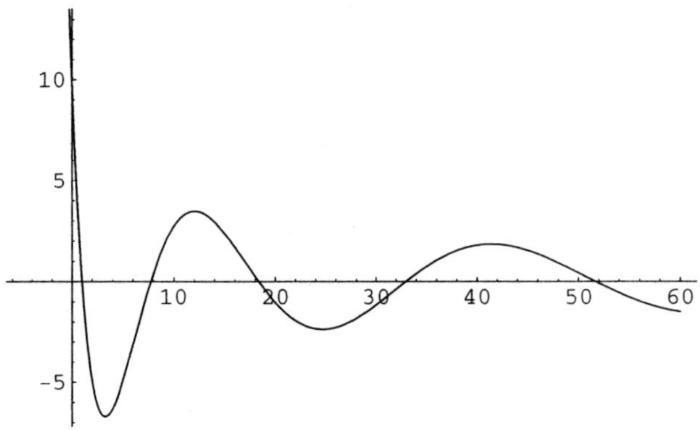

Fig. 8. Spectrum at $\sqrt{|\Lambda|}z_0 = \frac{\pi}{2} - 0.7$.

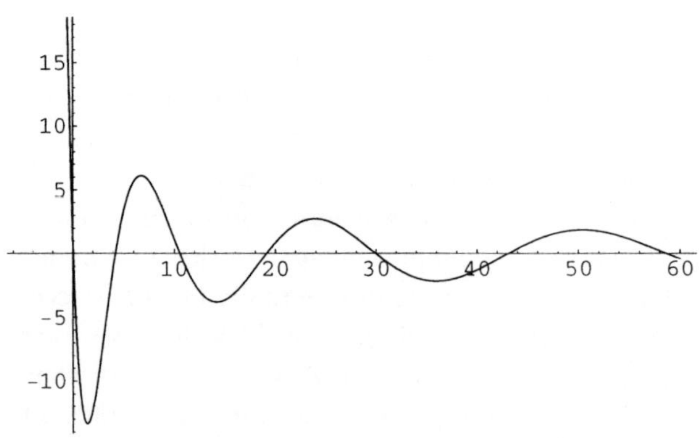

Fig. 9. Spectrum at $\sqrt{|\Lambda|}z_0 = \frac{\pi}{2} - 1.4$.

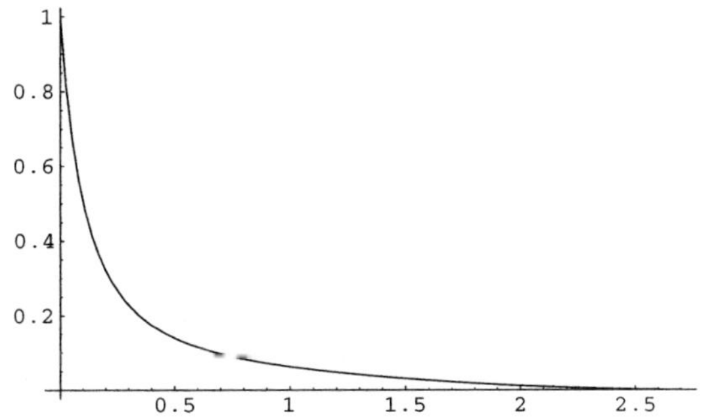

Fig. 10. Wavefunction of the almost massless mode, E=0.0419 at $\sqrt{|\Lambda|}z_0 = \frac{\pi}{2} - 1.4$.

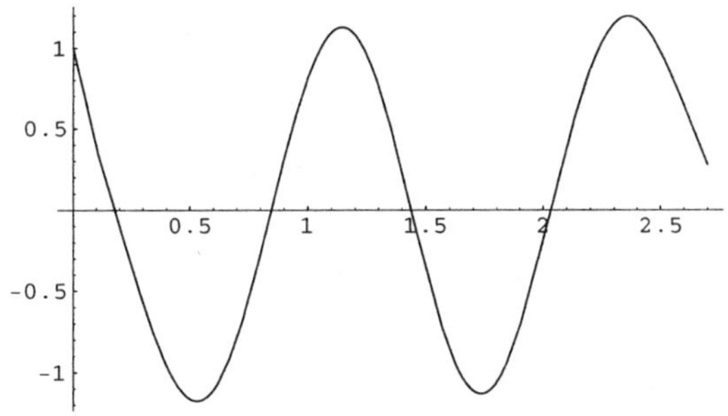

Fig. 11. Excited mode at $\sqrt{|\Lambda|}z_0 = \frac{\pi}{2} - 1.4$ and E=30.

died off anyway due to the gravitational potential of AdS. When we go to $\Lambda \to \frac{1}{L^2}$, which is the maximal value we can reach for very small tension branes, the mass of the almost zero mode goes like 4Λ and hence one should now see it contributing on the same footing as the 4d curvature effects. However since in this limit, the wavefunction is no longer bound to the brane and the almost zero mode has a mass comparable to the other excited modes, it really becomes one of the KK-modes and one no longer has four-dimensional gravity at all.

4. The non-TT Modes

So far we have only analyzed the spectrum of TT fluctuations. We have seen that in the AdS case, we get a massive spin 2 almost zero mode and the usual massive KK tower. This naively seems inconsistent, since the massive mode has extra polarization states that should contribute to the propagator. However, we expect a smooth flat space limit. In order to complete the analysis, we have to study the non- transverse-traceless modes as well. We will argue that although the additional modes can be gauged away up to a brane bending effect, they nonetheless contribute to the propagator, effectively reducing the degrees of freedom from 5 to 2. This will be discussed further in Ref. 10.

We first analyze the general form of a solution to the equations of motion and then compare this with the residual gauge transformations. We still write the perturbed metric as

$$ds^2 = e^{2A(r)}(g_{ij} + h_{ij})\, dx^i dx^j - dr^2\,, \tag{42}$$

in axial gauge

$$h_{\mu 5} = 0\,, \tag{43}$$

where $\mu = 0, 1, 2, 3, 5$, but do not restrict to TT components only. We first exhibit the general form of the solution to the bulk equations of motion.

The first order equations of motion require[e] vanishing of (where $h = g^{ij} h_{ij}$):

$$\begin{aligned}
55: &\quad (e^{2A} h')' \\
5i: &\quad -\frac{1}{2}\nabla_i h' + \frac{1}{2}\nabla^k h'_{ik} \\
ij: &\quad e^{2A}(\frac{1}{2}h''_{ij} + 2A'h'_{ij}) - \frac{1}{2}\nabla^2 h_{ij} + \frac{1}{2}g_{ij}e^{2A}A'h' - \\
&\quad \frac{1}{2}\nabla_i\nabla_j h + \frac{1}{2}(\nabla^k\nabla_i h_{jk} + \nabla^k\nabla_j h_{ik}) + 3\Lambda h_{ij}\,.
\end{aligned} \tag{44}$$

In order to manipulate these expressions, the following identities for covariant derivatives in dS and AdS are useful:

$$\begin{aligned}
R_{ijkl} =& \Lambda(g_{il}g_{jk} - g_{ik}g_{jl}) \\
R_{ij} = g^{jl}R_{ijkl} =& -3\Lambda g_{ij} \\
\nabla_i\nabla_j\nabla_k f - \nabla_j\nabla_i\nabla_k f =& \Lambda(g_{jk}\nabla_i - g_{ik}\nabla_j)f \\
\nabla_i\nabla_k\nabla^k f - \nabla^k\nabla_i\nabla_k f =& 3\Lambda\nabla_i f \\
\nabla_i\nabla_j\nabla^k\nabla_k f - \nabla^k\nabla_k\nabla_i\nabla_j f =& (8\Lambda\nabla_i\nabla_j - 2\Lambda g_{ij}\nabla^2)f \\
\nabla^k\nabla_i\nabla_j\nabla_k f - \nabla^k\nabla_k\nabla_i\nabla_j f =& (-\Lambda g_{ij}\nabla^2 f + \Lambda\nabla_i\nabla_j f)\,.
\end{aligned} \tag{45}$$

From the 55 equations it follows immediately that h_{ij} can be written as

$$h_{ij} = h^{TT}_{ij} + G\, f_{ij} + d_{ij} \tag{46}$$

where h^{TT}_{ij} are the TT components of the fluctuation which we studied in great detail before, f_{ij} and d_{ij} are functions of the 4d space-time only and G is defined as[f]

$$G = \int^r e^{-2A(\tilde{r})} d\tilde{r}. \tag{47}$$

[e] In the ij equations we used the zeroth order equations of motion to eliminate the terms involving the stress energy tensor on the right hand side of Einstein's equations as in Refs. 2,18.

[f] This of course only defines G up to a constant, which can be absorbed into d_{ij}. To be precise, we will always choose 0 as the lower bound on the integral.

Now from the $i5$ we find

$$\nabla_i f_k^k = \nabla^k f_{ik}. \tag{48}$$

Using as an ansatz

$$f_{ij} = \nabla_i \nabla_j \alpha - g_{ij}\beta \tag{49}$$

we see that (48) implies

$$f_{ij} = (\nabla_i \nabla_j - \Lambda g_{ij})\Phi. \tag{50}$$

Φ is what we would like to identify as the radion field. Plugging this into the ij equation, one finds after a tedious calculation using (45) and the following property of our warpfactors

$$A' + G\Lambda = -\frac{\lambda}{3} \tag{51}$$

that the full system is solved by

$$h_{ij} = h_{ij}^{TT} + 2G\,(\nabla_i \nabla_j - \Lambda g_{ij})\Phi - \frac{2}{3}\lambda g_{ij}\Phi \tag{52}$$

where Φ has to satisfy the following equation of motion:

$$\nabla^2 \Phi = 4\Lambda\Phi \tag{53}$$

and the TT modes h_{ij}^{TT} are determined by the analog quantum mechanics we analyzed before. The boundary condition at the brane demands

$$h'_{ij} = 0 = h_{ij}^{TT'} + 2e^{-2A}(\nabla_i \nabla_j - \Lambda g_{ij})\Phi. \tag{54}$$

Now let us proceed to analyze the residual gauge transformations that remain after fixing axial gauge. Under gauge transformations that take $x^\mu \to x^\mu + \xi^\mu$, the fluctuations transform as

$$\begin{aligned} h_{55} &\to h_{55} - 2e^{-2A}\xi^{5'} \\ h_{ij} &\to h_{ij} + (\nabla_i \xi_j + \nabla_j \xi_i) + 2A' g_{ij}\xi^5 \\ h_{i5} &\to h_{i5} - e^{-2A}\nabla_i \xi^5 + g_{ij}\xi^{j'} \end{aligned} \tag{55}$$

where we pulled out a factor of e^{2A} from the h_{55} and h_{i5} component as we did for the h_{ij} fluctuation in (42) and G is defined as in (47) above. In order to preserve the axial gauge (43) we require that h_{55} and h_{i5} stay zero, leaving us with the residual gauge transformations given by

$$\begin{aligned} \xi_5 &= \epsilon^5(x) \\ \xi_i &= G\nabla_i \epsilon^5 + \epsilon_i(x)\ . \end{aligned} \tag{56}$$

As in Ref. 8, the ϵ^i gauge transformations are needed in order to gauge away the longitudinal components of the graviton fluctuations, which we already neglected by our ansatz for h_{ij}. Using (51) one can see that the residual gauge transformations

(56) preserving axial gauge can indeed be used to gauge away the radion component Φ in the full solution (52) by choosing $\epsilon^5 = -\Phi$.

In the presence of sources on the brane, the ϵ_5 gauge transformation leads, as in Ref. 8, to a bent brane in the new gauge and to a brane bending contribution to the propagator, proportional to the propagator of a scalar field of mass 4Λ. This scalar is clearly not one of the physical modes, since we can gauge it away everywhere where there is no stress energy. Hence it does not correspond to a propagating ghost. In this respect our setup differs crucially from the quasilocalized gravity model of Ref. 19 that was analyzed in a very similar fashion in Refs. 20,21,22,23. There, due to the additional boundary condition on the second brane the radion cannot be gauged away[24] and represents a real ghost. In the critical tension limit, the bending contribution is precisely what is needed to smoothly match on to the standard massless 4d graviton propagator. The absence of a Veltman Van Dam discontinuity,[25,26] as the mass of the graviton goes to zero is consistent with the findings of Ref. 27[g] that states that no such discontinuity exists in AdS space as long as the mass m^2 goes to zero faster than Λ. This shouldn't come as a surprise, since we preserved the gauge invariance corresponding to 4d diffeomorphisms despite the fact that we are dealing with a massive graviton. As will be explained in Ref. 10, these additional gauge invariances can still be used to eliminate 3 of the polarization modes.

5. Conclusions

In this paper, we have demonstrated that localization of gravity applies, even for the AdS_4 brane for which there is no normalizable zero mode. Although expected from the vantage point of locality and holography, it is interesting that a massive graviton generates four-dimensional gravity.

Our result differs substantially from quasi-localization models, that also had a turn-around in the warp factor that modifies gravity on long distance scales. In the existing theories, there was a ghost when four-dimensional gravity is reproduced. Furthermore, those theories required unphysical matter in order to violate the C-theorem.

Our theory should have interesting implications for holography. A complete holographic description of the setup can be given in terms of a CFT living on the disk that is the left over part of the true AdS_5 boundary. In this description, all of the physics on the brane is reduced to correlators on the common boundary of the disk and AdS_4. This description is however not appropriate to study local physics on the brane. For latter purpose it should be more useful to split the bulk excitations into two sets, one dual to a CFT on the true boundary, the other one to a CFT on the brane, but the precise procedure is unclear. The mass of the almost zero mode is attributable to the long distance behavior of the warp factor which is sensitive to boundary physics; presumably the mass can be understood as a consequence of

[g]After completing this work, we became aware of Ref. 28, which obtains a similar result to Ref. 27.

the boundary CFT. Understanding how the theory on the brane is reproduced by CFT physics remains a challenge.

The much more general phenomenon of localization demonstrated in this paper might have other implications. Compactification of extra dimensional spaces is not essential to reproduce four-dimensional physics; a region of space which localizes gravity is sufficient. Furthermore, the current no-go theorems about localization in supergravity theories e.g. Refs. 30,31,32 relying on the asymptotic behavior of the geometry do not necessarily apply. It is important to know whether this physics can be realized in string theory.

Acknowledgments

We want to thank M. Aganagic, O. Aharony, N. Arkani-Hamed, A. Chamblin, O. DeWolfe, D. Freedman, M. Gremm, A. Katz, A. Miemiec, A. Naqvi, L. Rastelli, M. Schwartz, and B. Wald for useful discussions. We especially thank M. Porrati, both for discussions and for sharing his results.[27] We are also grateful to the organizers of the conferences at which much of this work has been previously reported.[33] Our work was supported by the U.S. Department of Energy under contracts # DE-FC02-94ER40818 and # DE-FG02-91ER4071.

References

1. L. Randall and R. Sundrum, *An Alternative to compactification*, *Phys. Rev. Lett.* **83**, 4690 (1999) [hep-th/9906064].
2. O. DeWolfe, D.Z. Freedman, S.S. Gubser, and A. Karch, *Modeling the fifth dimension with scalars and gravity*, *Phys. Rev.* **D62**, 046008 (2000) [hep-th/9909134].
3. N. Kaloper, *Bent domain walls as brane-worlds*, *Phys. Rev.* **D60**, 123506 (1999) [hep-th/9905210].
4. H.B. Kim and H.D. Kim, *Inflation and gauge hierarchy in Randall-Sundrum compactification*, *Phys. Rev.* **D61**, 064003 (2000) [hep-th/9909053].
5. T. Nihei, *Inflation in the five-dimensional universe with an orbifold extra dimension*, *Phys. Lett.* **B465**, 81 (1999) [hep-ph/9905487].
6. D.Z. Freedman, S.S. Gubser, K. Pilch, and N.P. Warner, *Renormalization group flows from holography supersymmetry and a c theorem*, hep-th/9904017.
7. L. Girardello, M. Petrini, M. Porrati, and A. Zaffaroni, *Novel local CFT and exact results on perturbations of $N = 4$ super Yang-Mills from AdS dynamics*, *JHEP* **9812**, 022 (1998) [hep-th/9810126].
8. S.B. Giddings, E. Katz, and L. Randall, *Linearized gravity in brane backgrounds*, *JHEP* **0003**, 023 (2000) [hep-th/0002091].
9. S.B. Giddings and E. Katz, *Effective theories and black hole production in warped compactifications*, hep-th/0009176.
10. A. Karch, E. Katz, and L. Randall, to appear.
11. C. Csaki, J. Erlich, T.J. Hollowood, and Y. Shirman, *Universal aspects of gravity localized on thick branes*, *Nucl. Phys.* **B581**, 309 (2000) [hep-th/0001033].
12. N. Alonso-Alberca, P. Meessen, and T. Ortin, *Supersymmetric brane-worlds*, *Phys. Lett.* **B482**, 400 (2000) [hep-th/0003248].
13. J. Garriga and M. Sasaki, *Brane-world creation and black holes*, *Phys. Rev.* **D62**, 043523 (2000) [hep-th/9912118].

14. O. Aharony, S.S. Gubser, J. Maldacena, H. Ooguri, and Y. Oz, *Large N field theories, string theory and gravity*, *Phys. Rept.* **323**, 183 (2000) [hep-th/9905111].
15. S. Ferrara, C. Fronsdal, and A. Zaffaroni, *On $N = 8$ supergravity on AdS(5) and $N = 4$ superconformal Yang-Mills theory*, *Nucl. Phys.* **B532**, 153 (1998) [hep-th/9802203].
16. A. Miemiec, *A Power Law for the lowest Eigenvalue in Quasilocalized Gravity*, to appear.
17. M.D. Schwartz, to appear.
18. O. DeWolfe and D.Z. Freedman, *Notes on fluctuations and correlation functions in holographic renormalization group flows*, hep-th/0002226.
19. R. Gregory, V.A. Rubakov, and S.M. Sibiryakov, *Opening up extra dimensions at ultra-large scales*, *Phys. Rev. Lett.* **84**, 5928 (2000) [hep-th/0002072].
20. C. Csaki, J. Erlich, and T.J. Hollowood, *Quasi-localization of gravity by resonant modes*, *Phys. Rev. Lett.* **84**, 5932 (2000) [hep-th/0002161].
21. C. Csaki, J. Erlich, T.J. Hollowood, and J. Terning, *Holographic RG and cosmology in theories with quasi-localized gravity*, hep-th/0003076.
22. G. Dvali, G. Gabadadze, and M. Porrati, *Metastable gravitons and infinite volume extra dimensions*, *Phys. Lett.* **B484**, 112 (2000) [hep-th/0002190].
23. G. Dvali, G. Gabadadze, and M. Porrati, *A comment on brane bending and ghosts in theories with infinite extra dimensions*, *Phys. Lett.* **B484**, 129 (2000) [hep-th/0003054].
24. L. Pilo, R. Rattazzi, and A. Zaffaroni, *The fate of the radion in models with metastable graviton*, *JHEP* **0007**, 056 (2000) [hep-th/0004028].
25. H. van Dam and M. Veltman, *Massive and massless Yang-Mills and gravitational fields*, *Nucl. Phys.* **B22**, 397 (1970).
26. V. Zahkarov, *JETP Lett.* **12**, 312 (1970).
27. M. Porrati, *No Van Dam-Veltman-Zakharov Discontinuity in AdS Space*, to appear.
28. I.I. Kogan, S. Mouslopoulos, and A. Papazoglou, *The $m \to 0$ limit for massive graviton in dS_4 and AdS_4. How to circumvent the van Dam-Veltman-Zakharov discontinuity*, hep-th/0011141.
29. A. Higuchi, *Forbidden mass range for spin-2 field theory in de Sitter space-time*, *Nucl. Phys.* **B282**, 397 (1987).
30. J. Maldacena and C. Nunez, *Supergravity description of field theories on curved manifolds and a no go theorem*, hep-th/0007018.
31. K. Behrndt and M. Cvetic, *Anti-de Sitter vacua of gauged supergravities with 8 supercharges*, *Phys. Rev.* **D61**, 101901 (2000) [hep-th/0001159].
32. R. Kallosh and A. Linde, *Supersymmetry and the brane world*, *JHEP* **0002**, 005 (2000) [hep-th/0001071].
33. Online proceedings of talks given at various conferences by L. Randall
`http://wwwth.cern.ch/susy2k/susy2ktalks/randall.ps.gz`,
`http://feynman.physics.lsa.umich.edu/strings2000/schedule.html#randall`,
`http://www.ictp.trieste.it/ highener/beyond4d.html`,
and A. Karch
`http://hep.physics.uoc.gr/string2000/program1.html`

COVARIANT QUANTIZATION OF THE SUPERSTRING

NATHAN BERKOVITS
Instituto de Física Teórica, Universidade Estadual Paulista
Rua Pamplona 145, 01405-900, São Paulo, SP, Brasil
E-mail: nberkovi@ift.unesp.br

After reviewing the Green-Schwarz superstring using the approach of Siegel, the superstring is covariantly quantized by constructing a BRST operator from the fermionic constraints and a bosonic pure spinor ghost variable. Physical massless vertex operators are constructed and, for the first time, N-point tree amplitudes are computed in a manifestly ten-dimensional super-Poincaré covariant manner. Quantization can be generalized to curved supergravity backgrounds and the vertex operator for fluctuations around $AdS_5 \times S^5$ is explicitly constructed.

1. Introduction

There are many motivations for covariantly quantizing the superstring. As in any theory, it is desirable to make all physical symmetries manifest in order to reduce the amount of calculations and simplify any cancellations coming from the symmetry. A second motivation comes from the desire to construct a quantizable sigma model action for the superstring in curved backgrounds with Ramond-Ramond flux. A third motivation is that a more symmetrical formulation of superstring theory may shed light on some of the mysteries of M-theory.

Although the light-cone Green-Schwarz (GS) formalism[1] has been useful for computing four-point tree-level and one-loop amplitudes, it is impractical for computing higher-point or higher-loop amplitudes because of dependence on the locations of the light-cone interaction points. Using the covariant RNS formalism, one can easily compute N-point tree-level and one-loop amplitudes involving bosons, but amplitudes involving fermions are complicated by the presence of spin fields.[2] Furthermore, lack of manifest spacetime supersymmetry in the RNS expressions complicates the analysis of finiteness properties and, for more than one-loop, leads to picture-changing problems.

For more than 15 years, a classical super-Poincaré covariant version of the GS formalism[3] has existed. But until the recent work described here, quantization problems have prevented this formalism from being used to compute non-vanishing scattering amplitudes. In this review, the covariant GS formalism is quantized by constructing a BRST operator from the fermionic constraints and a bosonic pure spinor ghost variable. After constructing physical massless vertex operators, N-point tree amplitudes are computed for the first time in a manifestly ten-dimensional super-Poincaré covariant manner.

In previous papers, this author has discussed quantization of the superstring compactified to four[4] or six[5] dimensions using "hybrid" variables which combine four or six-dimensional GS variables with a $c = 9$ or $c = 6$ superconformal field theory for the compactification. These hybrid formalisms have manifest four or six-dimensional super-Poincaré covariance, N=2 worldsheet supersymmetry, and are related to the usual RNS formalism by a field redefinition. Unfortunately, the precise relation between these hybrid formalisms and the new ten-dimensional super-Poincaré covariant formalism is still unclear, so this review will not discuss the hybrid formalisms.

In section 2 of this paper, the covariant GS formalism will be reviewed using the approach of Siegel[6] where the canonical momentum to θ^α is an independent variable. In section 3, it will be argued[7] that a bosonic pure spinor variable, λ^α, plays the role of the worldsheet ghost in this formalism, and a BRST operator is constructed out of the fermionic constraints and λ^α. In section 4, physical massless vertex operators will be constructed and, in section 5, they will be used to compute N-point tree amplitudes.[7,8] These amplitudes are manifestly super-Poincaré covariant and involve integration over an on-shell 'harmonic' superspace including five θ's and three λ's. In section 6, the cohomology of the BRST operator is shown to reproduce the light-cone GS spectrum.[9] In section 7, this quantization method is generalized to curved supergravity backgrounds[7] and the vertex operator is constructed for fluctuations around an $AdS_5 \times S^5$ background with Ramond-Ramond flux.[10] In section 8, some open problems and applications are discussed.

2. Review of GS Formalism using the Approach of Siegel

In conformal gauge, the classical covariant GS action for the heterotic superstring is[3]

$$S = \int d^2 z [\frac{1}{2}\Pi^m \bar{\Pi}_m + \frac{1}{4}\Pi_m \theta^\alpha \gamma^m_{\alpha\beta} \bar{\partial}\theta^\beta - \frac{1}{4}\bar{\Pi}_m \theta^\alpha \gamma^m_{\alpha\beta} \partial\theta^\beta] + S_R \tag{1}$$

where x^m and θ^α are the $d = 10$ worldsheet variables ($m = 0$ to 9, $\alpha = 1$ to 16), S_R describes the right-moving degrees of freedom for the $E_8 \times E_8$ or $SO(32)$ lattice, and $\Pi^m = \partial x^m + \frac{1}{2}\theta^\alpha \gamma^m_{\alpha\beta} \partial\theta^\beta$ and $\bar{\Pi}^m = \bar{\partial} x^m + \frac{1}{2}\theta^\alpha \gamma^m_{\alpha\beta} \bar{\partial}\theta^\beta$ are supersymmetric combinations of the momentum. Note that $\gamma^m_{\alpha\beta}$ and $\gamma^{m\,\alpha\beta}$ are 16×16 symmetric matrices satisfying $\gamma^{(m}_{\alpha\beta}\gamma^{n)\,\beta\gamma} = 2\eta^{mn}\delta^\gamma_\alpha$ and are the off-diagonal blocks in the Weyl representation of the 32×32 ten-dimensional Γ^m-matrices. In what follows, the right-moving degrees of freedom play no role and will be ignored. Also, all of the following remarks are easily generalized to the Type I and Type II superstrings.

Since the action of (1) is in conformal gauge, it needs to be supplemented with the Virasoro constraint $T = -\frac{1}{2}\Pi^m \Pi_m = 0$. Also, since the canonical momentum to θ^α does not appear in the action, one has the Dirac constraint $p_\alpha = \delta L / \delta \partial_0 \theta^\alpha = \frac{1}{2}(\Pi_m - \frac{1}{4}\theta\gamma_m \partial_1 \theta)(\gamma^m \theta)_\alpha$ where p_α is the canonical momentum to θ^α. If one defines

$$d_\alpha = p_\alpha - \frac{1}{2}(\Pi_m - \frac{1}{4}\theta\gamma_m \partial_1 \theta)(\gamma^m \theta)_\alpha, \tag{2}$$

one can use the canonical commutation relations to find $\{d_\alpha, d_\beta\} = -\gamma^m_{\alpha\beta}\Pi_m$, which implies (since $\Pi^m \Pi_m = 0$ is a constraint) that the sixteen Dirac constraints d_α have eight first-class components and eight second-class components. Since the anti-commutator of the second-class constraints is non-trivial (i.e. the anti-commutator is an operator Π^+ rather than a constant), standard Dirac quantization cannot be used since it would involve inverting an operator. So except in light-cone gauge (where the commutator becomes a constant), the covariant Green-Schwarz formalism cannot be easily quantized.

In 1986, Siegel suggested an alternative approach in which the canonical momentum to θ^α is an independent variable using the free-field action[6]

$$S = \int d^2 z[\frac{1}{2}\partial x^m \bar{\partial} x_m + p_\alpha \bar{\partial}\theta^\alpha]. \tag{3}$$

In this approach, Siegel attempted to replace the problematic constraints of the covariant GS action with some suitable set of first-class constraints constructed out of the supersymmetric objects (Π^m, d_α, $\partial\theta^\alpha$) where

$$d_\alpha = p_\alpha - \frac{1}{2}(\Pi^m - \frac{1}{4}\theta\gamma^m\partial\theta)(\gamma_m\theta)_\alpha \tag{4}$$

is defined as in (2) and is no longer constrained to vanish. The first-class constraints should include the Virasoro constraint $T = -\frac{1}{2}\Pi^m\Pi_m - d_\alpha\partial\theta^\alpha = -\frac{1}{2}\partial x^m\partial x_m - p_\alpha\partial\theta^\alpha$, and the first-class part of the d_α constraint, which is the irreducible part of $G^\alpha = \Pi^m(\gamma_m d)^\alpha$. To get to light-cone gauge, one also needs constraints such as $C^{mnp} = d_\alpha(\gamma^{mnp})^{\alpha\beta}d_\beta$ which is supposed to replace the second-class constraints in d_α. Although this approach was successfully used for quantizing the superparticle,[11] a set of constraints which closes at the quantum level and which reproduces the correct physical superstring spectrum was never found.

Nevertheless, the approach of Siegel has the advantage that all worldsheet fields are free which makes it trivial to compute the OPE's that

$$x^m(y)x^n(z) \to -2\eta^{mn}\log|y-z|, \quad p_\alpha(y)\theta^\beta(z) \to \delta_\alpha^\beta(y-z)^{-1}.$$

This gives some useful clues about the appropriate ghost degrees of freedom. Since $(\theta^\alpha, p_\alpha)$ contributes -32 to the conformal anomaly, the total matter contribution is -22 which is expected to be cancelled by a ghost contribution of $+22$. Furthermore, the spin contribution to the $SO(9,1)$ Lorentz currents in Siegel's approach is $M_{mn} = \frac{1}{2}p\gamma_{mn}\theta$, as compared with the spin contribution to the $SO(9,1)$ Lorentz currents in the RNS formalism which is $\psi_m\psi_n$. These two Lorentz currents satisfy similar OPE's except for the numerator in the double pole of M_{mn} with M_{mn}, which is $+4$ in Siegel's approach and $+1$ in the RNS formalism. This suggests that the worldsheet ghosts should have Lorentz currents which contribute -3 to the double pole.

3. Quantization using Pure Spinors

In fact, there exists an $SO(9,1)$ irreducible representation contributing $c = 22$ and with a -3 coefficient in the double pole of its Lorentz current.[7] This representation is a complex bosonic spinor λ^α satisfying the "pure spinor" condition that

$$\lambda^\alpha \gamma^m_{\alpha\beta} \lambda^\beta = 0 \tag{5}$$

for $m = 0$ to 9. Pure spinors have previously been used by Howe[12] for describing super-Yang-Mills and supergravity equations of motion as integrability conditions, and by Nilsson[13] as superspace auxiliary fields.

There are eleven independent complex degrees of freedom in λ^α, as can be seen by temporarily breaking $SO(9,1)$ to $SO(8)$ and solving the constraint of (5). If $\gamma^\pm = \frac{1}{2}(\gamma^0 \pm \gamma^9)$ and $\sigma^j_{a\dot a}$ are the $SO(8)$ Pauli matrices satisfying $\sigma^{(j}_{a\dot a}\sigma^{k)}_{a\dot b} = 2\delta^{jk}\delta_{\dot a\dot b}$ for $(j,a,\dot a) = 1$ to 8, then $\lambda\gamma^-\lambda = 0$ implies that $s^a = (\gamma^+\lambda)^a$ is a null complex $SO(8)$ spinor satisfying $s^a s^a = 0$. Furthermore, $\lambda\gamma^j\lambda = 0$ implies that $s^a \sigma^j_{a\dot a}(\gamma^-\lambda)^{\dot a} = 0$, which implies that $(\gamma^-\lambda)^{\dot a} = v^j(\sigma^j s)^{\dot a}$ for some complex vector v^j. $\lambda\gamma^+\lambda = 0$ gives no new constraints on s^a and v^j. But this parameterization of λ^α is invariant under the gauge transformation $\delta v^j = \epsilon^{\dot a}(\sigma^j s)^{\dot a}$, which allows one to gauge away half of the components of v^j. So a pure spinor λ^α can be parameterized by the seven complex components of a null s^a and the four remaining complex components of v^j.[9] After Wick rotation to $SO(10)$, one can alternatively describe λ^α as a complex scale parameter multiplying the complex ten-dimensional coset space $SO(10)/U(5)$.[7]

Using either the $SO(8)$ or $SO(10)/U(5)$ descriptions of λ^α, a free field action S_λ can be constructed out of the eleven left-moving unconstrained parameters and their canonical momenta with conformal anomaly $c = 22$. Although the unconstrained parameters do not transform covariantly under $SO(9,1)$, the only combinations which appear in the vertex operators are λ^α and its Lorentz current N_{mn}, which satisfy the manifestly SO(9,1) covariant OPE's

$$N_{mn}(y)\lambda^\alpha(z) \to \frac{\frac{1}{2}(\gamma^{mn}\lambda)^\alpha}{y-z}, \quad N_{kl}(y)N_{mn}(z) \to \frac{\eta_{m[l}N_{k]n} - \eta_{n[l}N_{k]m}}{y-z} - 3\frac{\eta_{k[n}\eta_{m]l}}{(y-z)^2}.$$

One can alternatively obtain the above OPE's of from an SO(9,1) WZW model of level -3, however, λ^a would not be a fundamental field in such an action.[14]

One still needs to define which states are physical in this Hilbert space. The physical state condition will be defined as ghost-number one states in the cohomology of the BRST-like operator[7]

$$Q = \int dz \lambda^\alpha(z) d_\alpha(z) \tag{6}$$

where λ^α carries ghost-number one and d_α is defined in (4). Since d_α satisfies the OPE $d_\alpha(y)d_\beta(z) \to -(y-z)^{-1}\gamma^m_{\alpha\beta}\Pi_m$, the pure spinor condition on λ^α implies that Q is nilpotent. Although it is a bit unusual that Q is constructed from second-class constraints, it will be shown in section 5 that its cohomology reproduces the correct light-cone GS spectrum.

4. Physical Massless Vertex Operators

Since there is no tachyon, the massless open superstring state is constructed from zero modes of the dimension-zero worldsheet fields x^m,θ^α, and λ^α. Since it has ghost-number one, it can be written as[7]

$$U = \lambda^\alpha A_\alpha(x,\theta) \tag{7}$$

where $A_\alpha(x,\theta)$ is a spinor superfield. Since $d_\alpha(y)A_\beta(x,\theta) \to (y-z)^{-1}D_\alpha A_\beta(x,\theta)$ where $D_\alpha = \frac{\partial}{\partial\theta^\alpha} + (\gamma^m\theta)_\alpha\partial_m$ is the superspace derivative, $QU = 0$ implies that $\lambda^\alpha\lambda^\beta D_\alpha A_\beta = 0$. But since $\lambda^\alpha\lambda^\beta = \frac{1}{32}(\gamma^{mnpqr})^{\alpha\beta}(\lambda\gamma_{mnpqr}\lambda)$, this implies that $(\gamma_{mnpqr})^{\alpha\beta}D_\alpha A_\beta = 0$ for every five-form direction $mnpqr$. Furthermore, the gauge invariance $\delta U = Q\Omega$ implies that A_α has the gauge transformation $\delta A_\alpha = D_\alpha\Omega$.

The equation $(\gamma_{mnpqr})^{\alpha\beta}D_\alpha A_\beta = 0$ and gauge invariance $\delta A_\alpha = D_\alpha\Omega$ describe the spinor gauge superfield for linearized on-shell $d = 10$ super-Yang-Mills.[15] It can be gauge-fixed to the form

$$A_\alpha(x,\theta) = a_m(x)(\gamma^m\theta)_\alpha + (\xi(x)\gamma_{mnp})_\alpha(\theta\gamma^{mnp}\theta) + \ldots \tag{8}$$

where $a_m(x)$ and $\xi^\alpha(x)$ are the on-shell gluon and gluino and all the components in ... are auxiliary fields which are related to a_m and ξ^α by equations of motion.

To compute scattering amplitudes, one also needs vertex operators in integrated form, $\int dzV$, where V is usually obtained from the unintegrated vertex operator U by anti-commuting with the b ghost. But since there is no natural candidate for the b ghost in this formalism, one needs to use an alternative method for obtaining V which is from the relation $[Q,V] = \partial U$.[16] Using this alternative method, one finds for the open superstring massless vertex operator that[7]

$$V = A_\alpha(x,\theta)\partial\theta^\alpha + A_m(x,\theta)\Pi^m + W^\alpha(x,\theta)d_\alpha + F^{mn}(x,\theta)N_{mn}, \tag{9}$$

where $A_m = \frac{1}{8}\gamma_m^{\alpha\beta}D_\alpha A_\beta$, $W^\alpha = \frac{1}{10}\gamma_m^{\alpha\beta}(D_\beta A^m - \partial^m A_\beta)$, and $F^{mn} = \partial^{[m}A^{n]}$. It will be useful to note that in components,

$$V = a_m(x)\partial x^m + \partial_{[m}a_{n]}(x)M^{mn} + \xi^\alpha(x)q_\alpha + O(\theta^2), \tag{10}$$

where $M^{mn} = \frac{1}{2}p\gamma^{mn}\theta + N^{mn}$ is the spin contribution to the Lorentz current and $q_\alpha = p_\alpha + \frac{1}{2}(\partial x^m + \frac{1}{12}\theta\gamma^m\theta)(\gamma_m\theta)_\alpha$ is the spacetime-supersymmetry current, so (10) closely resembles the RNS vertex operator[2] for the gluon and gluino. If one drops the $F^{mn}N_{mn}$ term, the vertex operator of (9) was suggested by Siegel[6] based on superspace arguments.

For the closed superstring, the massless vertex operator is $U = \lambda^\alpha\widehat{\lambda}^{\widehat{\beta}}A_{\alpha\widehat{\beta}}(x,\theta,\widehat{\theta})$ where $\widehat{\lambda}^{\widehat{\alpha}}$ and $\widehat{\theta}^{\widehat{\alpha}}$ are right-moving worldsheet fields and the chirality of the $\widehat{\alpha}$ index depends if the superstring is IIA or IIB. The physical state condition $QU = \widehat{Q}U = 0$ and gauge invariance $\delta U = Q\widehat{\Omega} + \widehat{Q}\Omega$ where $\widehat{Q}\widehat{\Omega} = Q\Omega = 0$ implies that[10]

$$\gamma^{\alpha\beta}_{mnpqr}D_\alpha A_{\beta\widehat{\gamma}} = \gamma^{\widehat{\alpha}\widehat{\gamma}}_{mnpqr}\widehat{D}_{\widehat{\alpha}}A_{\beta\widehat{\gamma}} = 0, \quad \delta A_{\alpha\widehat{\beta}} = D_\alpha\widehat{\Omega}_{\widehat{\beta}} + \widehat{D}_{\widehat{\beta}}\Omega_\alpha, \tag{11}$$

$$\gamma^{\alpha\beta}_{mnpqr} D_\alpha \Omega_\beta = \gamma^{\widehat{\widehat{\alpha}\widehat{\gamma}}}_{mnpqr} \widehat{D}_{\widehat{\alpha}} \widehat{\Omega}_{\widehat{\gamma}} = 0$$

for any five-form direction $mnpqr$, which are the linearized equations of motion and gauge invariances of the Type IIA or Type IIB supergravity multiplet. The integrated form of the closed superstring massless vertex operator is the left-right product of the open superstring vertex operator of (9) and will be used in section 7 for quantizing the superstring in a curved supergravity background.

5. Tree-Level Massless Scattering Amplitudes

As usual, the N-point tree-level open superstring scattering amplitude will be defined as the correlation function of 3 unintegrated vertex operators U_r and $N-3$ integrated vertex operators $\int dz V_r$. For massless external states, these vertex operators are given in (7) and (9).

The first step to evaluate the correlation function is to eliminate all worldsheet fields of non-zero dimension (i.e. ∂x^m, $\partial\theta^\alpha$, p_α and N^{mn}) by using their OPE's with other worldsheet fields and the fact that they vanish at $z \to \infty$. One then integrates over the x^m zero modes to get a Koba-Nielson type formula,

$$A = \int dz_4 \ldots dz_N \langle \lambda^\alpha \lambda^\beta \lambda^\gamma f_{\alpha\beta\gamma}(z_r, k_r, \eta_r, \theta) \rangle \tag{12}$$

where $\lambda^\alpha\lambda^\beta\lambda^\gamma$ comes from the three unintegrated vertex operators and $f_{\alpha\beta\gamma}$ is some function of the z_r's, the momenta k_r, the polarizations η_r, and the remaining θ zero modes.

One would like to define the correlation function $\langle \lambda^\alpha\lambda^\beta\lambda^\gamma f_{\alpha\beta\gamma}\rangle$ such that A is supersymmetric and gauge invariant. An obvious way to make A supersymmetric is to require that the correlation function vanishes unless all sixteen θ zero modes are present, but this gives the wrong answer by dimensional analysis. The correct answer comes from realizing that $Y = \lambda^\alpha\lambda^\beta\lambda^\gamma f_{\alpha\beta\gamma}$ satisfies the constraint $QY = 0$ since the external states are on-shell. Furthermore, gauge invariance implies that $\langle Y \rangle$ should vanish whenever $Y = Q\Omega$.

At zero momentum and ghost-number three, there is precisely one state in the cohomology of Q which is $(\lambda\gamma^m\theta)(\lambda\gamma^n\theta)(\lambda\gamma^p\theta)(\theta\gamma_{mnp}\theta)$. So if

$$f_{\alpha\beta\gamma}(\theta) = A_{\alpha\beta\gamma} + \theta^\delta B_{\alpha\beta\gamma\delta} + \ldots + (\gamma^m\theta)_\alpha(\gamma^n\theta)_\beta(\gamma^p\theta)_\gamma(\theta\gamma_{mnp}\theta)F + \ldots, \tag{13}$$

it is natural to define

$$\langle \lambda^\alpha\lambda^\beta\lambda^\gamma f_{\alpha\beta\gamma}(z_r, k_r, \eta_r, \theta)\rangle = F(z_r, k_r, \eta_r). \tag{14}$$

This definition is supersymmetric since $(\lambda\gamma^m\theta)(\lambda\gamma^n\theta)(\lambda\gamma^p\theta)(\theta\gamma_{mnp}\theta)\theta^\alpha$ is not annihilated by Q, and is gauge invariant since $(\lambda\gamma^m\theta)(\lambda\gamma^n\theta)(\lambda\gamma^p\theta)(\theta\gamma_{mnp}\theta) \neq Q\Omega$. Eq. (14) can be interpreted as integration over an on-shell harmonic superspace involving five θ's since $\langle\lambda^\alpha\lambda^\beta\lambda^\gamma f_{\alpha\beta\gamma}\rangle = \int (d\theta\gamma_m)^\alpha (d\theta\gamma_n)^\beta (d\theta\gamma_p)^\gamma (d\theta\gamma^{mnp} d\theta) f_{\alpha\beta\gamma}$.[7]

For three-point scattering, $A = \langle \lambda^a A^1_\alpha \lambda^\beta A^2_\beta \lambda^\gamma A^3_\gamma\rangle$, it is easy to check that the prescription of (14) reproduces the usual super-Yang-Mills cubic vertex. In the gauge of (8), each A_α contributes one, two or three θ's. If the five θ's are distributed as $(1,1,3)$, one gets the $a^1_m a^2_n \partial^{[m} a^{3n]}$ vertex, whereas if they are distributed as $(2,2,1)$, one gets the $(\xi^1\gamma^m\xi^2)a^3_m$ vertex. Together with Brenno Vallilo, it was proven that the above prescription agrees with the standard RNS prescription of reference [2] for N-point massless tree amplitudes involving up to four fermions.[8] The relation of (10) to the RNS massless vertex operator was used in this proof, and the restriction on the number of fermions comes from the need for different pictures in the RNS prescription.

6. Cohomology of Q

To compute the cohomology of $Q = \int dz \lambda^\alpha d_\alpha$, it is convenient to use the $SO(8)$ parameterization of λ^α in terms of s^a and v^j discussed in section 3. The gauge invariance $\delta v^j = \epsilon^{\dot a}(\sigma^j s)^{\dot a}$ leads to a new fermionic ghost anti-chiral spinor $t^{\dot a}$, and the invariance of the gauge parameter, $\delta\epsilon^{\dot a} = y^j(\sigma^j s)^{\dot a}$, leads to a new bosonic ghost-for-ghost vector $v^j_{(1)}$. These gauge invariances continue indefinitely until one has an infinite chain of bosonic vectors $v^j_{(n)}$ and fermionic anti-chiral spinors $t^j_{(n)}$ for $n=0$ to ∞ where $n=0$ labels the original vector and spinor.[9]

In terms of these left-moving dimension-zero worldsheet fields, $Q = \int dz s^a G^a$ where

$$G^a = (\gamma^- d)^a + \sigma^{a\dot a}_j [v^j_{(0)}(\gamma^+ d)^{\dot a} + \sum_{n=0}^{\infty}(w^j_{(n)} t^{\dot a}_{(n)} + v^j_{(n+1)} u^{\dot a}_{(n)})],$$

$(w^j_{(n)}, u^{\dot a}_{(n)})$ are the canonical momenta for $(v^j_{(n)}, t^{\dot a}_{(n)})$, and the infinite sum comes from the gauge invariances. Note that $Q^2 = 0$ since $s^a s^a = 0$ and $G^a(y)G^b(z) \to 2(y-z)^{-1}\delta^{ab}T(z)$ where

$$T = \frac{1}{2}\Pi^- + v^j\Pi^j + \frac{1}{2}v^j v^j \Pi^+ + t^{\dot a}_{(0)}(\gamma^+ d)^{\dot a} + \sum_{n=0}^{\infty}(v^j_{(n+1)} w^j_{(n)} + t^{\dot a}_{(n+1)} u^{\dot a}_{(n)}).$$

One can treat $s^a s^a = 0$ as a BRST constraint by defining $Q' = \int dz[s^a G^a - b s^a s^a + cT]$ where (b,c) are fermionic ghosts of dimension $(1,0)$. One can check that Q' is nilpotent with unconstrained s^a, and has cohomology equal to that of Q. Note that the algebra of G^a and T is not the usual $N=8$ superconformal algebra since, for example, T has dimension 1 and commutes with G^a. Nevertheless, since T and G^a are first-class constraints, they can be used to gauge-fix $x^+ = P^+\tau$ and $(\gamma^+\theta)^a = 0$, and solve for x^- and $(\gamma^- p)^a$.

The only remaining constraint is that physical operators must commute with the zero mode of T, which in this light-cone gauge is $T_0 + \frac{1}{2}P^-$ where $T_0 =$

$$\int dz[-\frac{1}{2}\theta\gamma^-\partial\theta + v^j_{(0)}\partial x^j + \frac{1}{2}v^j_{(0)}v^j_{(0)}P^+ + t^{\dot a}_{(0)}(\gamma^+ d)^{\dot a} + \sum_{n=0}^{\infty}(v^j_{(n+1)} w^j_{(n)} + t^{\dot a}_{(n+1)} u^{\dot a}_{(n)})]$$

and a Lorentz frame has been chosen in which $P^j = 0$ for $j = 1$ to 8. So physical operators are constructed from products of eigenvectors of T_0 whose $(mass)^2 =$

P^+P^- is equal to the sum of the eigenvalues multiplied by $-2P^+$. Since T_0 is quadratic, the bosonic and fermionic eigenvectors with eigenvalues N, a_N and b_N, can be expressed as linear combinations of the remaining light-cone variables,

$$a_N = \int d\sigma[f_N^j \partial x^j + \sum_{n=0}^{\infty}(g_{N(n)}^j v_{(n)}^j + h_{N(n)}^j w_{(n)}^j)],$$

$$b_N = \int d\sigma[j_N^{\dot a}(\gamma^+ p)^{\dot a} + k_N^{\dot a}(\gamma^- \theta)^{\dot a} + \sum_{n=0}^{\infty}(l_{N(n)}^{\dot a} t_{(n)}^{\dot a} + m_{N(n)}^{\dot a} u_{(n)}^{\dot a})].$$

If one requires that these eigenvectors satisfy the normalization condition that $\int d\sigma(f_N^j f_N^j + \sum_{n=0}^{\infty} g_{N(n)}^j h_{N(n)}^j)$ and $\int d\sigma(j_N^{\dot a} k_N^{\dot a} + \sum_{n=0}^{\infty} l_{N(n)}^{\dot a} m_{N(n)}^{\dot a})$ are finite, one finds that the only normalizable eigenvectors are the modes of

$$y^j = \partial x^j + \sum_{n=0}^{\infty}(P^+)^{-n-1}\partial^{n+1} w_{(n)}^j, \quad q^{\dot a} = (\gamma^+ p - \frac{1}{2}P^+\gamma^-\theta)^{\dot a} + \sum_{n=0}^{\infty}(P^+)^{-n-1}\partial^{n+1} u_{(n)}^{\dot a}.$$

These eight bosonic and eight fermionic eigenvectors generate the usual light-cone GS spectrum. So the T_0 constraint has imposed the second-class constraints as well as the mass-shell condition. It is interesting to note that an infinite set of fields has also been useful for treating the second-class constraints of the chiral boson[17] and the self-dual Type IIB four-form.[18]

7. Superstring in $AdS_5 \times S^5$ Background

To construct the superstring action in a curved Type II supergravity background, one adds the integrated form of the closed massless vertex operator of section 4 to the superstring action in a flat background, and covariantizes with respect to super-reparameterization invariance. As usual, one also needs to include a Fradkin-Tseytlin term for coupling the dilaton to the worldsheet curvature r. The resulting action is[7,19]

$$S = \int d^2 z[\frac{1}{2}\partial Y^M \bar\partial Y^N (G_{MN}(Y) + B_{MN}(Y)) \tag{15}$$

$$+\bar\partial Y^M \widetilde{d}_\alpha E_M^\alpha(Y) + \partial Y^M \widetilde{\widehat{d}}_{\widehat\alpha} E_M^{\widehat\alpha}(Y) + \widetilde{d}_\alpha \widetilde{\widehat{d}}_{\widehat\beta} F^{\alpha\widehat\beta}(Y) + \alpha' r\Phi(Y)] + S_\lambda + S_{\widehat\lambda}$$

where $Y^M = (x^m, \theta^\mu, \widehat{\theta^\mu})$ parameterizes the curved N=2 superspace background, the first line of (15) is the usual GS action, $\widetilde{d}_\alpha = d_\alpha + N_{mn}(\gamma^{mn}D)_\alpha$ and $\widetilde{\widehat{d}}_{\widehat\alpha} = \widehat{d}_{\widehat\alpha} + \widehat{N}_{mn}(\gamma^{mn}\widehat{D})_{\widehat\alpha}$ where D_α and $\widehat{D}_{\widehat\alpha}$ are $N = 2$ superspace derivatives which act on the background fields to their right, S_λ and $S_{\widehat\lambda}$ are the actions in a flat background for λ^α and $\widehat{\lambda}^{\widehat\alpha}$, and the lowest components of the superfields E_m^α and $E_m^{\widehat\alpha}$ are the gravitini, of $F^{\alpha\widehat\beta}$ are the Ramond-Ramond field strengths, and of Φ is the dilaton. It is convenient to treat d_α and $\widehat{d}_{\widehat\alpha}$ (instead of p_α and $\widehat{p}_{\widehat\alpha}$) as fundamental worldsheet fields when the background is curved. Although the above action does not have κ-symmetry, it is constructed such that $\lambda^\alpha d_\alpha$ and $\widehat{\lambda}^{\widehat\alpha}\widehat{d}_{\widehat\alpha}$ are holomorphic and anti-holomorphic currents when the background superfields are on-shell. One can therefore quantize the action using $Q = \int dz \lambda^\alpha d_\alpha$ and $\widehat{Q} = \int d\bar{z} \widehat{\lambda}^{\widehat{\alpha}} \widehat{d}_{\widehat{\alpha}}$ as in the flat case.

In the $AdS_5 \times S^5$ background with n units of Ramond-Ramond flux, $F^{\alpha\widehat{\beta}} = n\gamma_{01234}^{\alpha\widehat{\beta}} = n\delta^{\alpha\widehat{\beta}}$, so the term $n\delta^{\alpha\widehat{\beta}} d_\alpha \widehat{d}_{\widehat{\beta}}$ in (15) allows one to solve the equations of motion[20] for d_α and $\widehat{d}_{\widehat{\alpha}}$ in terms of $(x^m, \theta^\alpha, \widehat{\theta}^{\widehat{\alpha}})$. Plugging in the appropriate values for the $AdS_5 \times S^5$ background superfields, one obtains[7,21,22]

$$S = \frac{1}{n^2 g_s^2} \int d^2 z [\frac{1}{2}\eta_{cd} J^c \bar{J}^d + \frac{1}{4}\delta_{\alpha\widehat{\beta}}(3 J^{\widehat{\beta}} \bar{J}^\alpha - J^\alpha \bar{J}^{\widehat{\beta}}) + N_{cd} J^{cd} + \widehat{N}_{cd} \bar{J}^{cd}] + S_\lambda + S_{\widehat{\lambda}}$$

where $J^A = (g^{-1}\partial g)^A$ and $\bar{J}^A = (g^{-1}\bar{\partial} g)^A$ with $A = (a, \alpha, \widehat{\alpha}, cd)$ are left-invariant currents constructed from the coset supergroup $g(x, \theta, \widehat{\theta}) \in \frac{PSU(2,2|4)}{SO(4,1)\times SO(5)}$. This action is invariant under $g \to M g \Omega$ where M is a global $PSU(2,2|4)$ transformation and Ω is a local $SO(4,1) \times SO(5)$ transformation which also acts as a Lorentz rotation on the pure spinors λ^α and $\widehat{\lambda}^{\widehat{\alpha}}$. It has been checked to one-loop order[21,14] that S is conformally invariant and that $\lambda^\alpha d_\alpha = \delta_{\alpha\widehat{\beta}} \lambda^\alpha J^{\widehat{\beta}}$ and $\widehat{\lambda}^{\widehat{\alpha}} \widehat{d}_{\widehat{\alpha}} = \delta_{\alpha\widehat{\beta}} \widehat{\lambda}^{\widehat{\beta}} \bar{J}^\alpha$ are holomorphic and anti-holomorphic currents.

To construct the vertex operator $U = \lambda^\alpha \widehat{\lambda}^{\widehat{\beta}} A_{\alpha\widehat{\beta}}$ for fluctuations around the $AdS_5 \times S^5$ background, one needs to generalize the equations of motion and gauge invariances of (11) as was done in Ref. 23 for the $AdS_3 \times S^3$ background. The flat space equations of (11) can be generalized to $AdS_5 \times S^5$ by simply replacing D_α and $\widehat{D}_{\widehat{\alpha}}$ in (11) with $\nabla_\alpha = E_\alpha^M (\partial_M + \omega_M)$ and $\widehat{\nabla}_{\widehat{\alpha}} = E_{\widehat{\alpha}}^M (\partial_M + \omega_M)$ where $\omega_M^{cd} = (g^{-1}\partial_M g)^{cd}$ is the $SO(4,1) \times SO(5)$ spin connection and E_A^M is the super-vierbein obtained by inverting $E_M^A = (g^{-1}\partial_M g)^A$ for $A = (a, \alpha, \widehat{\alpha})$. The equations of motion and gauge invariances are still self-consistent since, although $\{\nabla_\alpha, \widehat{\nabla}_{\widehat{\beta}}\}$ is non-zero in the $AdS_5 \times S^5$ background, one can check that $\gamma_{mnpqr}^{\alpha\gamma} \{\nabla_\alpha, \widehat{\nabla}_{\widehat{\beta}}\} f_\gamma = 0$ and $\gamma_{mnpqr}^{\widehat{\beta}\widehat{\gamma}} \{\nabla_\alpha, \widehat{\nabla}_{\widehat{\beta}}\} \widehat{f}_{\widehat{\gamma}} = 0$ for any f_γ and $\widehat{f}_{\widehat{\gamma}}$ and for any five-form direction $mnpqr$.[10]

8. Open Problems and Applications

In this review, it was argued that pure spinors are the worldsheet ghosts of the GS formalism and that physical states are described by ghost-number one vertex operators in the cohomology of $Q = \int dz \lambda^\alpha d_\alpha$. Although these ideas were successfully used for computing N-point tree amplitudes, a more geometrical understanding would be useful for computing loop amplitudes where the ghosts play a more important role. In particular, one would like to understand how to formulate the action in a reparameterization-invariant manner and how the ghosts arise from the gauge-fixing procedure.

The relatively simple form of the vertex operator for fluctuations around the $AdS_5 \times S^5$ background suggests that it might be possible to compute tree amplitudes in this background. Although the currents J^A are not holomorphic since $\bar{\partial} J^A =$

$h^A_{BC} J^B \bar{J}^C$ where h^A_{BC} are constants,[24,20] conformal invariance together with the AdS isometries may be enough to imply their OPE's. Of course, even knowing these OPE's, one would still have to compute the zero mode contribution from $(x^m, \theta^\mu, \widehat{\theta^\mu})$ to the tree amplitude, which might be complicated in an $AdS_5 \times S^5$ background.[25]

Acknowledgments

I would like to thank Michael Bershadsky, Osvaldo Chandía, Warren Siegel, Brenno Vallilo, and Edward Witten for useful conversations, and CNPq grant 300256/94-9 and the Clay Mathematics Institute for partial financial support. This research was partially conducted during the period the author was employed by the Clay Mathematics Institute as a CMI Prize Fellow.

References

1. M.B. Green and J.H. Schwarz, *Supersymmetrical Dual String Theory*, *Nucl. Phys.* **B181**, 502 (1981).
2. D. Friedan, E. Martinec and S. Shenker, *Conformal Invariance, Supersymmetry and String Theory*, *Nucl. Phys.* **B271**, 93 (1986).
3. M.B. Green and J.H. Schwarz, *Covariant Description of Superstrings*, *Phys. Lett.* **B136**, 367 (1984).
4. N. Berkovits, *Covariant Quantization Of The Green-Schwarz Superstring in a Calabi-Yau Background*, *Nucl. Phys.* **B431**, 258 (1994) [hep-th/9404162].
5. N. Berkovits and C. Vafa, *N=4 Topological Strings*, *Nucl. Phys.* **B433**, 123 (1995) [hep-th/9407190].
6. W. Siegel, *Classical Superstring Mechanics*, *Nucl. Phys.* **B263**, 93 (1986).
7. N. Berkovits, *Super-Poincaré Covariant Quantization of the Superstring*, *JHEP* **0004**, 018 (2000) [hep-th/0001035].
8. N. Berkovits and B.C. Vallilo, *Consistency of Super-Poincaré Covariant Superstring Tree Amplitudes*, *JHEP* **0007**, 015 (2000) [hep-th/0004171].
9. N. Berkovits, *Cohomology in the Pure Spinor Formalism for the Superstring*, *JHEP* **0009**, 046 (2000) [hep-th/0006003].
10. N. Berkovits and Osvaldo Chandía, to appear.
11. F. Essler, M. Hatsuda, T. Kimura, E. Laenen, A. Mikovic, W. Siegel and J. Yamron, *Covariant Quantization of the First Ilk Superparticle*, *Nucl. Phys.* **B364**, 67 (1991).
12. P.S. Howe, *Pure Spinor Lines in Superspace and Ten-Dimensional Supersymmetric Theories*, *Phys. Lett.* **B258**, 141 (1991), Addendum *ibid* **B259**, 511 (1991); P.S. Howe, *Pure Spinors, Function Superspaces and Supergravity Theories in Ten Dimensions and Eleven Dimensions*, *Phys. Lett.* **B273**, 90 (1991).
13. B.E.W. Nilsson, *Pure Spinors as Auxiliary Fields in the Ten-Dimensional Supersymmetric Yang-Mills Theory*, *Class. Quant. Grav.* **3**, L41 (1986).
14. M. Bershadsky, private communication.
15. W. Siegel, *Superfields in Higher-Dimensional Spacetime*, *Phys. Lett.* **80B**, 220 (1979).
16. N. Berkovits, M. Hatsuda and W. Siegel, *The Big Picture*, *Nucl. Phys.* **B371**, 434 (1992) [hep-th/9108021].
17. B. McClain, Y.S. Wu and F. Yu, *Covariant Quantization of Chiral Bosons and OSp(1,1/2) Symmetry*, *Nucl. Phys.* **B343**, 689 (1990); C. Wotzasek, *The Wess-Zumino Term for Chiral Bosons*, *Phys. Rev. Lett.* **66**, 129 (1991).

18. N. Berkovits, *Manifest Electromagnetic Duality in Closed Superstring Field Theory, Phys. Lett.* **B388**, 743 (1996) [hep-th/9607070].
19. N. Berkovits and W. Siegel, *Superspace Effective Actions for 4D Compactifications of Heterotic and Type II Superstrings, Nucl. Phys.* **B462**, 213 (1996) [hep-th/9510106].
20. N. Berkovits, C. Vafa and E. Witten, *Conformal Field Theory of AdS Background with Ramond-Ramond Flux, JHEP* **9903**, 018 (1999) [hep-th/9902098].
21. N. Berkovits, M. Bershadsky, T. Hauer, S. Zhukov and B. Zwiebach, *Superstring Theory on $AdS_2 \times S^2$ as a Coset Supermanifold, Nucl. Phys.* **B567**, 61 (2000) [hep-th/9907200].
22. R. Metsaev and A. Tseytlin, *Type IIB superstring action in $AdS_5 \times S^5$ background, Nucl. Phys.* **B533**, 109 (1998) [hep-th/9805028].
23. L. Dolan and E. Witten, *Vertex Operators for $AdS_3 \times S^3$ Background with Ramond-Ramond Flux, JHEP* **9911**, 003 (1999) [hep-th/9910205].
24. M. Bershadsky, S. Zhukov and A. Zaintrob, *$PSL(n|n)$ Sigma Model as a Conformal Field Theory, Nucl. Phys.* **B559**, 205 (1999) [hep-th/9902180].
25. E. Witten, private communication.

VERTEX OPERATORS FOR AdS_3 WITH RAMOND BACKGROUND

LOUISE DOLAN
Department of Physics and Astronomy, University of North Carolina Chapel Hill, NC 27599-3255, USA

The computation of exact vertex operators for the Type IIB superstring in an $AdS_3 \times S^3$ background with Ramond-Ramond flux is described. The components of these vertex operators are shown to satisfy the supergravity linearized equations of motion for the six-dimensional (2,0) theory of a supergravity and tensor multiplet expanded around $AdS_3 \times S^3$ spacetime.

1. Introduction

The conjectured duality between M-theory or Type IIB string theory on anti-de Sitter (AdS) space and the conformal field theory on the boundary of AdS space may be useful in giving a controlled systematic approximation for strongly coupled gauge theories. Examples with maximal supersymmetry correspond to a set of p-branes whose near horizon geometry looks like $AdS_{p+2} \times S^{\mathcal{D}-p-2}$, where $\mathcal{D} = 10$ or 11 for branes in string or M-theory, and $p = 2, 3, 5$. The formulation of vertex operators and string theory tree amplitudes for the IIB superstring on $AdS_5 \times S^5$ will allow access to the dual conformal $SU(N)$ gauge field theory CFT_4 at large N, but *small* fixed 't Hooft coupling $x = g_{YM}^2 N$ in the dual picture, as $(g_{YM}^2 N)^{\frac{1}{2}}(4\pi)^{\frac{1}{2}} = R_{sph}^2/\alpha'$. Presently only the large N, and *large* fixed 't Hooft coupling x limit is accessible in the CFT, since only the supergravity limit ($\alpha' \to 0$) of the correlation functions of the AdS theory is known.

In this talk we discuss a case with non-maximal supersymmetry that is related to a system with a $D1$-brane and a $D5$-brane. Its quantizable worldsheet action[1] describes the IIB string on $AdS_3 \times S^3 \times M$ with background Ramond-Ramond flux, where M is T^4 or $K3$. The vertex operators for this model can be computed explicitly in the bulk.[2] Correlation functions constructed from these vertex operators, restricted to the boundary of AdS_3, would be those for a two-dimensional spacetime conformal field theory. We work to leading order in α', but because of the high degree of symmetry of the model, we expect our result for the vertex operators to be exact. Tree level n-point correlation functions for $n \geq 4$ presumably have α' corrections, since the worldsheet theory is not a free conformal field theory. But there may be sufficiently many symmetry currents to determine the tree level correlation functions exactly in α' as well.[3]

In terms of Berkovits-Vafa-Witten worldsheet variables, constraint equations for the vertex operators in the flat space $\mathcal{R}^6$ follow from the physical state conditions coming from an $N = 4$ superconformal algebra. We generalize[2] these constraint equations to $AdS_3 \times S^3$ for the vertex operators of the massless states that are independent of the compactification M. We then solve the constraints and identify the components of the vertex operators as supergravity fields, that satisfy the $D = 6$, $N = (2,0)$ theory[4] linearized around the $AdS_3 \times S^3$ background.

Recent work[5,6] by Berkovits discusses covariant ten-dimensional worldsheet variables and extends our analysis to vertex operators on AdS_5 in a spinor formulation.

2. Formulating Strings on AdS

In the Ramond-Neveu-Schwarz (RNS) formalism, the worldsheet action for strings on AdS space with background Ramond-Ramond flux involves 2d spin fields. These violate superconformal worldsheet symmetry, since the worldsheet supercurrents are not local with respect to the spin fields, and their presence makes the worldsheet theory difficult to understand.

For the Type IIB superstring on $AdS_3 \times S^3$ case, a sigma model[1] with conventional local interactions (no spin fields in the action) was found using the supergroup $PSU(2|2)$ as target, coupled to ghost fields ρ and σ. The spacetime symmetry group is $PSU(2|2) \times PSU(2|2)$, acting by left and right multiplication on the group manifold, *i.e.* by $g \to agb^{-1}$ where g is a $PSU(2|2)$-valued field, and $a, b \in PSU(2|2)$ are the symmetry group's Lie algebra elements. The supergroup is generated by the super Lie algebra with 12 bosonic generators forming a subalgebra $SO(4) \times SO(4)$ together with 16 odd generators. Hence our model has non-maximal supersymmetry with 16 supercharges.

The worldsheet field content generalizes the Berkovits-Vafa formalism which provides a manifest Lorentz covariant and supersymmetric quantization on R^6. Its six bosonic fields $x^p(z,\bar{z})$ contain both left- and right-moving modes. In addition there are left-moving fermi fields $\theta^a_L(z), p^a_L(z)$ of spins 0 and 1, together with ghosts $\sigma_L(z), \rho_L(z)$, and right-moving counterparts of all these left-moving fields. These variables allow Ramond-Ramond background fields to be incorporated without adding spin fields to the worldsheet action as follows: in the $AdS_3 \times S^3$ case, i.e. after adding RR background fields to the worldsheet action, one can integrate out the p's, so that the model has ordinary conformal fields x^p, θ^a, $\bar{\theta}^a$ (all now with both left- and right-moving components) as well as the ghosts. The $PSU(2|2)$-valued field g is given in terms of x, θ, and $\bar{\theta}$, which are identified as coordinates on the supergroup manifold. In addition, Type IIB on $AdS_3 \times S^3 \times M$ has worldsheet variables describing the compactification degrees of freedom on the four-dimensional space M. Their Virasoro currents have central charge $c = 6$, and will be labeled with a subscript C.

3. $N = 4$ Super Virasoro Generators

The holomorphic $N = 4$ superconformal generators with $c = 6$ are given for flat space by[1]

$$
\begin{aligned}
T &= -\tfrac{1}{2}\partial x^m \partial x_m - p_a \partial\theta^a - \tfrac{1}{2}\partial\rho\partial\rho - \tfrac{1}{2}\partial\sigma\partial\sigma + \partial^2(\rho + i\sigma) + T_C \\
G^+ &= -e^{-2\rho-i\sigma}(p)^4 + \tfrac{i}{2}e^{-\rho}p_a p_b \partial x^{ab} \\
&\quad + e^{i\sigma}(-\tfrac{1}{2}\partial x^m \partial x_m - p_a\partial\theta^a - \tfrac{1}{2}\partial(\rho+i\sigma)\partial(\rho+i\sigma) + \tfrac{1}{2}\partial^2(\rho+i\sigma)) + G_C^+ \\
G^- &= e^{-i\sigma} + G_C^- \\
J &= \partial(\rho + i\sigma) + J_C \\
\tilde{G}^+ &= e^{iH_C}\,[-e^{-3\rho-2i\sigma}(p)^4 + \tfrac{i}{2}e^{-2\rho-i\sigma}p_a p_b \partial x^{ab} \\
&\quad + e^{-\rho}(-\tfrac{1}{2}\partial x^m\partial x_m - p_a\partial\theta^a - \tfrac{1}{2}\partial(\rho+i\sigma)\partial(\rho+i\sigma) \\
&\quad + \tfrac{1}{2}\partial^2(\rho+i\sigma)] + e^{-\rho-i\sigma}\tilde{G}_C^- \\
J^+ &= e^{\rho+i\sigma}J_C^+ \\
J^- &= e^{-\rho-i\sigma}J_C^- \,.
\end{aligned}
\tag{1}
$$

These currents are given in terms of the left-moving bosons $\partial x^m, \rho, \sigma$, and the left-moving fermionic worldsheet fields p^a, θ^a, where $1 \le m \le 6, 1 \le a \le 4$. There are corresponding anti-holomorphic expressions. Both sets of generators are used to implement the physical state conditions on the vertex operators, a procedure which results in a set of string constraint equations.

4. String Constraint Equations for the Vertex Operators

The expansion of the massless vertex operator in terms of the worldsheet fields is

$$
V = \sum_{m,n=-\infty}^{\infty} e^{m(i\sigma+\rho)+n(i\bar{\sigma}+\bar{\rho})}\, V_{m,n}(x,\theta,\bar{\theta})\,. \tag{2}
$$

In flat space, the constraints from the left and right-moving worldsheet super Virasoro algebras are:

$$
\begin{aligned}
(\nabla)^4 V_{1,n} &= \nabla_a \nabla_b \partial^{ab} V_{1,n} = 0 \\
\tfrac{1}{6}\epsilon^{abcd}\,\nabla_b\nabla_c\nabla_d V_{1,n} &= -i\nabla_b\,\partial^{ab}V_{0,n} \\
\nabla_a\nabla_b V_{0,n} - \tfrac{i}{2}\epsilon_{abcd}\,\partial^{cd}\,V_{-1,n} &= 0\,, \qquad \nabla_a V_{-1,n} = 0\,;
\end{aligned}
\tag{3}
$$

$$
\begin{aligned}
\bar{\nabla}^4 V_{n,1} &= \bar{\nabla}_{\bar{a}}\bar{\nabla}_{\bar{b}}\bar{\partial}^{\bar{a}\bar{b}}V_{n,1} = 0 \\
\tfrac{1}{6}\epsilon^{\bar{a}\bar{b}\bar{c}\bar{d}}\bar{\nabla}_{\bar{b}}\bar{\nabla}_{\bar{c}}\bar{\nabla}_{\bar{d}}V_{n,1} &= -i\bar{\nabla}_{\bar{b}}\bar{\partial}^{\bar{a}\bar{b}}V_{n,0} \\
\bar{\nabla}_{\bar{a}}\bar{\nabla}_{\bar{b}}V_{n,0} - \tfrac{i}{2}\bar{\epsilon}_{\bar{a}\bar{b}\bar{c}\bar{d}}\,\bar{\partial}^{\bar{c}\bar{d}}\,V_{n,-1} &= 0\,, \qquad \bar{\nabla}_{\bar{a}}\,V_{n,-1} = 0
\end{aligned}
\tag{4}
$$

$$
\partial^p\partial_p V_{m,n} = 0 \tag{5}
$$

for $-1 \le m,n \le 1$, with the notation $\nabla_a = d/d\theta^a$, $\bar{\nabla}_{\bar{a}} = d/d\bar{\theta}^{\bar{a}}$, $\partial^{ab} = -\sigma^{pab}\,\partial_p$. These equations were derived in flat space by requiring the vertex operators to satisfy the physical state conditions

$$
G_0^- V = \tilde{G}_0^- V = \bar{G}_0^- V = \bar{\tilde{G}}_0^- V = T_0 V = \bar{T}_0 V = 0,
$$

$$J_0 V = \bar{J}_0 V = 0\,, \quad G_0^+ \tilde{G}_0^+ V = \bar{G}_0^+ \bar{\tilde{G}}_0^+ V = 0 \tag{6}$$

where T_n, $G_n^\pm$, $\tilde{G}_n^\pm$, $J_n, J_n^\pm$ and corresponding barred generators are the left and right $N = 4$ worldsheet superconformal generators. These conditions further imply $V_{m,n} = 0$ for $m > 1$ or $n > 1$ or $m < 1$ or $n < 1$, leaving nine non-zero components.

In $AdS_3 \times S^3$ space, we generalize these equations as follows:

$$\begin{aligned} F^4 V_{1,n} &= F_a\, F_b K^{ab} V_{1,n} = 0 \\ \tfrac{1}{6}\epsilon^{abcd}\, F_b\, F_c\, F_d V_{1,n} &= -iF_b\, K^{ab} V_{0,n} + 2iF^a V_{0,n} - E^a V_{-1,n} \\ F_a\, F_b\, V_{0,n} - \tfrac{i}{2}\epsilon_{abcd}\, K^{cd}\, V_{-1,n} &= 0\,, \qquad F_a\, V_{-1,n} = 0\,; \end{aligned} \tag{7}$$

$$\begin{aligned} \bar{F}^4 V_{n,1} &= \bar{F}_{\bar{a}} \bar{F}_{\bar{b}} \bar{K}^{\bar{a}\bar{b}} V_{n,1} = 0 \\ \tfrac{1}{6}\epsilon^{\bar{a}\bar{b}\bar{c}\bar{d}} \bar{F}_{\bar{b}} \bar{F}_{\bar{c}} \bar{F}_{\bar{d}} V_{n,1} &= -i\bar{F}_{\bar{b}} \bar{K}^{\bar{a}\bar{b}} V_{n,0} + 2i\bar{F}^{\bar{a}} V_{n,0} - \bar{E}^{\bar{a}} V_{n,-1} \\ \bar{F}_{\bar{a}} \bar{F}_{\bar{b}} V_{n,0} - \tfrac{i}{2}\bar{\epsilon}_{\bar{a}\bar{b}\bar{c}\bar{d}}\, \bar{K}^{\bar{c}\bar{d}}\, V_{n,-1} &= 0\,, \qquad \bar{F}_{\bar{a}}\, V_{n,-1} = 0. \end{aligned} \tag{8}$$

There is also a spin zero condition constructed from the Laplacian

$$(\, F_a\, E_a\, + \tfrac{1}{8}\epsilon_{abcd}\, K^{ab}\, K^{cd}\,)\, V_{n,m} = (\, \bar{F}_{\bar{a}}\, \bar{E}_{\bar{a}}\, + \tfrac{1}{8}\bar{\epsilon}_{\bar{a}\bar{b}\bar{c}\bar{d}}\, \bar{K}^{\bar{a}\bar{b}}\, K^{\bar{c}\bar{d}}\,)\, V_{n,m} = 0\,. \tag{9}$$

We derived[2] the curved space equations (7-9) by deforming the equations for the flat case (3-5), by requiring invariance under the $PSU(2|2)$ transformations (10) that replace the $D = 6$ super Poincare transformations of flat space. The Lie algebra of the supergroup $PSU(2|2)$ contains six even elements $K_{ab} \in SO(4)$ and eight odd ones E_a, F_a. They generate the infinitesimal symmetry transformations of the constraint equations:

$$\begin{aligned} \Delta_a^- V_{m,n} = F_a V_{m,n}\,, \quad \Delta_{ab} V_{m,n} = K_{ab} V_{m,n}\,, \quad \Delta_a^+ V_{1,n} = E_a V_{1,n}\,, \\ \Delta_a^+ V_{0,n} = E_a V_{0,n} + iF_a V_{1,n}\,, \quad \Delta_a^+ V_{-1,n} = E_a V_{-1,n} - iF_a V_{0,n}\,. \end{aligned} \tag{10}$$

We write E_a, F_a, and K_{ab} for the operators that represent the left action of e_a, f_a, and t_{ab} on g. In the above coordinates,

$$\begin{aligned} F_a &= \frac{d}{d\theta^a}\,, \qquad K_{ab} = -\theta_a \frac{d}{d\theta^b} + \theta_b \frac{d}{d\theta^a} + t_{Lab} \\ E_a &= \tfrac{1}{2}\epsilon_{abcd}\, \theta^b\, (t_L^{cd} - \theta^c \frac{d}{d\theta_d}) + h_{a\bar{b}} \frac{d}{d\bar{\theta}_{\bar{b}}}\,, \end{aligned} \tag{11}$$

where we have introduced an operator t_L that generates the left action of $SU(2) \times SU(2)$ on h alone, without acting on the θ's. Here

$$g = g(x, \theta\,, \bar{\theta}) \;= e^{\theta^a f_a}\, e^{\frac{1}{2}\sigma^{pcd} x_p t_{cd}}\, e^{\bar{\theta}^{\bar{a}} e_{\bar{a}}} = e^{\theta^a f_a}\, h(x)\, e^{\bar{\theta}^{\bar{a}} e_{\bar{a}}}\,, \tag{12}$$

$$t_{Lab}\, g = e^{\theta^a f_a}\, (-t_{ab})\, h(x)\, e^{\bar{\theta}^{\bar{a}} e_{\bar{a}}}\,, \tag{13}$$

and we found (11) by requiring $F_a g = f_a g$, $E_a g = e_a g$, $K_{ab} g = -t_{ab} g$. Similar expressions[2] hold for the right-acting generators $\bar{K}_{\bar{a}\bar{b}}$, $\bar{E}_{\bar{a}}$, and $\bar{F}_{\bar{a}}$.

5. Supersymmetry Algebras

In flat space, the $D = 6$ supersymmetry algebra for the left-movers is given by

$$\begin{aligned} \{q_a^+, q_c^-\} &= \tfrac{1}{2}\epsilon_{abcd}P^{cd} \\ [P_{ab}, P_{cd}] &= 0 = [P_{ab}, q_c^{\pm}] = \{q_a^+, q_b^+\} = \{q_a^-, q_b^-\} \end{aligned} \tag{14}$$

where $P_{ab} \equiv \delta_{ac}\delta_{bd}P^{cd}$ and

$$\begin{aligned} q_a^- &= \oint F_a(z) \\ q_a^+ &= \oint (e^{-\rho - i\sigma} F_a(z) + iE_a(z)) \\ P^{ab} &= \oint \partial x_m(z) \sigma^{mab}\,. \end{aligned} \tag{15}$$

$$\tag{16}$$

In flat space we have $F_a(z) = p_a(z)$ and $E_a(z) = \frac{1}{2}\epsilon_{abcd}\theta^b(z)\partial x_m(z)\sigma^{mcd}$. We distinguish between the currents and their zero moments E_a, F_a which together with P_{ab} also generate the flat space supersymmetry algebra

$$\begin{aligned} [P_{ab}, P_{cd}] &= 0 = [P_{ab}, F_c] = [P_{ab}, E_c]\,, \\ \{E_a, F_b\} &= \tfrac{1}{2}\epsilon_{abcd}P^{cd}\,, \qquad \{E_a, E_b\} = \{F_a, F_b\} = 0\,. \end{aligned} \tag{17}$$

On $AdS_3 \times S^3$, the Poincare supersymmetry algebra (17) is replaced by the $PSU(2|2)$ superalgebra

$$\begin{aligned} [K_{ab}, K_{cd}] &= \delta_{ac}K_{bd} - \delta_{ad}K_{bc} - \delta_{bc}K_{ad} + \delta_{bd}K_{ac} \\ [K_{ab}, E_c] &= \delta_{ac}E_b - \delta_{bc}E_a\,, \qquad [K_{ab}, F_c] = \delta_{ac}F_b - \delta_{bc}F_a \\ \{E_a, F_b\} &= \tfrac{1}{2}\epsilon_{abcd}K^{cd}\,, \quad \{E_a, E_b\} = 0 = \{F_a, F_b\}\,. \end{aligned} \tag{18}$$

The generators $q_a^{\pm}$, which generate the AdS transformations (10), still have a form similar to (15) but $E_a(z,\bar{z}), F_a(z,\bar{z})$, $K_{ab}(z,\bar{z})$ are no longer holomorphic and their zero moments with respect to z satisfy (18).

6. String Equations for the AdS Vertex Operator Components

The AdS supersymmetric constraints (7-9) imply

$$F_a\, F_b\, K^{ab} V_{1,1} = 0\,, \qquad \bar{F}_{\bar{a}} \bar{F}_{\bar{b}}\, \bar{K}^{\bar{a}\bar{b}} V_{1,1} \tag{19}$$

$$(F_a E_a + \tfrac{1}{8}\epsilon_{abcd} K^{ab} K^{cd}) V_{1,1} = (\bar{F}_{\bar{a}} \bar{E}_{\bar{a}} + \tfrac{1}{8}\bar{\epsilon}_{\bar{a}\bar{b}\bar{c}\bar{d}} \bar{K}^{\bar{a}\bar{b}} K^{\bar{c}\bar{d}}) V_{1,1} = 0 . \tag{20}$$

The vertex operators $V_{-1,1}, V_{1,-1}, V_{0,-1}, V_{-1,0}, V_{-1,-1}$ can be gauge fixed to zero, and therefore do not correspond to propagating degrees of freedom. Furthermore, this gauge symmetry can be used both to set to zero the components of $V_{1,1}$ with no θ's or no $\bar{\theta}$'s, and to gauge fix all components of $V_{0,1}, V_{1,0}, V_{0,0}$ that are independent of those of $V_{1,1}$. The physical degrees of freedom of the massless compactification independent vertex operators are thus described by a superfield

$$\begin{aligned} V_{1,1} \quad = \quad & \theta^a \bar{\theta}^{\bar{a}} V^{--}_{a\bar{a}} + \theta^a \theta^b \bar{\theta}^{\bar{a}} \sigma^m_{ab} \xi^-_{m\,\bar{a}} + \theta^a \theta^a \theta^{\bar{b}} \sigma^m_{\bar{a}\bar{b}} \xi^-_{m\,a} \\ + \quad & \theta^a \theta^b \bar{\theta}^{\bar{a}} \bar{\theta}^{\bar{b}} \sigma^m_{ab} \sigma^n_{\bar{a}\bar{b}} (g_{mn} + b_{mn} + \bar{g}_{mn}\phi) + \theta^a (\bar{\theta}^3)_{\bar{a}} A^{-+\,\bar{a}}_a + (\theta^3)_a \bar{\theta}^{\bar{a}} A^{+-\,a}_{\bar{a}} \\ + \quad & \theta^a \theta^b (\bar{\theta}^3)_{\bar{a}} \sigma^m_{ab} \bar{\chi}^{+\,\bar{a}}_m + (\theta^3)^a \bar{\theta}^{\bar{a}} \bar{\theta}^{\bar{b}} \sigma^m_{\bar{a}\bar{b}} \chi^{+\,a}_m + (\theta^3)_a (\bar{\theta}^3)_{\bar{a}} F^{++\,a\bar{a}} . \end{aligned} \tag{21}$$

This has the field content of $D = 6$, $N = (2,0)$ supergravity with one supergravity and one tensor multiplet. Further massless multiplets correspond to the compactification degrees of freedom. In flat space, the surviving constraint equations imply that the component fields Φ are all on shell massless fields, that is $\sum_{m=1}^{6} \partial^m \partial_m \Phi = 0$ as in (5), and in addition

$$\begin{aligned} \partial^m g_{mn} &= -\partial_n \phi , \quad \partial^m b_{mn} = 0 , \quad \partial^m \chi^{\pm b}_m = \partial^m \bar{\chi}^{\pm \bar{b}}_m = 0 \\ \partial_{ab} \chi^{\pm b}_m &= \partial_{\bar{a}\bar{b}} \chi^{\pm \bar{b}}_m = 0 , \quad \partial_{cb} F^{\pm\pm b\bar{a}} = \partial_{\bar{c}\bar{b}} F^{\pm\pm \bar{b}a} = 0 , \end{aligned} \tag{22}$$

where

$$\begin{aligned} F^{+-a\bar{a}} = \partial^{\bar{a}\bar{b}} A^{+-a}_{\bar{b}} , \qquad & F^{-+a\bar{a}} = \partial^{ab} A^{-+\bar{a}}_b , \quad F^{--a\bar{a}} = \partial^{ab} \partial^{\bar{a}\bar{b}} V^{--}_{b\bar{b}} \\ \chi^{-a}_m & = \partial^{ab} \xi^-_{mb} , \quad \bar{\chi}^{-\bar{a}}_m = \partial^{\bar{a}\bar{b}} \bar{\xi}^-_{m\bar{b}} . \end{aligned} \tag{23}$$

The equations of motion for the flat space vertex operator component fields describe $D = 6$, $N = (2,0)$ supergravity[4] expanded around the six-dimensional Minkowski metric.

In $AdS_3 \times S^3$ space corresponding gauge transformations reduce the number of degrees of freedom in a similar fashion, but the Laplacian must be replaced by the AdS Laplacian, and the constraints are likewise deformed. We focus on the vertex operator V_{11} that carries the physical degrees of freedom. We show the string constraint equations are equivalent to the $D = 6, N = (2,0)$ linearized supergravity equations expanded around the $AdS_3 \times S^3$ metric.

For the bosonic field components of the vertex operator the AdS constraint equations result in

$$\Box h^g{}_{\bar{a}}\, V^{--}_{ag} = -4\, \sigma^m_{ab}\, \sigma^n_{gh}\, \delta^{bh}\, h^g{}_{\bar{a}}\, G_{mn} \tag{24}$$

$$\Box h^g{}_{\bar{a}}\, h^h{}_{\bar{b}}\, \sigma^m_{ab}\, \sigma^n_{gh}\, G_{mn} = \tfrac{1}{4}\epsilon_{abce}\epsilon_{fghk}\, \delta^{ch}\, h^f{}_{\bar{a}}\, h^g{}_{\bar{b}}\, F^{++ek} \tag{25}$$

$$\Box h_g{}^{\bar{a}}\, F^{++ag} = 0 , \quad \Box h_g{}^{\bar{a}}\, A^{-+g}_a = 0 , \quad \Box h^g{}_{\bar{a}}\, A^{+-a}_g = 0 \tag{26}$$

$$\epsilon_{eacd}\, t_L^{cd}\, h^b{}_{\bar a}\, A_b^{+-a} = 0\,, \qquad \epsilon_{\bar e\bar b\bar c\bar d}\, t_R^{\bar c\bar d}\, h_a{}^{\bar a}\, A_{\bar a}^{-+\bar b} = 0 \tag{27}$$

$$\epsilon_{eacd}\, t_L^{cd}\, h_b{}^{\bar a}\, F^{++ab} = 0\,, \qquad \epsilon_{\bar e\bar b\bar c\bar d}\, t_R^{\bar c\bar d}\, h^a{}_{\bar a}\, F^{++\bar a\bar b} = 0 \tag{28}$$

$$t_L^{ab}\, h^g{}_{\bar a}\, h^h{}_{\bar b}\, \sigma^m_{ab}\, \sigma^n_{gh}\, G_{mn} = 0\,, \quad t_R^{\bar a\bar b}\, h_a{}^{\bar g}\, h_b{}^{\bar h}\, \sigma^m_{\bar g\bar h}\, \sigma^n_{\bar a\bar b}\, G_{mn} = 0\,. \tag{29}$$

We have expanded $G_{mn} = g_{mn} + b_{mn} + \bar g_{mn}\phi$. The $SO(4)$ Laplacian is $\Box \equiv \frac{1}{8}\epsilon_{abcd}\, t_L^{ab}\, t_L^{cd} = \frac{1}{8}\epsilon_{\bar a\bar b\bar c\bar d}\, t_R^{\bar a\bar b}\, t_R^{\bar c\bar d}$. In order to compare this with supergravity, we need to reexpress the above formulas containing the right- and left-invariant vielbeins $t_L^{ab}, t_R^{\bar a\bar b}$ in terms of covariant derivatives D_p on the group manifold. So we write

$$\mathcal{T}_L^{cd} \equiv -\sigma^{p\,cd}\, D_p\,, \qquad \mathcal{T}_R^{\bar c\bar d} \equiv \sigma^{p\,\bar c\bar d}\, D_p\,. \tag{30}$$

Acting on a scalar, $\mathcal{T}_L = t_L$ and $\mathcal{T}_R = t_R$, since both just act geometrically. But they differ in acting on fields that carry spinor or vector indices. For example, on spinor indices,

$$t_L^{ab} V_e = \mathcal{T}_L^{ab}\, V_e + \tfrac{1}{2}\delta_e^a\, \delta^{bc} V_c - \tfrac{1}{2}\delta_e^b\, \delta^{ac} V_c\,. \tag{31}$$

For $AdS_3 \times S^3$ we can write the Riemann tensor and the metric tensor as

$$\begin{aligned} \bar R_{mnp\tau} &= \tfrac{1}{4}\left(\bar g_{m\tau}\bar R_{np} + \bar g_{np}\bar R_{m\tau} - \bar g_{n\tau}\bar R_{mp} - \bar g_{mp}\bar R_{n\tau}\right) \\ \bar g_{mn} &= \tfrac{1}{2}\sigma_m^{ab}\, \sigma_{n\,ab}\,. \end{aligned} \tag{32}$$

The sigma matrices σ^{mab} satisfy the algebra $\sigma^{mab}\sigma^n_{ac} + \sigma^{nab}\sigma^m_{ac} = \eta^{mn}\delta^b_c$ in flat space, where η^{mn} is the six-dimensional Minkowski metric. Sigma matrices with lowered indices are defined by $\sigma^m_{ab} = \frac{1}{2}\epsilon_{abcd}\sigma^{mcd}$, although for other quantities indices are raised and lowered with δ^{ab}, so we distinguish σ^m_{ab} from $\delta_{ac}\,\delta_{bd}\,\sigma^{mcd}$. In curved space, η_{mn} is replaced by the $AdS_3 \times S^3$ metric $\bar g_{mn}$.

We then find from the string constraints that the six-dimensional metric field g_{rs}, the dilaton ϕ, and the two-form b_{rs} satisfy

$$\begin{aligned} \tfrac{1}{2}D^p D_p b_{rs} &= -\tfrac{1}{2}(\sigma_r\sigma^p\sigma^q)_{ab}\delta^{ab}\, D_p\,[\,g_{qs} + \bar g_{qs}\phi\,] + \tfrac{1}{2}(\sigma_s\sigma^p\sigma^q)_{ab}\delta^{ab} D_p\,[\,g_{qr} + \bar g_{qr}\phi\,] \\ &- \bar R_{\tau rs\lambda}\, b^{\tau\lambda} - \tfrac{1}{2}\bar R_r{}^\tau\, b_{\tau s} - \tfrac{1}{2}\bar R_s{}^\tau\, b_{r\tau} \\ &+ \tfrac{1}{4}F_{\text{asy}}^{++gh}\, \sigma_r^{ab}\sigma_s^{ef}\, \delta_{ah}\delta_{be}\delta_{gf} \end{aligned} \tag{33}$$

$$\begin{aligned} \tfrac{1}{2}D^p D_p(g_{rs} + \bar g_{rs}\phi) = &-\tfrac{1}{2}(\sigma_r\sigma^p\sigma^q)_{ab}\delta^{ab} D_p b_{qs} + \tfrac{1}{2}(\sigma_s\sigma^p\sigma^q)_{ab}\delta^{ab} D_p b_{rq} \\ &- \bar R_{\tau rs\lambda}(g^{\tau\lambda} + \bar g^{\tau\lambda}\phi) - \tfrac{1}{2}\bar R_r{}^\tau(g_{\tau s} + \bar g_{\tau s}\phi) \\ &- \tfrac{1}{2}\bar R_s{}^\tau(g_{r\tau} + \bar g_{r\tau}\phi) + \tfrac{1}{4}F_{\text{sym}}^{++gh}\sigma_{rga}\sigma_{shb}\delta^{ab}\,. \end{aligned} \tag{34}$$

This is the curved space version of the flat space zero Laplacian condition $\partial^p\partial_p b_{rs} = \partial^p\partial_p g_{rs} = \partial^p\partial_p\phi = 0$.

Four self-dual tensor and scalar pairs come from the string bispinor fields F^{++ab}, V_{ab}^{--}, , A_b^{+-a}, A_a^{-+b}. From the string constraint equations they satisfy

$$\sigma^p_{da}\, D_p\, F_{\text{asy}}^{++ab} = 0 \tag{35}$$

$$\tfrac{1}{4}\,[\,\delta^{Ba}\,\sigma^{r}_{ga}\,D_r\,F^{++gH}_{\text{sym}} \;-\; \delta^{Ha}\,\sigma^{r}_{ga}\,D_r\,F^{++gB}_{\text{sym}}\,] = -\tfrac{1}{4}\,\epsilon^{BH}{}_{cd}\,F^{++cd}_{\text{asy}} \tag{36}$$

We also find

$$\begin{aligned} &\tfrac{1}{2}\,D^p D_p\,V^{--}_{cd} \;-\tfrac{1}{2}\delta^{gh}\sigma^{p}_{ch}\,D_p\,V^{--}_{gd} \;+\tfrac{1}{2}\delta^{gh}\sigma^{p}_{dh}\,D_p\,V^{--}_{cg} \;+\tfrac{1}{4}\epsilon_{cd}{}^{gh}\,V^{--}_{gh} \\ = \;\; & -4\,\sigma^{m}_{ce}\,\sigma^{n}_{df}\,\delta^{ef}\,G_{mn}\,. \end{aligned} \tag{37}$$

The last constraints can be written as

$$\begin{aligned} &\epsilon_{eacd}\;\; t^{cd}_L\,h_b{}^{\bar a}\,F^{+-ab} = 0 \qquad && \epsilon_{\bar e\bar b\bar c\bar d}\,t^{\bar c\bar d}_R\,h^{a}{}_{\bar a}\,F^{+-\bar a\bar b} = 0 \\ &\epsilon_{eacd}\;\; t^{cd}_L\,h_b{}^{\bar a}\,F^{-+ab} = 0\,, && \epsilon_{\bar e\bar b\bar c\bar d}\,t^{\bar c\bar d}_R\,h^{a}{}_{\bar a}\,F^{-+\bar a\bar b} = 0 \end{aligned} \tag{38}$$

where

$$\begin{aligned} F^{+-a\bar a} &\equiv \delta^{\bar a\bar b}\,A^{+-a}_{\bar b} + t^{\bar a\bar b}_R\,A^{+-a}_{\bar b} \\ F^{-+a\bar a} &\equiv \delta^{ab}\,A^{-+\bar a}_{b} + t^{ab}_L\,A^{-+\bar a}_{b}\,, \end{aligned} \tag{39}$$

so F^{+-ab} and F^{-+ab} satisfy equations similar to those for F^{++ab}.

Independent conditions on the fermion fields are

$$\begin{aligned} &\Box h_a{}^{\bar g}\,\sigma^{m}_{\bar a\bar b}\,\xi^{-}_{m\bar g} = -\sigma^{m}_{\bar g\bar h}\,\epsilon_{\bar e\bar d\bar a\bar b}\,h_a{}^{\bar h}\,\delta^{\bar g\bar d}\,\bar\chi^{+\bar e}_{m} \\ &\Box h^{g}{}_{\bar a}\,\sigma^{m}_{ab}\,\bar\xi^{-}_{mg} = -\sigma^{m}_{gh}\,\epsilon_{edab}\,h^{h}{}_{\bar a}\,\delta^{gd}\,\chi^{+e}_{m} \\ &t_L^{ab} h^{g}{}_{\bar a}\,\sigma^{m}_{ab}\,\bar\xi^{-}_{mg} = 0\,, \qquad t^{\bar a\bar b}_R\,h_a^{\bar g}\,\sigma^{m}_{\bar a\bar b}\,\xi^{-}_{m\bar g} = 0 \\ &t_L^{ab}\sigma^{m}_{ab}\,h_g^{\bar a}\,\bar\chi^{+g}_{m} = 0\,, \qquad t^{\bar a\bar b}_R\,\sigma^{m}_{\bar a\bar b}\,h^{a}_{\bar g}\,\chi^{+\bar g}_{m} = 0 \\ &\epsilon_{deab}t^{ab}_L\,h^{g}{}_{\bar a}\,h^{h}{}_{\bar b}\,\sigma^{m}_{gh}\,\chi^{+e}_{m} = 0\,, \qquad \epsilon_{\bar d\bar e\bar a\bar b}\,t^{\bar a\bar b}_R\,h_a^{\bar g}\,h_b{}^{\bar h}\,\sigma^{m}_{\bar g\bar h}\bar\chi^{+\bar e}_{m} = 0\,. \end{aligned} \tag{40}$$

7. Comparison With Linearized *AdS* Supergravity Equations

We now show that the $AdS_3 \times S^3$ supersymmetric vertex operator constraint equations are equivalent to the linearized supergravity equations for the supergravity multiplet and one tensor multiplet of $D = 6$, $N = (2,0)$ supergravity[4] expanded around the $AdS_3 \times S^3$ metric and a self-dual three-form. We give the identification of the string vertex operator components in terms of the supergravity fields.

We will see that the two-form b_{mn} is a linear combination of *all* the oscillations corresponding to the five self-dual tensor fields and the anti-self-dual tensor field, *including the oscillation with non-vanishing background.* In flat space, b_{mn} corresponds to a state in the Neveu-Schwarz sector. In our curved space case, the string model describes vertex operators for AdS_3 background with Ramond-Ramond flux.

When matching the vertex operator component fields with the supergravity oscillations, we find that not only the bispinor V_{ab}^{--} (which is a Ramond-Ramond field in the flat space case), but also the tensor b_{mn} include supergravity oscillations with non-vanishing self-dual background.

The linearized supergravity equations are given by

$$D^p D_p \phi^i = \tfrac{2}{3}\, \bar{H}^i_{prs}\, g^{6prs} \tag{41}$$

$$\begin{aligned}
&\tfrac{1}{2}D^pD_p h_{rs} - \bar{R}_{\tau rs\lambda}h^{\tau\lambda} + \tfrac{1}{2}\bar{R}^{\tau}_{r}h_{\tau s} \\
&\quad + \tfrac{1}{2}\bar{R}^{\tau}_{s}h_{r\tau} - \tfrac{1}{2}D_sD^p h_{pr} - \tfrac{1}{2}D_sD^p h_{pr} + \tfrac{1}{2}D_rD_s h^p_p \\
&\quad = -\bar{H}^{i\ pq}_{r} g^i_{spq} - \bar{H}^{i\ pq}_{s} g^i_{rpq} + 2h^{pt}\bar{H}^{i\ q}_{rp}\bar{H}^i_{stq}
\end{aligned} \tag{42}$$

$$\begin{aligned}
D^p H_{prs} &= -2\,\bar{H}^i_{prs}\, D^p \phi^i \\
&+ B^i\,[-\bar{H}^{i\ pq}_{r} D_p\, h_{qs} + \bar{H}^{i\ pq}_{s} D_p\, h_{qr} + \bar{H}^{i\ \ q}_{rs} D^p h_{pq} - \tfrac{1}{2}\bar{H}^{i\ \ q}_{rs} D_q h^p{}_p]
\end{aligned} \tag{43}$$

where we have defined $H_{prs} \equiv g^6_{prs} + B^i\, g^i_{prs}$ as a combination of the supergravity exact forms $g^6 \equiv db^6, g^i \equiv db^i$, since we will equate this with the string field strength $H = db$. We will choose $B^1 = 2$. In zeroeth order, the equations are $\bar{R}_{rs} = -\bar{H}^i_{rpq}\,\bar{H}^{i\ pq}_{s}$.

We define the vertex operator components in terms of the supergravity fields $g^i_{prs}, g^6_{prs}, h_{rs}, \phi^i$, $1 \le i \le 5$, (and $2 \le I \le 5$) as

$$\begin{aligned}
H_{prs} &\equiv g^6_{prs} + 2\, g^1_{prs} + B^I\, g^I_{prs} \\
g_{rs} &\equiv h_{rs} - \tfrac{1}{6}\bar{g}_{rs}\, h^\lambda{}_\lambda \\
\phi &= -\tfrac{1}{3}\, h^\lambda{}_\lambda
\end{aligned}$$

$$\begin{aligned}
F^{++ab}_{\text{sym}} &= \tfrac{2}{3}(\sigma_p\sigma_r\sigma_s)^{ab}\, B^I\ g^I_{prs} + \delta^{ab}\,\phi^{++} \\
F^{++ab}_{\text{asy}} &= \sigma^{p\,ab}\, D_p\,\phi^{++} \\
\phi^{++} &= 4C^I\ \phi^I
\end{aligned} \tag{44}$$

which follows from choosing the graviton trace $h^\lambda{}_\lambda$ to satisfy $\phi^1 - h^\lambda{}_\lambda \equiv -2\,C^I\phi^I$, and we have used $H_{prs} \equiv \partial_p b_{rs} + \partial_r b_{sp} + \partial_s b_{pr}$.

The combinations $C^I\phi^I$ and $B^I g^I_{prs}$ reflect the $SO(4)_{\rm R}$ symmetry of the $D = 6, N = (2,0)$ theory on $AdS_3 \times S^3$. We relabel $C^I = C^I_{++}, B^I = B^I_{++}$. To define the remaining string components in terms of supergravity fields, we consider linearly independent quantities $C^I_\ell\phi^I$, $B^I_\ell g^I_{prs}$, $\ell = ++, +-, -+, --$.

$$\begin{aligned}
F^{+-ab}_{\text{sym}} &= \tfrac{2}{3}(\sigma_p\sigma_r\sigma_s)^{ab}\, B^I_{+-}\ g^I_{prs} + \delta^{ab}\,\phi^{+-} \\
F^{+-ab}_{\text{asy}} &= \sigma^{p\,ab}\, D_p\,\phi^{+-} \\
\phi^{+-} &= 4C^I_{+-}\ \phi^I \\
F^{-+ab}_{\text{sym}} &= \tfrac{2}{3}(\sigma_p\sigma_r\sigma_s)^{ab}\, B^I_{-+}\ g^I_{prs} + \delta^{ab}\,\phi^{-+} \\
F^{-+ab}_{\text{asy}} &= \sigma^{p\,ab}\, D_p\,\phi^{-+}
\end{aligned}$$

$$\phi^{-+} \;=\; 4C^I_{-+}\,\phi^I\,. \tag{45}$$

V^{--}_{ab} is given in terms of the fourth tensor/scalar pair $C^I_{--}\ \phi^I$, $B^I_{--}g^I_{mnp}$ through

$$D^p D_p V^{--}_{cd} - \delta^{gh}\sigma^p_{ch}\, D_p V^{--}_{gd} + \delta^{gh}\sigma^p_{dh}\, D_p V^{--}_{cg} + \tfrac{1}{2}\epsilon_{cd}{}^{gh}\, V^{--}_{gh} \;=\; -8\,\sigma^m_{ce}\,\sigma^n_{df}\,\delta^{ef}\, G_{mn}\,. \tag{46}$$

With these field definitions, the string constraint equations for the AdS_3 vertex operators reduce to the linearized supergravity equations for the vertex operator field components.

The fermion constraints imply the linearized *AdS* supergravity equations for the gravitinos and spinors, due to the above correspondence for the bosons and the supersymmetry of the two theories.

Acknowledgments

This work was supported in part by the U.S. Department of Energy, Grant No. DE-FG 05-85ER40219/Task A.

References

1. N. Berkovits, C. Vafa, and E. Witten, *Conformal Field Theory of AdS Background With Ramond-Ramond Flux*, *JHEP* **9903**, 018 (1999) [hep-th/9902098].
2. L. Dolan and E. Witten, *Vertex Operators for AdS3 Background With Ramond-Ramond Flux*, *JHEP* **9911**, 003 (1999) [hep-th/9910205].
3. M. Bershadsky, S. Zhukov, and A. Vaintrob, *PSL(N/N) Sigma Model As A Conformal Field Theory*, *Nucl. Phys.* **B559**, 205 (1999) [hep-th/9902180].
4. L. Romans, *Self-Duality for Interacting Fields*, *Nucl. Phys.* **B276**, 71 (1986).
5. N. Berkovits, *Super-Poincare Covariant Quantization of the Superstring*, *JHEP* **0004**, 018 (2000) [hep-th/0001035].
6. N. Berkovits, *Superstring Vertex Operators in an AdS_5 Background*, hep-th/0009168.

SUPERGRAVITY DESCRIPTION OF FIELD THEORIES ON CURVED MANIFOLDS AND A NO GO THEOREM

JUAN MALDACENA and CARLOS NUÑEZ
Jefferson Laboratory of Physics
Harvard University, Cambridge, MA 02138, USA

In the first part of this paper we find supergravity solutions corresponding to branes on worldvolumes of the form $R^d \times \Sigma$ where Σ is a Riemann surface. These theories arise when we wrap branes on holomorphic Riemann surfaces inside $K3$ or CY manifolds. In some cases the theory at low energies is a conformal field theory with two less dimensions. We find some non-singular supersymmetric compactifications of M-theory down to AdS_5. We also propose a criterion for permissible singularities in supergravity solutions.

In the second part of this paper, which can be read independently of the first, we show that there are no non-singular Randall-Sundrum or de-Sitter compactifications for large class of gravity theories.

1. Introduction

In the first part of this paper we study the large N limit of branes wrapped on non-trivial cycles. More precisely we consider a $d+2$ dimensional field theory defined on a space of the form $R^d \times \Sigma_g$ where Σ_g is a genus g Riemann surface. These theories reduce to d dimensional field theories at low energies, at energies small compared to the inverse size of the Riemann surface. We consider supergravity solutions that describe the flow between the $d+2$ dimensional field theory and the d dimensional field theory. Our $d+2$ dimensional field theories are superconformal and have the maximum amount of supersymmetry. More precisely we consider $\mathcal{N}=4$ super-Yang-Mills in four dimensions and the $(0,2)$ theory in six dimensions. These will give rise to two and four dimensional field theories respectively. These field theories will have less supersymmetry. We consider situations where the supersymmetry gets reduced to $1/2$ or $1/4$ of the original supersymmetry of the $d+2$ dimensional theory. The amount of supersymmetry preserved depends on how the two dimensional surface Σ_g is embedded in a higher dimensional space. In the field theory limit, these different embedding possibilities translate into different normal bundles and therefore different external $SO(n)$ gauge fields on the worldvolume theory, where n is the number of directions normal to the brane. There are several cases where the resulting d dimensional field theory is conformal. In those cases we find an AdS_{d+1} geometry in the IR. These geometries in the IR have an interest of its own and provide new AdS/CFT examples. In particular, we find new $\mathcal{N}=2$ and $\mathcal{N}=1$ superconformal field theories in four dimensions that arise from wrapping M5 branes on negatively curved Riemann surfaces (surfaces with $g>1$).

They are dual to AdS_5 warped compactifications of eleven dimensional supergravity. These are AdS_5 fibrations over a six dimensional space, the five dimensional warp factor depends on the six dimensional coordinates.[a] Similarly we find some compactifications of IIB string theory to AdS_3 that preserve (2,2) supersymmetry.

The basic technique that we use to find the solutions is the following. The field theories on branes are twisted theories.[4] What this means is that together with the coupling to the curvature of the brane worldvolume there is a coupling to an external $SO(n)$ gauge field, if n is the number of transverse directions to the brane. So we have a field theory on a curved space coupled to a $SO(n)$ gauge field. In supergravity this translates into boundary conditions at the boundary of AdS_{d+3} for the metric and the $SO(n)$ gauge fields. Fortunately these are modes of gauged supergravity, so that we can consider just the $d+3$ dimensional gauged supergravity equations. We further consider only constant curvature Riemann surfaces and simple embedding of the spin connection into the $SO(n)$ connection. A more general analysis is left for the future.

Some of these supergravity solutions can be viewed as compactifications of $d+3$ dimensional gauged supergravities on $AdS_{d+1} \times \Sigma_{g>1}$ with magnetic fluxes on Σ. They are similar in spirit to the solutions in Refs. 5,6,7,8,9.

Since we have obtained a large family of AdS_5 compactifications it is natural to ask if we could find a smooth Randall-Sundrum[10] compactification based on these. In section we show that it is not possible to find a smooth Randall-Sundrum compactification of usual supergravity theories. In fact we prove something even more general, it is not possible to find warped compactifications which have only singularities where the warp factor is non-increasing as we approach the singularity. We also show that there are no deSitter compactifications. This section is self contained and can be read independently of the rest of this paper. For other no go theorems for supersymmetric RS compactifications of 5d gauged supergravity see Refs. 11,12,13,14.

There are many papers in the literature which consider compactifications similar to the ones considered here.[15,6,7,8,9,16,17,18,19,20,21,22] While this paper was in preparation the paper Ref. 23 appeared which has some overlap with ours and presents another interesting method, based on Ref. 3, for obtaining a large class of solutions corresponding to branes on Riemann surfaces.

This paper is organized as follows. We first describe the general idea behind twisted field theories and how we plan to find the gravity solutions. In section 3 we consider $\mathcal{N}=4$ super Yang Mills on $R^2 \times \Sigma$. We consider two possible twisting which preserve $(4,4)$ and $(2,2)$ supersymmetry in 1+1 dimensions. In section four we consider the (0,2) superconformal field theory that lives on M5 branes compactified on spaces of the form $R^4 \times \Sigma$. We consider again two cases, preserving $\mathcal{N}=2,1$

[a]Related compactifications were found in Refs. 1,2, but in their case they the warp factor was going to zero at some points in the internal space. In our case the metric is completely smooth. In Ref. 3 similar configurations were studied and the conditions for preserved supersymmetry were studied in great generality and some solutions were found corresponding to intersecting branes with a certain amount of "smearing" .in one of the transverse dimensions.

supersymmetry in four dimensions. In section we discuss a criterion for allowed singularities in gravity theories that are dual to field theories. In section we discuss the absence of certain warped compactifications or de Sitter solutions in a class of gravity theories, including 11d supergravity, IIB, IIA and massive IIA.

In the appendix we give some more details on our calculations.

2. General Idea

If we start with a supersymmetric field theory and we put it on a curved manifold Ω then, in general, we will break supersymmetry since we will not have a covariantly constant spinor, obeying $(\partial_\mu + \omega_\mu)\epsilon = 0$. If the supersymmetric theory has a global R-symmetry, then we can couple the theory to an external gauge field that couples to the R-symmetry current. If we choose the external gauge field to be equal to the spin connection $A_\mu = \omega_\mu$ (the precise meaning of this equation will be explained below) then we see that we can find a covariantly constant spinor since $\partial_\mu + \omega_\mu - A_\mu \epsilon = \partial_\mu \epsilon$, so that we can just consider a constant spinor. The resulting theory is a so called "twisted" theory, since we can view the the coupling to the external gauge field as effectively changing the spins of all fields. The supersymmetry parameter becomes a scalar. Though it might sound like a contrived way of preserving supersymmetry, it is precisely the way that branes wrapping on non-trivial cycles in M-theory or string theory compactifications manage to preserve some supersymmetries.[4] In this case Ω is the worldvolume geometry of the cycle and the external gauge field takes into account the fact that the directions normal to the cycle form a non-trivial bundle, the normal bundle, and A_μ is the connection on this normal bundle. The condition that the cycle preserves some supersymmetry then boils down to the condition that the spin connection is equal to the gauge connection. So if we are interested in understanding field theories arising on branes wrapping non-trivial cycles, then we will have to study these twisted theories. Let us first clarify the nature of the limit in which we decouple this twisted field theory from the full original string theory. We consider a brane wrapping a cycle and we take the decoupling limit $l_p \to 0$ as in Ref. 24 keeping the volume of the cycle fixed. In this limit we get a field theory on the brane that is twisted as above, due to the non-trivial embedding of the cycle in the ambient space. If the theory on the brane is conformal before we wrap it, this is all we have to do. If we have a D-p-brane with $p \neq 3$ then we also should scale the string coupling as in the flat case.[25] Notice that in this limit the theory is not sensitive to the global geometry of the spacetime where the cycle is embedded. The reason is that a finite fluctuation of the scalar field parameterizing transverse displacements corresponds to an infinitesimal fluctuation in the position of the brane. In other words, in the scaling limit we are considering, the size of the Calabi Yau (or any other space where the brane is embedded) is fixed, while the typical fluctuation of the position of the brane goes to zero as l_p goes to zero. This of course does not imply that the scalars associated to motions of the brane will have definite expectation values. Whether they do or not will depend on the number of non-compact dimensions of the theory. We will discuss this more later. Let us consider some examples. Suppose we have type IIB string theory on $R^6 \times K3$. We can wrap a D3 brane on an S^2 inside the $K3$ manifold leaving two non-compact directions. The worldvolume is then $\Omega = R^2 \times S^2$. In this case, the spin connection is in a $U(1)$ subgroup of the tangent group $SO(3,1)$, since the curvature is purely in the S^2 directions. The brane has 6 normal directions. Two of them are in the directions of $K3$ that are normal to the S^2. They will form a non-trivial $U(1)$ normal bundle. It turns out that for holomorphically embedded spheres, the spin connection is essentially equal to the connection on the $U(1)$ normal bundle. The other four normal directions are totally flat, with no gauge field. This gauge connection will break the $SO(6)$ R-symmetry group into $U(1) \times SO(4)$. Out of the 16 spinors that generate the supersymmetries of $\mathcal{N} = 4$ Yang Mills, there will be only half for which the chirality on S^2 and the $U(1)$ charges are correlated so that the spin connection and gauge connection cancel. In terms of the two dimensional low energy theory on R^2 the theory has $(4,4)$ supersymmetry. These kind of theories were considered in Refs. 26,27. Below we find the supergravity solutions associated to these twisted field theories. For this particular example we find that there is a family of supergravity solutions and that they all seem to have a singularity in the IR. We argue later that this singularity is associated to the IR properties of the brane theory and that only the singularities of an allowable type, according to the criterion in section or the one in Ref. 28, produce the right physics. We will consider however some other examples, involving branes wrapping negatively curved Riemann surfaces, for which there are non-singular solutions which in the IR have an *AdS* form. For example if we wrap a $D3$ brane on a genus $g > 1$ Riemann surface times R^2 with a particular normal bundle we specify below then the supergravity solution interpolates between an AdS_5 region close to the boundary and an AdS_3 region in the IR, corresponding to the fact that the theory in the UV is just $3+1$ dimensional SYM and it flows to a 1+1 dimensional conformal field theory. So these are examples of flows "across dimensions", the four dimensional conformal field theory flows to a two dimensional conformal field theory.

We only consider cases where the genus of the Riemann surface Σ_g is $g \neq 1$, since in the case of $g = 1$ the constant curvature metric is just the metric on a flat T^2 and the theory is identical to the untwisted theory.

We will find similar examples for $M5$ branes wrapping on negatively curved Riemann surfaces. These will give us a new family of examples of four dimensional conformal field theories and their associated AdS_5 compactifications of M-theory. The case analyzed in Refs. 1,29 would probably arise as a singular limit of the smooth solutions analyzed here. In principle it should also be possible to find the solutions in our paper using the methods of Refs. 3,23 but we will not pursue that here. In Horava-Witten M-theory compactifications to 3+1 dimensions we could have M5 branes wrapping on Riemann surfaces in the CY manifold at some points in moduli space. Our analysis implies that coincident branes will typically give rise to conformal field theories. More accurately, that they give rise to conformal field theories in the large N limit and the large CY volume limit. Finite N effects, or finite volume effects of the CY manifold could destroy conformality. Since having a conformal field theory, or a theory with a logarithmically running coupling would give a natural explanation of the gauge hierarchy problem, it is interesting that these arise quite naturally in these compactifications. One should also be careful with this conclusion because in Horava-Witten compactifications we will also generically have some fluxes of the four form field strength. These might induce relevant perturbations of the field theory, as in Ref. 30.

Since we are dealing with the theory of coincident M5 branes we find it hard to give a purely four dimensional field theory description of the theory in the IR. It looks like it should be possible to say what it is more precisely, but leave this to the future. Dimensional reducing the M5 solutions we get D4 solutions corresponding to D4 branes wrapped on Riemann surfaces. In this case we can state more precisely what the 3 dimensional field theories are.

In this paper we will consider Riemann surfaces with constant curvature. We think of them as H_2/Γ where Γ is a discrete subgroup of $SL(2,R)$. Notice that for our purposes, which is to find a precise sugra solution, the precise metric on the Riemann surface *does* matter. One can find solutions for non-constant curvatures by the methods of this paper or by the methods of Ref. 23 where a large family of solutions was found.

3. Twisted 4d N=4 SYM

The possible twistings of $N = 4$ SYM were considered in Ref. 26. These twistings are distinguished by different ways of embedding the spin connection in the $SU(4)$ R-symmetry group. In principle we could consider the most general case by the techniques of this paper. But in order to keep formulas simple we will concentrate on four dimensional spaces of the form $R^2 \times \Sigma_2$ where Σ_2 is a two dimensional manifold. In this case the spin connection is in a U(1) subgroup. So, different twistings of the theory are defined by different embedding of this $U(1)$ in $SU(4) = SO(6)$. We will consider two cases, one which preserves (4,4) supersymmetry and one preserving (2,2) supersymmetry.

Throughout this section we work in units where the five dimensional AdS_5 radius is one. In order to restore the radius dependence of the solutions we just multiply the five or ten dimensional metrics we will write by $R^2_{AdS_5} = \sqrt{4\pi g_s N}\alpha'$.

3.1. *Twists preserving* (4,4) *susy*

The first twisting that we consider corresponds to picking a $U(1)$ in SO(6) in such a way that we break $SO(6) \to SO(2) \times SO(4)$, where the first $SO(2)$ is the $U(1)$ that we are picking inside $SO(6)$. In other words, if we think of $SO(6)$ as acting on six coordinates ϕ^I, then the $U(1)$ is the group of rotations in the 12 plane. This twisting turns out to be exactly the same as the one considered in Ref. 26[b] and the

[b]In Ref. 26 more general manifolds were considered, but it was noted that if the four dimensional manifold is Kahler (as in our case), then only the $U(1)$ mentioned above has a non-trivial gauge connection

resulting field theories were analyzed in Ref. 27. In order to explain more clearly how this works, let us consider a field ϕ in the Yang-Mills Lagrangian, with spin s under the $SO(2)$ spin connection on Σ_2 and $U(1)$ charge q. The Lagrangian is obtained from the flat Lagrangian, by replacing ordinary derivatives of ϕ in the Σ directions by covariant derivatives

$$\mathcal{D}_\mu \phi = (\partial_\mu + si\omega_\mu + iqA_\mu)\phi \tag{1}$$

where $\omega_\mu = \epsilon_{ab}\omega_\mu^{ab}/2$ with ω_μ^{ab} the usual spin connection. If the metric on Σ_2 is $ds^2 = e^{2h}(dx^2 + dy^2)$ then the spin connection is $\omega_\mu = \epsilon_{\mu\nu}\partial_\nu h$, and we also have $A_\mu = \omega_\mu$. This implies that spinors with $s = -q$ can be covariantly constant. In fact they are actually constant, the twisting effectively made them scalars. In this situation the $SO(1,3) \times SO(6)$ symmetry group of the tangent and normal bundles is decomposed as $SO(1,1) \times SO(2)_\Sigma \times U(1) \times SU(2)_L \times SU(2)_R$ and the preserved spinors transform in the $(+, \pm, \mp, 1, 2)$ and $(-, \pm, \mp, 2, 1)$ representations. So we have (4,4) supersymmetry, in $1 + 1$ dimensional notation. Let us consider some examples where these theories arise. Suppose we have a compactification of the form $R^6 \times K3$. Then we can wrap a $D3$ on a holomorphic Riemann surface inside $K3$ and we obtain a field theory on the $D3$ which is of the above from. The $SO(4) = SU(2)_L \times SU(2)_R$ symmetry of the field theory is the rotational symmetry in the four directions in R^6 that are orthogonal to the brane. More precisely, the limit in which we get the above field theory is a limit where we take $\alpha' \to 0$ keeping the size of the Riemann surface (and the $K3$) fixed. Therefore, this limit corresponds to a large volume compactification, large in string units. In this limit the directions normal to the Riemann surface become effectively non-compact, since the brane explores only an infinitesimal neighborhood of the surface. In other words, a field ϕ parameterizing fluctuations of the position of the brane is related to the actual displacement by $r = \alpha'\phi$ so that the actual displacement goes to zero as $\alpha' \to 0$. We can then take a further low energy limit where we consider energies much smaller than the size of the Riemann surface. In this limit we obtain a two dimensional effective theory, which in the IR is a (4,4) superconformal theory.[27] With the order of limits that we took, this conformal field theory has a non-compact target space since we are only exploring a neighborhood of the Riemann surface. The fact that the target space is non-compact implies that this CFT does not have a well defined vacuum state. We would have obtained a different theory if we had taken the size of the Riemann surface and K3 fixed in string units. In that case the target space would have been essentially compact,[c] and quantum fluctuations would have explored the whole moduli space of Riemann surfaces. The resulting theory would be (a T-dual version of) the familiar $D1 - D5$ system, if the genus of the Riemann surface is $g > 1$.

Let us describe in more detail about the Lagrangian of these theories. The full Lagrangian can be found in Ref. 26. Here we give only some terms that are rele-

[c] It still might still have some non-compact directions via effects such as the ones discussed in Ref. 31.

vant in comparing with gravity solutions. The only parts of the Lagrangian that are different from the usual $\mathcal{N} = 4$ Lagrangian are the terms involving covariant derivatives along Σ or fields that are charged under the $U(1)$ part of the normal bundle that has a non-zero gauge connection. Let us consider the part of the Lagrangian involving the two twisted scalar fields. These are the two scalar fields parameterizing fluctuations of the surface in the two normal directions which transform under $U(1)$. Let us arrange these two fields into a complex field $\mathcal{Z} = X^1 + iX^2$. The quadratic terms in the Lagrangian involving these fields are

$$S = \int Tr\{|\mathcal{D}_z \mathcal{Z}|^2 + |\mathcal{D}_{\bar{z}} \mathcal{Z}|^2 + \frac{1}{4} R |\mathcal{Z}|^2\} \tag{2}$$

The first two terms are the terms we would obviously expect from (1) and the last term is a curvature coupling that was derived from supersymmetry in Ref. 26, but in the case of single D-brane it can also be obtained directly by expanding the Nambu action for a brane on this surface. The other scalar fields ϕ^I, $I = 1, ..., 4$ have a simple Lagrangian of the form $(\partial\phi^I)^2$. Note that integrating by parts the first term in (2) we can recast it as the second term up to a commutator $[\mathcal{D}_z, \mathcal{D}_{\bar{z}}]$. This commutator precisely cancels the curvature term in (2). We can choose a holomorphic basis for the normal bundle so that $\mathcal{D}_{\bar{z}} \mathcal{Z} = \partial_{\bar{z}} \mathcal{Z}$. This implies that holomorphic sections of the normal bundle $\mathcal{Z}(z)$ are solutions of the equations of motion and describe configurations with the same energy as the original configuration. This is precisely what we expect since any holomorphic deformation of the surface preserves supersymmetry. Of course, whether these deformations exist globally on the surface or not depends on the global aspects of the geometry and the normal bundle.

Let us look for the supergravity dual of these field theories. Since these are just $\mathcal{N} = 4$ 3+1 dimensional SYM theories on some particular backgrounds, we expect to be able to find the gravity dual by starting with $AdS_5 \times S^5$ and changing the asymptotic boundary conditions to reflect the fact that the theory is defined on $R^2 \times \Sigma$ and is coupled to an $SO(6)$ gauge field. This is easy to achieve. We impose that at the boundary of AdS_5 the metric behaves as

$$ds^2 \sim \frac{-dt^2 + dz^2 + e^{2h}(dx^2 + dy^2) + dr^2}{r^2} \quad \text{for small } r \tag{3}$$

where $ds^2_\Sigma = e^{2h(x,y)}(dx^2 + dy^2)$ is the metric of the two dimensional surface. Similarly we impose that the AdS_5 SO(6) gauge fields are asymptotic to the corresponding field theory values. This translates into a condition on components of the metric with one index in AdS_5 and one index on S^5. In other words we require $g_{\phi\alpha} \sim A_\alpha \sim \epsilon_\alpha^{\ \beta} \partial_\beta h$ near the boundary. These two conditions are rather obvious. A bit less obvious is the fact that we also need to turn on an operator in the **20** of SO(6). This becomes apparent once we look at the curvature coupling in (2) and realize that that coupling is not present for the other four scalar fields. This means that the operator

$$\mathcal{O}_2 = Tr[\frac{2}{3}|\mathcal{Z}|^2 - \frac{1}{3}({\phi^1}^2 + ... + {\phi^4}^2)] \tag{4}$$

is turned on. We also have the singlet operator turned on, with the expected coefficient proportional to the scalar curvature.[32] So we can rewrite the curvature coupling in (2) as

$$S = \frac{1}{2}\int R\left(\frac{1}{6}(|\mathcal{Z}|^2 + {\phi^1}^2 + ... + {\phi^4}^2) + \frac{1}{2}\mathcal{O}_2\right) \tag{5}$$

Fortunately all the operators that are turned on correspond to fields in the five dimensional gauged supergravity multiplet. This is a general feature for these twisted theories, even in the most general curved backgrounds. In order to find the gravity solutions we can therefore consider just the five dimensional gauged supergravity equations. If we were interested in the most general twisted theory we would have to use the full $\mathcal{N} = 8$ gauged supergravity of Ref. 33. However, in our case the connection is in $U(1)$ so we can further use a $U(1)$ truncation of the equations of the form considered in Ref. 34. Furthermore, in Ref. 34 one can find formulas to express the ten dimensional solution given any solution of the truncated five dimensional equations. We will consider therefore a theory involving the five dimensional metric, a $U(1)$ gauge field and a scalar field φ which is dual to the operator $\mathcal{O}_2$ appearing above. In order to find supersymmetric solutions we look at the supersymmetry variation equations of the fermionic fields. These can be read of from Refs. 35,6 as explained in the appendix.

$$\begin{aligned}\frac{1}{\sqrt{6}}\delta\lambda &= -\frac{1}{24}e^{-2\varphi}\Gamma^{\mu\nu}F_{\mu\nu}\epsilon - \frac{i}{4}\Gamma^{\mu}\partial_\mu\varphi\epsilon + \frac{i}{6}(e^{2\varphi} - e^{-\varphi})\epsilon \\ \delta\psi_\mu &= \mathcal{D}_\mu(\omega)\epsilon + \frac{i}{24}e^{-2\varphi}(\Gamma_\mu^{\nu\rho} - 4\delta_\mu^\nu\Gamma^\rho)F_{\nu\rho}\epsilon + \frac{1}{6}\Gamma_\mu(2e^{-\varphi} + e^{2\varphi})\epsilon - \frac{i}{2}A_\mu\epsilon\end{aligned} \tag{6}$$

In order to solve this equation we make the following ansatz for the metric

$$ds^2 = e^{2f}(dr^2 + dz^2 - dt^2) + \frac{e^{2g}}{y^2}(dx^2 + dy^2) \tag{7}$$

where f and g are functions of r to be determined. For simplicity we are considering constant curvature Riemann surfaces of genus $g > 1$.[d] When $r \to 0$ the boundary conditions are $f(r) \sim g(r) \sim -log(r)$. The gauge field is $A_x = 1/y$. This value comes from demanding equality with the spin connection. Setting to zero the supersymmetry variations we obtain, see the appendix,

$$\begin{aligned} g' &= -e^f[\frac{1}{3}(2e^{-\varphi} + e^{2\varphi}) - \frac{1}{3}e^{-2g-2\varphi}] \\ f' &= -\frac{1}{6}e^f[2(2e^{-\varphi} + e^{2\varphi}) + e^{-2g-2\varphi}] \\ \varphi' &= \frac{1}{3}e^f[2(-e^{-\varphi} + e^{2\varphi}) + e^{-2g-2\varphi}] \end{aligned} \tag{8}$$

[d]The reader should not be confused by the fact that g denotes both the function $g(r)$ appearing in (7) and the genus.

The solution is given by

$$
\begin{aligned}
e^{-3\varphi} &= 1 + e^{-2\rho}[\frac{1}{2}\log(e^{2\rho} - \frac{1}{2}) + C_1] \\
e^{2g} &= e^{2\rho}e^{-\varphi} \\
e^{2f} &= (e^{2\rho} - \frac{1}{2})e^{-\varphi} \\
(\frac{dr}{d\rho})^2 e^{2f} &= \frac{e^{4\rho}}{(e^{2\rho} - \frac{1}{2})^2} e^{2\varphi}
\end{aligned}
\tag{9}
$$

Notice that using the last line we can rewrite the metric in terms of the new radial coordinate ρ, $\rho \to +\infty$ corresponds to the boundary. C_1 is an integration constant, there is of course another trivial integration constant which amounts to shifting f by a constant. Note that once we solve the equations for H_2 we can quotient the solution by a subgroup Γ of $SL(2,R)$ that produces the Riemann surface $\Sigma_g = H^2/\Gamma$. This group also acts in the $U(1)$ that we are twisting. That implies that the spinor parameter that generates the preserved supersymmetry transforms like a scalar under $SL(2,R)$ transformations and it therefore survives the quotienting procedure.

We see that there is a singularity of the metric at $\rho = \rho^*$ where $e^{-3\varphi(\rho_*)} = 0$. This singularity is qualitatively the same regardless of the value of C_1. These singularities are related to regions in the Coulomb or Higgs branches of the theory. Let us explore this a bit more. If we expand the first term in equation (9) for large ρ we find that $\varphi \sim -e^{-2\rho}\rho - e^{-2\rho}C_1$. The logarithmic term is related to the fact that we are inserting the operator $\mathcal{O}_2$ as in (5) the subleading term can be thought of as the expectation value of this operator.[36] We would like to relate the sign of C_1 to the sign of the expectation value of $\mathcal{O}_2$. We can do that by looking at Ref. 37 where configurations in $\mathcal{N} = 4$ super Yang-Mills with both expectation values for $\mathcal{O}_2$ were considered. We find from their paper that $C_1 \sim \langle \mathcal{O}_2 \rangle$. From the explicit expression for the operator $\mathcal{O}_2$ (4) we see that configurations with $\langle \mathcal{O}_2 \rangle > 0$ correspond to Higgs branch configurations while ones where $\langle \mathcal{O}_2 \rangle < 0$ correspond to the Coulomb branch. Since we also have the operator inserted, it is a bit hard to separate its expectation value, so we should trust this criterion only for $|C_1|$ large. The behavior at the singularity is very similar for both signs of C_1. These singularities look very much like the singularities in the Coulomb branch, where branes are distributed on an S^3 as in Ref. 37, even though we expect that negative values of C_1 should correspond to the Higgs branch. This seems to be related to the Higgs-Coulomb correspondence as in Refs. 38,39,40 where it was shown that some singularities in the Higgs branch look very similar to the near region of the Coulomb branch. What seems surprising, when we compare this system to the familiar D1-D5 system, is that we do not find any solution that has an AdS_3 region in the IR. We think that this is related to the fact that the (4,4) superconformal field theory that we are dealing with here has a non-compact target space. This non-compactness is different in nature from the one that arises at special points in moduli space for the D1-D5 system.[31] In the D1-D5 system the singularities are small instanton singularities which require tuning of order Q_5 parameters out of the order $Q_1 Q_5$ parameters in instanton moduli in order to reach them. In our (4,4) we have another kind of non-compactness of the moduli space. This non-compactness is related to the possibility of moving the branes in the directions orthogonal to the Riemann surface. In other words, we are saying that the non-compactness of the Higgs branch in this case is very similar to the non-compactness we would have if we were to consider the D1-D5 system but with the internal space replaced by R^4 instead of T^4 and a finite number Q_1 of one branes. It was shown in Ref. 41 that that system has no AdS_3 region.

Let us find the theory we expect at low energies from the field theory point of view. It was argued in Ref. 4 that this (4,4) superconformal field theory was a conformal field theory whose target space is Hitchin space. Its central charge is $c = 6N^2(g-1)$. This moduli space is non-compact and that seems to be preventing the AdS_3 solutions. Another way to say what the low energy theory is; is the following. Suppose we add two more dimensions along the brane so that we have a D5 brane wrapped on Σ_g, $g > 1$. Then we have an $\mathcal{N} = 2$ theory in four dimensions. This four dimensional theory is a $U(N)$ theory with g adjoint hypers. A way to see this is to go to weak coupling and calculate explicitly the number of low energy modes and their susy properties, see Ref. 42 for a related discussion. We can calculate that number using an index argument. Supersymmetry then determines the form of the Lagrangian. This theory has a Higgs branch of real dimension $4N^2(g-1)$. We can now dimensionally reduce two of the dimensions of the D5 brane to make it a D3 brane. So in the IR we have a CFT which is a sigma model whose target space is the Higgs branch of the gauge theory. A way of thinking about a point in this Higgs branch was given in Ref. 43. The N coincident branes are moved and their intersections "blown up" in such a way that we have a single surface which now has $2N^2(g-1)$ geometric moduli and the same number of $U(1)$ worldvolume gauge field Wilson lines. Of course, in the $\alpha' \to 0$ limit that we are taking this single surface has a thickness smaller than α' and only explores a small neighborhood of the original surface but not the global structure of the manifold where the surface Σ is embedded.

The singularities in (9) are allowed under the criterion given in section which demands that g_{00} does not increase as we approach the singularity or the one given in Ref. 28, which demands that the potential for the scalars should be bounded above in the solution. See the appendix for the explicit computation of the potential.

It is also possible to find the solution for the case that the branes are wrapped on $R^2 \times S^2$. Since we can go from the metric in H_2, $ds^2 = d\theta^2 + \sinh^2\theta d\psi^2$, to minus the metric on S^2 by taking $\theta \to i\theta$ we can find the equations for the branes on $R^2 \times S^2$ by formally replacing $e^{2g} \to -e^{2g}$ on the sphere by taking $\theta \to i\theta$ in

(8). Then we get a solution similar to (9) that reads

$$
\begin{aligned}
e^{-3\varphi} &= 1 + e^{-2\rho}[-\frac{1}{2}\log(e^{2\rho}+\frac{1}{2}) + C_1] \\
e^{2g} &= e^{2\rho}e^{-\varphi} \\
e^{2f} &= (e^{2\rho}+\frac{1}{2})e^{-\varphi} \\
(\frac{dr}{d\rho})^2 e^{2f} &= \frac{e^{4\rho}}{(e^{2\rho}+\frac{1}{2})^2}e^{2\varphi}
\end{aligned}
\tag{10}
$$

Again we interpret C_1 as giving the expectation value of the operator $\mathcal{O}_2$. In this case something interesting happens. For large enough negative values of C_1 we have a singularity in the IR that looks very similar to that analyzed for the case of H^2. For positive enough values of C_1, corresponding to positive values of $\langle \mathcal{O}_2 \rangle$, we find that the singularity is *not* allowed by the criterion in section or Gubser's criterion.[28] This is very fortunate because we do not expect this field theory to have a Higgs branch. This is due to the fact that the scalars normal to the sphere have no zero modes as can be seen from the fact that (2) is positive for non-zero values of $|\mathcal{Z}|$ if $R > 0$. The reduction of $N = 4$ SYM on S^2 was studied in Ref. 44. This theory does have a Coulomb branch and indeed we see that in the supergravity solution.

When we talk about Coulomb and Higgs branches through this section we should remember that vacuum expectation values for massless fields are not well defined in two dimensions. So we should interpret these solutions as describing some semi-classical states were some aspects do not change very quickly with time. Once the singularities get resolved by the full theory, by taking into account the IR degrees of freedom that give rise to the singularities then we expect that the solution would change slowly with time as the vevs of massless fields drift through moduli space.

3.2. *Twists preserving (2,2) supersymmetry*

Another way in which we can embed the $U(1)$ group of the spin connection is the following. Consider the $SO(2)^3$ subgroup of $SO(6)$ which corresponds to rotations within three orthogonal planes. Let us call these three generators T_i, $i = 1, 2, 3$. We can take the generator $T = \frac{1}{2}T_2 + \frac{1}{2}T_3$. This breaks $SO(6) \to SO(2) \times SO(2) \times SU(2)$. Under this choice of subgroup the scalars split into two for which the normal bundle is trivial and four for which the normal bundle is such that they effectively become spinors. Another way to state this choice is to consider the splitting of $SO(6) \to SO(2) \times SO(4)$ that we considered in the previous section but now we take the gauge connection in $U(1) \subset SU(2)_L \subset SO(4)$.

It can be shown that a Riemann surface in a CY manifold typically has a normal bundle which is topologically like the one we are considering here.[45] We expect then that if we wrap a D3 brane on this Riemann surface we get a theory that is similar to the theories that we are considering here. In these more general theories the normal bundle does not generically have the additional $SU(2)$ symmetry mentioned above.

Let us consider the bosonic Lagrangian for this theory. Let us call $\mathcal{W}_1, \mathcal{W}_2$ the two complex fields parameterizing the four directions where the gauge field is nonzero. This Lagrangian will be similar to (2) but the coefficient of the curvature term will be different. In principle we should get these terms by demanding supersymmetry. However we will find them by demanding holomorphicity, as in the discussion below (2). The Lagrangian is then

$$S = \int Tr\{\sum_i |\mathcal{D}_z \mathcal{W}_i|^2 + |\mathcal{D}_{\bar{z}} \mathcal{W}_i|^2 + \frac{1}{8} R|\mathcal{W}_i|^2\} \tag{11}$$

Again the remaining two untwisted scalars have an action of the form $\int(\partial\phi)^2$. The presence of the curvature term in (11) implies, as in the discussion around (5), that we are turning on the operator in the **20**. It turns out to be the same operator that we were turning on before but with a different coefficient.

When we consider the IR limit of this theory, at energies smaller than the size of the Riemann surface again we expect to find a conformal field theory. We can compute the central charge of this theory by an anomaly argument. When we reduce the theory to two dimensions we find left and right moving massless fermions with different transformation properties under the R-symmetry group. Let us classify the massless fermions according to their transformation properties under the $SO(2) \times SO(2) \times SO(2)$ subgroup of $SO(6)$. These fermions come from the four dimensional gauginos. There are four possibilities for their charges: (A) (+,+,+), (B) (-,-,+), (C) (-,+,-) (D) (+,-,-) . These have positive chirality in $SO(6)$, their complex conjugates have negative chirality in $SO(6)$. Now we can calculate the number of fermion zero modes on Σ_g via an index theorem. Let us do the calculation in the weak coupling limit where we have essentially N^2 free fields. These are fermions which are coupled to an external gauge field, which is the gauge field implementing the twisting, whose generator was defined as T above. The charges for the fermions fields are $t_A = 1/2$, $t_B = -1/2$, $t_C = t_D = 0$. The index theorem says that the difference between the number of zero modes on Σ_g with positive and negative chirality is

$$n^+ - n^- = \frac{1}{2\pi} t \int_\Sigma F = \frac{t}{2\pi} \int_\Sigma R = 2t(g-1) \; , \tag{12}$$

where we used that the gauge connection is the same as the spin connection. Since the ten-dimensional chiralities of the gauginos are positive, and we have taken them to have positive $SO(6)$ chirality, then we conclude that the chiralities in $1+1$ dimensions are the same as those in Σ. So we conclude that the difference between the number of left and right movers $n_R - n_L$ with given charges is also given by (12). Now it is useful to define $U(1)_L$ and $U(1)_R$ symmetries that will become the $U(1)_L$ and $U(1)_R$ symmetries of the CFT in the IR. In order to do this we note that the right-moving preserved supercharge has charges as those of fermions in A, while the left-moving one has charges as those in B. So we define $T_R = \frac{1}{2}T_1 + \frac{1}{2}T_2 + T_3$ and $T_L = -\frac{1}{2}T_1 - \frac{1}{2}T_2 + T_3$ with this definition we see that the supersymmetry parameter generating the right moving supersymmetries has charge $(q_R, q_L) = (1,0)$ and vice versa for the left moving one. These symmetries are anomalous and their anomalies are given by the usual formulas $k_R = \sum_i (n_R^i - n_L^i)(q_R^i)^2$ where the sum runs over all fermions in the theory. Using (12) we get $k_L = k_R = N^2(g-1)$. Since the anomalies should be the same at weak and strong coupling, and since the (2,2) superconformal algebra links the anomaly to the central charge we can compute the central charge in the field theory to be

$$c_L = c_R = 3N^2(g-1) \ . \tag{13}$$

We can find the supergravity solution by the method we explained above. We start by writing the supersymmetry variation equations, see the appendix,

$$g' = -\frac{1}{3}e^f[(2e^{-\varphi} + e^{2\varphi}) - e^{-2g}e^{\varphi}] \tag{14}$$

$$f' = -\frac{e^f}{6}[2(2e^{-\varphi} + e^{2\varphi}) + e^{-2g}e^{\varphi}] \tag{15}$$

$$\varphi' = \frac{2}{3}e^f[(-e^{-\varphi} + e^{2\varphi}) - \frac{1}{4}e^{-2g}e^{\varphi}] \tag{16}$$

We could not solve these equations completely. However we could partially solve them by finding an analytic expression relating g and φ

$$e^{2g+\varphi} = e^{2g-2\varphi} + \frac{g+2\varphi}{2} + C_1 \tag{17}$$

When $C_1 = 1/4$ the solution ends at the fixed point where $g' = \varphi' = 0$. This corresponds to an AdS_3 region where the Riemann surface has constant size. We can see from the equations (14) that this happens when $e^{2g} = 2^{-\frac{4}{3}}$ and $e^{3\varphi} = 2$. We also have then $e^{2f} = \frac{1}{2^{\frac{4}{3}}r^2}$. The solution interpolates between AdS_5 and $AdS_3 \times \Sigma$. We can calculate the central charge from the supergravity solution by calculating the effective three dimensional Newton's constant and then use the formula $c = 3R_{ads}/(2G_N^{(3)})$. In order to do this it is useful first to calculate the five dimensional Newton's constant and then reduce from five to three dimensions. We are quotienting H_2 so that we have a finite volume Riemann surface. Since the surface has constant curvature we can relate its volume to the radius of curvature of H_2 and the genus. After we do this we get a central charge which agrees with (13). See the appendix for more details.

Once we know the five dimensional metric and gauge fields it is easy to lift up the solutions to ten dimensions using the formulas in Ref. 34. The ten dimensional metric in the AdS_3 region, where φ and g are constant, reads

$$ds^2 = 2^{-\frac{3}{2}}\sqrt{\Delta}ds_3^2 + \frac{2^{-\frac{3}{2}}}{\sqrt{\Delta}}\Bigg[\Delta\frac{dx^2+dy^2}{y^2} + 2\Delta d\theta^2 \\ + 4\cos^2\theta(d\psi^2 + \sin^2\psi(d\phi_1 + \frac{1}{2y}dx)^2 + \cos^2\psi(d\phi_2 + \frac{1}{2y}dx)^2) + 2\sin^2\theta d\phi_3^2\Bigg] \tag{18}$$

where we have defined;

$$ds_3^2 = \frac{(dr^2 - dt^2 + dz^2)}{r^2} \tag{19}$$

$$\Delta = 1 + \sin^2\theta \tag{20}$$

the angles $\theta, \psi, \phi_i, i : 1, 2, 3$ were introduced in order to parametrize the five sphere. The expression for the self dual five form F_5 can also be obtained from the formulas in Ref. 34. The metric (18) has $SO(2) \times SO(2) \times SU(2)$ isometries as expected. The $SU(2)$ isometry is not so obvious in the coordinates in (18) but can be made more manifest by writing the original S^3 parametrized by ψ, ϕ_1, ϕ_2 as the Hopf fibration on S^2. Then the gauge field is only appearing in the fiber and the $SU(2)$ isometries are those of S^2.

It is interesting to note that there are other solutions like the one above, where we allow a second scalar field φ_2. This second scalar field is associated in the original four dimensional field theory to the operator $\mathcal{O}'_2 \sim |\mathcal{W}_1|^2 - |\mathcal{W}_2|^2$. In the IR region they describe marginal deformations of the theory with $\varphi_2 = 0$. This marginal deformation breaks the $SU(2)$ isometry mentioned above to $SO(2)$. In the AdS_3 region the fields take values that can be parametrized in term of φ_2 as (see the appendix for details)

$$e^{3\varphi_1} = 2\cosh\varphi_2 \qquad e^{2g} = \frac{e^{2\varphi_1}}{4} \qquad f = -\log[e^{2\varphi_1} r] \tag{21}$$

It is easy to see that the supergravity central charge is the same as above (13).

There are more twistings of the D3 brane theory that we could consider. In the appendix we discuss more general cases. In some of those cases the spins of fields after twisting are not integer of half integer, so it is not clear to us that the theories really make sense.

Some twisted solutions of five dimensional gauged supergravities were considered in Refs. 6,7. Those solutions had the restriction that the scalar fields had to be constant in the flow from five to three dimensions. This restriction seems hard to obey if we insist in not having fields with fractional spins, at least for the case of a D3 brane.

4. Twists of the M5 Brane Theory

In this section we consider $M5$ branes wrapped on two dimensional Riemann surfaces Σ_g times R^4. We take the limit $l_p \to 0$ keeping the size of the Riemann surface fixed so that we obtain the $(0,2)$ six-dimensional superconformal theory on $R^4 \times \Sigma_g$. We will consider two possible ways of embedding the spin connection in the $SO(5)$ R-symmetry group of the $(0,2)$ SCFT. In the first we preserve two supersymmetries and in the second we preserve one supersymmetry (in $d = 4$ notation). It is convenient to use a $U(1) \times U(1)$ truncation of the $SO(5)$ gauged supergravity theory. If we denote by 1,...,5 the directions orthogonal to the 5-brane, then these two $U(1)$'s rotate the directions 12 and 34 respectively. This truncation was studied in Ref. 34

and the supersymmetry variation was studied in Ref. 46. In both cases we will make the following ansatz for the metric

$$ds^2 = e^{2f(r)}(dr^2 + du^2 + dv^2 + dz^2 - dt^2) + \frac{e^{2g(r)}}{y^2}(dx^2 + dy^2) \qquad (22)$$

The boundary condition is as in the previous case that $g \sim f \sim -log(r)$, for $r \to 0$.

Throughout this section we work in units where the radius of AdS_7 is one. We can restore the dependence on this parameter by multiplying the seven or ten dimensional metric by the radius of AdS_7 which is given by $R^2_{AdS_7} = 4(\pi N)^{2/3} l_p^3$. [e]

4.1. $\mathcal{N} = 2$ *case*

In this section we consider configurations preserving $\mathcal{N} = 2$ supersymmetry in four dimensions. The field content and the Lagrangian of the $d = 7$ gauged supergravity, together with the relevant supersymmetry transformations are described in detail in Ref. 46 and in the appendix.

In order to get an idea of what the field theory looks like let us first consider a single M5 brane. In this case, in analogy with the D3 brane case we expect that there is an operator $\mathcal{O}_4 = |\mathcal{Z}|^2 - 2/3(\mathcal{W}_3^2 + \mathcal{W}_4^2 + \mathcal{W}_5^2)$ where $\mathcal{Z} = \mathcal{W}_1 + i\mathcal{W}_2$ is the scalar that we are twisting and $\mathcal{W}_i$ are the five scalar fields describing fluctuation of the M5 brane in the five orthogonal dimensions. The presence of this operator can be seen by expanding directly the Nambu action for the brane. We can calculate its coefficient as we did for the D3 brane case around (2), (5). In the case of multiple coincident branes we also have an operator of dimension 4, with the same $SO(5)$ transformation properties which is turned on. So we expect that a supergravity field, dual to this operator, will have a non-trivial boundary condition at infinity.

In order to get an idea of what the field theory looks like, let us consider first the case of a single M5 brane. If it is wrapped on S^2 then we only get four massless modes, the three scalar fields in the directions that are untwisted and the component of the B field on S^2. These form a single four-dimensional hypermultiplet. This theory therefore has only a "higgs" branch. In the case that the M5 wraps a genus $g > 1$ Riemann surface then we get the same hypermultiplet that we were getting above plus g vector multiplets. These vector multiplets correspond to modes of the B field with one index along a non-contractible cycle in the Riemann surface and one index in four dimensions. So we get a $U(1)^g$ theory with one neutral hypermultiplet. The scalars in the vector multiplet correspond to modes of the twisted scalar field $\mathcal{Z}$ and they represent deformations of the Riemann surface in the normal directions that are twisted.

It is less clear what the resulting four dimensional gauge theory is when we have N coincident branes, since we do not have an explicit Lagrangian we could use in six dimensions to derive the four dimensional gauge theory. It seems possible to give a DLCQ definition of the theory in the spirit of Refs. 47,48,49. Of course,

[e] In conventions were the eleven dimensional Newton's constant is $G_N^{11} = 16\pi^7 l_p^9$. [24]

it would be very nice to give a direct definition of the theory. As a step in that direction we compactify one of the worldvolume directions on a circle so that we get a D4 brane wrapped on Σ. Then the low energy theory is a three dimensional $U(N)$ gauge theory with 8 supercharges. In the case of S^2 get get a pure gauge theory with only a Coulomb branch while in the $g > 1$ case we also get g adjoint hypers. Notice that the four dimensional theory we want to find should be such that when we reduce it on a circle the Higgs branch of the 4d theory should become the Coulomb branch of the 3d theory and the Higgs branch of the 3d theory results from the Coulomb branch of the 4d theory as in Ref. 50. After dimensional reduction to three dimensions Higgs and Coulomb branches would be exchanged.[51] In the case of S^2 we could say a bit more about this four dimensional low energy theory. When we go down to three dimensions we get a $U(N)$ field theory. The Coulomb branch of this theory is given by the moduli space of N monopoles in $SU(2)$.[50,52] Three dimensional mirror symmetry exchanges this Coulomb branch with the Higgs branch of the mirror. So the four dimensional theory is a theory whose Higgs branch is the same as the moduli space of N monopoles of $SU(2)$. This space is smooth and has no singularities. So the four dimensional theory is just this sigma model with only hypermultiplets.[f] It is harder to say what the four dimensional field theory is in the case of $g > 1$. Again we can say fairly easily what the Higgs branch of the four dimensional theory is. It is the same as the Coulomb branch of the 3d theory. It is less clear what the Coulomb branch of the four dimensional theory is. What we know is that upon reducing to three dimensions and doubling the variables, as in Ref. 50 we should find the Higgs branch of the D4 theory. The dimensionality of this Higgs branch is what we analyzed for the D3 brane in section 3.1, it has dimension $4N^2(g-1)$. So we expect that the Coulomb branch of the four dimensional theory has dimension $2N^2(g-1)$. This is suggestive of a very large gauge group of the form $U(N)^{(g-1)N}$.[g] Below we will find, from a supergravity analysis, that the theory has a conformal fixed point where the effective number of degrees of freedom goes as $N^3(g-1)$, which is in rough agreement with the type of gauge group we expect.

Let us now turn to the supergravity solutions. Setting to zero the supersymmetry variations we get

$$\begin{aligned} g' &= -\, e^{f}[e^{2\lambda} - \kappa\frac{e^{-2g}}{4}e^{3\lambda}] + \frac{1}{2}\lambda' \\ \lambda' &= -\, \frac{2}{5}e^{f}[2(e^{-3\lambda} - e^{2\lambda}) + \kappa\frac{1}{4}e^{-2g+3\lambda}] \\ f' &= -\, e^{f+2\lambda} + \frac{1}{2}\lambda' \end{aligned} \tag{23}$$

[f]We thank M. Douglas for suggesting this possibility.
[g]For a related discussion see Ref. 1.

where $\kappa = 1$ for H^2 and $\kappa = -1$ for S^2. The general solution of the equations is

$$
\begin{aligned}
e^{5\lambda} &= \frac{e^{2\rho} + \kappa\frac{1}{2} + C_1 e^{-2\rho}}{e^{2\rho} + \kappa\frac{1}{4}} \\
e^{2g} &= e^{\lambda}(e^{2\rho} + \kappa\frac{1}{4}) \\
e^{2f} &= C_2 e^{2\rho} e^{\lambda} \\
e^{2f}\left(\frac{dr}{d\rho}\right)^2 &= g_{\rho\rho} = e^{-4\lambda}
\end{aligned}
\tag{24}
$$

We have chosen a new variable ρ as implicit in the last equation. $\rho \to +\infty$ corresponds to the boundary of AdS_7. C_2 is a trivial integration constant that can be absorbed by rescaling the four dimensional coordinates. It is related to the size of the surface where we are wrapping the six dimensional theory. In the case $C_1 = 0$ $\kappa = 1$ we find that the solution interpolates between AdS_7 and $AdS_5 \times \Sigma_{g>1}$. This is telling us that the IR dynamics of the $(0,2)$ six dimensional theory on $\Sigma_{g>1} \times R^4$ is given by a four dimensional superconformal field theory. The solution in the IR has the fixed point values

$$
e^{5\lambda} = 2\,, \qquad e^{2g-\lambda} = \frac{1}{4}\,, \qquad e^{f+2\lambda} = \frac{1}{r}\,. \tag{25}
$$

The sign of the expectation value of the operator $\mathcal{O}_4$ is again given by the sign of C_1. A positive value of C_1 therefore corresponds to a positive expectation value for the twisted scalars and therefore to the Coulomb branch. A negative value of C_1 corresponds to an expectation value for the scalars that are untwisted and therefore to the Higgs branch. However for the case of S^2 we do not expect a Coulomb branch since the twisted scalars cannot have an expectation value. Fortunately either Gubser's criterion or our criterion in section rules out this case but allows all the other cases. So in the case of H^2 we can have both signs on C_1 and in the case of S^2 we find that $C_1 \leq \frac{1}{16}$. The fact that we have $1/16$ instead of zero is due to the fact that it is difficult to disentangle the expectation value from the insertion of the operator in supergravity so that our field theory expectation are easy to check only for large values of $|C_1|$, presumably a more correct analysis valid also for small values of C_1 would still agree with field theory expectations.

Using the formulas in Ref. 34 we can uplift the solution to eleven dimensions. Here we give only the form of the solution in its IR fixed point, the AdS_5 region.

$$
\begin{aligned}
ds^2_{11} &= \frac{1}{2}\Delta^{1/3} ds^2_{AdS_5} + \frac{\Delta^{-\frac{2}{3}}}{4}\left[\Delta\frac{(dx^2+dy^2)}{y^2} + \right. \\
&\left. + \Delta d\theta^2 + \cos^2\theta(d\psi + \sin^2\psi d\phi_1^2) + 2\sin^2\theta(d\phi_2 + \frac{dx}{y})^2\right] \\
\Delta &= (1+\cos^2\theta)
\end{aligned}
\tag{26}
$$

Here, the angles $\theta, \psi, \phi_1, \phi_2$ parametrize the four sphere (before the twisting). $ds^2_{AdS_5}$ denotes a unit radius Anti-de-Sitter metric. Note that (26) has $SU(2)\times U(1)$

isometries as required by $\mathcal{N}=2$ superconformal invariance. An explicit expression for the four form gauge field can be written following section 4 of Ref. 34.

This is giving non-singular AdS_5 warped compactifications of M theory. These compactifications are completely smooth. Singular compactifications to AdS_5 were given in Refs. 1,2, see however Refs. 3,23.

It is also possible to find the supergravity duals of the D4 brane theories wrapped on Σ that we discussed above by compactifying the M5 brane solution along a longitudinal direction and then using the standard reduction from eleven dimensional supergravity to type IIA supergravity. We write this solution explicitly in the appendix. In the case of S^2 this solution is related to pure 3d SYM with 8 supercharges. This is the same system that was explored in Ref. 53. In Ref. 53 the starting point was a D6 brane wrapping K3. At low energies this reduces to pure SYM in 3d. A puzzling aspect of the solutions in Ref. 53 was that the K3 remained in the IR geometry, at least at first sight. If we take a limit of the solution in Ref. 53 where the volume of K3 goes to zero, then we can T-dualize and obtain a large K3 and the D6 brane becomes a D4 brane wrapping S^2 which is what we have here. In the IR region of our solution with $C_1 = 1/16$ we find that the volume of S^2 shrinks to zero much faster than g_{00} so we expect that KK excitations in this S^2 decouple from the low energy theory. More precisely, dual of large N pure SYM in 3d would be the region close to the singularity. It is possible that there is a better gravity description for the IR region than the one we explored here, or it could be that the theory only admits a string theory dual but not a weakly coupled gravity dual. The excision of the singularity in Ref. 53 amounts to our criterion that the singularity is admissible by the criterion in section . In fact it was argued in Ref. 53 that it was necessary to move the branes into the Coulomb branch of the 3d theory to remove the singularities.

4.2. $\mathcal{N}=1$ *case*

In this section we consider the case where we embed the spin connection in both of the $U(1)$ factors described above. When we wrap an M5 brane on a Riemann surface in a CY space we typically get normal bundles which are topologically equivalent to this one. The resulting four dimensional field theory has 4 supercharges. It seems hard to say explicitly what this four dimensional theory is. Again if we compactify one circle and we go to three dimensions we can say more. In that case we have a D4 brane wrapped on the Riemann surface. In the S^2 case we find just pure $U(N)$ gauge theory with 4 supercharges.

Let us turn to the supergravity analysis. The equations that result from setting to zero the supersymmetry variations are (see the appendix),

$$\begin{aligned} f' &= -\,e^{4\phi+f} + \phi' \\ g' &= -\,e^{f}\,(e^{4\phi} - \kappa\frac{1}{4}e^{\phi-2g}) + \phi' \\ \phi' &= -\,\frac{1}{5}e^{f}\,(4(e^{-\phi} - e^{4\phi}) + \kappa\frac{1}{4}e^{\phi-2g}) \end{aligned} \tag{27}$$

where $\kappa = 1$ for H^2 and $\kappa = -1$ for S^2. In the case of H^2, this set of equations have solutions for constant values of ϕ, g given by

$$e^{5\phi} = \frac{4}{3} , \qquad e^{-2g-3\phi} = 4 , \qquad e^{f(r)} = e^{-4\phi}\frac{1}{r} . \tag{28}$$

The corresponding eleven dimensional metric is smooth and can be obtained from Ref. 34. Here we give only the AdS_5 region of the solution which can be found using the IR values (28)

$$ds^2_{11} = \Delta^{\frac{1}{3}} ds^2_7 + \frac{\Delta^{-\frac{2}{3}}}{4}[e^{-4\phi}d\mu_0^2 + e^{\phi}(d\mu_1^2 + d\mu_2^2 + \mu_1^2(d\phi_1 + \frac{1}{2y}dx)^2 + \mu_2^2(d\phi_2 + \frac{1}{2y}dx)^2)]$$
$$\Delta = e^{4\phi}\mu_0^2 + e^{-\phi}(\mu_1^2 + \mu_2^2) \tag{29}$$

where μ_i parameterize an S^2, $\mu_0^2 + \mu_1^2 + \mu_2^2 = 1$ and the seven dimensional metric is as in (22) with the values given in (28).

It would be nice to determine whether these AdS_5 compactifications of M-theory are stable under quantum corrections. In other words, it would be interesting to determine whether these field theories are exactly conformal of if they are non-conformal when we take into account $1/N$ corrections, as the ones in Ref. 54.

5. A Criterion for Allowed Singularities

In this section we discuss a proposal for a criterion that would tell us if a supergravity singularity in the IR region of a geometry describing a field theory is allowed or not. The strong form of the final criterion is

The g_{00} component of the metric should not increase as we approach the singularity

we will also discuss a weak form of the criterion

The g_{00} component of the metric should be bounded above.

In what follows we explain this criterion and give a heuristic motivation for it. When we are trying to find supergravity solutions that are dual to field theories we typically encounter singular solutions. These singularities do not necessarily mean that the solutions are wrong, they might be telling us that the supergravity description is failing and that we should go to a dual description. This is the case for $D-p$-branes for $p \neq 3$.[25] It is clear however that not all singularities are allowed. For example a solution like negative mass Schwarzschild should not be allowed since it would imply that the energy is not bounded below. We would like to propose a necessary criterion for weeding out unphysical singularities. Our criterion applies for the IR regions of supergravity backgrounds which are dual to some field theory. If we want to interpret some region as being dual to the IR of some field theory we expect that g_{00} should decrease so that fixed proper energy excitations correspond to lower and lower energy excitations from the point of view of coordinate time, which is the same as field theory time. So our criterion will be that g_{00} should not increase as we approach the singularity. In particular it should not go to infinity. In many cases we can approach the singularity in various directions in the internal manifold. We require that g_{00} does not increase as we approach the singularity along any direction in the internal manifold. Note that it makes sense to talk about g_{00} since we are talking about a field theory with a time translation isometry, which is generated by a Killing vector of the dual geometry, so we are choosing coordinates so that this vector is $\frac{\partial}{\partial t}$.

This criterion certainly forbids negative mass Schwarzschild singularities where $g_{00} \to \infty$. There are some singularities where g_{00} stays constant, like orbifold singularities. We should certainly allow those. Or course the full string theory will then tell us whether it is really allowed or not. There are some singularities where g_{00} increases as one approaches the singularity but stays bounded. A D8 brane has a singularity of this form and it looks like it should be allowed in the full string theory but it seems that it should not be allowed in a region that we want to interpret as the IR of some field theory. But in order to be sure we are not over-restrictive we could allow these and we get the weak form criterion stated in the beginning of this section. If we rule these out we get the first. A diverging g_{00} implies that the singularity is repulsive, massive particles are repelled from the singularity. In these singularities a finite proper energy excitation will have very high energy from the point of view of the field theory. It seems to violate the UV/IR correspondence. It is also hard to see how these singularities can form from gravitational collapse since they are repulsive. For a concrete example where an example of a repulsive singularity is discussed see Ref. 53. Note that the c theorem in Ref. 55 does not imply that g_{00} always decreases. In Ref. 55 it was proven that the metric of gauged supergravity decreases, but the full ten or eleven dimensional g_{00} factor includes a warp factor and this warp factor could increase even though the metric appearing in the gauged supergravity action is decreasing. The physical redshift factor is given by the ten or eleven dimensional metric, as long as the supergravity solution is valid. In string theory one should use the Einstein metric for this analysis.

In Ref. 28 it was proposed that admissible singularities are those that can be generalized to finite temperature. The criterion proposed here is easy to check and it can be applied directly to the ten or eleven dimensional metric and does not need to use the gauged supergravity form of the potential. In the cases analyzed here it gives the same results as Gubser's criterion[28] we suspect that the boundedness of the gauged supergravity potential will ensure that g_{00} is bounded but we did not prove it.

Note that this criterion is valid for regions that we want to interpret as the IR of a gauge theory. It is certainly violated when we approach the boundary of AdS or the boundaries of the geometries describing some D-p-branes.[25] In fact it is necessary to put boundary conditions for fields in the boundary of AdS. These boundary conditions are interpreted as defining the details (operator insertions) of the field theory dual to the given background.

6. A No Go Theorem

In this section, which is to great extent disconnected with the rest of the paper, we present an argument saying that there are no non-singular wrapped compactifications in a large class of supergravity theories. Our main assumption will be that the potential for scalar fields is non-positive. The massive type IIA case is treated separately. We do not need to use the equations of motion of the matter fields. Then we will show that there is no non-singular warped compactification to R^d or de-Sitter space dS^d, $d \geq 2$ with finite d dimensional Newton's constant. We will do this for general d, but the reader interested in real world applications might want to take $d = 4$. The no go theorem remains true even in the case that we allow singularities such as those allowed by the strong form of the criterion in section . These are singularities that we *might* be able to interpret as arising from the IR dynamics of some field theory. The argument is quite general and only relies on the equation of motion for the warp factor and does not rely on supersymmetry. There are ways to evade this argument which involve including higher derivative corrections to the supergravity equations, or starting from a theory that already has a positive cosmological constant. There are other no go theorems for supersymmetric RS models using five dimensional gauged supergravity.[13,12,11,14] These are complementary to our arguments, since the five dimensional potential could be positive. If the five dimensional gauged supergravity arises from a large volume compactification then we could apply our results but the arguments in Refs. 13,12,11,14 also cover 5d theories which cannot be interpreted as large volume compactifications.

We consider a D dimensional gravity theory, with $D > 2$, compactified down to d dimensions. We denote by $M, N, L, ..$ the D dimensional indices. We denote by $\nu, \nu, \rho, ...$ the d dimensional indices and by m, n, l, the $D - d$ dimensional indices. We will assume that the D dimensional gravity theory satisfies the following conditions.

- The gravity action does not contain higher curvature corrections.
- The potential is non-positive, $V \leq 0$. This condition in not obeyed in massive IIA supergravity which has a positive cosmological constant so we treat that case separately in . V could be just a negative cosmological constant or it can depend on the scalars but it cannot be positive (at least in the range of values of scalar fields that is explored in the solution under consideration).
- The theory contains massless fields with positive kinetic terms. These massless fields have field strengths which are n forms, $F_{i_1,..,i_n}$. For $n = 1$ we have scalar fields, $n = 2$ Maxwell fields (these could be non-abelian, as long as the metric on the group is positive definite so that the kinetic terms are positive), etc. We consider $n < D$, if $n = D$ it would give a contribution similar to a potential and we go back to the previous assumption.
- The d dimensional effective Newton's constant is finite.

We start by writing out Einstein's equations in D dimensions

$$R_{MN} = T_{MN} - \frac{1}{D-2} g_{MN} T^L_L \tag{30}$$

Notice that in (30) we neglected higher derivative corrections. We write the metric as

$$ds_D^2 = \Omega^2(y)\left(dx_d^2 + \hat{g}_{mn}dy^n dy^m\right) \tag{31}$$

where $dx_d^2 = \eta_{\mu\nu}dx^\mu dx^\nu$ where η is the metric of the d dimensional space which is either Minkowski or de-Sitter space. Now we calculate the $R_{\mu\nu}$ components of the D dimensional metric and we find that Einstein's equations imply

$$R_{\mu\nu} = R_{\mu\nu}(\eta) - \eta_{\mu\nu}\left(\hat{\nabla}^2 \log\Omega + (D-2)(\hat{\nabla}\log\Omega)^2\right) = T_{\mu\nu} - \frac{1}{D-2}\Omega^2\eta_{\mu\nu}T^L_{\ L} \tag{32}$$

where the hat denotes covariant derivatives and contraction of indices with respect to the metric $\hat{g}$. Taking the trace over η on both sides we find

$$\hat{\nabla}^2\log\Omega + (D-2)(\hat{\nabla}\log\Omega)^2 = \frac{1}{(D-2)\Omega^{D-2}}\nabla^2\Omega^{D-2} = R(\eta) + \Omega^2(-T^\mu_{\ \mu} + \frac{d}{D-2}T^L_{\ L}) \tag{33}$$

where in the term involving the stress tensor on the right hand side we contract the indices with the D dimensional metric and $R(\eta)$ is the curvature of the d dimensional metric η. We will now proceed to prove that the term in the right hand side involving the stress tensor is non-negative.

6.1. $\tilde{T} \geq 0$

The stress tensor will be the sum of the contributions to the stress tensor of the various massless fields. We will consider each contribution individually since they are all adding up to the total stress tensor. Let us define

$$\tilde{T} \equiv -T^\mu_{\ \mu} + \frac{d}{D-2}T^L_{\ L} \tag{34}$$

We want to show that all contributions to $\tilde{T}$ are non-negative. Let us first consider the potential term. We will not keep track of irrelevant positive numerical constants. The stress tensor is

$$T_{MN} \sim -Vg_{MN}\ , \qquad \tilde{T} \sim -V\frac{2d}{D-2} \geq 0 \tag{35}$$

if $V < 0$ as assumed. Now let us consider the n form field strengths. Their stress tensors are

$$\begin{aligned} T_{MN} =& F_{ML_1..L_{n-1}}F_N^{\ L_1..L_{n-1}} - \frac{1}{2n}g_{MN}F^2 \\ \tilde{T} =& -F_{\mu L_1..L_{n-1}}F^{\mu L_1..L_{n-1}} + \frac{d}{D-2}(1-\frac{1}{n})F^2 \end{aligned} \tag{36}$$

In principle we could have functions of scalar fields multiplying these expressions, as we have in some supergravity theories, and we could also have many types of n form fields. We will not indicate these explicitly but it is obvious how to extend the following arguments to those cases. The space time indices of non-vanishing components of F could be completely along the internal dimensions or, if $n \geq d$, they could have d out of n indices along the d dimensions and the rest along the internal dimensions. Other possibilities do not preserve the isometries of R^d or dS^d. In constructing $\tilde{T}$ these two types of components will make separate contributions. We will therefore consider them independently and show that each of them is positive. So let us first consider the part of F with all indices internal. Then we have that $F^2 \geq 0$ and we see from (36) that we have a positive contribution. For all $n > 1$ forms this contribution is strictly positive if we have a non-vanishing field strength, but for $n = 1$ the contribution is zero even if we have a non-vanishing field strength. Now we consider the part of the field strength with components along the d-dimensional space. The difference between the term that contains a trace over the μ index and the others is that we are choosing a particular order of contractions of the indices comparing the two we find that

$$F_{\mu L_1..L_{n-1}} F^{\mu L_1..L_{n-1}} = \frac{d}{n} F^2.$$

Then we find that

$$\tilde{T} = -F^2 \frac{d(D-1-n)}{n(D-2)} \geq 0 \tag{37}$$

Where we used that $F^2 < 0$ since we are considering temporal components of F. We have also used that we are considering $n \leq D-1$.

6.2. *Condition on the warp factor*

Multiplying (33) by a power of Ω and using that $\tilde{T} \geq 0$ we conclude that

$$\hat{\Omega}^{(D-2)} \nabla^2 \Omega^{(D-2)} \geq 0 \tag{38}$$

with equality holding only if the right hand side of (33) is zero so that the d dimensional space is Minkowski space. Remember that the d dimensional Newton constant is given by

$$\frac{1}{G_N^d} \sim \int d^d y \sqrt{\hat{g}} \Omega^{(D-2)} \tag{39}$$

We are assuming that this Newton constant is finite.

Let us first assume that Ω is bounded below and above in the internal manifold. In that case the internal manifold should be compact. Integrating (38) over the compact internal space by parts we conclude that $\int d^(D-d)y\sqrt{\hat{g}}(\hat{\nabla}\Omega^{(D-2)})^2 \leq 0$ which is possible only if Ω is constant. In that case we conclude that the right hand side of (33) is zero, so that we cannot have a deSitter space and the only n forms that we can be turned on are the $n = 1, D-1$ forms.

As discussed in section we expect that singularities where Ω diverges should not be allowed. So we conclude that Ω is bounded above. Now suppose that we have regions where $\Omega \to 0$ or we allow singularities obeying the strong form of the criterion in section , which says that g_{00} should not increase as we approach the singularity. In this case we can define a region $\mathcal{R}$ which leaves out the singularities and such that $\Omega > \epsilon$ in $\mathcal{R}$ for a suitably small ϵ. By our assumptions about the singularities it is clear that we can choose $\mathcal{R}$ so that $\nabla\Omega$ is either zero or pointing inwards at the boundary of $\mathcal{R}$.

Now we can integrate (38) by parts in region $\mathcal{R}$, and we get

$$\int_{\mathcal{R}_\epsilon} \left(\hat{\nabla}\Omega^{(D-2)}\right)^2 \leq - \int_{\partial\mathcal{R}_\epsilon} (\vec{n}\nabla\Omega^{(D-2)})\Omega^{(D-2)} \leq 0 \tag{40}$$

where we used the assumption that Ω was non-increasing at the boundary. Again we conclude that the warp factor has to be constant and that a R-S or deSitter compactification of this type is not allowed.[h]

In summary we have proven that given our assumptions there are no compactifications to deSitter space or Randall-Sundrum compactifications where the only possible singularities are such that the warp factor Ω going to zero at the singularity. As we explained in section , these are the singularities which *might* have a field theory interpretation.

The fact that de-Sitter is not allowed *does not* imply that there are no expanding universe solutions in large volume compactifications, it only says that there are no homogeneous $SO(1, d+1)$ invariant de-Sitter solutions. So we could have solutions where some scalar fields are time dependent.[i]

The most natural question is whether there are ways to evade this argument. For that we note that once we include higher derivative corrections to the gravity action, like the ones present in string theory or M-theory, then the positivity argument does not hold and there can be warped compactifications.[58,59,60,61,62,63] For example, in heterotic compactifications or Horava Witten compactifications[64] we can have warp factors but we need crucially the higher derivative terms which modify the Bianchi identity for the three or four form field strengths respectively. The same can be said about type I examples where we have orientifolds. These higher derivative terms are crucial to get important physical aspects of string compactifications. All we are saying is that these stringy corrections to the gravity equations are also crucial if one wants a deSitter of Randall Sundrum compactification.[10] Of course, it is very interesting to study these solutions more precisely. It is possible to evade the arguments in this section by by allowing potentials for scalar fields that are positive, as explicitly demonstrated in.[65,66]

6.3. *No Minkowski or de Sitter compactifications of massive IIA sugra*

In this section we show that there are no compactification of massive IIA on smooth manifolds without boundaries down to R^d or dS^d. The equations of motion for the

---

[h] Note that the solutions in Ref. 56 do not obey the condition that the higher dimensional g_{00} goes to zero at the singularity.

[i] The same comment applies to the statement in Ref. 57 that their compactifications have no de-Sitter or Anti-de-Sitter solutions. We expect to find also expanding or contracting universes with time dependent scalar.[57]

metric and the dilaton in massive IIA supergravity are given by[67]

$$R_{MN} = \frac{m^2}{16} e^{-5\phi} g_{MN} + 2\partial_M\phi\partial_N\phi + e^{2\phi}(G_M^{\ PQ} G_{NPQ} - \frac{1}{12} g_{MN} G^2) + \\ + 2m^2 e^{-3\phi}(B_M^{\ P} B_{NP} - \frac{1}{16} g_{MN} B^2) + \frac{1}{3} e^{-\phi}(F_M^{\ PQR} F_{NPQR} - \frac{3}{32} g_{MN} F^2) \tag{41}$$

$$0 = \Box\phi + \frac{5m^2}{8} e^{-5\phi} + \frac{1}{48} e^{-\phi} F^2 - \frac{1}{6} e^{2\phi} G^2 + \frac{3m^2}{4} e^{-3\phi} B^2 \tag{42}$$

where we have defined the square of a tensor of n indices as $H^2 = H^{M_1 \ldots M_n} H_{M_1 \ldots M_n}$. We will not need the precise definition of the tensors B, G and F in (41)(42) but we will use that they are real and antisymmetric. We assume $d \geq 2$. We make an ansatz for the metric as in (31). Following the steps that lead to (33) we find

$$\frac{1}{(D-2)\Omega^D} \nabla^2 \Omega^{D-2} = \Omega^{-2} R(\eta) + d\left[-\frac{m^2}{16} e^{-5\phi} + \frac{1}{12} e^{2\phi} G_e^2 + \frac{m^2}{8} e^{-3\phi} B_e^2 + \frac{1}{32} e^{-\phi} F_e^2 \right] \\ - d\left[\theta(3-d)\frac{1}{4} e^{2\phi} G_l^2 + \theta(2-d)\frac{7m^2}{8} B_l^2 + \theta(4-d)\frac{5}{96} e^{-\phi} F_l^2 \right] \tag{43}$$

where we have denoted by H_e^2 the square of a tensor with components purely in the internal dimensions and by H_l^2 the square of the tensor with components along the d dimensional spacetime directions. These can only appear if the rank of the tensor is bigger or equal to d that is the reason we have factors of $\theta(n-d)$ where $\theta(x) = 1$ for $x \geq 0$ and zero otherwise. Now notice that

$$\Box\phi = \frac{1}{\Omega^D} \hat{\nabla}_m \left(\Omega^{D-2} \hat{g}^{mn} \partial_n \phi \right) \tag{44}$$

We then see that if we multiply (43) by $10\Omega^D$ and add (44) times $d\Omega^D$ we get

$$10\Omega^D \times (43) + d\Omega^D \times (42) = \frac{10}{D-2} \hat{\nabla}^2 \Omega^{D-2} + d\hat{\nabla}_m \left(\Omega^{D-2} \hat{g}^{mn} \partial_n \phi \right) = \\ = \Omega^{D-2} R(\eta) + + d\Omega^D \left[\frac{2}{3} e^{2\phi} G_e^2 + 2m^2 e^{-3\phi} B_e^2 + \frac{1}{3} e^{-\phi} F_e^2 - \right. \\ \left. -\theta(3-d)\frac{8}{3} e^{2\phi} G_l^2 - \theta(2-d) 8m^2 B_l^2 - \theta(4-d)\frac{1}{2} e^{-\phi} F_l^2 \right] \tag{45}$$

We see that the right hand side of (45) is positive since $H_e^2 > 0$ and $H_l^2 < 0$ for all n forms.

If we have a compact internal manifold then we can integrate (45)over the manifold. We get zero from the left hand side since we have a total derivative. On the right hand side we get a non-zero result unless $R(\eta) = 0$ and $B = F = G = 0$. Now that we know this we can integrate just the equation for the dilaton (42) over the internal manifold and we get a contradiction. So we find that there are no non-singular compactifications (over a compact internal manifold) to either de Sitter or Minkowski space. If we assume that we have regions where the warp factor could go to zero but the dilaton stays constant, as we expect in an AdS region. Then we also get a contradiction by following steps similar to those in the previous subsection and including boundary terms for the conformal factor. This excludes compactifications of the RS type. In Ref. 67 there are several examples of compactifications to Anti-de-Sitter manifolds.

Obviously there are compactifications of massive IIA to Minkowski space on spacetimes with boundaries where the conformal factor is not decreasing as we approach the boundaries as explicitly demonstrated in Ref. 68. It would be nice to see if we can get compactifications to de Sitter space in cases where the compact manifold has boundaries.

Acknowledgments

We would like to thank M. Bershadsky, M. Douglas, J. Gomis, K. Hori, C. Johnson, A. Kapustin, A. Strominger and C. Vafa for discussions.

The research of C.N. was supported by CONICET and Fundacion Antorchas. The research of J.M. was supported in part by DOE grant DE-FGO2-91ER40654, NSF grant PHY-9513835, the Sloan Foundation and the David and Lucile Packard Foundation.

References

1. M. Alishasha and Y. Oz, *Phys. Lett.* **B465**, 136 (1999).
2. M. Cvetic, H. Lu, C. Pope and J. Vazquez-Poritz, hep-th/0005246.
3. A. Fayyazuddin and D.J. Smith, *JHEP* **9904**, 030 (1999).
4. M. Bershadsky, C. Vafa and V. Sadov, *Nucl. Phys.* **B463**, 420 (1996).
5. D.Z. Freedman and B. Zwiebach, *Nucl. Phys.* **B237**, 573 (1984).
6. A.H. Chamseddine and W.A. Sabra, *Phys. Lett.* **B477**, 329 (2000).
7. D. Klemm and W.A. Sabra, *Phys. Rev.* **D62**, 024003 (2000).
8. S. Cacciatori, D. Klemm, W.A. Sabra and D. Zanon, *Nucl. Phys.* **B587**, 277 (2000) [hep-th/0004077].
9. A.H. Chamseddine and W.A. Sabra, *Phys. Lett.* **B485**, 301 (2000) [hep-th/0003213].
10. L. Randall and R. Sundrum, *Phys. Rev. Lett.* **83**, 4690 (1999).
11. R. Kallosh and A. Linde, *JHEP* **0002**, 005 (2000).
12. K. Behrndt and M. Cvetic, *JHEP* **0006**, 015 (2000).
13. M. Wijnholt and S. Zhukov, hep-th/9912002.
14. G.W. Gibbons and N.D. Lambert, *Phys. Lett.* **B488**, 90 (2000) [hep-th/0003197].
15. G. Papadopoulos, J.G. Russo and A.A. Tseytlin, *Class. Quant. Grav.* **17**, 1713 (2000).
16. H.-S. Yang, hep-th/9902128.
17. Y. Kiem and D. Park, *Phys. Lett.* **B450**, 41 (1999).
18. D. Youm, hep-th/9902208.
19. A. Loewy, *Phys. Lett.* **B463**, 41 (1999).
20. A. Gomberoff, D. Kastor, D. Marolf and J. Traschen, *Phys. Rev.* **D61**, 024012 (2000).
21. D. Youm, hep-th/9902208.
22. S.A. Cherkis, hep-th/9906203.
23. A. Fayyazuddin and D.J. Smith, *JHEP* **0010**, 023 (2000) [hep-th/0006060].
24. J. Maldacena, *Adv. Theor. Math. Phys.* **2**, 231 (1998).

25. N. Itzhaki, J.M. Maldacena, J. Sonnenschein and S. Yankielowicz, *Phys. Rev.* **D58**, 046004 (1998).
26. C. Vafa and E. Witten, *Nucl. Phys.* **B431**, 3 (1994).
27. M. Bershadsky, A. Johansen, V. Sadov and C. Vafa, *Nucl. Phys.* **B448**, 166 (1995).
28. S.S. Gubser, hep-th/0002160.
29. Y. Oz, hep-th/0004009.
30. J. Polchinski and M.J. Strassler, hep-th/0003136.
31. N. Seiberg and E. Witten, *JHEP* **9904**, 017 (1999). .
32. E. Witten, *Adv. Theor. Math. Phys.* **2**, 253 (1998).
33. M. Gunaydin, L.J. Romans and N.P. Warner, *Phys. Lett.* **B154**, 268 (1985).
34. M. Cvetic *et al.*, *Nucl. Phys.* **B558**, 96 (1999).
35. M. Gunaydin, G. Sierra and P.K. Townsend, *Nucl. Phys.* **B253**, 573 (1985).
36. V. Balasubramanian, P. Kraus and A. Lawrence, *Phys. Rev.* **D59**, 046003 (1999).
37. D.Z. Freedman, S.S. Gubser, K. Pilch and N.P. Warner, *JHEP* **0007**, 038 (2000) [hep-th/9906194].
38. E. Witten, *JHEP* **9707**, 003 (1997).
39. M. Berkooz and H. Verlinde, *JHEP* **9911**, 037 (1999).
40. O. Aharony and M. Berkooz, *JHEP* **9910**, 030 (1999).
41. D. Marolf and A. Peet, *Phys. Rev.* **D60**, 105007 (1999).
42. S. Katz, D.R. Morrison and M.R. Plesser, *Nucl. Phys.* **B477**, 105 (1996).
43. C.G. Callan and J.M. Maldacena, *Nucl. Phys.* **B513**, 198 (1998).
44. A. Imaanpur *Nucl. Phys.* **B551**, 467 (1999).
45. C. Vafa, private communication.
46. J.T. Liu and R. Minasian, *Phys. Lett.* **B457**, 39 (1999).
47. O. Aharony, M. Berkooz, S. Kachru, N. Seiberg and E. Silverstein, *Adv. Theor. Math. Phys.* **1**, 148 (1998).
48. O. Aharony, M. Berkooz and N. Seiberg, *Adv. Theor. Math. Phys.* **2**, 119 (1998).
49. A. Kapustin and S. Sethi, *Adv. Theor. Math. Phys.* **2**, 571 (1998).
50. N. Seiberg and E. Witten, hep-th/9607163.
51. K. Intriligator and N. Seiberg, *Phys. Lett.* **B387**, 513 (1996).
52. A. Hanany and E. Witten, *Nucl. Phys.* **B492**, 152 (1997).
53. C.V. Johnson, A.W. Peet and J. Polchinski, *Phys. Rev.* **D61**, 086001 (2000).
54. S. Kachru and E. Silverstein, *Phys. Rev. Lett.* **80**, 4855 (1998).
55. D.Z. Freedman, S.S. Gubser, K. Pilch and N.P. Warner, hep-th/9904017.
56. M. Cvetic, H. Lu and C.N. Pope, *Class. Quant. Grav.* **17**, 4867 (2000) [hep-th/0001002].
57. N. Arkani-Hamed, S. Dimopoulos, N. Kaloper, and R. Sundrum, *Phys. Lett.* **B480**, 193 (2000).
58. A. Strominger, *Nucl. Phys.* **B274**, 253 (1986).
59. E. Witten, *Nucl. Phys.* **B471**, 135 (1996).
60. K. Dasgupta, G. Rajesh and S. Sethi, *JHEP* **9908**, 023 (1999).
61. C.S. Chan, P.L. Paul and H. Verlinde, *Nucl. Phys.* **B581**, 156 (2000) [hep-th/0003236].
62. K. Behrndt and S. Gukov, *Nucl. Phys.* **B580**, 225 (2000).
63. P. Mayr, *JHEP* **0011**, 013 (2000) [hep-th/0006204].
64. P. Horava and E. Witten, *Nucl. Phys.* **B460**, 506 (1996).
65. M. Cvetic and H.H. Soleng, *Phys. Rept.* **282**, 159 (1997).
66. O. DeWolfe, D.Z. Freedman, S.S. Gubser and A. Karch, *Phys. Rev.* **D62**, 046008 (2000) [hep-th/9909134].
67. L.J. Romans, *Phys. Lett.* **B169**, 374 (1986).
68. J. Polchinski and E. Witten, *Nucl. Phys.* **B460**, 525 (1996).
69. A. Chamblin, R. Emparan, C.V. Johnson and R.C. Myers, *Phys. Rev.* **D60**, 064018 (1999).

70. O. Aharony, S.S. Gubser, J. Maldacena, H. Ooguri, and Y. Oz, *Phys. Rept.* **323**, 183 (2000).
71. M. Henningson and K. Skenderis, *JHEP* **9807**, 023 (1998).
72. M.J. Duff, *Class. Quant. Grav.* **11**, 1387 (1994).

Appendix A

Here we discuss some details in obtaining the supergravity solutions.

A.1. *D=5*

We start with IIB supergravity in $AdS_5 \times S^5$. It is believed that we can obtain a consistent truncation to $\mathcal{N} = 8$ supergravity. $\mathcal{N} = 8$ supergravity involves an $SO(6)$ gauge field. We can further truncate this theory to supergravity theories involving a smaller gauge group. The advantage of considering these truncations is a simplification in the equations of motion. The truncation that we used is a truncation to a supergravity theory in five dimensions with three $U(1)$ gauge fields and two scalar fields described in Refs. 69,34. This is an $\mathcal{N} = 2$ supergravity theory in five dimensions with two vector multiplets. The third gauge field is the graviphoton. The three $U(1)$ gauge fields are associated to rotations on the 12, 34 and 56 planes respectively, where 1-6 are directions orthogonal to the D3 brane.

The Lagrangian for this theory is given by[34,6]

$$L = R - \frac{1}{2}(\partial_\mu \phi_1)^2 - \frac{1}{2}(\partial_\mu \phi_2)^2 + 4\sum_i^3 e^{\alpha_i} - \frac{1}{4}(\sum_i^3 e^{2\alpha_i} {F^i}_{\mu\nu}{}^2) + \frac{1}{4}\epsilon^{\mu\nu\alpha\beta\rho} {F^1}_{\mu\nu} {F^2}_{\alpha\beta} {A^3}_\rho \tag{A.1}$$

where we have defined

$$\alpha_1 = \frac{\phi_1}{\sqrt{6}} + \frac{\phi_2}{\sqrt{2}}, \quad \alpha_2 = \frac{\phi_1}{\sqrt{6}} - \frac{\phi_2}{\sqrt{2}}, \quad \alpha_3 = \frac{-2\phi_1}{\sqrt{6}}, \tag{A.2}$$

We work in units where $R_{AdS_5} = 1$ for the usual AdS_5 solution, related to $\mathcal{N} = 4$ SYM. The corresponding supersymmetry transformations are

$$\begin{aligned} \delta\lambda_i &= (\frac{3}{8}\partial_i X_I \Gamma^{\mu\nu} F^I_{\mu\nu} - \frac{i}{2} g_{ij} \Gamma^\mu \partial_\mu \phi^j + \frac{3i}{2} V_I \partial_i X^I)\epsilon \\ \delta\psi_\mu &= (D_\mu + \frac{i}{8} X_I (\Gamma_\mu^{\nu\rho} - 4\delta_\mu^\nu \Gamma^\rho) F^I_{\nu\rho} + \frac{1}{2}\Gamma_\mu X^I V_I - \frac{3i}{2} V_I A^I_\mu)\epsilon \end{aligned} \tag{A.3}$$

we have defined

$$X^{(1)} = e^{-\frac{\phi_1}{\sqrt{6}} - \frac{\phi_2}{\sqrt{2}}} \qquad X^{(2)} = e^{-\frac{\phi_1}{\sqrt{6}} + \frac{\phi_2}{\sqrt{2}}} \qquad X^{(3)} = e^{2\frac{\phi_1}{\sqrt{6}}} \tag{A.4}$$

$$X_{(1)} = \frac{1}{3} e^{\frac{\phi_1}{\sqrt{6}} + \frac{\phi_2}{\sqrt{2}}} \qquad X_{(2)} = \frac{1}{3} e^{\frac{\phi_1}{\sqrt{6}} - \frac{\phi_2}{\sqrt{2}}} \qquad X_{(3)} = \frac{1}{3} e^{-2\frac{\phi_1}{\sqrt{6}}} \tag{A.5}$$

$$g_{ij} = \frac{1}{2}\delta_{ij} \qquad V_I = \frac{1}{3} \tag{A.6}$$

For the metric (7), we have the spin connection

$$\omega_{\hat{x}}^{xy} = -\frac{1}{y} \quad \omega_{\hat{x}}^{xr} = \omega_{\hat{y}}^{yr} = \frac{e^{g-f}g'}{y}, \quad \omega_{\hat{z}}^{zr} = \omega_{\hat{t}}^{tr} = f', \tag{A.7}$$

(here the hatted index is curved and the others are flat) and with the choice of gauge fields

$$A_x^{(1)} = \frac{a}{y} \quad A_x^{(2)} = \frac{b}{y} \quad A_x^{(1)} = \frac{c}{y} \tag{A.8}$$

we have the equations

$$\partial_x\epsilon + \frac{1}{2}\omega_{\hat{x}}^{xy}\Gamma_{xy}\epsilon + \frac{1}{2}\omega_{\hat{x}}^{xr}\Gamma_{xr}\epsilon - \frac{i}{2}X_I F_{xy}^I y e^{-g}\Gamma_y\epsilon + \frac{e^g}{2y}X^I V_I \Gamma_x\epsilon - \frac{i}{2}(A_x^{(1)} + A_x^{(2)} + A_x^{(3)})\epsilon = (\tag{A.9}$$

$$\partial_z\epsilon + \frac{1}{2}\omega_{\hat{z}}^{zr}\Gamma_{zr}\epsilon + \frac{i}{4}X_I F_{xy}^I e^{-2g}y^2\Gamma_{zxy}\epsilon + \frac{1}{2}e^f X^I V_I \gamma_z\epsilon = 0 \tag{A.10}$$

$$\frac{3}{4}\partial_{\phi 1}X_I F_{xy}^I e^{-2g}y^2\Gamma_{xy}\epsilon - \frac{i}{4}e^{-f}\phi_1'\Gamma_r\epsilon + \frac{i}{2}\partial_{\phi 1}(X^{(1)} + X^{(2)} + X^{(3)})\epsilon = 0 \tag{A.11}$$

$$\frac{3}{4}\partial_{\phi 2}X_I F_{xy}^I e^{-2g}y^2\Gamma_{xy}\epsilon - \frac{i}{4}e^{-f}\phi_2'\Gamma_r\epsilon + \frac{i}{2}\partial_{\phi 2}(X^{(1)} + X^{(2)} + X^{(3)})\epsilon = 0 \tag{A.12}$$

and another equation describing the radial dependence of the spinor that we do not write here. We impose the following condition on the spinors

$$\Gamma_{xy}\epsilon = -i\beta\epsilon \qquad \Gamma_r\epsilon = \eta\epsilon \tag{A.13}$$

where $\beta = \pm 1$, $\eta = \pm 1$. By doing simply parity transformations we set $\beta = \eta = 1$. We obtain, from eq. (A.9)

$$1 = (a + b + c) \tag{A.14}$$

$$g' = -2e^f[\frac{1}{6}(X^{(1)} + X^{(2)} + X^{(3)}) - \frac{1}{2}X_I a^I e^{-2g}] \tag{A.15}$$

Equations (A.10)(A.11) (A.12) reduce to

$$f' = -2e^f[\frac{1}{6}(X^{(1)} + X^{(2)} + X^{(3)}) + \frac{1}{2}X_I a^I e^{-2g}] \tag{A.16}$$

$$\phi_1' = 4e^f[\frac{1}{2}\partial_{\phi 1}(X^{(1)} + X^{(2)} + X^{(3)}) - \frac{3}{4}\partial_{\phi 1}X_I a^I e^{-2g}] \tag{A.17}$$

$$\phi_2' = 4e^f[\frac{1}{2}\partial_{\phi 2}(X^{(1)} + X^{(2)} + X^{(3)}) - \frac{3}{4}\partial_{\phi 2}X_I a^I e^{-2g}] \tag{A.18}$$

It is easy to see that, the choice

$$\phi_2 = 0, a = b \tag{A.19}$$

solves (A.18). Using the explicit expressions for the scalar fields X^I and defining $\varphi = \frac{\phi}{\sqrt{6}}$ we obtain

$$g' = -\frac{1}{3}e^{f-g}[e^g(2e^{-\varphi} + e^{2\varphi}) - e^{-g}(2ae^{\varphi} + ce^{-2\varphi})] \tag{A.20}$$

$$f' = -\frac{1}{6}[2e^f(2e^{-\varphi} + e^{2\varphi}) + e^{f-2g}(2ae^{\varphi} + ce^{-2\varphi})] \tag{A.21}$$

$$\phi' = \frac{4}{\sqrt{6}}e^f[(-e^{-\varphi} + e^{2\varphi}) - \frac{1}{2}e^{-2g}(ae^{\varphi} - ce^{-2\varphi})] \tag{A.22}$$

$$2a + c = 1 \tag{A.23}$$

In the case $a = b = 0$; $c = 1$ we obtain eqs. (8), while for $a = b = 1/2$ we obtain eqs. (14) - (16) .

It can be seen that solutions with constant g, ϕ can be obtained

$$\begin{aligned} a &= b\ , \qquad c = 1 - 2a\ , \qquad e^{\frac{\phi}{\sqrt{6}}} = (4 - \frac{1}{a})^{\frac{1}{3}} \\ e^{-2g} &= \frac{(4 - \frac{1}{a})^{\frac{1}{3}}}{a} = \frac{e^{\varphi}}{a} \\ e^f &= e_0^f\frac{1}{r}\ , \qquad e_0^f = \frac{2a(4 - \frac{1}{a})^{\frac{1}{3}}}{(6a - 1)} \end{aligned} \tag{A.24}$$

A simple analysis of the eqs. (A.20)-(A.23) near $r = 0$, shows that

$$\begin{aligned} g(r) &= -Log(r) + \frac{7(2a + c)}{36}r^2 + ... \\ f(r) &= -Log(r) - \frac{(2a + c)}{18}r^2 + ... \\ \phi(r) &= -\frac{(a - c)}{3}r^2 Log(r) + ... \end{aligned} \tag{A.25}$$

The last equation shows that the operator dual to the field ϕ is turned on. The field ϕ is dual to an operator of dimension $\Delta = 2$. For this dimension the two solutions of the wave equation go as $\phi \sim r^2$, $\phi \sim r^2 \log(r)$. The second solution is the non-normalizable mode associated to the insertion of an operator. We see that in both cases analyzed in this paper the operator ϕ is turned on. It is interesting that there is also a special solution where $a = b = c = 1/3$ where the operator ϕ is not turned on. The field $\phi = 0$ in the whole solution. This solution is of the form of the solutions analyzed in Ref. 7. It would be interesting to see if this solution really makes sense since some fields acquire fractional spins with this twisting. In this case the metric has the form[7]

$$ds^2 = e^{-g}(3e^{2g} - 1)^{\frac{3}{2}}(-dt^2 + dz^2) + \frac{9}{(3 - e^{-2g})^2}dg^2 + \frac{e^{2g}}{y^2}(dx^2 + dy^2) \tag{A.26}$$

We now give some details on the computation of the central charge for the solutions of the form (A.24) from the supergravity side. We have

$$c = \frac{3R_{AdS3}}{2G_N^3} \tag{A.27}$$

and

$$\frac{1}{G_N^3} = \frac{Vol_7}{G_N^{10}} = \frac{2N^2}{\pi}Vol H_2 \tag{A.28}$$

where we used that in the units used in our paper where $R_{AdS_5} = 1$ we have that $G_N^{10} = \frac{\pi^4}{2N^2}$ and that the volume of a unit radius five-sphere is π^3.

In the case of the solutions (A.24) $R_{AdS3} = e^{f_0}$. We can calculate the volume of a constant curvature Riemann surface as follows

$$\int_{\Sigma_2} \sqrt{g}R = 8\pi(1-g)\ ; \qquad R = -2e^{-2g} \quad \rightarrow \quad \int_{\Sigma} \sqrt{g} = 4\pi(g-1)e^{2g} \quad \text{(A.29)}$$

$$c = 12N^2(g-1)e^{2g+f_0} = \frac{24a^2N^2(g-1)}{6a-1} \quad \text{(A.30)}$$

for the case $a = \frac{1}{2}$ we reproduce (13) .

As we pointed out in the paragraph before eq. (21) there are other solutions for the case in which we excite two gauge fields $A_x^{(1)} = \frac{a}{y}$, $A_x^{(2)} = \frac{b}{y}$ and we keep both scalar fields ϕ_1, ϕ_2 nonvanishing.

In that case the eqs. to solve read

$$g' = -\frac{1}{3}e^f[(e^{-\frac{\phi_1}{\sqrt{6}}-\frac{\phi_2}{\sqrt{2}}} + e^{-\frac{\phi_1}{\sqrt{6}}+\frac{\phi_2}{\sqrt{2}}} + e^{2\frac{\phi_1}{\sqrt{6}}}) - e^{-2g}(ae^{\frac{\phi_1}{\sqrt{6}}+\frac{\phi_2}{\sqrt{2}}} + be^{\frac{\phi_1}{\sqrt{6}}-\frac{\phi_2}{\sqrt{2}}} + ce^{-2\frac{\phi_1}{\sqrt{6}}})$$
$$\text{(A.31)}$$

$$f' = -\frac{1}{6}e^f[2(e^{-\frac{\phi_1}{\sqrt{6}}-\frac{\phi_2}{\sqrt{2}}} + e^{-\frac{\phi_1}{\sqrt{6}}+\frac{\phi_2}{\sqrt{2}}} + e^{2\frac{\phi_1}{\sqrt{6}}}) + e^{-2g}(ae^{\frac{\phi_1}{\sqrt{6}}+\frac{\phi_2}{\sqrt{2}}} + be^{\frac{\phi_1}{\sqrt{6}}-\frac{\phi_2}{\sqrt{2}}} + ce^{-2\frac{\phi_1}{\sqrt{6}}})$$
$$\text{(A.32)}$$

$$\phi_1' = \frac{1}{\sqrt{6}}e^f[2(-e^{-\frac{\phi_1}{\sqrt{6}}-\frac{\phi_2}{\sqrt{2}}} - e^{-\frac{\phi_1}{\sqrt{6}}+\frac{\phi_2}{\sqrt{2}}} + 2e^{2\frac{\phi_1}{\sqrt{6}}}) - e^{-2g}(ae^{\frac{\phi_1}{\sqrt{6}}+\frac{\phi_2}{\sqrt{2}}} + be^{\frac{\phi_1}{\sqrt{6}}-\frac{\phi_2}{\sqrt{2}}} - 2ce^{-2\frac{\phi_1}{\sqrt{6}}})$$
$$\text{(A.33)}$$

$$\phi_2' = \frac{1}{\sqrt{2}}e^f[2(-e^{-\frac{\phi_1}{\sqrt{6}}-\frac{\phi_2}{\sqrt{2}}} + e^{-\frac{\phi_1}{\sqrt{6}}+\frac{\phi_2}{\sqrt{2}}}) - e^{-2g}(ae^{\frac{\phi_1}{\sqrt{6}}+\frac{\phi_2}{\sqrt{2}}} - be^{\frac{\phi_1}{\sqrt{6}}-\frac{\phi_2}{\sqrt{2}}}) \quad \text{(A.34)}$$

A constant solution to these equations, can be obtained iff $a = b = \frac{1}{2}$ and reads, after redefining $\varphi_1 = \frac{\phi_1}{\sqrt{6}}$ and $\varphi_2 = \frac{\phi_2}{\sqrt{2}}$

$$f[r] = -Log[e^{2\varphi_1}r], \qquad e^{2g} = \frac{e^{2\varphi_1}}{4}, \qquad 2\cosh\varphi_2 = e^{3\varphi_1} \quad \text{(A.35)}$$

We end this subsection by noting that according to Ref. 34 the $g_{00}^{(10)}$ component of the ten dimensional metric is related to the five dimensional $g_{00}^{(5)}$ metric appearing in (A.1) through

$$g_{00}^{(10)} = Wg_{00}^{(5)}\ , \qquad W^2 = \left(e^{-\frac{\phi_1}{\sqrt{6}}-\frac{\phi_2}{\sqrt{2}}}\sin^2\psi + e^{-\frac{\phi_1}{\sqrt{6}}+\frac{\phi_2}{\sqrt{2}}}\cos^2\psi\right)\cos^2\theta + e^{2\frac{\phi_1}{\sqrt{6}}}\sin^2\theta\ .$$
$$\text{(A.36)}$$

A.2. *Effective Potential*

Here we consider the effective action for the configurations studied above. Start by considering the Action

$$S = \int d^5x\sqrt{G}[R - \frac{1}{2}\partial\phi^2 - V - \frac{1}{4}\Sigma e^{2\alpha_i}F^2_{\mu\nu,i}] \tag{A.37}$$

with α_i as in (A.2) and

$$V = -4\sum_i^3 e^{\alpha_i}. \tag{A.38}$$

Now consider a compactification to three dimensions on the space

$$ds_5^2 = ds_3^2(r,z,t) + \frac{e^{2g}}{y^2}(dx^2 + dy^2) \qquad F_{x,y,i} = c_i/y^2, \qquad \phi_2 = 0 \tag{A.39}$$

where we are thinking that we will quotient the H^2 space to obtain a finite volume region. We get

$$S \sim \int d^3x\sqrt{g_3}e^{2g}[R(g_3) - \frac{1}{2}g^{rr}(\partial_r\phi)^2 - 2e^{-2g} - V - \frac{1}{2}\Sigma e^{2\alpha_i}c_i^2 e^{-4g}]\,. \tag{A.40}$$

Transforming to the Einstein frame

$$g_3 = e^{-4g}\hat{g}_3 \tag{A.41}$$

we find

$$S = \int d^3x\sqrt{\hat{g}_3}[R(\hat{g}_3) - \frac{1}{2}g^{rr}(\partial_r\phi)^2 - V_{eff}] \tag{A.42}$$

where,

$$V_{eff} = e^{-4g}(\pm 2e^{-2g} + V + \frac{e^{-4g}}{2}(e^{2\frac{\phi}{\sqrt{6}}}a^2 + e^{2\frac{\phi}{\sqrt{6}}}b^2 + e^{-4\frac{\phi}{\sqrt{6}}}c^2)) \tag{A.43}$$

where $+$ is for genus $g > 1$ and $-$ is for S^2.

A.3. *D=7*

Again in this case it is is convenient to choose a truncation of $SO(5)$ gauged supergravity to a gauged supergravity containing only two $U(1)$ gauge fields. These two $U(1)$ fields correspond to rotations in the 12 and 34 planes, where 1-5 are directions orthogonal to the M5 brane. We use Ref. 46 for the supersymmetry transformations and Ref. 34 for lifting up the solutions to eleven dimensions. The effective Lagrangian is[46]

$$\begin{aligned} L = R &- 5\partial(\lambda_1+\lambda_2)^2 - \partial(\lambda_1-\lambda_2)^2 - e^{-4\lambda_1}F^2_{\mu\nu,1} - e^{-4\lambda_2}F^2_{\mu\nu,2} \\ &- \frac{1}{2}m^2(-8e^{2(\lambda_1+\lambda_2)} - 4e^{-2\lambda_1-4\lambda_2} - 4e^{-4\lambda_1-2\lambda_2} + e^{-8(\lambda_1+\lambda_2)}) + L[C_3 = 0] \end{aligned} \tag{A.44}$$

Again we will work in units where $R_{AdS_7} = 1$ for the usual M5 solution. In these units the radius of S^4 is $1/2$ and $m = 2$ above.

The supersymmetry transformations can be found in Ref. 46,

$$\delta\psi_\mu = [\nabla_\mu + k/2(A^{(1)}_\mu\Gamma^{12} + A^{(2)}_\mu\Gamma^{34}) + \frac{m}{4}e^{-4(\lambda_1+\lambda_2)}\gamma_\mu + \frac{1}{2}\gamma_\mu\gamma^\nu\partial_\nu(\lambda_1+\lambda_2) + \frac{1}{2}\gamma^\nu(e^{-2\lambda_1}F_{\mu\nu,1}\Gamma^{12} + e^{-2\lambda_2}F_{\mu\nu,2}\Gamma^{34})]\epsilon \tag{A.45}$$

$$\delta\lambda^{(1)} = [\frac{m}{4}(e^{2\lambda_1} - e^{-4(\lambda_1+\lambda_2)}) - \frac{1}{4}\gamma^\mu\partial_\mu(3\lambda_1+2\lambda_2) - \frac{1}{8}\gamma^{\mu\nu}e^{-2\lambda_1}F_{\mu\nu,1}\Gamma^{12}]\epsilon \tag{A.46}$$

$$\delta\lambda^{(2)} = [\frac{m}{4}(e^{2\lambda_2} - e^{-4(\lambda_1+\lambda_2)}) - \frac{1}{4}\gamma^\mu\partial_\mu(2\lambda_1+3\lambda_2) - \frac{1}{8}\gamma^{\mu\nu}e^{-2\lambda_2}F_{\mu\nu,2}\Gamma^{34}]\epsilon \tag{A.47}$$

Where the spin connection components, for the metric considered in the text (22) are given by

$$\omega^{u,r}_{\hat{u}} = f' \qquad \omega^{xr}_{\hat{x}} = \omega^{yr}_{\hat{y}} = \frac{e^{g-f}g'}{y} \qquad \omega^{xy}_{\hat{x}} = -\frac{1}{y} \tag{A.48}$$

Here the hatted indices are curved while the unhatted one are flat.

The vanishing gravitino and gaugino equations read

$$\partial_{\hat{u}}\epsilon + \frac{\omega^{u,r}_{\hat{u}}\gamma_{ur}}{2}\epsilon + \frac{m}{4}e^{-4(\lambda_1+\lambda_2)+f}\gamma_u\epsilon + \frac{\lambda_1'+\lambda_2'}{2}\gamma_{ur}\epsilon = 0 \tag{A.49}$$

$$[\partial_{\hat{x}}\epsilon + \frac{1}{2}\omega^{xy}_{\hat{x}}\gamma_{xy}\epsilon + \frac{k}{2}(A^{(1)}_{\hat{x}}\Gamma^{12} + A^{(2)}_{\hat{x}}\Gamma^{34})\epsilon + \frac{1}{2}\omega^{xr}_{\hat{x}}\gamma_{xr}\epsilon + \frac{m}{4y}e^{-4(\lambda_1+\lambda_2)+g}\gamma_x\epsilon$$
$$+\frac{e^{g-f}(\lambda_1'+\lambda_2')}{2y}\gamma_{xr} + \frac{ye^{-g}}{2}(e^{-2\lambda_1}F^{(1)}_{xy} + e^{-2\lambda_2}F^{(2)}_{xy})\gamma_y\epsilon = 0 \tag{A.50}$$

$$m(e^{2\lambda_1} - e^{-4(\lambda_1+\lambda_2)})\epsilon - e^{-f}(3\lambda_1'+2\lambda_2')\gamma_r\epsilon - e^{-2g-2\lambda_1}y^2F^{(1)}_{xy}\gamma_{xy}\Gamma^{(12)}\epsilon = 0 \tag{A.51}$$

$$m(e^{2\lambda_1} - e^{-4(\lambda_1+\lambda_2)})\epsilon - e^{-f}(2\lambda_1'+3\lambda_2')\gamma_r\epsilon - e^{-2g-2\lambda_2}y^2F^{(2)}_{xy}\gamma_{xy}\Gamma^{(34)}\epsilon = 0 \tag{A.52}$$

$$\partial_{\hat{r}}\epsilon + +\frac{m}{4}e^{-4(\lambda_1+\lambda_2)+f}\gamma_r\epsilon + \frac{\lambda_1'+\lambda_2'}{2}\epsilon = 0 \tag{A.53}$$

Imposing the conditions on the spinor

$$\gamma_r\epsilon = \epsilon, \quad ,\gamma_{xy}\epsilon = i\epsilon, \quad \Gamma^{(12)}\epsilon = i\epsilon, \quad \Gamma^{(34)}\epsilon = i\epsilon, \quad \partial_{u,y,t,z,x,w}\epsilon = 0, \tag{A.54}$$

Different choices of signs in (A.54) are related by simple parity transformations. we choose $\eta = \beta = \alpha = \gamma = 1$ and evaluating over the configurations $A^{(1)}_x = \frac{a}{y}$ $A^{(2)}_x = \frac{b}{y}$

$$\begin{aligned} f' &= -e^{f-4(\lambda_1+\lambda_2)} - \lambda_1' - \lambda_2' \\ g' &= -e^f\left[\frac{1}{2}e^{-4(\lambda_1+\lambda_2)} - \kappa e^{-2g}(ae^{-2\lambda_1} + be^{-2\lambda_2})\right] - (\lambda_1'+\lambda_2') \\ 3\lambda_1' + 2\lambda_2' &= e^f\left[2(e^{2\lambda_1} - e^{-4(\lambda_1+\lambda_2)}) + \kappa a e^{-2g-2\lambda_1}\right] \\ 3\lambda_2' + 2\lambda_1' &= e^f\left[2(e^{2\lambda_2} - e^{-4(\lambda_1+\lambda_2)}) + \kappa b e^{-2g-2\lambda_2}\right] \\ 1 &= 4(a+b) \end{aligned} \tag{A.55}$$

where $\kappa = 1$ for H^2 and $\kappa = -1$ for S^2. In the $\mathcal{N} = 2$ configurations we set $a = 1/4$, $b = 0$, $3\lambda_2 + 2\lambda_1 = 0$ and we define $\lambda = \lambda_2$. Then we get the equations (23). For $\mathcal{N} = 1$ configurations we choose $a = b = 1/8$ and $\lambda_1 = \lambda_2 = -\phi/2$ we get the equations (27).

We can also calculate the supergravity "central charge" of the four dimensional conformal field theory, which is given by

$$c = \frac{\pi R^3_{AdS_5}}{8G^5_N} \tag{A.56}$$

in a normalization where $c = N^2/4$ for $\mathcal{N} = 4$ $U(N)$ SYM. In our case, a calculation parallel to that done for the D3 brane gives

$$c = \frac{8}{3}(g-1)e^{2g+3f_0}N^3 \tag{A.57}$$

where f_0 is the constant piece in f defined as $f = f_0 - \log r$ where f, r are those of the ansatz (22). From (25) and (28) we can see that we get $e^{2g+3f_0} = 1/8$ for the $\mathcal{N} = 2$ case and $e^{2g+3f_0} = 27/2^8$ for the $\mathcal{N} = 1$ case. In the $\mathcal{N} = 2$ case we can calculate the effective number of vector multiplets as follows. We know that the conformal anomaly coefficients a, c (see Refs. 70,71) are equal to leading order in N for theories that have a gravity dual. This implies that this theory has the same number of vector and hypermultiplets. Then from the formulas for a, c in Ref. 72 and (A.56) we find $n_V = n_H = \frac{4}{3}(g-1)N^3$.

Let us finish this appendix by writing the expression for the warp factor that is obtained when uplifting this solutions to M theory[34]

$$g^{(11)}_{00} = W g^{(7)}_{00}\,, \qquad W^3 = e^{-4(\lambda_1+\lambda_2)}\mu_0^2 + e^{2\lambda_2}\mu_1^2 + e^{2\lambda_1}\mu_2^2 \tag{A.58}$$

where μ_i parametrize S^2, $\mu_0^2 + \mu_1^2 + \mu_2^2 = 1$.

A.4. *D4 brane solutions*

Here we consider the solutions that we get when we compactify a direction along the M5 worldvolume on S^1 and we reduce the 11 dimensional solution to a 10 dimensional solution describing a D4 brane wrapped on the Riemann surface.

We find that the dilaton and ten dimensional metric are

$$\begin{aligned} e^{2\phi} &= W e^{3f} \\ ds^2_{str} &= e^{2\phi/3}\, ds^2_{11}\big|_{10} \\ W^2 &= e^{2\lambda}\cos^2\theta + e^{-3\lambda}\sin^2\theta \end{aligned} \tag{A.59}$$

where the right hand side of the second line denotes the eleven dimensional metric along the ten directions of the type IIA solution, the factor of the dilaton produces the string metric in ten dimensions. The functions λ, f that appear here are those of the solution (24). In principle we could also find the various components of the four form field strength of type IIA from that of the 11 dimensional solution. Of course we can do the same with the other solution we found, the one with $\mathcal{N} = 1$ supersymmetry in four dimensions.

BLACK HOLE THERMODYNAMICS FROM CALCULATIONS IN STRONGLY COUPLED GAUGE THEORY

DANIEL KABAT
Department of Physics, Columbia University
New York, NY 10027, USA

GILAD LIFSCHYTZ
Department of Physics, Princeton University
Princeton, NJ 08544, USA

DAVID LOWE
Department of Physics, Brown University
Providence, RI 02912, USA

We develop an approximation scheme for the quantum mechanics of N D0-branes at finite temperature in the 't Hooft large-N limit. The entropy of the quantum mechanics calculated using this approximation agrees well with the Bekenstein-Hawking entropy of a ten-dimensional non-extremal black hole with 0-brane charge. This result is in accord with the duality conjectured by Itzhaki, Maldacena, Sonnenschein and Yankielowicz. Our approximation scheme provides a model for the density matrix which describes a black hole in the strongly-coupled quantum mechanics.

1. Introduction

Properties of black holes in quantum theories of gravity have intrigued physicists for many years. In recent years much progress has been made in understanding some of these properties using string theory. However because string theory is usually formulated perturbatively about a given spacetime background, it has been difficult to obtained a unified and complete description of black hole physics. In the past few years nonperturbative formulations of string theory have been proposed in terms of large N gauge theories. These formulations give new hope for understanding some of these fundamental issues.

In particular, Maldacena's conjecture gives a promising arena to answer some of these questions. Maldacena's conjecture[1] relates string theories in anti-de Sitter backgrounds to conformal field theories in a large-N limit. However there is no free lunch. In this framework, seemingly obvious questions on the string theory side are hard to formulate on the gauge theory side. Moreover, in the range of parameters where semi-classical string theory may be used to construct black hole geometries, the dual gauge theory is strongly coupled. This makes understanding black hole states very difficult.

The version of Maldacena duality to be considered here relates black holes in ten dimensions with 0-brane charge to supersymmetric gauged $SU(N)$ quantum mechanics with sixteen supercharges.[2] The black holes in question have a non-zero Hawking temperature T. The dual quantum mechanics is to be taken at the same finite temperature. The black hole has a free energy, which arises from its Bekenstein-Hawking entropy.

$$\beta F = -2.52\, N^2 \left(\frac{T}{(g_{YM}^2 N)^{1/3}} \right)^{1.80} \tag{1}$$

Duality predicts that the quantum mechanics should have the same free energy. The supergravity description is expected to be valid when the curvature and the dilaton are small near the black hole horizon. This regime corresponds to the 't Hooft large N limit of the quantum mechanics, when the dimensionless effective coupling $g_{YM}^2 N/T^3$ is large.

In this article we describe a set of approximations that can be applied to the quantum mechanics in the regime of interest. Using these techniques we calculate the finite temperature partition function of the quantum mechanics. Over a certain range of temperature our results can be well fit by a power law,

$$\beta F \approx \text{const.} - 2.0\, N^2 \left(\frac{T}{(g_{YM}^2 N)^{1/3}} \right)^{1.8} . \tag{2}$$

This is in quite good agreement with the black hole prediction (1). Although in the present paper we are primarily interested in the thermodynamics of the quantum mechanics, our approximation scheme should also be useful for addressing questions about the spacetime structure of the black hole, perhaps along the lines of Refs. 3,4.

The basic idea is to treat the $\mathcal{O}(N^2)$ degrees of freedom of the quantum mechanics as statistically independent, using a type of mean field approximation. This assumption is motivated by the overall N^2 dependence of the free energy (1). The approximation involves constructing a trial action S_0 from the full action S. All quantities can then be systematically computed as an expansion in powers of $S - S_0$. The terms in the trial action are fixed by solving a truncated version of the Schwinger-Dyson equations of the quantum mechanics. This procedure can be viewed as re-summing an infinite number of Feynman diagrams. Since we are interested in large-N behavior, we will only resum planar diagrams. Thus the crucial test of our approximation is to obtain the correct dependence on the effective dimensionless coupling $g_{YM}^2 N/T^3$.

The sort of approximation that we are considering has several attractive features, which we regard as *a priori* reasons to believe that it captures some of the essential physics of the strongly-coupled quantum mechanics.

- As mentioned above, the approximation automatically respects 't Hooft large-N counting.

- The approximation partially respects the symmetries of the problem. Our trial action will have $\mathcal{N} = 2$ supersymmetry and $SO(2) \times SO(7)$ rotational symmetry (out of the underlying $\mathcal{N} = 16$ supersymmetry and $SO(9)$ rotational symmetry).

- The approximation is non-perturbative in the Yang-Mills coupling constant.

Before continuing there is one other issue we would like to comment on. The partition function of the full quantum mechanics contains an infrared divergence from the regions in moduli space when the D0-branes are far apart. This leads to a divergent contribution to the entropy with an overall coefficient $\mathcal{O}(N)$. From the supergravity point of view, this corresponds to a thermal gas of gravitons. This divergence may be regulated by putting the system in a finite box. The black hole entropy which is $\mathcal{O}(N^2)$ can easily be made to dominate over the $\mathcal{O}(N)$ contribution. Our mean field approximation computes the $\mathcal{O}(N^2)$ piece by design, and this infrared divergence does not make an appearance.

2. Gaussian Approximation for 0-brane Quantum Mechanics

In this section we sketch the application of the Gaussian approximation[5] to gauged $SU(N)$ supersymmetric quantum mechanic with sixteen supercharges. Further details will appear in Ref. 6. To avoid explicit breaking of supersymmetry by the approximation, we adopt an unconstrained superfield formulation. This ensures that we recover exact supersymmetry in the zero temperature limit. We will use an $\mathcal{N} = 2$ superspace; for more details on our notation see Ref. 5.

$\mathcal{N} = 2$ superspace has an $SO(2)$ R-symmetry, with spinor indices $\alpha, \beta = 1, 2$ and vector indices $i, j = 1, 2$. The $\mathcal{N} = 16$ vector multiplet decomposes into a set of real scalar superfields Φ_a transforming in the **7** of G_2, plus a gauge superfield Γ_α. Thus only an $SO(2) \times G_2$ subgroup of the underlying $SO(9)$ R-symmetry is manifest. Φ has components ϕ, ψ_α and f, while Γ_α has components A_0, X^i, χ_α, λ_α and d. We impose the supersymmetric gauge condition $D^\alpha \Gamma_\alpha = 0$. This sets $\partial A_0 / \partial t = 0$, $d = 0$ and $\lambda = \frac{1}{2} \partial \chi / \partial t$. This is a convenient gauge fixing, as all the remaining degrees of freedom transform homogeneously under gauge transformations. This feature helps make the approximation compatible with Ward identities.[6]

We are interested in the finite temperature behavior of the quantum mechanics. As usual we compactify the Euclidean time coordinate on a circle of circumference β, which is identified with the inverse temperature, and expand the fields in Matsubara frequencies. For example we write

$$X^i(\tau) = \frac{1}{\sqrt{\beta}} \sum_{l \in \mathbb{Z}} X_l^i e^{i 2\pi l \tau / \beta} .$$

Note that in Euclidean space the zero mode of the gauge field, which we denote A_{00}, survives as a physical degree of freedom (fluctuations in A_0 are eliminated by our gauge condition).

Our strategy is to construct a trial action S_0, which we use as an approximation to the full action $S = S_{\text{sym}} + S_{\text{ghost}}$. The free energy can then be calculated as an expansion in powers of $S - S_0$; note that such an expansion is non-perturbative in the Yang-Mills coupling. Throughout this paper we use the leading-order expression for the free energy in this scheme,

$$\beta F \approx \beta F_0 + \langle S - S_0 \rangle_0 \,. \tag{3}$$

Here βF_0 is the free energy of the trial action, and $\langle \cdot \rangle_0$ denotes an expectation value computed using S_0.

We make the following ansatz for the trial action:

$$S_0 = -\frac{N}{\lambda}\mathrm{Tr}(U + U^\dagger) + \sum_{l,i} \frac{1}{2\sigma_l^2}\mathrm{Tr}(X_l^i X_{-l}^i) + \sum_{l,a} \frac{1}{2\Delta_l^2}\mathrm{Tr}(\phi_l^a \phi_{-l}^a) + \cdots \,. \tag{4}$$

Here all fields (except the gauge field) appear in Gaussian form. The gauge field must be treated in a special way, owing to its periodicity properties. To do this we have introduced the timelike Wilson loop operator U, which can be expressed in terms of the gauge zero mode.

$$U = Pe^{i\int_0^\beta A_0} = e^{i\sqrt{\beta}A_{00}}$$

This makes it manifest that at finite temperature $A_0 \sim A_0 + 2\pi/\beta$ is periodic. As a trial action for the gauge field we have adopted the unitary one plaquette model action. As λ varies the trial action goes through a Gross-Witten phase transition at $\lambda = 2$.[7]

The key step in the approximation is to find a closed set of "gap" equations for the dressed propagators σ_l^2, Δ_l^2, ... appearing in (4). Again, the gauge field must be treated as a special case. All other propagators are obtained by demanding stationarity of the quantity[a]

$$\beta F_0 + \langle S - S_0 \rangle_0 - \frac{1}{2}\langle S_{III}^2 \rangle_{\mathsf{C},0} \,. \tag{5}$$

Here S_{III} refers to cubic terms in the original SYM plus ghost action, and the subscript C denotes a connected correlation function. Up to contributions from the gauge field, it can be shown this procedure correctly re-sums all one-loop self-energy corrections to the propagators.

The gap equation for λ is obtained from the Schwinger-Dyson equation for $\langle U \rangle$ that arises from the change of variables $U \to gU$ with $g \in SU(N)$. Demanding that this equation hold with respect to the one-plaquette measure yields

$$\langle \mathrm{Tr} U \rangle_0 = \frac{1}{\beta^{3/2}} \mathrm{Tr} \left\langle U \left(\frac{\delta S}{\delta A_{00}} - \frac{1}{2}\frac{\delta S_{III}^2}{\delta A_{00}} \right) \right\rangle_{C,0} \,. \tag{6}$$

This equation resums one-loop corrections to the Wilson loop, in the same sense that (5) resums one-loop corrections to the propagators. At large N the terms on the right hand side factorize into a gauge field correlator times matter field correlators; the terms involving the gauge fields may be computed using the results of Ref. 7.

[a]This quantity can be identified with the two-loop 2PI effective action of Ref. 8.

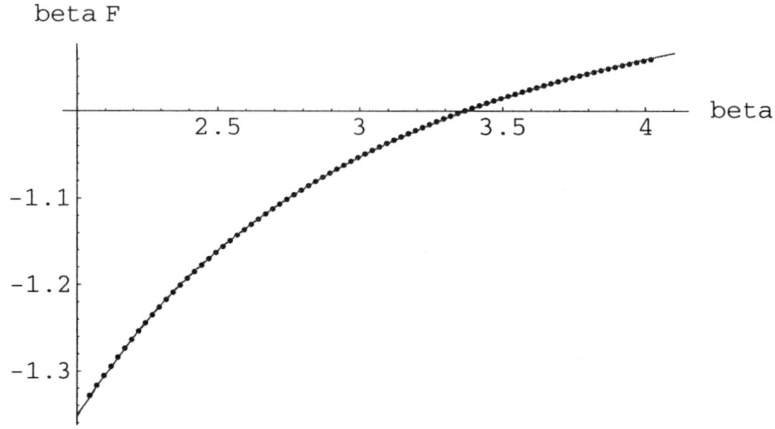

Fig. 1. The solid curve is the power law fit (8) for βF. The data points are calculated from numerical solutions to the gap equations.

3. Numerical Results

The gap equations can be solved numerically, using the methods discussed in Appendix B of Ref. 5. The basic strategy is to start at high temperature, where the gap equations can be solved semi-analytically, then use Newton-Raphson to solve the gap equations at a sequence of successively lower temperatures. As mentioned in the introduction, g_{YM}^2 and N only appear in the combination $g_{YM}^2 N/T^3$. Henceforth we choose units for T which effectively scales $g_{YM}^2 N$ to 1.

In principle the resulting Gaussian action contains a great deal of information about correlation functions in the quantum mechanics. But in this section we will just concentrate on the behavior of three basic quantities: the free energy, the Wilson loop, and the mean size of the state.

At high temperature, where the gauge theory is weakly coupled, we find that the free energy of the system is

$$\beta F = 6 \log \beta + \mathcal{O}(1)\,. \tag{7}$$

This result can be obtained analytically: the gap equations are dominated by the bosonic zero modes, and the free energy is dominated by βF_0.

In general, for a weakly-coupled theory in $0+1$ dimensions, one would expect the free energy to behave like $\log \beta$. But note that, even though the gauge theory is weakly coupled at high temperature, the perturbation series is afflicted with IR divergences. Thus, to determine the coefficient of the logarithm (which depends on the value of the dynamically generated IR cutoff) one must re-sum part of the perturbation series.

As the temperature is lowered the behavior of the free energy changes: at $\beta \approx 0.8$ we find that it begins to roll over and fall off as a non-trivial power of the temperature. In the range $2 < \beta < 4$ the numerical results for the free energy are well fit by

$$\beta F \approx -0.78 - 2.0\,\beta^{-1.8}\,. \tag{8}$$

This fit to the numerical results is illustrated in Fig. 1. Note that supersymmetry is crucial in making such power-law behavior possible. Without supersymmetry the free energy would behave as $\beta F \approx \beta E_0$ in the low temperature regime $\beta > 1$, where E_0 is the ground state energy of the system.

We obtained (8) by performing a Levenberg-Marquardt nonlinear least-squares fit to 80 numerical calculations of the free energy, carried out in the temperature range $2 \leq \beta \leq 4$. To estimate the uncertainty in the best fit parameters we varied the window of β over which the fit was performed. Fitting over the ranges $1.5 \leq \beta \leq 3.5$ and $2.5 \leq \beta \leq 4.5$ leads to uncertainties: $-0.78 \pm {}^{0.04}_{0.02}$, $-2.0 \pm {}^{0.1}_{0.3}$ and $-1.8 \pm {}^{0.1}_{0.2}$.

It is quite remarkable that the power law (8) is in excellent agreement with the semiclassical black hole prediction

$$\beta F = -2.52\,\beta^{-1.80}\,. \tag{9}$$

(An additive constant appears in the Gaussian approximation for the free energy. We will generally ignore this 'ground state degeneracy', since it seems to be an artifact of the Gaussian approximation when applied to systems with a continuous spectrum. Similar behavior was noted in Ref. 5.)

As we go to still lower temperatures, we find that the energy $\partial(\beta F)/\partial\beta$ calculated in the Gaussian approximation begins to drop below the energy of the black hole. In fact the Gaussian energy becomes negative around $\beta = 7.3$. Ultimately, as $\beta \to \infty$, the Gaussian energy does asymptote to zero, as required by the $\mathcal{N} = 2$ supersymmetry which is manifest in the approximation. But a negative energy is clearly problematic.

Fortunately, we can be rather precise about exactly where the approximation is going wrong: the difficulty is with the Schwinger-Dyson gap equation we have been using to fix the value of the one-plaquette coupling λ. Although we do not know how to write down a better gap equation for λ, we can give a *prescription* for fixing λ, that will allow us to obtain reasonable results at much lower values of the temperature. This may be regarded either as a check on our understanding of why the approximation is breaking down, or as a way of building a model for the black hole that can be used at lower temperatures. Our prescription for fixing λ is simply that, when $\beta > 4$, we choose λ so that the free energy is given by (8).[b]

The energy $E = \partial(\beta F)/\partial\beta$ calculated with this prescription is shown in Fig. 2, and the behavior of λ is shown in Fig. 3. Note that λ increases monotonically with β. A Gross-Witten phase transition takes place when $\lambda = 2$; this value is reached at $\beta = 9.45$. Thus a phase transition takes place as the system moves into the supergravity regime.[5]

[b] The numerical results we present in the rest of the paper were actually obtained by fitting $\beta F = -0.7857 - 2.050\beta^{-1.858}$ for $\beta > 4$.

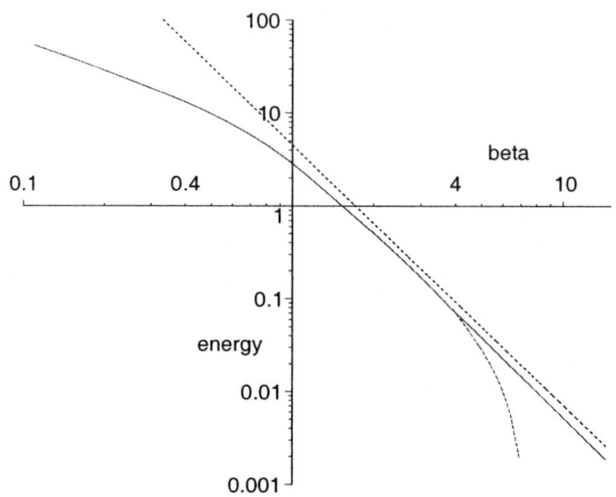

Fig. 2. Energy vs. β. For $\beta > 4$ fixing λ by fitting βF to a power law leads to the solid line, while the Schwinger-Dyson gap equation for lambda leads to the lower dashed line. The upper dashed line is the semiclassical energy of the black hole.

By adopting the prescription of fitting βF to a power law, we cannot say anything about the order of the phase transition. But note that, as shown in Fig. 3, the values of λ obtained from this prescription are fairly close to the results obtained from the Schwinger-Dyson equation for the Wilson loop, at least up to $\beta = 13$. (A relatively small change in λ has a large effect on the energy, due to the delicate supersymmetric cancellations that enter in computing βF.) If one takes the Schwinger-Dyson result for λ seriously, then the Gross-Witten transition occurs at $\beta = 10.63$, and is second order (the second derivative of the free energy drops by 0.15 in crossing the transition).

Our prescription for choosing λ begins to break down around $\beta = 14$, as we find that λ rapidly diverges as β approaches 14.5. By itself, this is not necessarily a problem: infinite λ simply means that the Wilson loop is uniformly distributed over $U(N)$. But unfortunately, we do not have a good prescription for continuing past this temperature. Evidently some of the other gap equations (not just the gap equation for λ) start to break down at this point. Note that this breakdown does not occur until well into the strong coupling regime, as an inverse temperature $\beta = 14.5$ corresponds to an effective gauge coupling $\beta^3 = 3 \times 10^3$.

Finally, let us comment on the average 'size' of the state. In our approximation the scalar fields $X^i(\tau)$ and $\phi^a(\tau)$ are Gaussian random matrices, and their eigenvalues obey a Wigner semi-circle distribution. We can define the size of the state

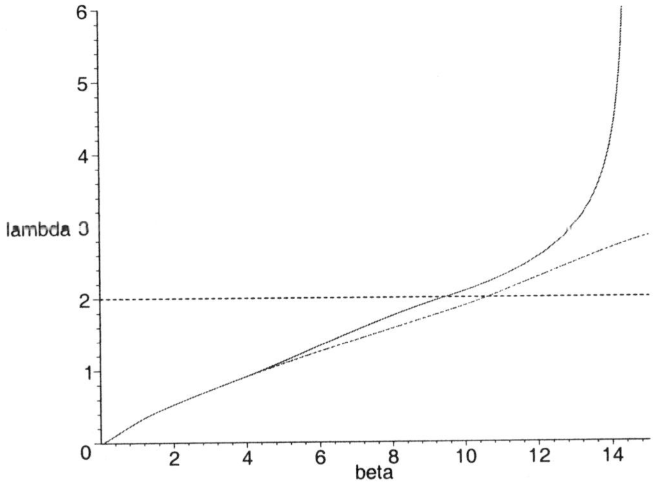

Fig. 3. The one-plaquette coupling λ vs. β. The Gross-Witten transition occurs when $\lambda = 2$. For $\beta < 4$ we use the Schwinger-Dyson gap equation to determine λ. For $\beta > 4$ the Schwinger-Dyson gap equation gives the dashed line, while fitting βF to a power law gives the solid line.

in terms of the quantities

$$\begin{aligned} R^2_{\text{scalar}} &= \frac{1}{N}\text{Tr}\langle\left(X^i(\tau)\right)^2\rangle_0 \\ R^2_{\text{gauge}} &= \frac{1}{N}\text{Tr}\langle(\phi^a(\tau))^2\rangle_0 \,. \end{aligned} \tag{10}$$

The radius of the Wigner semi-circle, given by $2\sqrt{R^2}$, is shown in Fig. 4. Note that the radius stays fairly constant in the region corresponding to the black hole. However, because the superfield formalism we are using does not respect the full $SO(9)$ invariance, the radius measured in the scalar multiplet directions is not the same as the radius measured in the gauge multiplet directions. The ratio of the two radii,

$$\left(R^2_{\text{scalar}}/R^2_{\text{gauge}}\right)^{1/2} = 2.1 \qquad \text{at } \beta = 14\,,$$

can be regarded as a measure of $SO(9)$ breaking.

In Fig. 4 we have also plotted the Schwarzschild radius of the black hole, given by[c] $U_0/2\pi = 1.76\ \beta^{-2/5}$. Note that, as the temperature decreases, the Schwarzschild radius becomes much smaller than the radius of the eigenvalue distributions. It seems appropriate to identify the radius of the eigenvalue distributions with the size of the region $U \ll (g^2_{\text{YM}}N)^{1/3}$ in which 10-dimensional supergravity is valid.[2] The Gaussian approximation should provide a good laboratory for studying the

[c]Our Higgs fields are related to radial position by $X = r/2\pi\alpha'$, while Ref. 2 sets $U = r/\alpha'$.

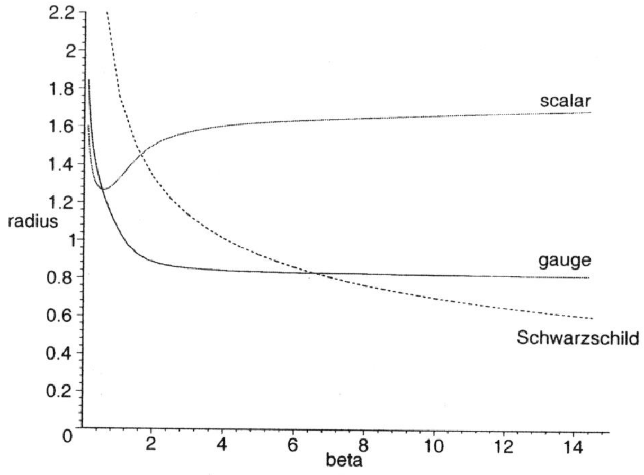

Fig. 4. Range of eigenvalues (radius of the Wigner semi-circle) vs. β. The upper solid curve is for the scalar fields in the scalar multiplets; the lower solid curve is for the scalar fields in the gauge multiplet. The dashed curve is the Schwarzschild radius of the black hole. These results were calculated with βF fit to a power law for $\beta > 4$.

way in which the horizon of the black hole can be detected in the dual quantum mechanics, perhaps along the lines suggested in Refs. 3,4.

Acknowledgments

DK wishes to thank Rutgers University and New York University for hospitality, and Emil Martinec and Mark Stern for valuable discussions. The work of DK is supported by the DOE under contract DE-FG02-90ER40542 and by the generosity of Martin and Helen Chooljian. GL would like to thank the Aspen Center for Physics for hospitality, and Vipul Periwal for useful discussions. The work of GL is supported by the NSF under grant PHY-98-02484. D.L. wishes to thank the Aspen Center for Physics, the Abdus Salam International Centre for Theoretical Physics and the Erwin Schroedinger Institute program in Duality, String Theory and M-Theory for hospitality during the course of this research. The research of D.L. is supported in part by DOE grant DE-FE0291ER40688-Task A.

References

1. J. Maldacena, *The large N limit of superconformal field theories and supergravity*, *Adv. Theor. Math. Phys.* **2**, 231 (1998) [hep-th/9711200].
2. N. Itzhaki, J.M. Maldacena, J. Sonnenschein and S. Yankielowicz, *Supergravity and the large N limit of theories with sixteen supercharges*, *Phys. Rev.* **D58**, 046004 (1998) [hep-th/9802042].
3. D. Kabat and G. Lifschytz, *Tachyons and black hole horizons in gauge theory*, *JHEP* **9812**, 002 (1998) [hep-th/9806214].

4. D. Kabat and G. Lifschytz, *Gauge theory origins of supergravity causal structure*, *JHEP* **9905**, 005 (1999) [hep-th/9902073].
5. D. Kabat and G. Lifschytz, *Approximations for strongly-coupled supersymmetric quantum mechanics*, *Nucl. Phys.* **B571**, 419 (2000) [hep-th/9910001].
6. D. Kabat, G. Lifschytz and D.A. Lowe, to appear.
7. D. Gross and E. Witten, *Possible third-order phase transition in the large-N lattice gauge theory*, *Phys. Rev.* **D21**, 446 (1980).
8. J. Cornwall, R. Jackiw and E. Tomboulis, *Effective action for composite operators*, *Phys. Rev.* **D10**, 2428 (1974).

D-BRANE PHYSICS AT LOW ENERGIES*

I. ANTONIADIS†
CERN, Theory Division, CH–1211 Geneva 23, Switzerland

1. Introduction

In how many dimensions do we live? Could they be more than the four we are aware of? If so, why don't we see the other dimensions? Is there a way to detect them?. While the possibility of extra-dimensions has been considered by physicists for long time, a compelling reason for their existence has arisen with string theory. It seems that a quantum theory of gravity requires that we live in more than four dimensions, probably in ten or eleven dimensions. The remaining (space-like) six or seven dimensions are hidden to us: observed particles do not propagate in them. The theory does not tell us yet why four and only four have been accessible to us. However, it predicts that this is only a low-energy effect: with increasing energy, particles which propagate in a higher dimensional space could be produced. What is the value of the needed high energy scale? could it be just close by, at reach of near future experiments?

Another scale which appears in our attempts to answer the previous questions is related to the extended nature of fundamental objects. It is the scale at which internal degrees of freedom are excited. In string theory this scale M_s is related to the string tension and sets the mass of the first heavy oscillation mode. The point-like behavior of known particles as observed at present colliders allows to conclude that M_s has to be higher than a few hundred GeV. It is the aim of these lectures as to provide a short summary of the present status of research on extra-dimensions and string-like sub-structure of matter.

---

*This work was partly supported by the European Commission under TMR contract ERBFMRX-CT96-0090.

†On leave from *Centre de Physique Théorique, Ecole Polytechnique, 91128 Palaiseau, France* (unité mixte du CNRS et de l'EP, UMR 7644).

2. Early Motivation for Large Extra Dimensions

Attempts to construct a consistent theory for quantum gravity have lead only to one candidate: string theory. The only vacua of string theory free of any pathologies are supersymmetric. Not being observed in nature, supersymmetry should be broken.

In contrast to ordinary supergravity, where supersymmetry breaking can be introduced at an arbitrary scale, through for instance the gravitino, gaugini and other soft masses, in string theory this is not possible (perturbatively). The only way to break supersymmetry at a scale hierarchically smaller than the (heterotic) string scale is by introducing a large compactification radius whose size is set by the breaking scale. This has to be therefore of the order of a few TeV in order to protect the gauge hierarchy.[1] This is one of the very few general predictions of perturbative (heterotic) string theory that leads to the spectacular prediction of the possible existence of extra dimensions accessible to future accelerators.[2] The main theoretical problem is though that the heterotic string coupling becomes necessarily strong.

The strong coupling problem can be understood from the effective field theory point of view from the fact that at energies higher than the compactification scale, the KK excitations of gauge bosons and other Standard Model particles will start being produced and contribute to various physical amplitudes. Their multiplicity turns very rapidly the logarithmic evolution of gauge couplings into a power dependence,[3] invalidating the perturbative description, as expected in a higher dimensional non-renormalizable gauge theory. A possible way to avoid this problem is to impose conditions which prevent the power corrections to low-energy couplings.[2] For gauge couplings, this implies the vanishing of the corresponding β-functions, which is the case for instance when the KK modes are organized in multiplets of $N = 4$ supersymmetry, containing for every massive spin-1 excitation, 2 Dirac fermions and 6 scalars. Examples of such models are provided by orbifolds with no $N = 2$ sectors with respect to the large compact coordinate(s).

The simplest example of a one-dimensional orbifold is an interval of length πR, or equivalently S^1/Z_2 with Z_2 the coordinate inversion. The Hilbert space is composed of the untwisted sector, obtained by the Z_2-projection of the toroidal states, and of the twisted sector which is localized at the two end-points of the interval, fixed under the Z_2 transformations. This sector is chiral and can thus naturally contain quarks and leptons, while gauge fields propagate in the (5d) bulk.

Similar conditions should be imposed to Yukawa's and in principle to higher (non-renormalizable) effective couplings in order to ensure a soft ultraviolet (UV) behavior above the compactification scale. We now know that the problem of strong coupling can be addressed using string S-dualities which invert the string coupling and relate a strongly coupled theory with a weakly coupled one. For instance, as we will discuss below, the strongly coupled heterotic theory with one large dimension is described by a weakly coupled type IIB theory with a tension at intermediate energies $(Rl_H)^{-1/2} \simeq 10^{11}$ GeV.[4] Furthermore, non-abelian gauge interactions emerge from tensionless strings[5] whose effective theory describes a higher-dimensional nontrivial infrared fixed point of the renormalization group.[6] This theory incorporates all conditions to low-energy couplings that guarantee a smooth UV behavior above the compactification scale. In particular, one recovers that KK modes of gauge bosons form $N = 4$ supermultiplets, while matter fields are localized in four dimensions. It is remarkable that the main features of these models were captured already in the context of the heterotic string despite its strong coupling.[2]

In the case of two or more large dimensions, the strongly coupled heterotic string is described by a weakly coupled type IIA or type I/I$'$ theory.[4] Moreover, the tension of the dual string becomes of the order or even lower than the compactification scale. In fact, as it will become clear in the following, the string tension becomes an arbitrary parameter.[7] It can be anywhere below the Planck scale and as low as a few TeV.[8] The main advantage of having the string tension at the TeV, besides its obvious experimental interest, is that it offers an automatic protection to the gauge hierarchy, alternative to low-energy supersymmetry or technicolor.[9,10,11]

3. Low-Scale Strings

In ten dimensions, superstring theory has two parameters: a mass (or length) scale M_s ($l_s = M_s^{-1}$), and a dimensionless string coupling g_s given by the vacuum expectation value (VEV) of the dilaton field $e^{<\phi>} = g_s$ on which we impose the weakly coupled condition $g_s < 1$. Compactification to lower dimensions introduces other parameters describing for instance volumes and shapes of the internal space. The D-dimensional compactification volume V_D will always be chosen to be bigger than unity in string units, $V_D \geq l_s^D$. This choice can always be done by appropriate T-duality transformations which inverts the compactification radius. To illustrate this duality let us consider a string vacuum with a $d_{\|}$-brane on which the standard model gauge bosons are localized. There are three type of strings:

- Closed strings have masses given by

$$M^2_{closed} = \sum_{i=1}^{d_{\|}} \frac{n_i^2}{R^2_{\|i}} + \sum_{j=1}^{d_\perp=9-d_{\|}} \frac{n_j^2}{R^2_{\perp i}} + \sum_{i=1}^{d_{\|}} \frac{m_i^2 R^2_{\|i}}{l_s^4} + \sum_{j=1}^{d_\perp=9-d_{\|}} \frac{m_j^2 R^2_{\perp j}}{l_s^4} + \frac{N}{l_s^2}, \quad (1)$$

- open strings with both ends on the $d_{\|}$-brane with masses

$$M^2_{DD} = \sum_{i=1}^{d_{\|}} \frac{n_i^2}{R^2_{\|i}} + \sum_{j=1}^{d_\perp=9-d_{\|}} \frac{m_j^2 R^2_{\perp j}}{l_s^4} + \frac{N}{l_s^2}, \quad (2)$$

- open strings with one ends on the $d_{\|}$-brane and another on a $d'_{\|}$-brane intersecting along $d_{\|} \bigcap d'_{\|}$ dimensions, for which the mass formula reads

$$M^2_{DD'} = \sum_{i \in d_{\|} \bigcap d'_{\|}} \frac{n_i^2}{R^2_{\|i}} + \sum_{j \in 9-(d_{\|} \bigcup d'_{\|})} \frac{m_i^2 R^2_{\perp i}}{l_s^4} + \frac{N}{l_s^2}, \quad (3)$$

where n_i, m_i and N are integer numbers. Note that the later have neither KK excitations ($p = \frac{m}{R}$) nor winding modes ($w = \frac{nR}{l_s^2}$) along the directions $(d_{\|} \bigcup d'_{\|}) - (d_{\|} \bigcap d'_{\|})$ in which they are localized.

T-duality not only exchanges Kaluza-Klein (KK) momenta $p = \frac{m}{R}$ with string winding modes $w = \frac{nR}{l_s^2}$, but also rescales the string coupling:

$$R \to \frac{l_s^2}{R} \qquad g_s \to g_s \frac{l_s}{R}, \tag{4}$$

so that the lower-dimensional coupling $g_s\sqrt{l_s/R}$ remains invariant. When R is smaller than the string scale, the winding modes become very light, while T-duality trades them as KK momenta in terms of the dual radius $\tilde{R} \equiv l_s^2/R$. The enhancement of the string coupling is then due to their multiplicity which diverges in the limit $R \to 0$ (or $\tilde{R} \to \infty$).

Upon compactification in $D = 4$ dimensions, these parameters determine the values at the string scale of the four-dimensional (4d) Planck mass (or length) M_p ($l_p = M_p^{-1}$) and gauge coupling g_{YM} that for phenomenological purposes should have the correct strength magnitude. For instance, generically the four-dimensional Planck mass can be expressed as:

$$M_p^2 \equiv f_p \frac{(M_s^6 V_6)}{g_s^2} M_s^2 , \tag{5}$$

where V_6 is the six-dimensional internal volume felt by gravitational interactions while the four-dimensional gauge coupling can be written as

$$\frac{1}{g_{YM}^2} \equiv f_{YM} \frac{(M_s^d V_d)}{g_s^q} , \tag{6}$$

where V_d is the d-dimensional internal volume felt by gauge interactions, and the coefficients f_p, f_{YM} have been computed for known classical string vacua. In the lowest order approximation, they are moduli-independent $\mathcal{O}(1)$ constants.[a]

In the past, weakly coupled heterotic strings were providing the most promising framework for phenomenological applications. In this case, the standard model was considered as descending from the ten-dimensional E_8 gauge symmetry, and we have $V_d = V_6$, $d = 6$ and $q = 2$. Taking the ratio of the two equations, one finds $\frac{M_s^2}{M_p^2} = \frac{f_{YM}}{f_p} g_{YM}^2 \sim g_{YM}^2$. Requiring $g_{YM} \sim \mathcal{O}(1)$, it was concluded that both the string scale M_s and the compactification scale $R^{-1} \equiv V_6^{-1/6}$ had to lie just below the Planck scale, at energies $\sim 10^{18}$GeV far out of reach of any near future experiment.[2,9]

The situation changed during recent years when it was discovered that string theory provides classical solutions (vacua) where gauge degrees of freedom live on subspaces i.e. $d < D$ along with the possibility of $p \neq q$. For instance, while $D = 6$

[a]Below we will often simplify the discussion by taking $f_p = f_{YM} = 1$

and $p = 2$, $(d, q) = (d, 1)$ in type I and $(d, q) = (2, 0)$ in type II or weakly coupled heterotic strings with small instantons. In these cases, it is an easy exercise to check that both the string and compactification scales can be made arbitrarily low.

The possibility of decreasing the string scale offers new insights on the physics beyond the standard model. For instance, a string scale at energies as low as TeV, would in addition to the plethora of experimental signatures, provides a solution to the problem of gauge hierarchy alternative to supersymmetry or technicolor. The hierarchy in gauge symmetry versus fundamental (cut-off) scales is then nullified as the two are of the same order.[9,10] Another possibility[12,13] is an intermediate scale which then identifies the string scale with natural scales where some new physics is expected, as for instance the scale of supersymmetry breaking in a hidden sector, the Peccei-Quinn axion physics, the neutrino see-saw scale etc. For instance, in a generic brane configuration, there might be a non-supersymmetric brane (as an anti-brane) which is located far away from the supersymmetric brane on which the standard model fields are localized. In this case supersymmetry is broken on the far brane at M_s and if communicated through gravity, the scale of supersymmetry breaking on our brane will be of order M_s^2/M_p. Requiring the latter to be in the TeV range implies a string scale at intermediate energies.

We review below the different possible realizations of low scale string theories.

3.1. *Type I/I′ string theory and D-branes*

Type I/I′ is a ten-dimensional theory of closed and open unoriented strings. Closed strings describe gravity, while gauge interactions are described by open strings whose ends are confined to propagate on p-dimensional sub-spaces defined as Dp-branes. The internal space has 6 compactified dimensions, $p-3$ longitudinal and $9-p$ transverse to the Dp-brane.

The gauge and gravitational interactions appear at different order in string loops perturbation theory, leading to different powers of g_s in the corresponding effective action:

$$S_I = \int d^{10}x \frac{1}{g_s^2 l_s^8} \mathcal{R} + \int d^{p+1}x \frac{1}{g_s l_s^{p-3}} F^2 \,, \tag{7}$$

The $1/g_s$ factor in front of the gauge kinetic terms corresponds to the lowest order open string diagram represented by a disk.

Upon compactification in four dimensions, the Planck length and gauge couplings are given to leading order by

$$\frac{1}{l_p^2} = \frac{V_{\|} V_{\perp}}{g_s^2 l_s^8} \,, \qquad \frac{1}{g_{YM}^2} = \frac{V_{\|}}{g_s l_s^{p-3}} \,, \tag{8}$$

where $V_{\|}$ ($V_{\perp}$) denotes the compactification volume longitudinal (transverse) to the Dp-brane. From the second relation above, it follows that the requirements of weak coupling $g_{YM} \sim \mathcal{O}(1)$, $g_s < 1$ imply that the size of the longitudinal space must be of order of the string length ($V_{\|} \sim l_s^{p-3}$), while the transverse volume $V_{\perp}$ remains unrestricted. Using the longitudinal volume in string units $v_{\|} \gtrsim 1$, and assuming an isotropic transverse space of $n = 9 - p$ compact dimensions of radius $R_\perp$, we can rewrite these relations as:

$$M_p^2 = \frac{1}{g_{YM}^4 v_\parallel} M_s^{2+n} R_\perp^n \,, \qquad g_s = g_{YM}^2 v_\parallel \,. \tag{9}$$

From the relations (9), it follows that the type I/I′ string scale can be chosen hierarchically smaller than the Planck mass at the expense of introducing extra large transverse dimensions that are felt only by the gravitationally interacting light states, while keeping the string coupling weak.[10] The weakness of 4d gravity compared to gauge interactions (ratio M_W/M_p) is then attributed to the largeness of the transverse space $R_\perp/l_s$.

An important property of these models is that gravity becomes (4+n)-dimensional with a strength comparable to those of gauge interactions at the string scale. The first relation of Eq. (9) can be understood as a consequence of the (4+n)-dimensional Gauss law for gravity, with

$$G_N^{(4+n)} = g_{YM}^4 l_s^{2+n} v_\parallel \tag{10}$$

the Newton's constant in $4+n$ dimensions.

Taking the type I string scale M_s to be at 1 TeV, one finds a size for the transverse dimensions $R_\perp$ varying from 10^8 km, .1 mm (10^{-3} eV), down to .1 fermi (10 MeV) for $n = 1, 2$, or 6 large dimensions, respectively. This shows that while $d_\perp = 1$ is excluded, $d_\perp \geq 2$ are allowed by present experimental bounds on gravitational forces.[14]

3.1.1. *Type II theories*

We proceed now with discussion of the relations (5) and (6) for the case of models derived from compactifications of Type II strings. For simplicity, we shall restrict ourselves to four-dimensional compactifications of type II on $K3 \times T^2$, yielding $N = 4$ supersymmetry. Calabi-Yau manifolds that lead to $N = 2$ supersymmetry can be obtained by replacing T^2 by a "base" two-sphere over which $K3$ varies while more interesting phenomenological models with $N = 1$ supersymmetry can be obtained by a freely acting orbifold, although the most general $N = 1$ compactifications would require F-theory on Calabi-Yau fourfolds.

In type IIA non-abelian gauge symmetries arise in six dimensions from D2-branes wrapped around non-trivial vanishing 2-cycles of a singular $K3$. The gauge kinetic terms are independent of the string coupling g_s and the corresponding effective action is:

$$S_{IIA} = \int d^{10}x \frac{1}{g_s^2 l_s^8} \mathcal{R} + \int d^6x \frac{1}{l_s^2} F^2 \,. \tag{11}$$

Upon compactification on a two-torus T^2 of size V_{T^2} to four dimensions, the gauge couplings are determined by V_{T^2}, while the Planck mass is also controlled by V_{K3} and g_s:

$$\frac{1}{g_{YM}^2} = \frac{V_{T^2}}{l_s^2} \qquad \frac{1}{l_p^2} = \frac{V_{T_2} V_{K3}}{g_s^2 l_s^8} = \frac{V_{K3}}{g_s^2 l_s^6} \frac{1}{g_{YM}^2} \,. \tag{12}$$

Therefore the area of T^2 should be of order l_s^2, while both g_s and V_{K3} can be used to separate the Planck mass from the string scale:[8,4]

$$M_s = M_p g_{YM} \frac{g_s l_s^2}{\sqrt{V_{K3}}} , \tag{13}$$

Taking $M_s \sim M_W$, with M_W the weak scale, the hierarchy between the electroweak and the Planck scales could be now obtained with a choice of string-size internal manifold and an ultra-weak coupling $g_s = 10^{-14}$.[4] As a result, gravity remains weak even at the string scale where the corresponding string interactions are suppressed by the tiny string coupling, or equivalently by the 4d Planck mass. The main observable effects in particle accelerators are the production of KK excitations along the two TeV dimensions of T^2 with gauge interactions.

In a way similar to the case of Type I/I' strings, one can instead produce a hierarchy of scalesM_s/M_p by keeping g_s of order unity and allowing some of the $K3$ (transverse) directions to be large. This corresponds to $V_{K3}/l_s^4 \sim 10^{28}$, implying a fermi size for the four $K3$ compact dimensions. Alternatively, one could play with both parameters g_s and V_{K3}.

An interesting possibility to mention is that it is possible to satisfy Eq. (12) while taking one direction much bigger than the string scale and the other much smaller. For instance, in the case of a rectangular torus of radii r and R, $V_{T^2} = rR \sim l_s^2$ with $r \gg l_s \gg R$. This can be treated by performing a T-duality (4) along R to type IIB: $R \to l_s^2/R$ and $g_s \to \tilde{g}_s = g_s l_s/R$ with $l_s = \tilde{l}_s$. One thus obtains:

$$\frac{1}{g_{YM}^2} = \frac{r}{R} \qquad \frac{1}{l_p^2} = \frac{V_{T_2} V_{K3}}{\tilde{g}_s^2 \tilde{l}_s^8} = \frac{R^2 V_{K3}}{\tilde{g}_s^2 \tilde{l}_s^6} \frac{1}{g_{YM}^2} . \tag{14}$$

which shows that the gauge couplings are now determined by the ratio of the two radii, or in general by the shape of T^2, while the Planck mass is controlled by its size.

Since T^2 is felt by gauge interactions, its size cannot be larger than $\mathcal{O}(\mathrm{TeV}^{-1})$ implying that in a scenario where $R \gg \tilde{l}_s$, the type IIB string scale should be much larger than TeV. The condition of weakly coupled ten (and six) dimensional type II theory implies $\tilde{M}_s \lesssim \sqrt{M_p/R}$, so that the largest value for the string tension, when $R \sim 1\mathrm{TeV}^{-1}$, is an intermediate scale $\sim 10^{11}$ GeV when the string coupling is of order unity. In the energy range between the KK scale $1/R$ and the type IIB string scale, one has an effective 6d theory without gravity at a non-trivial superconformal fixed point described by a tensionless string.[5,6] This is because in type IIB gauge symmetries still arise non-perturbatively from vanishing 2-cycles of $K3$, but take the form of tensionless strings in 6 dimensions, given by D3-branes wrapped on the vanishing cycles. Only after further compactification does this theory reduce to a standard gauge theory, whose coupling involves the shape rather than the volume of the two-torus, as described above. Since the type IIB coupling is of order unity, gravity becomes strong at the type IIB string scale and the main experimental signals at TeV energies are similar to those of type IIA models with tiny string coupling i.e. production of KK excitations of gauge degrees of freedom.

Similar constructions can be also realized in the context of the heterotic string when the standard model is embedded in non-perturbative gauge group arising from small instantons. In this case, the heterotic string scale can also be lowered in the TeV region.[15]

3.2. *Relation between type I/I′ and type II – heterotic*

The type I/I′ and type II models discussed above describe particular strongly coupled heterotic vacua with large dimensions.[16,4] Somewhat surprisingly, it turns out that all these TeV scale string models can be recovered as different strongly coupled decompactification limits, assuming only one large scale in addition to the Planck-size heterotic tension. More precisely, let us consider the heterotic string compactified on a six-torus with k large dimensions of radius $R \gg l_s$ and $6-k$ string-size dimensions. Applying the standard duality maps, it is simple to show that the type I′ theory with n transverse dimensions provides a weakly coupled dual description for the heterotic string only with $k = 4, 5, 6$ large dimensions, since otherwise the remaining T-dualities needed to obtain volumes above the resulting string scale lead to strongly-coupled descriptions. $k = 4$ is described by $n = 2$, $k = 6$ (for the $SO(32)$ gauge group) is described by $n = 6$, while for $n = 5$ one finds a type I′ model with 5 large transverse dimensions and one extra-large. The case $k = 4$ is particularly interesting: the heterotic string with 4 large dimensions at a TeV is described by a perturbative type I′ theory with the string scale at a TeV and gauge interactions confined to D7-branes with two transverse dimensions of millimeter size, T-dual to the two string-size heterotic coordinates. On the other hand, the type II theory provides a weakly coupled description for $k = 1, 2, 3, 4$ and $k = 6$ (for $E_8 \times E_8$). In particular, $k = 1$ is described by type IIB with string tension at intermediate energies, $k = 2$ is described by type IIA with tension and all compactification radii at a TeV and an infinitesimal coupling $g_s \sim l_s/R$, while for $k = 3$ all four (transverse) $K3$ directions are extra large.

4. Theoretical Implications

We will now focus on some theoretical implications of the low scale string scenario. Unless explicitly stated otherwise, we will restrict ourselves to the context of type I strings.

4.1. *U.V./ I.R. correspondence*

In addition to the open strings describing the gauge degrees of freedom, consistency of string theory requires the presence of closed strings associated with gravitons and different kind of moduli fields m_a. There are two types of extended objects: D-branes and orientifolds. The former are hypersurfaces on which open strings end while the latter are hypersurfaces located at fixed points when acting simultaneously with a Z_2 parity on the transverse space and world-sheet coordinates.

Closed strings can be emitted by D-branes and orientifolds, the lowest order diagrams being described by a cylindric topology. In this way D-branes and orientifolds appear as to lowest order classical point-like sources in the transverse space. For weak type-I string coupling this can be described by a Lagrangian of the form

$$\int d^n x_\perp \left[\frac{1}{g_s^2}(\partial_{x_\perp} m_a)^2 + \frac{1}{g_s}\sum_s f_s(m_a)\delta(x_\perp - x_{\perp s})\right], \tag{15}$$

where $x_{\perp s}$ is the location of the source s (D-branes and orientifolds) while $f_s(m_a)$ encodes the coupling of this source to the moduli m_a. As a result while m_a have constant values in the four-dimensional space, their expectation values will generically vary as a function of the transverse coordinates $x_\perp$ of the n directions with size $\sim R_\perp$ large compared to the string length l_s.

Solving the classical equation of motion for m_a in (15) leads to contributions to the parameters (couplings) on the brane of the low energy effective action given by a sum of Green's functions of the form:[11]

$$\frac{1}{V_\perp} \sum_{|p_\perp|<M_s} \frac{1}{p_\perp^2} F(\vec{p}_\perp), \tag{16}$$

where $V_\perp = {R_\perp}^{d_\perp}$ is the volume of the transverse space, $\vec{p}_\perp = (m_1/R_\perp \cdots m_{d_\perp}/R_\perp)$ is the transverse momentum exchanged by the massless closed string, $F(\vec{p}_\perp)$ are the Fourier-transformed to momentum space of derivatives of $f_s(m_a)$. An explicit expression can be given in the simple case of toroidal compactification with vanishing antisymmetric tensor, where the global tadpole cancelation fixes the number of D-branes to be 32:

$$F(\vec{p}_\perp) \sim \left(32\prod_{i=1}^{d_\perp}\frac{1+(-)^{m_i}}{2} - 2\sum_{a=1}^{16}\cos(\vec{p}_\perp\vec{x}_a)\right), \tag{17}$$

where $\vec{p}_\perp = (m_1/R_\perp \cdots m_{d_\perp}/R_\perp)$, the orientifolds are located at the corners of the cell $[0, \pi R_\perp]^{d_\perp}$ and are responsible for the first term in (17), and $\pm\vec{x}_a$ are the transverse positions of the 32 D-branes (corresponding to Wilson lines of the T-dual picture) responsible of the second term.

In a compact space where flux lines can not escape to infinity, the Gauss-law implies that the total charge, thus global tadpoles, should vanish $F(0) = 0$ while local tadpoles may not vanish $F(\vec{p}_\perp) \neq 0$ for $\vec{p} \neq 0$. In that case, obtained for generic positions of the D-branes, the tadpole contribution (16) leads to the following behavior in the large radius limit for the moduli m_a:

$$m_a(x_{\perp s}) \sim \begin{cases} O(R_\perp M_s) & \text{for} \quad d_\perp = 1 \\ O(\ln R_\perp M_s) & \text{for} \quad d_\perp = 2 \\ O(1) & \text{for} \quad d_\perp > 2 \end{cases}, \tag{18}$$

which is dictated by the large-distance behavior of the two-point Green function in the $d_\perp$-dimensional transverse space.

There are some important implications of these results:

- The tree-level exchange diagram of a closed string can also be seen as one-loop exchange of open strings. While from the former point of view, a long cylinder represents an infrared limit where one computes the effect of exchanging light closed strings at long distances, in the second point of view the same diagram is conformally mapped to an annulus describing the one-loop running in the ultraviolet limit of very heavy open strings streching between the two boundaries of the cylinder. Thus, from the brane gauge theory point of view, there are ultraviolet effects that are not cut-off by the string scale M_s but instead by the winding mode scale $R_\perp M_s^2$.
- In the case of one large dimension $d_\perp = 1$, the corrections are linear in $R_\perp$. Such correction appears for instance for the dilaton field which sits in front of gauge kinetic terms, that drive the theory rapidly to a strong coupling singularity and, thus, forbid the size of the transverse space to become much larger than the string length. It is possible to avoid such large corrections if the tadpoles cancel locally. This happens when D-branes are equally distributed at the two fixed points of the orientifold.
- The case $d_\perp = 2$ is particularly attractive because it allows the effective couplings of the brane theory to depend logarithmically on the size of the transverse space, or equivalently on M_p, exactly as in the case of softly broken supersymmetry at M_s. Both higher derivative and higher string loop corrections to the bulk supergravity Lagrangian are expected to be small for slowly (logarithmically) varying moduli. The *classical* equations of motion of the effective 2d supergravity in the transverse space are analogous to the renormalization group equations used to resum large corrections to the effective field theory parameters with appropriate boundary conditions.

It turns out that low-scale type II theories with infinitesimal string coupling share many common properties with type I′ when $d_\perp = 2$.[4] In fact, the limit of vanishing coupling does not exist due to subtleties related to the singular character of the compactification manifold and to the non perturbative origin of gauge symmetries. In general, there are corrections depending logarithmically on the string coupling, similarly to the case of type I′ strings with 2 transverse dimensions.

4.2. *Unification*

One of the main succes of low-energy supersymmetry is that the three gauge couplings of the Standard Model, when extrapolated at high energies assuming the particle content of its $N = 1$ minimal supersymmetric extension (MSSM), meet at an energy scale $M_{\rm GUT} \simeq 2\times 10^{16}$ GeV. This running is described at the the one-loop level by:

$$\frac{1}{g_a^2(\mu)} = \frac{1}{g^2} + \frac{b_a}{4\pi} \ln \frac{M_{\rm GUT}^2}{\mu^2}\,, \tag{19}$$

where μ is the energy scale and a denotes the 3 gauge group factors of the Standard Model $SU(3) \times SU(2) \times U(1)$. Note that even in the absence of any GUT group, if one requires keeping unification of all gauge couplings then the string relations we discussed in section *3* suggest that the gauge theories arise from the same kind of branes. Decreasing the string scale below energies of order M_{GUT} is expected to cut-off the runing of the couplings before they meet and thus spoils the unification. Is there a way to reconcile the apparent unification with a low string scale?

One possibility is to use power-law running that may accelerate unification in an energy region where the theory becomes higher dimensional.[17] There are two remarks to be made on this approach: (i) the result is very sensitive (power-like) to the initial conditions and thus to string threshold corrections, in contrast to the usual unification based on logarithmic evolution, (ii) only the case of one extra-dimension appears to lead to power-like corrections in type I models.

In fact the one-loop corrected gauge couplings in $N = 1$ orientifolds are given by the following expression:[18]

$$\frac{1}{g_a^2(\mu)} = \frac{1}{g^2} + s_a m + \frac{b_a}{4\pi} \ln \frac{M_I^2}{\mu^2} - \sum_{i=1}^{3} \frac{b_{a,i}^{N=2}}{4\pi} \left\{ \ln T_i + f(U_i) \right\} , \tag{20}$$

where the first two terms in the r.h.s. correspond to the tree-level (disk) contribution and the remaining ones are the one-loop (genus-1) corrections. Here, we assumed that all gauge group factors correspond to the same type of D-branes, so that gauge couplings are the same to lowest order (given by g). m denotes a combination of the twisted moduli, whose VEVs blow-up the orbifold singularities and allow the transition to smooth (Calabi-Yau) manifolds. However, in all known examples, these VEVs are fixed to $m = 0$ from the vanishing of the D-terms of anomalous $U(1)$'s.

As expected, the one-loop corrections contain an infrared divergence, regulated by the low-energy scale μ, that produces the usual 4d running controlled by the $N = 1$ beta-functions b_a. The last sum displays the string threshold corrections that receive contributions only from $N = 2$ sectors, controlled by the corresponding $N = 2$ beta-functions $b_{a,i}^{N=2}$. They depend on the geometric moduli T_i and U_i, parameterizing the size and complex structure of the three internal compactification planes. In the simplest case of a rectangular torus of radii R_1 and R_2, $T = R_1 R_2 / l_s^2$ and $U = R_1/R_2$. The function $f(U) = \ln \left(\mathrm{Re} U |\eta(iU)|^4 \right)$ with η the Dedekind-eta function; for large U, $f(U)$ grows linearly with U. Thus, from expression (20), it follows that when $R_1 \sim R_2$, there are logarithmic corrections (as explained for transverse directions to the brane for the previous section) $\sim \ln(R_1/l_s)$, while when $R_1 > R_2$, the corrections grow linearly as R_1/R_2. Note that in both cases, the corrections are proportional to the $N = 2$ β-functions and there no power law corrections in the case of more than one large compact dimensions.

Obviously, unification based on logarithmic evolution requires the two (transverse) radii to be much larger than the string length, while power-low unification can happen either when there is one longitudinal dimension a bit larger than the string scale ($R_1/R_2 \sim R_{\parallel}/l_s$ keeping $g_s < 1$), or when one transverse direction is bigger than the rest of the bulk.

The most advantageous possibility is to obtain large logarithmic thresholds depending on two large dimensions transverse to the brane ($d_\perp = 2$). One hopes that such logarithmic corrections may restore the "old" unification picture with a GUT scale given by the winding scale, which for millimeter-size dimensions has the correct order of magnitude.[19,11,20] In this way, the running due to a large desert in energies is replaced by an effective running due to a "large desert" in transverse distances from our world-brane. However, the logarithmic contributions are model dependent[18] and at present there is no compelling explicit realization of this idea.

4.3. *Supersymmetry breaking and scales hierarchy*

When decreasing the string scale, the question of hierarchy of scales i.e. of why the Planck mass is much bigger than the weak scale, is translated into the question of why there are transverse dimensions much larger than the string scale, or why the string coupling is very small. For instance for a string scale in the TeV range, From Eq. (9) in type I/I′ strings, the required hierarchy $R_\perp/l_I$ varies from 10^{15} to 10^5, when the number of extra dimensions in the bulk varies from $n = 2$ to $n = 6$, respectively, while in type II strings with no large dimensions, the required value of the string coupling g_{II} is 10^{-14}. There are two issues that one needs to address:

- We have seen in section 4.1 that although the string scale is very low, there might be large quantum corrections that arise, dependending on the size of the large dimensions transverse to the brane. This is as if the UV cutoff of the effective field theory on the brane is not the string scale but the winding scale $R_\perp M_I^2$, dual to the large transverse dimensions and which can be much larger than the string scale. In particular such correction could spoil the nullification of gauge hierarchy that remain the main theoretical motivation of TeV scale strings.
- Another important issue is to understand the dynamical question on the origin of the hierarchy.

TeV scale strings offer a solution to the technical (at least) aspect of gauge hierarchy without the need of supersymmetry, provided there is no effective propagation of bulk fields in a single transverse dimension, or else closed string tadpoles should cancel locally. The case of $d_\perp = 2$ leads to a logarithmic dependence of the effective potential on $R_\perp/l_s$ which allows the possible radiative generation of the hierarechy between $R_\perp$ and l_s as for no-scale models. Moreover, it is interesting to notice that the ultraviolet behavior of the theory is very similar with the one with soft supersymmetry breaking at $M_s \sim TeV$. It is then natural to ask the question whether there is any motivation leftover for supersymmetry or not. This bring us to the problems of the stability of the new hierarchy and of the cosmological constant.[10]

In fact, in a non-supersymmetric string theory, the bulk energy density behaves generically as $\Lambda_{\text{bulk}} \sim M_s^{4+n}$, where n is the number of transverse dimensions much larger than the string length. In the type I/I′ context, this induces a cosmological constant on our world-brane which is enhanced by the volume of the transverse space $V_\perp \sim R_\perp^n$. When expressed in terms of the 4d parameters using the type I/I′ mass-relation (9), it is translated to a quadratically dependent contribution on the Planck mass:

$$\Lambda_{\text{brane}} \sim M_I^{4+n} R_\perp^n \sim M_I^2 M_p^2 \,, \tag{21}$$

where we used $s = I$. This contribution is in fact the analogue of the quadratic divergent term $\mathrm{Str}\mathcal{M}^2$ in softly broken supersymmetric theories, with M_I playing the role of the supersymmetry breaking scale.

The brane energy density (21) is far above the (low) string scale M_I and in general destabilizes the hierarchy that one tries to enforce. One way out is to resort to special models with broken supersymmetry and vanishing or exponentially small cosmological constant.[21] Alternatively, one could conceive a different scenario, with supersymmetry broken primordially on our world-brane maximally, i.e. at the string scale which is of order of a few TeV. In this case the brane cosmological constant would be, by construction, $\mathcal{O}(M_I^4)$, while the bulk would only be affected by gravitationally suppressed radiative corrections and thus would be almost supersymmetric.[10,22] In particular, one would expect the gravitino and other soft masses in the bulk to be extremely small $O(M_I^2/M_p)$. In this case, the cosmological constant induced in the bulk would be

$$\Lambda_{\text{bulk}} \sim M_I^4 / R_\perp^n \sim M_I^{6+n} / M_p^2 \,, \tag{22}$$

i.e. of order $(10 \text{ MeV})^6$ for $n = 2$ and $M_I \simeq 1$ TeV. The scenario of brane supersymmetry breaking is also required in models with a string scale at intermediate energies $\sim 10^{11}$ GeV (or lower), discussed in the beginning of section *3*. It can occur for instance on a brane distant from our world and is then mediated to us by gravitational (or gauge) interactions.

In the absence of gravity, brane supersymmetry breaking can occur in a non-BPS system of D-branes. The simplest examples are based on orientifold projections of type IIB, in which some of the orientifold 5-planes have opposite charge, requiring an open string sector living on anti-D5 branes in order to cancel the RR (Ramond-Ramond) charge. As a result, supersymmetry is broken on the intersection of D9 and anti-D5 branes that coincides with the world volume of the latter. The simplest construction of this type is a T^4/Z_2 orientifold with a flip of the Ω-projection (world-sheet parity) in the twisted orbifold sector. It turns out that several orientifold models, where tadpole conditions do not admit naive supersymmetric solutions, can be defined by introducing non-supersymmetric open sector containing anti-D-branes. A typical example of this type is the ordinary $Z_2 \times Z_2$ orientifold with discrete torsion.

The resulting models are chiral, anomaly-free, with vanishing RR tadpoles and no tachyons in their spectrum.[22] Supersymmetry is broken at the string scale on a collection of anti-D5 branes while, to lowest order, the closed string bulk and the other branes are supersymmetric. In higher orders, supersymmetry breaking is of course mediated to the remaining sectors, but is suppressed by the size of the transverse space or by the distance from the brane where supersymmetry breaking primarily occurred. The models contain in general uncancelled NS (Neveu-Schwarz) tadpoles reflecting the existence of a tree-level potential for the NS moduli, which is localized on the (non-supersymmetric) world volume of the anti-D5 branes.

As a result, this scenario implies the absence of supersymmetry on our world-brane but its presence in the bulk, a millimeter away! The bulk supergravity is needed to guarantee the stability of gauge hierarchy against large gravitational quantum radiative corrections.

References

1. I. Antoniadis, C. Bachas, D. Lewellen and T. Tomaras, *Phys. Lett.* **B207**, 441 (1988).
2. I. Antoniadis, *Phys. Lett.* **B246**, 377 (1990).
3. T. Taylor and G. Veneziano, *Phys. Lett.* **B212**, 147 (1988).
4. I. Antoniadis and B. Pioline, *Nucl. Phys.* **B550**, 41 (1999) [hep-th/9902055].
5. E. Witten, *Proceedings of Strings 95*, hep-th/9507121; A. Strominger, *Phys. Lett.* **B383**, 44 (1996); **B475**, 94 (1996).
6. N. Seiberg, *Phys. Lett.* **B390**, 169 (1997).
7. E. Witten, *Nucl. Phys.* **B471**, 135 (1996).
8. J.D. Lykken, *Phys. Rev.* **D54**, 3693 (1996); G. Shiu and S.-H.H. Tye, *Phys. Rev.* **D58**, 106007 (1998).
9. N. Arkani-Hamed, S. Dimopoulos and G. Dvali, *Phys. Lett.* **B429**, 263 (1998).
10. I. Antoniadis, N. Arkani-Hamed, S. Dimopoulos and G. Dvali, *Phys. Lett.* **B436**, 263 (1998).
11. I. Antoniadis, C. Bachas, *Phys. Lett.* **B450**, 99 (83).
12. K. Benakli, *Phys. Rev.* **D60**, 104002 (1999).
13. C.P. Burgess, L.E. Ibáñez, F. Quevedo, *Phys. Lett.* **B429**, 257 (1999).
14. See *e.g.* J.C. Long, H.W. Chan and J.C. Price, *Nucl. Phys.* **B539**, 23 (1999).
15. K. Benakli and Y. Oz, *Phys. Lett.* **B472**, 83 (2000).
16. I. Antoniadis and M. Quirós, *Phys. Lett.* **B392**, 61 (1997).
17. K.R. Dienes, E. Dudas and T. Gherghetta, *Phys. Lett.* **B436**, 55 (1998); *Nucl. Phys.* **B537**, 47 (1999).
18. I. Antoniadis, C. Bachas and E. Dudas, *Nucl. Phys.* **B560**, 93 (1999).
19. C. Bachas, *JHEP* **9811**, 23 (1998).
20. N. Arkani-Hamed, S. Dimopoulos and J. March-Russell, hep-th/9908146.
21. S. Kachru and E. Silverstein, *JHEP* **9811**, 1 (1998) [hep-th/9810129]; J. Harvey, *Phys. Rev.* **D59**, 26002 (1999); R. Blumenhagen and L. Görlich, *Nucl. Phys.* **B551**, 601 (1999) [hep-th/9812158]; C. Angelantonj, I. Antoniadis and K. Foerger, *Nucl. Phys.* **B555**, 116 (1999) [hep-th/9904092].
22. I. Antoniadis, E. Dudas and A. Sagnotti, *Phys. Lett.* **B464**, 38 (1999) [hep-th/9908023]; G. Aldazabal and A.M. Uranga, *JHEP* **9910**, 024 (1999) [hep-th/9908072].

D-BRANES IN THE BACKGROUND OF NS FIVEBRANES

SHMUEL ELITZUR[a], AMIT GIVEON[a], DAVID KUTASOV[b],
ELIEZER RABINOVICI[a] and GOR SARKISSISAN[a]
[a] *Racah Institute of Physics, The Hebrew University*
Jerusalem 91904, Israel
[b] *Department of Physics, University of Chicago*
5640 S. Ellis Av., Chicago, IL 60637, USA

We study the dynamics of D-branes in the near-horizon geometry of NS fivebranes. This leads to a holographically dual description of the physics of D-branes ending on and/or intersecting $NS5$-branes. We use it to verify some properties of such D-branes which were deduced indirectly in the past, and discuss some instabilities of non-supersymmetric brane configurations. Our construction also describes vacua of Little String Theory which are dual to open plus closed string theory in asymptotically linear dilaton spacetimes.

1. Introduction

The work discussed is based on the results appearing in hep-th/0005052. In the last few years it was found that embedding various supersymmetric gauge theories in string theory, as the low energy worldvolume dynamics on branes, provides an efficient tool for studying many aspects of the vacuum structure and properties of BPS states in these theories. One class of constructions (reviewed in Ref. 1) involves systems of D-branes ending on and/or intersecting $NS5$-branes. For applications to gauge theory one is typically interested in taking the weak coupling limit $g_s \to 0$ (as well as the low energy limit). In this limit one might expect the system to be amenable to a perturbative worldsheet treatment, but the presence of the $NS5$-branes and branes ending on branes complicate the analysis.

In the absence of a derivation of the properties of D-branes interacting with $NS5$-branes from first principles, some of their low energy properties were postulated in the past based on symmetry considerations and consistency conditions. One of the main purposes of this talk is to derive some of these properties by a direct worldsheet study of D-branes in the vicinity of $NS5$-branes.

In the analysis, we will use the improved understanding of the dynamics of $NS5$-branes achieved in the last few years. It is now believed that in the weak coupling limit (but not necessarily at low energies) fivebranes decouple from gravity and other bulk string modes and give rise to a rich non-gravitational theory, Little String Theory (LST).[2] In Ref. 3 it was proposed to study LST using holography. The theory on a stack of $NS5$-branes was argued to be holographically dual to string theory in the near-horizon geometry of the fivebranes. Many properties of LST can be understood by performing computations in string theory in this geometry. In particular, we will find below that it is an efficient way to study properties of

D-branes ending on or intersecting $NS5$-branes.

From the point of view of LST, D-branes in the vicinity of $NS5$-branes give rise to new particle and extended object states in the theory, as well as new vacua, typically with reduced supersymmetry (when the branes are space-filling). Another motivation of this work is to understand the non-perturbative spectrum and dynamics of extended objects in LST, and more generally study the interplay between the non-trivial dynamics on fivebranes and the worldvolume physics of D-branes in their vicinity.

The plan of the talk is the following. In section 2 we review some facts regarding the relevant brane configurations and the physical issues involved. In particular, we describe the conjectured properties of these configurations that we will try to verify. Section 3 is a review of the near-horizon geometry of $NS5$-branes (the CHS geometry[4]) and its holographic relation to LST. We also describe a modification of this geometry corresponding to fivebranes positioned at equal distances around a circle, which plays a role in the analysis. In section 4 we review some facts about D-branes in flat space and on a three-sphere (or $SU(2)$ WZW model). In section 5 we study D-branes in the CHS geometry as well as its regularized version. We verify some of the properties described in section 2, and describe additional features which follow from our analysis.

2. Some Properties of Brane Configurations

One class of brane constructions, that gave rise to many insights into gauge dynamics, involves D-branes suspended between $NS5$-branes. The construction realizes four dimensional $N = 1$ supersymmetric gauge theory with gauge group $G = U(N_c)$ and N_f chiral superfields in the fundamental representation (more precisely, N_F fundamentals Q^i, $i = 1, \cdots, N_F$, and N_F anti-fundamentals $\tilde{Q}_i$). This will help introduce the issues that will be discussed later.

Consider a configuration[5] of N_c $D4$-branes stretched between an $NS5$-brane and an $NS5'$-brane separated by a distance L along the x^6 direction (Fig. 1). At distances greater than L, the five dimensional theory on the $D4$-branes reduces to a four dimensional theory with $N = 1$ supersymmetry. The boundary conditions provided by the fivebranes imply that out of all the massless degrees of freedom on the fourbranes (described by $4-4$ strings), only the $4d$ $N = 1$ vector multiplets for $G = U(N_c)$ survive. To decouple the gauge dynamics from the complications of string theory one takes the limit $g_s \to 0$, $L/l_s \to 0$, with the four dimensional gauge coupling $g^2 = g_s l_s/L$ held fixed.[1]

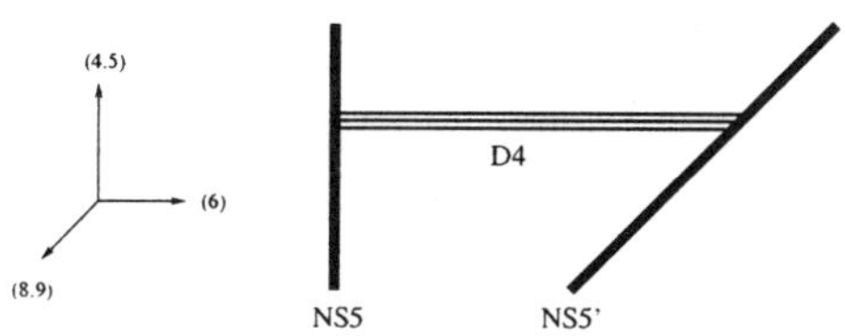

Fig. 1.

One can add matter in the fundamental representation of G in one of the two ways illustrated in Fig. 2 (which, as explained in Ref. 6, are related). One is to place N_f $D6$-branes between the fivebranes (Fig. 2a). The $4-6$ strings give N_f fundamentals of $U(N_c)$, Q^i, $\tilde{Q}_i$, whose mass is proportional to the separation between the sixbranes and the fourbranes in (x^4, x^5). Alternatively, one can add to the configuration N_f $D4$-branes stretching from the $NS5$-brane to infinity (Fig. 2b). In this case, the N_f fundamentals of $U(N_c)$ should arise from $4-4$ strings stretched between the two kinds of fourbranes. Their mass is the separation between the fourbranes in (x^4, x^5).

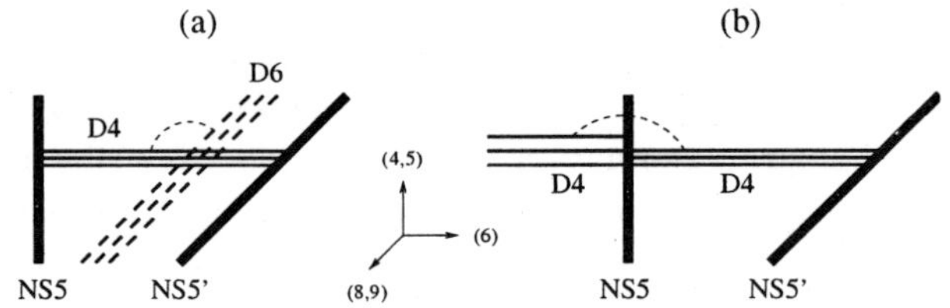

Fig. 2.

In the configuration of Fig. 2a, the low energy physics should be independent of the locations of the $D6$-branes along the interval between the fivebranes. In Ref. 7 it was pointed out that a particularly natural location for the sixbranes is at the same value of (x^4, x^5, x^6) as the $NS5'$-brane. In this case the $NS5'$-brane is embedded in the $D6$-branes; it divides them into two disconnected parts (Fig. 3).

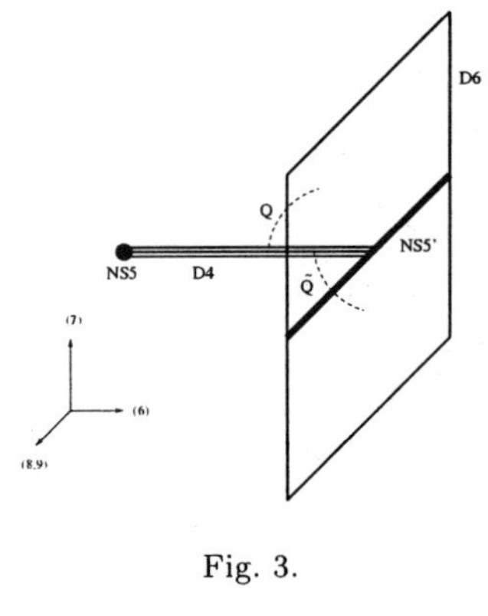

Fig. 3.

Consequently, the configuration has two separate $U(N_f)$ symmetries, acting on the two semi-infinite sixbranes, and one may attempt to interpret them as the $U(N_f) \times U(N_f)$ global symmetry of $N = 1$ SQCD, under which Q and $\tilde{Q}$ transform as $(N_F, 1)$ and $(1, \overline{N}_F)$, respectively. This would imply that $4-6$ strings connecting the $D4$-branes to the upper part of the $D6$-branes give rise at low energies to the N_f chiral multiplets in the fundamental of $U(N_c)$, Q^i, while those that connect the fourbranes to the lower part of the sixbranes give the N_f chiral superfields in the anti-fundamental of $U(N_c)$, $\tilde{Q}_i$, as indicated in Fig. 3.

The low-lying excitations of the brane configurations discussed above can be divided into two classes: those that are bound to one of the fivebranes, and those that are not. In this talk we will analyze the properties of the first class of excitations. It includes the following:

- (a) $4-6$ strings connecting N_c $D4$-branes to N_F $D6'$-branes, all of which end on an $NS5$-brane[a] (Fig. 4a). The prediction is that they give rise to a chiral spectrum: a chiral superfield Q in the (N_c, N_F) of $U(N_c) \times U(N_F)$.
- (b) $4-4$ strings connecting fourbranes ending on an $NS5$-brane from opposite sides (Fig. 4b). They should give rise to chiral superfields, Q in the (N_c, N_F) and $\tilde{Q}$ in the $(\overline{N}_c, \overline{N}_F)$, or hypermultiplets $(Q, \tilde{Q})$.
- (c) $4-6$ strings connecting $D4$-branes ending on $NS5$-branes to $D6$-branes intersecting the fivebranes (Fig. 4c). They should also give rise to hypermultiplets $(Q, \tilde{Q})$.

We verify the predictions (a), (b), (c). In this talk we will describe the ingredients leading to the verification of the chiral symmetry.

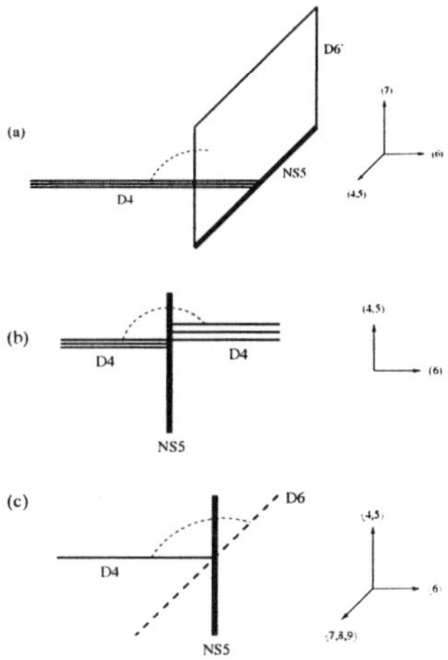

Fig. 4.

3. The Near-Horizon Geometry of $NS5$-branes and Holography

The background fields around a stack of k parallel $NS5$-branes are:[4]

$$e^{2(\Phi-\Phi_0)} = 1 + \sum_{j=1}^{k} \frac{l_s^2}{|\vec{x} - \vec{x}_j|^2}$$

$$G_{IJ} = e^{2(\Phi-\Phi_0)} \delta_{IJ}$$

[a]The configuration of Fig. 4a can be obtained from that of Fig. 3 by exchanging $(x^4, x^5) \leftrightarrow (x^8, x^9)$ and removing some branes.

$$G_{\mu\nu} = \eta_{\mu\nu}$$
$$H_{IJK} = -\epsilon_{IJKL}\partial^L \Phi \tag{1}$$

$I, J, K, L = 6, 7, 8, 9$ label the directions transverse to the fivebranes. $\mu, \nu = 0, 1, \cdots, 5$ are the directions along the brane. $\{\vec{x}_j\}$ are the locations of the fivebranes in $\vec{x} = (x^6, \cdots, x^9)$. H is the field strength of the NS-NS B-field; G, Φ are the metric and dilaton, respectively.

The background (1) interpolates between flat ten dimensional spacetime far from the fivebranes, and a near-horizon region, in which the 1 on the right hand side of the first line of (1) can be neglected (Fig. 5).

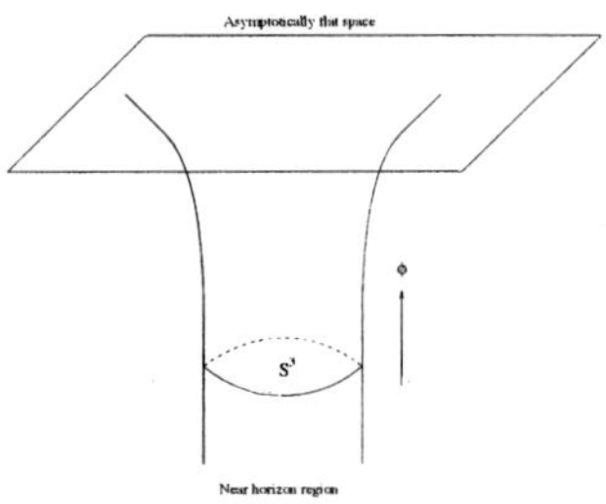

Fig. 5.

String propagation in the near-horizon geometry can be described by an exact worldsheet Conformal Field Theory (CFT). The target space is

$$R^{5,1} \times R_\phi \times SU(2) \tag{2}$$

R_ϕ corresponds to the radial direction $r = |\vec{x}|$:

$$\phi = \frac{1}{Q} \log \frac{|\vec{x}|^2}{kl_s^2}$$
$$\Phi = -\frac{Q}{2}\phi \tag{3}$$

where we set $\Phi_0 = 0$ by rescaling $\vec{x}$. Q is related to the number of fivebranes via

$$Q = \sqrt{\frac{2}{k}}. \tag{4}$$

Equation (3) implies infinite string coupling at the origin. To allow a perturbative treatment this background should be modified by adding a Liouville term which shields this singular point. This modification is equivalent to replacing $R_\phi \times SU(2)$ by the cigar shaped coset space leading to the background:

$$R^{5,1} \times \frac{SL(2,R)_k}{U(1)} \times \frac{SU(2)_k}{U(1)} \tag{5}$$

Since the background (5) can be made arbitrarily weakly coupled, one can use it to compute correlation functions in LST in a weak coupling regime in its moduli space of vacua.[14] Consider the (NS,NS) sector of the theory. The observables are primaries of the $N = 1$ superconformal algebra with scaling dimension $(h, \bar{h}) = (\frac{1}{2}, \frac{1}{2})$. By analyzing correlation functions of $V_{j;m,\bar{m}}$, one finds that poles in correlators correspond to discrete representations of $SL(2)$

$$|m| = j + n, \quad |\bar{m}| = j + \bar{n}; \quad n, \bar{n} = 1, 2, 3, \cdots \tag{6}$$

This leads to a discrete spectrum of states in LST, which exhibits Hagedorn growth at high energy.

4. Some Properties of D-branes

In this section we review some properties of D-branes in flat space,[16,17] and on S^3, [18,19,20] in preparation for our discussion of D-branes in the CHS geometry (2), and its regularized version (5).

4.1. *4-6 strings in flat space*

Later, we analyze $4-6$ strings connecting $D4$-branes and $D6'$-branes, both of which end on a stack of $NS5$-branes (Fig. 4a).

Consider an open string, one of whose ends is on a $D4$-brane. The other end is on a $D6'$ brane. The vertex operator describing the (bosonic) ground state of a $4-6'$ string stretched between a $D4$-brane at $(x^8, x^9) = (0, 0)$ and a $D6'$-brane at $(x^8, x^9) = (a, b)$,

$$V = e^{-\varphi} \sigma_{4567} S_{4567} e^{\frac{i}{\pi}(a(x_L^8 - x_R^8) + b(x_L^9 - x_R^9))} e^{ik_\mu x^\mu} \tag{7}$$

where k_μ ($\mu = 0, 1, 2, 3$) is the $4d$ spacetime momentum, σ_{4567} is a dimension $\frac{1}{4}$ operator which twists the coordinates $x^{4,5,6,7}$ from Dirichlet to Neumann boundary conditions, S_{4567} is a dimension $\frac{1}{4}$ spin operator for $\psi_{4,5,6,7}$ and φ is the bosonized superconformal ghost. The vertex operator (7) is written in the -1 picture; one can check that the coefficient of $e^{-\varphi}$ in (7) is an $N = 1$ superconformal primary, which is a necessary condition for its BRST invariance. The requirement that it has worldsheet dimension $1/2$, which is also necessary for BRST invariance, implies that the mass squared of the ground state of the $4 - 6'$ string is $-k_\mu^2 = \frac{1}{\pi^2}(a^2 + b^2)$, as one would expect. (we work in a convention $\alpha' = 1/2$, in which the scalars x are canonically normalized on the boundary and the tension of the fundamental string is $T = 1/\pi$). In particular, when the $D6'$-brane intersects the $D4$-brane, i.e. when $a = b = 0$, this mass vanishes. The vertex operator (7) describes a particle which transforms as a scalar under $3 + 1$ dimensional Lorentz rotations. The spin field S_{4567} has 4 components, half of which are projected out by the GSO projection, so (7) actually describes two real scalar particles.

4.2. *D-branes on the $SU(2)$ group manifold*

We next turn to some facts regarding D-branes on a group manifold G, focusing on the case $G = SU(2)$. In the absence of D-branes, the WZW model has an affine

$G_L \times G_R$ symmetry. If $g(z, \overline{z})$ is a map from the worldsheet to the group G, the symmetry acts on it as:

$$g \to h_L(z) g h_R(\overline{z}) \tag{8}$$

If the worldsheet has a boundary, there is a relation between left-moving and right-moving modes, and the $G_L \times G_R$ symmetry is broken. One can still preserve some diagonal symmetry G, say the symmetry

$$g \to h g h^{-1} \tag{9}$$

corresponding to $h_L = h_R^{-1} = h$ in (8). The presence of this symmetry constrains the boundary conditions that can be placed on g. Allowing $g(\text{boundary}) = f$ for some $f \in G$ we must also allow $g(\text{boundary}) = hfh^{-1}$ for every $h \in G$. This means that g on the boundary takes value in the conjugacy class containing f. For $G = SU(2)$, conjugacy classes are parametrized by a single angle θ, $0 \leq \theta \leq \pi$, corresponding to the choice $f = \exp(i\theta\sigma_3)$. Thinking of $SU(2)$ as the group of three dimensional rotations, the conjugacy class C_θ is the set of all rotations of angle 2θ about any axis. In addition not any value of θ gives rise to a consistent model. One finds that the possible conjugacy classes on which a boundary state can live are quantized, the corresponding θ must satisfy

$$\theta = 2\pi \frac{j}{k} \tag{10}$$

with j integer or half integer satisfying $0 \leq j \leq \frac{k}{2}$. It is also possible for group element to lie at the boundary in the conjugacy class shifted by arbitrary element l:

$$g(\text{boundary}) \in \mathrm{C}_\theta \mathrm{l}. \tag{11}$$

This choice preserves another diagonal subgroup of (8): $h_R = l^{-1} h_L^{-1} l$.

5. D-branes in the Near-Horizon Geometry of $NS5$-branes

After assembling the necessary tools, we are now ready to study the physics of the configurations of Fig. 4. consider the configuration of Fig. 4a. A stack of N_c $D4$-branes ends from the left, *i.e.* from negative x^6, on k coincident $NS5$-branes. N_F $D6'$-branes end on the $NS5$-branes from above (positive x^7). From the point of view of the geometry (1) (Fig. 5), the D-branes extend into the CHS throat, as indicated in Fig. 6a.

The $D4$-branes intersect the three-sphere at the point $x^7 = x^8 = x^9 = 0$; the $D6'$-branes at $x^6 = x^8 = x^9 = 0$ (see Fig. 6b). Thus, they correspond to the boundary states $g|_{\text{boundary}} = 1$ and $i\sigma_3$, respectively. The $D4$-branes are described by the boundary state with $\theta = 0$ and $l = 1$, while the $D6'$-branes correspond to a transformed state, with $\theta = 0$ and $l = \exp(i\pi\sigma_3/2) = i\sigma_3$. The $SU(2)$ currents J^a satisfy the boundary conditions for strings ending on the $D4$ and $D6'$-branes, respectively. In order to preserve worldsheet supersymmetry one has to impose analogous boundary conditions on the fermions. For example, for a string ending on the $D4$-branes one has

$$\chi^a = \overline{\chi}^a \tag{12}$$

while for a boundary on a $D6'$-brane the χ^a satisfy an analog of the sixbranes extend into the throat, the boundary conditions on ϕ are Neumann.

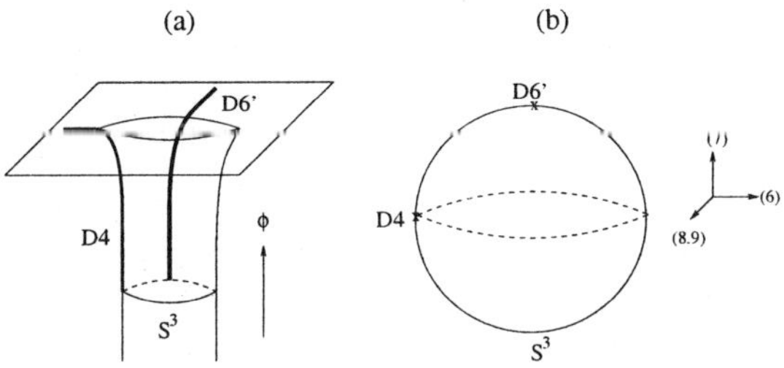

Fig. 6.

We would like to construct the vertex operator for emitting the lowest lying $4-6'$ string in the geometry of Fig. 6, i.e. generalize (7) to the fivebrane near-horizon geometry. Some parts of the discussion leading to (7) are unchanged. In particular, the geometry is the same as there in $(x^0, x^1, x^2, x^3, x^4, x^5)$. The presence of ϕ allows also a contribution $\exp(\beta\phi)$ to the vertex operator. Thus, the analog of (7) for this case has the form

$$V = e^{-\varphi}\sigma_{45}S_{45}e^{ik_\mu x^\mu}e^{\beta\phi}V_2 \tag{13}$$

where V_2 is the contribution of the $SU(2)$ group manifold to the vertex operator, to which we turn next.

The $4-6'$ vertex operator V_2 changes the worldsheet boundary conditions from $g = 1$ to $g = l = \exp(i\alpha\sigma_3/2)$. The $D6'$-brane corresponds to $\alpha = \pi$, but it is instructive to discuss the general case, in which the angle between the $D4$ and $D6'$-branes is $\alpha/2$ (see Fig. 7). V_2 turns on to be an operator of $U(1)$ charge $-\frac{k}{2}\frac{\alpha}{2\pi}$ and dimension $-\frac{k}{4}\left(\frac{\alpha}{2\pi}\right)^2$.

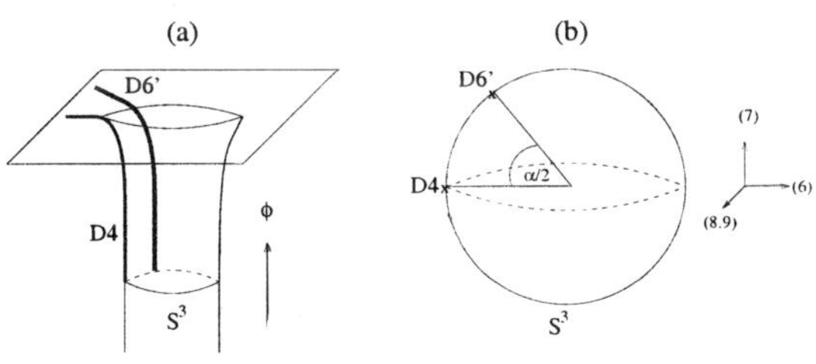

Fig. 7.

We find that the lowest lying open string connecting the $D4$ and $D6'$-branes in the configuration of Fig. 7 is described by the vertex operator

$$V_{4-6'}^{+} = e^{-\varphi}\sigma_{45}S_{45}e^{ik_\mu x^\mu}V_{jm} \tag{14}$$

with

$$\frac{k_\mu^2}{2} + \frac{m^2 - j(j+1)}{k} = \frac{1}{4} \tag{15}$$

and

$$m = -\frac{k}{2}\frac{\alpha}{2\pi} \tag{16}$$

To compute the spectrum of low-lying normalizable excitations of the $4-6'_+$ string, one computes the two point function of the operators (14) on the disk. This amplitude exhibits first order poles in j. Using the mass-shell condition (15), these can be interpreted as poles in k_μ^2; they correspond to on-shell particles in four dimensions, created from the vacuum by the operator (14).

The calculation of this two point function is very similar to its closed string analog, and is described in appendix A of the original paper. The result is, as in the closed string case, a series of poles corresponding to the discrete representations of $SL(2)$, (6). The lowest lying state corresponds to $n = 1$, i.e. $j = |m| - 1$. Plugging this together with (16) into (15) we find that the mass of the lowest lying normalizable state of the $4-6'_+$ string is

$$M^2(\alpha) = \frac{1}{2}(\frac{\alpha}{\pi} - 1) \tag{17}$$

In particular, we find that as expected, for $\alpha = \pi$ the lowest lying state is massless. The vertex operator which creates this massless particle from the vacuum is

$$V^+_{4-6'}(k_\mu^2 = 0) = e^{-\varphi}\sigma_{45}S_{45}e^{ik_\mu x^\mu}V_{\frac{k}{4}-1,-\frac{k}{4}} \tag{18}$$

The degeneracy works out as well. The chiral structure is recaptured. This validates both the approximate picture and the LST methods used.
We note that for $\alpha < \pi$ the lowest lying state (17) is tachyonic; the stable vacuum is obtained by its condensation. This process has a very natural interpretation from the point of view of brane theory, which also makes it clear what is the endpoint of the condensation. For $\alpha \neq \pm\pi$, the configuration of Fig. 7 is not supersymmetric, hence stability needs to be checked. For $\alpha < \pi$ the configuration of Fig. 7 can reduce its energy by having the $D4$-brane slide away from the $NS5$-brane so that it ends on the $D6'$-brane instead (Fig. 8). The resulting vacuum is stable. For $\alpha > \pi$, the configuration of Figs. 7, 8 is stable under small deformations. Indeed, the lowest lying open string state is massive in this case (17).

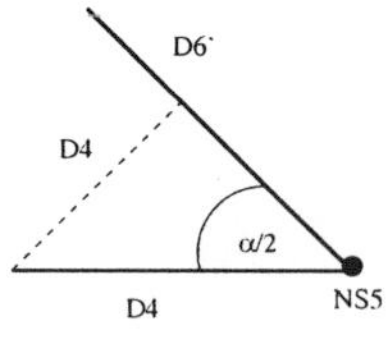

Fig. 8.

From the point of view of the brane configurations describing four dimensional gauge theories like that of Fig. 3, one can change α from π in a number of ways.

One is to change the relative position of the $NS5$ and $NS5'$-branes in x^7 by the amount Δx^7. In the gauge theory on the $D4$-branes this corresponds to turning on a Fayet-Iliopoulos D-term. Depending on the sign of the FI term, either Q or $\tilde{Q}$ should condense to minimize the D-term potential

$$V_D \sim (Q^\dagger Q - \tilde{Q}^\dagger \tilde{Q} - r)^2 \tag{19}$$

where r is proportional to Δx^7. The gauge theory analysis is nicely reproduced by our string theory considerations[b] $\Delta x^7 > 0$ corresponds to $0 < \alpha < \pi$. In this case the ground state of the $4 - 6_+$ string is tachyonic, Q condenses and the $D4$-branes detach from the $NS5'$-branes and attach to the $D6$-branes. $\Delta x^7 < 0$ corresponds to $\alpha > \pi$ for the $4 - 6_+$ string which is hence massive, but since $0 < -\alpha < \pi$ for the $4 - 6_-$ strings ($\tilde{Q}$), a similar process of condensation to the above occurs for them.

Another way of changing α is to tilt the $D6$-branes by some angle in the (x^6, x^7) plane, which must lead to a potential similar to (19). The calculations in the cases (b) and (c) mentioned in the introduction can be found in the original paper.

Acknowledgments

This research was supported in part by NSF grant #PHY94-07194 and by the Israel Academy of Sciences and Humanities – Centers of Excellence Program. The work of A.G. and E.R. is also supported in part by BSF – American-Israel Bi-National Science Foundation. D.K. is supported in part by DOE grant #DE-FG02-90ER40560.

References

1. A. Giveon and D. Kutasov, *Rev. Mod. Phys.* **71**, 983 (1999) [hep-th/9802067].
2. N. Seiberg, *Phys. Lett.* **408B**, 98 (1997) [hep-th/9705221]; R. Dijkgraaf, E. Verlinde, H. Verlinde, *Nucl. Phys.* **B486**, 89 (1997) [hep-th/9604055]; *Nucl. Phys.* **B506**, 121 (1997) [hep-th/9704018]; Micha Berkooz, Moshe Rozali, Nathan Seiberg, *Phys. Lett.* **B408**, 105 (1997) [hep-th/9704089].
3. O. Aharony, M. Berkooz, D. Kutasov and N. Seiberg, *JHEP* **9810**, 004 (1998) [hep-th/9808149].
4. C. Callan, J. Harvey and A. Strominger, hep-th/9112030.
5. S. Elitzur, A. Giveon and D. Kutasov, *Phys. Lett.* **B400**, 269 (1997) [hep-th/9702014].
6. A. Hanany and E. Witten, *Nucl. Phys.* **B492**, 152 (1997) [hep-th/9611230].
7. J.H. Brodie and A. Hanany, *Nucl. Phys.* **B506**, 157 (1997) [hep-th/974043].
8. A. Hanany and K. Hori, *Nucl. Phys.* **B513**, 119 (1998) [hep-th/9707192].
9. K. Landsteiner, E. Lopez and D.Lowe, *JHEP* **9802**, 007 (1998) [hep-th/9801002].
10. S. Elitzur, A. Giveon, D. Kutasov and D. Tsabar, *Nucl. Phys.* **B524**, 251 (1998) [hep-th/9801020].
11. I. Brunner, A. Hanany, A. Karch and D. Lust, *Nucl. Phys.* **B528**, 197 (1998) [hep-th/9801017].
12. H. Ooguri and C. Vafa, *Nucl. Phys.* **B463**, 55 (1996) [hep-th/9511164].

[b] Both in gauge theory and in string theory, in the context of the full configuration of Fig. 3 the foregoing discussion is valid for $N_f \geq N_c$.

13. A. Giveon, D. Kutasov and O. Pelc, *JHEP* **9910**, 035 (1999) [hep-th/9907178].
14. A. Giveon and D. Kutasov, *JHEP* **9910**, 034 (1999) [hep-th/9909110]; *JHEP* **0001**, 023 (2000) [hep-th/9911039].
15. J. Teschner, *Nucl. Phys.* **B546**, 390 (1999) [hep-th/9712256]; *Nucl. Phys.* **B546**, 369 (1999) [hep-th/9712258]; *Nucl. Phys.* **B571**, 555 (2000) [hep-th/9906215]; V.A. Fateev, A.B. Zamolodchikov and Al.B. Zamolodchikov, unpublished.
16. J. Polchinski, S. Chaudhuri and C.V. Johnson, hep-th/9602052.
17. A. Hashimoto, *Nucl. Phys.* **B496**, 243 (1997) [hep-th/9608127].
18. C. Klimcik and P. Severa, *Nucl. Phys.* **B488**, 653 (1997).
19. A. Alekseev and V. Schomerus, *Phys. Rev.* **D60**, 061901 (1999) [hep-th/9812193].
20. K. Gawedzki, hep-th/9904145.

DOMAIN WALL WORLD(S)

M. CVETIČ*
Department of Physics and Astronomy, University of Pennsylvania
Philadelphia, PA 19104, USA

Gravitational properties of domain walls in fundamental theory and their implications for the trapping of gravity are reviewed. In particular, the difficulties to embed gravity trapping configurations within gauged supergravity is reviewed and the status of the domain walls obtained via the breathing mode of sphere reduced Type IIB supergravity is presented.

Over the past few years domain walls have played an important role both from the point of view of the AdS/CFT correspondence, i.e. by shedding light on the renormalization group flow and bound state spectra of strongly coupled field theories, as well as from the point of view of the brane world scenarios, such as Randall-Sundrum scenario[1] for localization of gravity on domain walls in five-dimensions. In this contribution, however, we shall focus on gravitational properties of domain walls, their implications for the brane world scenarios and their implementation in the fundamental theory.

This presentation is based on the work with K. Behrndt,[2,3] along with the parallel work by Kallosh and Linde and Shmakova,[4,5] leading to the "no-go" theorems for the implementation of the Randall-Sundrum scenario, when one employs massless modes of gauged supergravity. (For subsequent developments see, e.g., Refs. 6,7,8.) The focal part, however, will be based on the work with H. Lü and C. Pope[9,10] where the massive-breathing mode of sphere reduced gauged supergravity was proposed as the candidate field for the gravity trapping domain walls. For subsequent related works see Ref. 11 and in particular Ref. 12 as well as the contributions to these proceedings by S. de Alwis, J. Liu and K. Stelle. In addition, there the origin of the Randall-Sundrum solution[13] based on S^1/Z_2 topology and two branes sources was addressed in detail.

The outline of this contribution is the following: (i) We shall first review the properties of supergravity domain walls, identifying the flat (Minkowski) walls with BPS saturated ones and the bent (deSitter and anti-deSitter) ones with those whose energy density is either larger or smaller than that of the BPS saturated ones. (This part of the review is primarily based on a much earlier work on domain walls in four-dimensional N=1 supergravity, initiated in Ref. 14, its generalizations to non-BPS solutions were given in Ref. 15 and reviewed in Ref. 16; for a recent work on a generalization of the analysis to D-dimensions see Ref. 17 and references therein.)

*Based on the talk given at Strings'2000.

(ii) We shall then turn to the discussion of the properties of domain walls in N=2 D=5 gauged supergravity (with vector multiplets), leading to the "no-go" theorem for the implementation of the Randall-Sundrum scenario. (iii) As the last topic we turn to the status of the breathing mode domain walls, where the breathing mode parameterizes the volume of the compactified Einstein space (with the sphere as the most symmetric example). While these walls may provide the candidate gravity trapping solutions, we also mention difficulties with the interpreting of the delta function sources for infinitely thin domain wall within Type IIB string theory.

1. Domain Walls in N=1 Supergravity

Since the scalar potential in supergravity theories is of the restricted form it is natural to expect that the nature of domain walls is of special form as well. The bosonic Lagrangian of D=4 N=1 supergravity theory takes the form:

$$\mathcal{L} = g_{A\bar{B}}\partial_\mu\Phi^A\partial^\mu\Phi^{\bar{B}} - V - \frac{R}{8\pi M_{pl}^2}\,, \tag{1}$$

where the scalar potential (for the gauge neutral fields) is of the form:

$$V = g^{A\bar{B}}\partial_A\hat{W}\partial_{\bar{B}}\hat{W} - \frac{3}{M_{pl}^2}\hat{W}^2\,. \tag{2}$$

Here the Kähler invariant quantity $\hat{W}$ takes the form: $\hat{W} = \zeta|W|e^{\frac{K}{2M_{pl}^2}}$, where $\zeta = \pm 1$ and ζ can change the sign only when W goes through the zero value. Here W and K are the superpotential and Kähler potential, respectively. Supersymmetric extrema are at $\partial_A\hat{W} = 0$, and correspond to either a zero cosmological constant, i.e. $\Lambda \equiv V_{ext} = 0$ when $\hat{W}_{ext} = 0$, or a negative cosmological constant, i.e. $\Lambda \equiv V_{ext} < 0$, when $\hat{W}_{ext} \neq 0$.

The surprising result[14] in the study of domain walls in supergravity theory is that the static (flat) domain wall solutions between supersymmetric extrema do exist irrespective of the fact that the actual value of the cosmological constants of the two isolated extrema may be different. Such solutions turn out to be BPS ones: they satisfy the Killing spinor equations: $\delta\psi^\alpha = 0$, and $\delta\psi_\mu^\alpha = 0$ for the supersymmetric variation of the scalar-field super-partner and gravitino, respectively. These equations are solved with the static Ansatz for the scalar field Φ (only one chosen for simplicity) and the following conformally flat metric Ansatz:[a]

$$ds^2 = A(z)(-dt^2 + dx^2 + dy^2 + dz^2) \equiv \mathcal{A}(\tilde{z})(-dt^2 + dx^2 + dy^2) + d\tilde{z}^2\,, \tag{3}$$

leading to the following coupled first order equations of motion:[14]

$$\begin{aligned} \partial_z\Phi^A &= 2g^{A\bar{B}}\partial_{\bar{B}}\hat{W}\,, \\ M_{pl}^2\partial_z \log\mathcal{A} &= -2\hat{W}\,, \end{aligned} \tag{4}$$

[a] We shall describe the domain wall metric interchangeably in terms of the conformal factor A or the warp factor $\mathcal{A}$.

and the energy density of the wall takes the following form:

$$\sigma_{BPS} = 2(\hat{W}_+ - \hat{W}_-) = 2M_{pl}(\zeta_+\sqrt{-\frac{\Lambda_+}{3}} - \zeta_-\sqrt{-\frac{\lambda_-}{3}})\,. \tag{5}$$

The asymptotic behavior of the solution is the following:

$$\begin{aligned} z \to \pm\infty: \quad \Phi &\to \Phi_\pm + e^{2\partial_\Phi^2 \hat{W}_\pm z}; \quad \{\partial_\Phi^2 \hat{W}_- > 0, \partial_\Phi^2 \hat{W}_+ < 0\}, \\ \mathcal{A} &\to e^{-\frac{2\hat{W}_\pm z}{M_{pl}^2}}\,. \end{aligned} \tag{6}$$

Note that sign$(\hat{W}_\pm)$ determines the asymptotic behavior of the metric. It turns out that the necessary condition for the exponential fall-off of the metric warp factor $\mathcal{A}$ is that at the supersymmetric extrema the potential $V(\Phi_\pm)$ has the minimum.

A typical Z_2 symmetric example corresponds to the following choice:

$$W = \sqrt{\lambda}\Phi(\frac{\Phi^2}{2} - \eta^2), \quad K = \Phi\Phi^*\,, \tag{7}$$

where the kink solution (4) interpolates between two supersymmetric anti-deSitter minima with the same cosmological constant ($\hat{W}_+ = -\hat{W}_- \equiv \hat{W}_0 = M_{pl}\sqrt{-\frac{\Lambda}{3}}$). The metric conformal factor $A(z)$ (3) falls-off on either side as $M_{pl}^4(\hat{W}_0 z)^{-2}$, and the energy density of the wall is $\sigma_{BPS} = 2\times 2\hat{W}_0 = 4M_{pl}\sqrt{-\frac{\Lambda}{3}}$.

The thin wall limit of the Z_2 symmetric BPS solution is achieved by taking the limit $\lambda \to \infty$, $\eta \to 0$, while $\lambda\eta^3$ remains fixed. In this particular case the superpotential $\hat{W}$ approaches the step function, while the potential and the scalar kinetic energy term approach the delta function. This leads to the thin wall effective (supersymmetric) Lagrangian with the delta function source that precisely reproduces the thin- wall flat domain-wall solution, discussed by Randall-Sundrum in D=5. (For further discussions for a supersymmetric implementation of the effective Lagrangian with the delta function sources see Refs. 18,19,20,21,22.)

In the case of broken supersymmetry, the walls become bent. In the thin wall analysis of walls that can assume the Z_2 symmetric limit, the space-time internal to the wall is either deSitter (when $\sigma > \sigma_{BPS}$) or anti-deSitter (when $\sigma < \sigma_{BPS}$). The conformal factor (3) on the other hand takes the form:[15]

$$\begin{aligned} A(z) &= M_{pl}^2\beta^2\left[\sqrt{-\frac{\Lambda_\pm}{3}}\sinh(\beta(z+z_\pm))\right]^{-2}, \text{ for } \sigma > \sigma_{BPS}\,, \\ A(z) &= M_{pl}^2\beta^2\left[\sqrt{-\frac{\Lambda_\pm}{3}}\cos(\beta(z+z_\pm))\right]^{-2}, \text{ for } \sigma < \sigma_{BPS}\,. \end{aligned} \tag{8}$$

where $\beta^2 \sim |\Lambda_{wall}|$ specifies the cosmological constant on the bent wall for respective anti-deSitter and deSitter space-times on the wall. The constants $z_\pm$ are appropriately chosen so that at $z = 0$ the conformal factor is $A(z)$ is normalized to 1. In the former case the conformal factor falls-off even faster then in the BPS case

($|z| \to \infty$ corresponds to the cosmological horizons), while for the latter case the conformal factor, while first decreasing it turns around at intermediate distances ($|z| \to \infty$ is the space-time boundary). This latter case allows for the possibility of quasi-localized gravity on the wall.[23]

While we have briefly reviewed the BPS and non-BPS thin walls for supergravity theory in four dimensions, it was shown in Ref. 17 that the global and local space-time structure of co-dimension one objects in D dimensions is universal, only that the role of the cosmological constant factor $\sqrt{-\frac{\Lambda_\pm}{3}}$ is replaced by $\sqrt{-\frac{(D-2)\Lambda_\pm}{2(D-1)}}$. In particular, for domain walls in D=5, this factor becomes $\sqrt{-\frac{3\Lambda_\pm}{8}}$.

2. Domain Walls in D=5 Gauged Supergravity

A natural question to be asked is to identify the origin of gravity trapping solutions in fundamental (M-)theory, as initiated in Ref. 24. In particular, can such domain walls arise in an effective five-dimensional theory, that can be obtained as a compactification of, e.g., M-/string theory on Einstein-Sasaki spaces? Since such compactifications are expected to produce an effective gauged supergravity theory, we now turn to the study of domain walls in five-dimensional N=2 gauged supergravity. For the sake of simplicity, we shall focus on D=5 N=2 gauged supergravity with Abelian ($U(1)_R$)-gauging and with the vector supermultiplets, only.[25]

The bosonic sector is of the constrained form, with the vector supermultiplets X^I (real, neutral fields) subject to the following condition:

$$F = \sum_{IKJ} C_{IJK} X^I X^J X^K = 1\,, \tag{9}$$

which can be solved for the physical scalar fields Φ^A. The potential of $U(1)$ gauged supergravity is also of the constrained form:

$$W = \sum_I h_I X^I\,. \tag{10}$$

The bosonic part of the Lagrangian nevertheless takes an analogous form as in the case of D=4 N=1 supergravity (We have now set M_{pl}=1.):

$$\mathcal{L} = g_{AB}\partial_\mu\Phi^A\partial^\mu\Phi^B - V + R\,, \tag{11}$$

where the metric and the potential are of the form:

$$g_{AB} = \frac{1}{2}(\partial_I\partial_J F)\partial_A X^I \partial_B X^J; \;\; V = g^{AB}\partial_A W \partial_B W - \frac{4}{3}W^2\,. \tag{12}$$

Again, the Killing spinor equations, corresponding to the supersymmetric domain wall solutions, reduce to the equations of the analogous type:

$$\begin{aligned} \partial_z \Phi^A &= \mp 3 g^{AB}\partial_B W\,, \\ \partial_z \log \mathcal{A} &= \pm 2W\,. \end{aligned} \tag{13}$$

However, due to the fact that W and g_{AB} are of the constrained form, they satisfy a relationship,[25] that takes the following form at the supersymmetric extrema:

$$\partial_A \partial_B W_{ext} = \tfrac{2}{3}[g_{AB} W]_{ext} \,. \tag{14}$$

Expanding the Killing spinor equations around the supersymmetric minima $\partial_\Phi W|_\pm = 0$, as $\Phi^A = \Phi^A_\pm + \delta\Phi^A$ and using the relationship (14), yields the following asymptotic form of these equations:

$$\begin{aligned} \partial_z(\delta\Phi^A) &= +W_\pm \delta\Phi^A \,, \\ \partial_z(\log \mathcal{A}) &= \pm W_\pm \,. \end{aligned} \tag{15}$$

The condition for the asymptotic kink solution requires $\mathrm{sign}(W_+) = -\mathrm{sign}(W_-)$, and then the conspiracy of signs in the above equations (15) necessarily requires that the kink solution has the metric factor that *grows* exponentially on either side of the wall. Thus, these solutions are not relevant for the trapping of gravity; those are typical domain walls relevant for AdS/CFT correspondence. In addition, the constraint (14) implies that supersymmetric extrema with $W_{ext} > 0$ are necessarily the minima of the superpotential (for g_{AB}-positive definite), while supersymmetric extrema with $W_{ext} < 0$ are maxima. Therefore, the kink necessarily has to cross the singular region in the superpotential manifold and thus the solution is generically singular.

Further studies of more general solutions reveal that neither inclusion of non-Abelian tensor multiplets,[6,5] nor inclusion of hypermultiplets[7,8] allow for non-singular domain walls that would have a fall-off metric conformal factor on both sides of the wall. (For a related no-go theorem see Ref. 26.)

3. Breathing Mode and Gravity Trapping Domain Walls

In this section we shall review a framework within gauged supergravity theories that has a chance of implementing a variant of the Randall-Sundrum scenario. (For related work see, e.g., Ref. 27.) Recall that in order to obtain Z_2 symmetric domain-wall solutions of the Randall-Sundrum scenario, the gauged supergravity potential would have to have two isolated supersymmetric *minima*. Since the potentials for the massless scalar fields in a gauged supergravity generically do not have this feature, we now turn to an alternative proposal[9] to include other scalar fields that do not lie in the massless supermultiplet.

We shall focus on the special classes of gauged supergravities that arise from sphere reductions of M-theory or string theory, with particular emphasis on the $D = 5$ case. For examples in the Kaluza-Klein reduction of Type IIB string theory on a five-sphere (S^5), there will be an infinite tower of massive supermultiplets in addition to the massless multiplet, and so one could consider the potentials for one or more of the massive scalar fields. In general, one cannot focus attention on a single such field in isolation, on account of its couplings to other fields. However, in certain special cases a consistent truncation to a single massive scalar can be performed.

One such example is the "breathing mode" that parameterises the overall volume of the compactifying S^5. (Unlike the breathing mode in a toroidal reduction, which is massless, the breathing mode in a spherical reduction is a member of a massive supermultiplet.)

The scalar potentials for the breathing-mode scalars in various Kaluza-Klein spherical reductions were studied in Ref. 28. Although the breathing mode is a member of a massive multiplet, the truncation is nonetheless consistent since it is a singlet under the isometry group of the internal sphere. (It would not in general be consistent to turn on a finite subset of other fields as well.)

The resulting D-dimensional Lagrangians all turn out to have the following form:

$$\mathcal{L}_D = R - \tfrac{1}{2}(\partial\phi)^2 - V\,, \tag{16}$$

where the potential is given by[28]

$$V = \tfrac{1}{2}g^2\Big(\frac{1}{a_1^2}\,e^{a_1\Phi} - \frac{1}{a_1 a_2}\,e^{a_2\Phi}\Big)\,. \tag{17}$$

The positive constants a_1 and a_2 are given by

$$a_1^2 = \frac{4}{N} + \frac{2(D-1)}{D-2}\,, \qquad a_1\,a_2 = \frac{2(D-1)}{D-2}\,, \tag{18}$$

where Type IIB reduction on S^5 corresponds to $N=1$, $D=5$. Since $a_1 > a_2 > 0$, the potential has a minimum at $\phi = 0$, corresponding to the self-dual point where the volume of the five-sphere and the radius of the AdS$_5$ are equal (in appropriate units). In addition, this potential can be cast in the standard supersymmetric form:

$$V = \Big(\frac{\partial W}{\partial\phi}\Big)^2 - \frac{D-1}{2(D-2)}\,W^2\,, \tag{19}$$

where

$$W = \sqrt{\frac{N}{2}}\,g\,\Big(\frac{1}{a_1}\,e^{a_1\Phi/2} - \frac{1}{a_2}\,e^{a_2\Phi/2}\Big)\,. \tag{20}$$

Thus, there is a domain wall solution that can be obtained in terms of the coupled first-order differential equations (15). Solving for ϕ and A one finds the result:

$$A^{\frac{D-1}{2}} = c\,\frac{\partial W}{\partial\Phi}\,e^{-\frac{1}{2}(a_1+a_2)\Phi}\,, \tag{21}$$

where c is an integration constant and the solution for Φ is given by:

$$z - z_0 = \frac{4}{a_2\,g\,\sqrt{N}}\,e^{-\frac{1}{2}a_2\,\Phi}\,{}_2F_1\Big[\frac{a_2}{a_2-a_1}, 1, 1+\frac{a_2}{a_2-a_1}; e^{\frac{1}{2}(a_1-a_2)\,\Phi}\Big]\,. \tag{22}$$

The solutions above *have two different branches.* In one branch, ϕ runs from 0 to $+\infty$, with z running from $z=-\infty$ to $z=0$, where we have chosen the integration constant z_0 to yield the following result:

$$\begin{aligned} e^{-\frac{1}{2}a_1\,\phi} &\sim -\tfrac{1}{4}a_1\,\sqrt{N}\,g\,z\,,\\ A^{\frac{(D-1)}{2}} &\sim c\sqrt{\frac{N}{8}}\,g\,e^{-\frac{1}{2}a_2\,\phi} \sim c\sqrt{\frac{N}{8}}\,g\,\Big(-\tfrac{1}{4}a_1\,\sqrt{N}\,g\,z\Big)^{\frac{a_2}{a_1}}\,. \end{aligned} \tag{23}$$

In this branch, when the coordinate z reaches its limit at $z = 0$, the metric factor therefore goes to zero, and there is a power-law naked curvature singularity. (Note that in this regime the solution extends into large positive values of the potential (17) with a large cost to the energy density of the wall, and it thus terminates at a finite value of the transverse coordinate.) As z approaches $-\infty$, the metric asymptotically approaches the AdS space-time, described in horospherical coordinates with $z \to -\infty$ corresponding to the Cauchy horizon. Note that on that side of the wall the gravity is repulsive and provides "one half" of the Randall-Sundrum wall.

The study of the gravitational fluctuating modes, internal to the wall, reduces to the study of the Schrödinger equation, whose potential[10] has an attractive, singular region near the naked singularity. Nevertheless the spectrum has energy levels bounded from below. However, the boundary conditions at the naked singularity exclude the massless normalizable mode (corresponding to the four-dimensional graviton). On the other hand corrections (of the order of the inverse string scale), that would smooth out the naked singularity, would in turn provide the non-singular attractive Schrödinger potential with precisely one normalizable massless state. Further investigations to identify the origin of the smoothing out of such singularities within the string theory context is needed.

In the second branch, ϕ runs from 0 to $-\infty$, while z runs from $z = -\infty$ to $z = +\infty$. The behaviour of the solution near $z = -\infty$ is the same as in the branch discussed previously, with the metric approaching asymptotically AdS. As z approaches $+\infty$, the solution becomes

$$\begin{aligned} e^{-\frac{1}{2}a_2\,\phi} &\sim \tfrac{1}{4}a_2\,\sqrt{N}\,g\,z\,, \\ A^{\frac{(D-1)}{2}} &\sim -c\sqrt{\frac{N}{8}}\,g\,e^{-\frac{1}{2}a_1\,\phi} \sim -c\sqrt{\frac{N}{8}}\,g\left(\tfrac{1}{4}a_2\,\sqrt{N}\,g\,z\right)^{\frac{a_1}{a_2}}. \end{aligned} \qquad (24)$$

(The constant c is negative in this case.) This side describes one-side of a supersymmetric *dilatonic* domain wall. Interestingly, it has no curvature singularity; as z tends to $+\infty$ the curvature falls off as $1/z^2$, while the diverging dilaton $\phi \to -\infty$ approaches the weak coupling limit. Unfortunately, the dilatonic vacuum side does not provide for the gravity trapping solution.

Thus within a pure field-theoretic framework, i.e. by employing only the breathing-mode scalar field to construct the domain wall solution, one did not fully succeed in constructing domain wall solutions that would allow for trapping of gravity on the wall, though the first branch would provide for such a scenario if the mechanism to smooth out the naked singularity within string theory existed. (For a possible related mechanism see Ref. 29.)

In Ref. 9 it was therefore proposed to add a delta-function source to the effective Lagrangian. In this case, the diverging behaviour of the dilaton is cut-off by a delta-function source at some finite value of z, say $z = z_*$ and the solution for $z > z_*$ can be replaced by a reflection of the solution for $z < z_*$. Now the solution is Z_2 symmetric and the metric factor $A(z)$ falls-off on either side of the infinitely thin wall, supported by the delta function source, thus reproducing the Randall-Sundrum scenario. The origin of such a delta function source within Type IIB supergravity was further explored in Ref. 30.

Acknowledgments

I would like to thank K. Behrndt, H. Lü and C. Pope for many discussions and enjoyable collaboration on topics presented in this contribution. The work is supported by U.S. Department of Energy Grant No. DOE-EY-76-02-3071 and by the NATO Linkage grant 97061.

References

1. L. Randall and R. Sundrum, *An alternative to compactification*, *Phys. Rev. Lett.* **83**, 4690 (1999) [hep-th/9906064].
2. K. Behrndt and M. Cvetič, *Supersymmetric domain-wall world from $D = 5$ simple gauged supergravity*, *Phys. Lett.* **B475**, 253 (2000) [hep-th/9909058].
3. K. Behrndt and M. Cvetič, *Anti-de Sitter vacua of gauged supergravities with 8 supercharges*, *Phys. Rev.* **D61**, 101901 (2000) [hep-th/0001159].
4. R. Kallosh, A. Linde and M. Shmakova, *Supersymmetric multiple basin attractors*, *JHEP* **9911**, 010 (1999) [hep-th/9910021].
5. R. Kallosh and A. Linde, *Supersymmetry and the brane world*, *JHEP* **0002**, 005 (2000) [hep-th/0001071].
6. M. Gunaydin and M. Zagermann, *The vacua of 5d, $N = 2$ gauged Yang-Mills/Einstein/tensor supergravity: Abelian case*, *Phys. Rev.* **D62**, 044028 (2000) [hep-th/0002228].
7. A. Ceresole and G. Dall'Agata, *General matter coupled $N = 2$, $D = 5$ gauged supergravity*, *Nucl. Phys.* **B585**, 143 (2000) [hep-th/0004111].
8. K. Behrndt, C. Herrmann, J. Louis and S. Thomas, *Domain walls in five dimensional supergravity with non-trivial hypermultiplets*, hep-th/0008112.
9. M. Cvetič, H. Lü and C.N. Pope, *Domain walls and massive gauged supergravity potentials*, *Class. Quant. Grav.* **17**, 4867 (2000) [hep-th/0001002].
10. M. Cvetič, H. Lü and C.N. Pope, *Localized gravity in the singular domain wall background?*, hep-th/0002054.
11. S.P. de Alwis, *Brane world scenarios and the cosmological constant*, hep-th/0002174.
12. M.J. Duff, J.T. Liu and K.S. Stelle, *A supersymmetric type IIB Randall-Sundrum realization*, hep-th/0007120.
13. L. Randall and R. Sundrum, *A large mass hierarchy from a small extra dimension*, *Phys. Rev. Lett.* **83**, 3370 (1999) [hep-ph/9905221].
14. M. Cvetič, S. Griffies, and S.-J. Rey, *Static domain walls in $N = 1$ supergravity*, *Nucl. Phys.* **B381**, 301 (1992) [hep-th/9201997].
15. M. Cvetič, S. Griffies, and H.H. Soleng, *Local and global gravitational aspects of domain wall space-times*, *Phys. Rev.* **D48**, 2613 (1993) [hep-th/9306005].
16. M. Cvetič and H.H. Soleng, *Supergravity domain walls*, *Phys. Rept.* **282**, 159 (1997) [hep-th/9604090].
17. M. Cvetič and J. Wang, *Vacuum domain walls in D-dimensions: Local and global space-time structure*, *Phys. Rev.* **D61**, 124020 (2000) [hep-th/9912187].
18. T. Gherghetta and A. Pomarol, *Bulk Fields and Supersymmetry in a Slice of AdS*, *Nucl. Phys.* **B586**, 141 (2000) [hep-ph/0003129].
19. R. Altendorfer, J. Bagger and D. Nemeschansky, *Supersymmetric Randall-Sundrum scenario*, hep-th/0003117.

20. A. Falkowski, Z. Lalak and S. Pokorski, *Supersymmetrizing branes with bulk in five-dimensional supergravity*, *Phys. Lett.* **B491**, 172 (2000) [hep-th/0004093].
21. E. Bergshoeff, R. Kallosh and A. Van Proeyen, *Supersymmetry in singular spaces*, *JHEP* **0010**, 033 (2000) [hep-th/0007044].
22. M. Cvetič and S.-C. Yoon, work in progress.
23. A. Karch and L. Randall, *Locally localized gravity*, hep-th/0011156.
24. H. Verlinde, *Holography and compactification*, *Nucl. Phys.* **B580**, 264 (2000) [hep-th/9906182].
25. M. Gunaydin, G. Sierra and P.K. Townsend, *More On $D = 5$ Maxwell-Einstein Supergravity: Symmetric Spaces And Kinks*, *Class. Quant. Grav.* **3**, 763 (1986).
26. J. Maldacena and C. Nunez, *Supergravity description of field theories on curved manifolds and a no go theorem*, hep-th/0007018.
27. C.S. Chan, P.L. Paul and H. Verlinde, *A note on warped string compactification*, *Nucl. Phys.* **B581**, 156 (2000) [hep-th/0003236].
28. M.S. Bremer, M.J. Duff, H. Lü, C.N. Pope and K.S. Stelle, *Instanton cosmology and domain walls from M-theory and string theory*, *Nucl. Phys.* **B543**, 321 (1999) [hep-th/9807051].
29. C.V. Johnson, A.W. Peet and J. Polchinski, *Gauge theory and the excision of repulson singularities*, *Phys. Rev.* **D61**, 086001 (2000) [hep-th/9911161].
30. M. Cvetič, M.J. Duff, J.T. Liu, H. Lü, C.N. Pope and K.S. Stelle, *Randall-Sundrum brane tensions*, hep-th/0011167.

LIGHT CONE SUPERSTRINGS IN ADS SPACE

A.A. TSEYTLIN*
Physics Department, The Ohio State University
Columbus, OH 43210–1106, USA
E-mail: tseytlin.1@osu.edu

We discuss light-cone gauge description of type IIB Green-Schwarz superstring in $AdS_5 \times S^5$ with a hope to make progress towards understanding spectrum of this theory. As in flat space, fixing light-cone gauge consists of two steps: (i) fixing kappa symmetry in such a way that the fermionic part of the action does not depend on x^-; (ii) fixing 2-d reparametrizations by $x^+ = \tau$ and a condition on 2-d metric. In curved AdS space the latter cannot be the standard conformal gauge and breaks manifest 2-d Lorentz invariance. It is natural, therefore, to work in phase-space framework, imposing the GGRT light-cone gauge conditions $x^+ = \tau$, $P^+ =$const. We obtain the resulting light-cone superstring Hamiltonian.

1. Introduction

One lesson of developments in the last 4 years is that string theory duals of ($\mathcal{N} = 4, 2, 1, 0$) supersymmetric gauge theories should be fermionic strings in 4+1+more dimensions propagating in curved space with Ramond-Ramond backgrounds (see Refs. 1,2 and, e.g., Refs. 3,4,5 for reviews and references). It is important to make progress in understanding properties of such string theories.

The basic most symmetric example is (large N) $\mathcal{N}$=4 Super Yang-Mills theory dual to (weakly coupled) type IIB superstring theory in $AdS_5 \times S^5$ space with Ramond-Ramond flux.[2] Though this background has a lot of symmetry, solving the corresponding string theory appears to be a complicated problem. The commonly used procedure, in the known exactly solvable cases, is to start with a string action, solve the classical string equations, then quantize the theory, find the string spectrum, vertex operators, scattering amplitudes, etc. Even the first steps in this program are nontrivial in the $AdS_5 \times S^5$ case.

In contrast to other known cases, here fermionic degrees of freedom play crucial role and cannot be ignored from the start. The bosonic part of the string sigma model is the sum of the two symmetric space chiral models – $SO(2,4)/SO(1,5)$ and $SO(6)/SO(5)$. These are not conformally invariant ($R_{mn} \neq 0$), and it is the fermionic R-R couplings ($\bar{\theta}\gamma^{\cdots}\theta\partial x \partial x F_{5\ldots}$) that "glue" the AdS_5 and S^5 bosons together ensuring the vanishing of the 2-d beta-function ($R_{mn} - (F_5^2)_{mn} = 0$).

The presence of a curved R-R background indicates that one should use the manifestly supersymmetric Green-Schwarz[6] description of superstring. Finding an explicit expression for curved space GS action $S(x,\theta)$ is difficult in general (one

*Also at Imperial College, London and Lebedev Institute, Moscow

needs to know the component expansion of the background supergravity superfields which enter the formal expression[7] for the action). In the present $AdS_5 \times S^5$ case, this technical problem has a nice geometrical solution[8] based on viewing string as moving on the supercoset $PSU(2,2|4)/[SO(1,4) \times SO(5)]$ which replaces the flat superspace (10-d super Poincare)$/SO(1,9)$ in original GS construction.

The resulting action, though explicitly known,[8,9,10] looks highly nonlinear containing terms of many orders in θ. This is, however, an illusion of complexity: the fermionic part of the action simplifies dramatically in proper κ-symmetry gauges – it becomes quadratic and quartic in θ only.[11,12,13] Its structure is similar to that of the flat space GS action in a covariant κ-symmetry gauge which is also quartic in fermions.[a]

If one interprets the $AdS_5 \times S^5$ supergroup $PSU(2,2|4)$ as the $\mathcal{N} = 4$ superconformal group in 4 dimensions, it is natural to split the fermionic generators into 4 standard supergenerators Q_i and 4 special conformal supergenerators S_i (we suppress the 4-d spinor indices). The associated superstring coordinates will be denoted as θ_i and η_i (we use fermionic parametrization of Refs. 14,15). The 4-d Lorentz covariant κ-symmetry "S-gauge[14]" (which leads to an action equivalent to the one in Refs. 11,12) corresponds to setting all η_i to zero. The resulting superstring Lagrangian written in "4+6" parametrization of $AdS_5 \times S^5$ where $(R=1)$

$$ds^2 = Y^2 dx^a dx^a + Y^{-2} dY^M dY^M \ , \tag{1}$$

or, equivalently, $(a = 0, ..., 3,\ M = 1, ..., 6,\ Y^M = e^{\phi} u^M)$

$$ds^2 = e^{2\phi} dx^a dx^a + d\phi^2 + du^M du^M \ , \qquad u^M u^M = 1 \ , \tag{2}$$

has the following simple structure[14 b]

$$L = -\frac{1}{2}\sqrt{g}\Big(Y^2[\partial_\mu x^a - (i\theta_{\mathsf{a}i}\sigma^{a\mathsf{a}\dot{\mathsf{b}}}\partial_\mu\theta^i_{\dot{\mathsf{b}}} + h.c.)]^2 + Y^{-2}\partial^\mu Y^M \partial_\mu Y^M\Big)$$
$$- \big(i\epsilon^{\mu\nu}\partial_\mu Y^M \theta^{\dot{\mathsf{a}}i}\rho^M_{ij}\partial_\nu\theta^j_{\dot{\mathsf{a}}} + h.c.\big) \ . \tag{3}$$

This covariant κ-symmetry gauge fixed action is well-defined and useful for developing perturbation theory near classical "long" string configurations ending at the boundary of AdS_5,[13,16,17] e.g., the ones appearing in Wilson loop computations.[18] However, the kinetic term of the fermions $\sim \partial x \theta \partial \theta$ is degenerate for "short" strings,[13] and thus this action is not directly applicable for computing the spectrum of "short" closed string in the bulk of $AdS_5 \times S^5$.

[a]For example, in $\theta^1 = \theta^2$ gauge the flat space type IIB GS action has the following structure: $S \sim \int [(\partial x)^2 + \partial x \bar\theta \gamma \partial\theta + (\bar\theta\partial\theta)^2]$.

[b]The actions in Refs. 11,12 have isomorphic form, corresponding to a specific choice of the 10-d Dirac matrix representation. We use "chiral" representations for the 4-d and 6-d Dirac matrices, $\gamma^a = \begin{pmatrix} 0 & \sigma^a \\ \bar\sigma^a & 0 \end{pmatrix}$, $\gamma^M = \begin{pmatrix} 0 & \rho^{M\,ij} \\ \rho^M_{ij} & 0 \end{pmatrix}$, with $(\rho^M)^{ij} \equiv -(\rho^M_{ij})^*$, $i,j = 1,2,3,4$. $\mathsf{a}, \dot{\mathsf{b}} = 1,2$ are the $sl(2,C)$ (4-d spinor) indices and $\theta^\dagger_{\mathsf{a}i} = -\theta^i_{\dot{\mathsf{a}}}$, $\theta^{\mathsf{a}\dagger}_i = \theta^{\dot{\mathsf{a}}i}$. The 10-d spinors are split as (θ^i, θ_i).

In order to avoid this degeneracy problem, it is natural to try to follow the same approach which worked remarkably well in flat space:[19,6] use light-cone gauge (for an alternative covariant approach to quantization of GS action see Ref. 20). In flat space superstring light-cone gauge fixing consists of the two steps: (i) fermionic light-cone gauge choice, i.e. fixing the κ-symmetry by $\gamma^+\theta^I = 0$; (ii) bosonic light-cone gauge choice, i.e. using the conformal gauge[c] $\sqrt{g}g^{\mu\nu} = \eta^{\mu\nu}$ and fixing the residual conformal diffeomorphism symmetry by $x^+(\tau,\sigma) = p^+\tau$. Fixing the fermionic light-cone gauge already produces a substantial simplification of the flat-space GS action: it becomes quadratic in θ. Similarly, in $AdS_5 \times S^5$ the first task should be to find a light-cone κ-symmetry gauge in which the fermion kinetic term becomes $\partial x^+ \theta \partial \theta$, i.e. involves only one combination – x^+ – of 4-d coordinates, so that the non-degeneracy of the kinetic term for the fermions will not depend on a choice of a specific string background in transverse directions. To simplify the fermion kinetic term further one should then try to choose the light-cone bosonic gauge $x^+ = \tau$.

Fixing the fermionic light-cone gauge was discused in detail in Ref. 14,[d] and the fixing the bosonic part of light-cone gauge and derivation of the resulting light-cone Hamiltonian was described in Ref. 15. Our aim below is to review these results.

The metric (2) corresponds to the AdS space in the Poincaré coordinate patch. One needs to use the Poincaré coordinates to have a null isometry in the bulk and at the boundary (the boundary should have $R^{1,3}$ topology). The AdS/CFT duality suggests that since the boundary SYM theory in $R^{1,3}$ has a well-defined light-cone gauge description,[22] it should be possible to fix some analog of a light-cone gauge for the dual string theory as well.

The SYM theory does not admit a manifestly $\mathcal{N} = 4$ supersymmetric Lorentz-covariant description, but has a simple superspace description in the light-cone gauge $A^+ = 0$.[22] It is based on a single chiral superfield $\Phi(x,\theta) = A(x) + \theta^i\psi_i(x) + \ldots$ where $A = A_1 + \mathrm{i}A_2$ represents the transverse components of the gauge field and ψ_i its fermionic partner which transforms under the fundamental representation of R-symmetry group $SU(4)$. In addition to the standard light-cone supersymmetry (shifts of θ), the light-cone superspace SYM action $S[\Phi]$ has also a non-linear superconformal symmetry. This suggests that it may be possible to formulate the bulk string theory in a way which is naturally related to the light-cone form of the boundary SYM theory. In particular, it may be useful to split the fermionic string coordinates into the two parts with manifest $SU(4) \simeq SO(6)$ transformation properties which will be the counterparts of the linearly realized Poincaré supersymmetry supercharges Q_i and the nonlinearly realized conformal supersymmetry supercharges S_i of the SYM theory.

As was shown in Refs. 23,24,25, field theories in AdS space (in particular, type IIB supergravity) admit a simple light-cone description. There exists a light-cone Hamiltonian for a superparticle in $AdS_5 \times S^5$ which may be used to formulate AdS/CFT correspondence between chiral primary states of SYM theory and super-

---

[c] We use Minkowski signature 2-d world sheet metric $g_{\mu\nu}$ with $g \equiv -\det g_{\mu\nu}$.

[d] Some previous work in this direction using different approach was reported in Ref. 21.

gravity states of the bulk theory in the light-cone gauge. This indicates that the full superstring theory in $AdS_5 \times S^5$ should also have a kind of light-cone gauge formulation, which may be useful in the context of the AdS/CFT correspondence. The string light-cone Hamiltonian should reduce to the flat space one in the $R \to \infty$ limit and to the superparticle Hamiltonian[25] in the zero slope limit.

This is the motivation behind our light-cone gauge fixing program,[14,15] in spite of the fact that there is no globally well-defined null Killing vector in AdS space (its norm proportional to $e^{2\phi}$ vanishes at the horizon $\phi = -\infty$ of the Poincaré patch). We use a formal approach, assuming that the degeneracy of the light-cone description reflected in the $e^{2\phi} \to 0$ singularity of the resulting light-cone gauge fixed action should have some physical resolution, e.g., the form of the light-cone Hamiltonian may suggest how the wave functions should be defined in this region.

2. Superstring Action in Light-Cone Kappa Symmetry Gauge

In Ref. 14 it was shown how the light-cone type κ-symmetry gauge can be fixed in the original action of Ref. 8 so that the resulting action has indeed the required form: only ∂x^+ appears in the fermionic kinetic term. In contrast to the Lorentz covariant "S-gauge" where 16 fermions η_i are set equal to zero, the light-cone gauge used in Ref. 14 corresponds to setting to zero "half" of the 16 θ_i and "half" of the 16 η_i ("half" is defined with respect to $SO(1,1)$ rotations in the light-cone directions). The remaining fermions (which will be again denoted by θ_i, η_i) from now on are simply 4+4 complex anticommuting variables carrying no extra Lorentz spinor indices. For comparison, the flat space GS action in the light-cone gauge ($\gamma^+\theta = 0$),[19] written in a similar parametrization of the 16 fermionic coordinates, has the following structure:

$$\mathcal{L} = -\frac{1}{2}\sqrt{g}(\partial_\mu x)^2 - \left[\frac{\mathrm{i}}{2}\sqrt{g}\partial^\mu x^+(\theta\partial_\mu\theta + \eta\partial_\mu\eta) - \epsilon^{\mu\nu}\partial_\mu x^+\eta\partial_\nu\theta + h.c.\right] . \tag{4}$$

Note that θ's and η's enter diagonally in the kinetic term, but they are mixed in the WZ term. This form of the original GS Lagrangian is the flat space limit of the light-cone $AdS_5 \times S^5$ Lagrangian of Ref. 14 (cf. (3))

$$\begin{aligned}\mathcal{L} = &-\sqrt{g}\Big[Y^2(\partial^\mu x^+\partial_\mu x^- + \partial^\mu x\partial_\mu \bar{x}) + \frac{1}{2}Y^{-2}(\partial_\mu Y^M + \mathrm{i}\eta\rho^{MN}\eta Y^N Y^2\partial_\mu x^+)^2\Big] \\ &- \frac{\mathrm{i}}{2}\sqrt{g}g^{\mu\nu}Y^2\partial_\mu x^+\Big[\theta^i\partial_\nu\theta_i + \theta_i\partial_\nu\theta^i + \eta^i\partial_\nu\eta_i + \eta_i\partial_\nu\eta^i + \mathrm{i}Y^2\partial_\nu x^+(\eta^2)^2\Big] \\ &+ \epsilon^{\mu\nu}|Y|\partial_\mu x^+\Big[\eta^i\rho^M_{ij}Y^M(\partial_\nu\theta^j - \mathrm{i}\sqrt{2}|Y|\eta^j\partial_\nu x) + h.c.\Big] .\end{aligned} \tag{5}$$

We decomposed x^a into the light-cone and 2 complex coordinates: $(x^+, x^-, x, \bar{x})$, $x^\pm = \frac{1}{\sqrt{2}}(x^3 \pm x^0)$; $x, \bar{x} = \frac{1}{\sqrt{2}}(x^1 \pm \mathrm{i}x^2)$.[e] Written in terms of the radial direction ϕ of AdS_5 and unit 6-d vector u^M parametrizing S^5 this Lagrangian becomes ($Y^M =$

[e] We use the following notation: 4-d indices are $a, b = 0, 1, 2, 3$; $SO(6)$ indices are $M, N = 1, ..., 6$; $SU(4)$ indices are $i, j = 1, 2, 3, 4$; 2-d indices are $\mu, \nu = 0$. The matrices ρ^M (off-diagonal blocks of

$e^{\phi}u^M,\ |Y| = e^{\phi})^{14}$

$$\mathcal{L} = -\sqrt{g}\Big[e^{2\phi}(\partial^\mu x^+\partial_\mu x^- + \partial^\mu x\partial_\mu \bar{x}) + \frac{1}{2}\partial^\mu\phi\partial_\mu\phi + \frac{1}{2}(\partial_\mu u^M + \mathrm{i}\eta\rho^{MN}\eta u^N e^{2\phi}\partial_\mu x^+)^2\Big]$$
$$- \frac{\mathrm{i}}{2}\sqrt{g}g^{\mu\nu}e^{2\phi}\partial_\mu x^+\Big[\theta^i\partial_\nu\theta_i + \theta_i\partial_\nu\theta^i + \eta^i\partial_\nu\eta_i + \eta_i\partial_\nu\eta^i + \mathrm{i}e^{2\phi}\partial_\nu x^+(\eta^2)^2\Big]$$
$$+ \epsilon^{\mu\nu}e^{2\phi}\partial_\mu x^+\Big[\eta^i\rho^M_{ij}u^M(\partial_\nu\theta^j - \mathrm{i}\sqrt{2}e^{\phi}\eta^j\partial_\nu x) + h.c.\Big]\ . \tag{6}$$

This action has several important properties:

(a) It contains x^- only in the bosonic part and only linearly (in $\partial x^+\partial x^-$ term), and the fermion kinetic terms are multiplied by the derivative of x^+ only. It is the crucial property of this light-cone κ-symmetry gauge fixed form of the action that the fermion kinetic term involves the derivative of only *one* space-time direction – x^+, i.e. its (non)degeneracy does not depend on transverse string profile. One expects, therefore, that after one fixes the bosonic light-cone gauge, it should this action should be a well defined starting point for quantizing the theory in the "short" string sector.

(b) The fermionic κ-symmetry light-cone gauge we used reduces the 32 fermionic coordinates θ^I_α (two left Majorana-Weyl 10-d spinors) to 16 physical Grassmann variables: "linear" θ^i and "nonlinear" η^i and their hermitian conjugates θ_i and η_i, which transform in fundamental representations of $SU(4)$. The superconformal algebra $psu(2,2|4)$ implies that these are related to the Poincaré and the conformal supersymmetry in the light-cone gauge description of the boundary theory. The action and symmetry generators have simple (quadratic) dependence on θ^i, but complicated (quartic) dependence on η^i. The action is symmetric under shifting $\theta \to \theta + \epsilon$ (supplemented by a shift of x^-). It is this symmetry that is responsible for the fact that the theory is linear in θ, i.e. that there is no quartic terms in θ.

(c) The fact that the action has only quadratic and quartic fermionic terms has to do with special symmetries of the $AdS_5 \times S^5$ background (covariantly constant curvature and 5-form field strength). The presence of the η^4 term reflects the curvature of the background.[f] As follows from the discussion in Ref. 8, the 'extra' $O(\eta^2)$ terms should have the interpretation of the couplings to the R-R 5-form background.

(d) The $AdS_5 \times S^5$ superstring action in general depends on two parameters: the scale (equal radii) R of $AdS_5 \times S^5$ and the inverse string tension or α'. Restoring the dependence on R one finds that in the flat space limit $R \to \infty$ the quartic fermionic

6-d Dirac matrices in chiral representation) satisfy: $\rho^M_{ij} = -\rho^M_{ji}$, $(\rho^M)^{il}\rho^N_{lj} + (\rho^N)^{il}\rho^M_{lj} = 2\delta^{MN}\delta^i_j$, $(\rho^M)^{ij} \equiv -(\rho^M_{ij})^*$, and $\rho^{MN} \equiv \rho^{[M}\rho^{\dagger N]}$, i.e. $\rho^{MN i}{}_j = \frac{1}{2}(\rho^M)^{il}\rho^N_{lj} - (M \leftrightarrow N)$. Also, $\theta^\dagger_i = \theta^i$, $\eta^\dagger_i = \eta^i$, $\theta^2 \equiv \theta^i\theta_i$, $\eta^2 \equiv \eta^i\eta_i$.

[f] Note that the light-cone gauge GS action in a curved space of the form $R^{1,1} \times M^8$ with generic NS-NS and R-R backgrounds[26] (reconstructed from the light-cone flat space GS vertex operators) contains, in general, higher than quartic fermionic terms, multiplied by higher derivatives of the background fields. This light-cone GS action has quartic fermionic term[27,26] involving the curvature tensor $R_{....}\partial x^+\partial x^+(\bar\theta\gamma^{-\cdot\cdot}\theta)(\bar\theta\gamma^{-\cdot\cdot}\theta) \sim R_{....}(p^+)^2(\bar\theta\gamma^{-\cdot\cdot}\theta)(\bar\theta\gamma^{-\cdot\cdot}\theta)$ which is similar to the one present in the NSR string action (i.e. in the standard 2-d supersymmetric sigma model).

term goes away, while the kinetic term reduces to the flat space light-cone GS action after representing each of the two $SO(8)$ spinors in terms of the two $SU(4)$ spinors.

3. Fixing the Bosonic Light-Cone Gauge

As a next step to quantization of the theory one would like, as in the flat case, to eliminate the ∂x^+-factors from the fermion kinetic terms in (6). In flat space this was possible by choosing the bosonic light-cone gauge. In the Polyakov formulation this may be done by fixing the conformal gauge $\sqrt{g}g^{\mu\nu} = \eta^{\mu\nu}$, and then noting that since the resulting equation $\partial^2 x^+ = 0$ has the general solution $x^+(\tau,\sigma) = f(\tau-\sigma) + h(\tau+\sigma)$, one can fix the residual conformal diffeomorphism symmetry on the plane by choosing $x^+(\tau,\sigma) = p^+\tau$.

Let us first not make any explicit gauge choice and consider the superstring path integral assuming that there is no sources for x^-. The linear dependence of the action (5),(6) on x^- allows us to integrate over x^- explicitly: we get δ-function constraint imposing the equation of motion $\partial_\mu(\sqrt{g}g^{\mu\nu}e^{2\phi}\partial_\nu x^+) = 0$ for x^+, which is formally solved by setting

$$\sqrt{g}g^{\mu\nu}e^{2\phi}\partial_\nu x^+ = \epsilon^{\mu\nu}\partial_\nu f \ , \quad \text{i.e.} \qquad e^{2\phi}\partial_\mu x^+ = g_{\mu\nu}\frac{\epsilon^{\nu\lambda}}{\sqrt{g}}\partial_\lambda f \ . \tag{7}$$

where $f(\tau,\sigma)$ is an arbitrary function. Since our action (6) depends on x^+ only through $e^{2\phi}\partial x^+$, we are then able to integrate over x^+ as well, eliminating it in favor of the function f. The resulting fermion kinetic term is then non-degenerate (for a properly chosen f), and may be interpreted as an action of 2-d fermions in curved 2-d geometry determined by f and $g_{\mu\nu}$ (cf. Refs. 17,14 and refs. there).

To proceed further, one would like to fix 2-d diffeomorphisms by a condition on $g_{\mu\nu}$ and a condition on f corresponding to the usual light-cone gauge $x^+ = \tau$. An important observation[28,29,14] is that in the case of the AdS type curved spaces, the bosonic light-cone gauge $x^+ = \tau$ can not be combined with the standard conformal gauge $\sqrt{g}g^{\mu\nu} = \eta^{\mu\nu}$ (imposing the conformal gauge one in general is unable to solve the equation for x^+ by $x^+ = \tau$). Instead, one needs to impose a condition on $g_{\mu\nu}$ that breaks the manifest 2-d Lorentz symmetry and leads to a rather non-standard string action, with all terms coupled to the radial function ϕ of AdS_5 space.

A consistent gauge choice is[14]

$$e^{2\phi}\sqrt{g}g^{00} = -1 \ , \qquad x^+ = p^+\tau \ . \tag{8}$$

Indeed, the equation for x^+ is then satisfied identically. This choice is equivalent to $f = \sigma$ in (7). A closely related alternative, originally suggested by Polyakov,[29] is a modification of the conformal gauge $\sqrt{g}g^{ab} = \text{diag}(-e^{-2\phi}, e^{2\phi})$ which is also consistent with the light-cone gauge condition $x^+ = \tau$.

The resulting action has AdS_5 radial direction ϕ factors coupled differently to ∂_0 and ∂_1 derivative terms, and the S^5 part of the action is also coupled to ϕ. The absence of manifest 2-d Lorentz symmetry suggests that it is natural to use the *phase space* formulation of the light-cone gauge fixed theory. The coordinate space Polyakov approach based on (8) is equivalent to the phase space approach based on fixing the diffeomorphisms by $x^+ = p^+\tau,\ P^+ =$const.[15] In general, the original GGRT phase space approach[30] to light-cone gauge fixing is directly applicable in the present curved space case.

To illustrate the derivation of phase space Lagrangian let us consider first the example of a classical bosonic string in AdS_5 ($2\pi\alpha' = 1,\ R = 1$)

$$\mathcal{L} = -h^{\mu\nu}\left[\partial_\mu x^+\partial_\nu x^- + \frac{1}{2}\partial_\mu x\partial_\nu\bar{x} + e^{-2\phi}\partial_\mu\phi\partial_\nu\phi\right] , \qquad h^{\mu\nu} \equiv \sqrt{g}g^{\mu\nu}e^{2\phi} . \quad (9)$$

Introducing the momenta $\mathcal{P}^a$ for the light-cone coordinates (x^+, x^-) and the two transverse coordinates $(x, \bar{x})$ and the momentum π for the radial direction ϕ we get

$$\begin{aligned}\mathcal{L} &= \dot{x}_\perp\mathcal{P}_\perp + \dot{\phi}\pi + \dot{x}^+\mathcal{P}^- + \dot{x}^-\mathcal{P}^+ \\ &+ \frac{1}{2h^{00}}\left[(\mathcal{P}_\perp^2 + 2\mathcal{P}^+\mathcal{P}^-) + e^{4\phi}(\acute{x}_\perp^2 + 2\acute{x}^+\acute{x}^-) + e^{2\phi}(\pi^2 + \acute{\phi}^2)\right] \\ &+ \frac{h^{01}}{h^{00}}(\acute{x}_\perp\mathcal{P}_\perp + \acute{\phi}\pi + \acute{x}^+\mathcal{P}^- + \acute{x}^-\mathcal{P}^+) , \end{aligned} \quad (10)$$

where $1/h^{00}$ and h^{01}/h^{00} play the role of the Lagrange multipliers imposing two constraints. Choosing the light-cone gauge $x^+ = \tau$, $\mathcal{P}^+ = p^+ =$ const and integrating over $\mathcal{P}^-$ we get the relation

$$h^{00} = -p^+ , \quad (11)$$

which is equivalent (up to a rescaling) to the condition on the metric in (8). The expression for $\mathcal{P}^-$ follows from the $1/h^{00}$ constraint after using the h^{01}/h^{00} constraint. The resulting light-cone gauge Hamiltonian density is[15]

$$\mathcal{H} = -\mathcal{P}^- = \frac{1}{2p^+}\left[\mathcal{P}_\perp^2 + e^{4\phi}\acute{x}_\perp^2 + e^{2\phi}(\pi^2 + \acute{\phi}^2)\right] . \quad (12)$$

As usual, the coordinate x^- does not appear in the Hamiltonian, but is determined from the reparametrization constraint

$$p^+\acute{x}^- + \mathcal{P}_\perp\acute{x}_\perp + \acute{\phi}\pi = 0 . \quad (13)$$

For closed string, the integral of this constraint over σ constrains the state space to the subspace invariant under σ translations.

For comparison, for string in AdS_3 described by $SL(2,R)$ WZW model (in the standard Gauss parametrization)

$$\mathcal{L} = -\sqrt{g}g^{\mu\nu}(e^{2\phi}\partial_\mu x^+\partial_\nu x^- + \frac{1}{2}\partial_\mu\phi\partial_\nu\phi) + \epsilon^{\mu\nu}e^{2\phi}\partial_\mu x^+\partial_\nu x^- , \quad (14)$$

one finds

$$\mathcal{H} = \frac{e^{2\phi}}{2p^+}(\pi + \acute{\phi})^2 , \qquad p^+\acute{x}^- + \pi\acute{\phi} = 0 . \quad (15)$$

4. Light Cone Superstring Hamiltonian

Repeating the same procedure for the superstring action in (6) we get[15 g]

$$\mathcal{L} = \mathcal{P}_\perp \dot{x}_\perp + \pi\dot{\phi} + \mathcal{P}_M \dot{u}^M + \frac{\mathrm{i}}{2}p^+(\theta^i\dot{\theta}_i + \eta^i\dot{\eta}_i + \theta_i\dot{\theta}^i + \eta_i\dot{\eta}^i) + \mathcal{P}^-$$

$$- \frac{h^{01}}{p^+}\Big[p^+\acute{x}^- + \mathcal{P}_\perp \acute{x}_\perp + \pi\acute{\phi} + \mathcal{P}_M \acute{u}^M + \frac{\mathrm{i}}{2}p^+(\theta^i\acute{\theta}_i + \eta^i\acute{\eta}_i + \theta_i\acute{\theta}^i + \eta_i\acute{\eta}^i)\Big] , \quad (16)$$

where the light-cone Hamiltonian density is ($\mathcal{H} = -\mathcal{P}^-$)

$$\mathcal{H} = \frac{1}{2p^+}\Big(\mathcal{P}_\perp^2 + e^{4\phi}\acute{x}_\perp^2 + e^{2\phi}[\pi^2 + \acute{\phi}^2 + (l^i{}_j)^2 + \acute{u}^M\acute{u}^M + p^{+2}(\eta^2)^2 + 4p^+\eta_i l^i{}_j\eta^j]\Big)$$

$$- e^{2\phi}\eta^i\rho^M_{ij}u^M(\acute{\theta}^j - \mathrm{i}\sqrt{2}e^\phi\eta^j\acute{x}) - e^{2\phi}\eta_i\rho^{Mij}u^M(\acute{\theta}_j + \mathrm{i}\sqrt{2}e^\phi\eta_j\acute{\bar{x}}) , \quad (17)$$

and h^{01} imposes the reparametrization constraint. Here $\mathcal{P}^M$ is the momentum corresponding to the unit 6-vector u^M, $u^M\mathcal{P}^M = 0$, and $l^i{}_j \equiv \frac{\mathrm{i}}{2}(\rho^{MN})^i{}_j u^M\mathcal{P}^N$, so that $(l^i{}_j)^2 \equiv l^i{}_j l^j{}_i = \mathcal{P}^M\mathcal{P}^M$. In the particle theory limit $l^i{}_j$ has the interpretation of the $SU(4)$ angular momentum operator[25] and (17) reduces to the (classical limit of) light-cone Hamiltonian for a superparticle in $AdS_5 \times S^5$ space[25]

$$\mathcal{H} = \frac{1}{2p^+}\Big[\mathcal{P}_\perp^2 + e^\phi\pi e^\phi\pi + e^{2\phi}(X - \frac{1}{4})\Big] , \quad (18)$$

$$X \equiv (l^i{}_j)^2 + (p^+\eta^2 - 2)^2 + 4p^+\eta_i l^i{}_j\eta^j . \quad (19)$$

The analogous expressions for the string Lagrangian and Hamiltonian written in terms of 6 Cartesian coordinates Y^M and the associated momentum $\mathcal{P}_M$, i.e. corresponding to the Lagrangian (5) are

$$\mathcal{L} = \mathcal{P}_\perp \dot{x}_\perp + \mathcal{P}_M\dot{Y}^M + \frac{\mathrm{i}}{2}p^+(\theta^i\dot{\theta}_i + \eta^i\dot{\eta}_i + \theta_i\dot{\theta}^i + \eta_i\dot{\eta}^i) - \mathcal{H} , \quad (20)$$

$$\mathcal{H} = \frac{1}{2p^+}\Big(\mathcal{P}_\perp^2 + Y^4\acute{x}_\perp^2 + Y^4\mathcal{P}_M\mathcal{P}_M + \acute{Y}^M\acute{Y}^M$$

$$+ Y^2[p^{+2}(\eta^2)^2 + 2\mathrm{i}p^+\eta\rho^{MN}\eta Y_M\mathcal{P}_N]\Big) - |Y|Y^M\Big[\eta\rho^M(\acute{\theta} - \mathrm{i}\sqrt{2}|Y|\eta\acute{x}) + h.c.\Big] , \quad (21)$$

$$p^+\acute{x}^- + \mathcal{P}_\perp\acute{x}_\perp + \mathcal{P}_M\acute{Y}^M + \frac{\mathrm{i}}{2}p^+(\theta^i\acute{\theta}_i + \eta^i\acute{\eta}_i + \theta_i\acute{\theta}^i + \eta_i\acute{\eta}^i) = 0 . \quad (22)$$

The fact that in the particle theory limit the string Hamiltonian (17) or (21) reduces[15] to the superparticle Hamiltonian[25] implies that the "massless" (zero-mode) spectrum of the superstring coincides indeed with the spectrum of type IIB supergravity compactified on S^5. The vertex operators for the supergravity states are then obtained by solving the linearized supergravity equations (or 1-st quantized superparticle state equations) in $AdS_5 \times S^5$ background.

[g] We explicitly introduce momenta only for the bosons. The odd (fermionic) part of the phase space may be viewed as represented by θ^i, η^i considered as fermionic coordinates and θ_i, η_i considered as fermionic momenta.

To restore the dependence on R and α' one needs to make the replacements: $e^{\phi} \to e^{\phi/R}$, $u^M \to Ru^M$, $\mathcal{P} \to T^{-1/2}\mathcal{P}$, $x' \to T^{1/2}x'$, etc., where $T = (2\pi\alpha')^{-1}$. It may be interesting to study the limiting cases of the parameters, like $\alpha'/R^2 \to \infty$ for fixed momenta, etc. $R \to \infty$ gives of course the standard flat space light-cone superstring Hamiltonian (in the specific parametrization of fermions we are using, see section 2).

For generic values of parameters the Hamiltonian (17),(21) looks quite non-linear. Further progress may depend on a possibility of making a transformation to some new variables which may allow one to solve for the string theory spectrum. Such transformation should involve fermions in an essential way. The phase space formulation seems a good starting point for searching for such a transformation since the bosonic momenta are already nontrivially dependent on the fermions. One potentially interesting idea is to try to introduce twistor-like variables (see Ref. 31 for some previous discussions of twistors in *AdS* space).

Acknowledgments

This work was supported in part by the DOE grant DE-FG02-91ER-40690 and the INTAS project 991590. I am grateful to R.R. Metsaev and C.B. Thorn for collaboration on the work described in this contribution.

References

1. A.M. Polyakov, *String theory as a universal language*, hep-th/0006132; *The wall of the cave*, *Int. J. Mod. Phys.* **A14**, 645 (1999) [hep-th/9809057]; *String theory and quark confinement*, *Nucl. Phys. Proc. Suppl.* **68**, 1 (1998) [hep-th/9711002].
2. J. Maldacena, *The large-N limit of superconformal field theories and supergravity*, *Adv. Theor. Math. Phys.* **2**, 231 (1998) [hep-th/9711200]; S.S. Gubser, I.R. Klebanov and A.M. Polyakov, *Gauge theory correlators from non-critical string theory*, *Phys. Lett.* **B428**, 105 (1998) [hep-th/9802109]; E. Witten, *Anti-de Sitter space and holography*, *Adv. Theor. Math. Phys.* **2**, 253 (1998) [hep-th/9802150].
3. I.R. Klebanov and M.J. Strassler, *Supergravity and a confining gauge theory: Duality cascades and (χ)SB-resolution of naked singularities*, *JHEP* **0008**, 052 (2000) [hep-th/0007191]. J.M. Maldacena and C. Nunez, *Towards the large n limit of pure $N = 1$ super Yang Mills*, hep-th/0008001.
4. I.R. Klebanov and A.A. Tseytlin, *D-branes and dual gauge theories in type 0 strings*, *Nucl. Phys.* **B546**, 155 (1999) [hep-th/9811035].
5. I.R. Klebanov, *TASI lectures: Introduction to the AdS/CFT correspondence*, hep-th/0009139.
6. M.B. Green and J.H. Schwarz, *Covariant description of superstrings*, *Phys. Lett.* **B136**, 367 (1984); *Nucl. Phys.* **B243**, 285 (1984).
7. M.T. Grisaru, P. Howe, L. Mezincescu, B. Nilsson and P.K. Townsend, *N=2 Superstrings In A Supergravity Background*, *Phys. Lett.* **B162**, 116 (1985).
8. R.R. Metsaev and A.A. Tseytlin, *Type IIB superstring action in $AdS_5 \times S^5$ background*, *Nucl. Phys.* **B533**, 109 (1998) [hep-th/9805028].
9. R. Kallosh, J. Rahmfeld and A. Rajaraman, *Near horizon superspace*, *JHEP* **9809**, 002 (1998) [hep-th/9805217].

10. R.R. Metsaev and A.A. Tseytlin, *Supersymmetric D3 brane action in $AdS_5 \times S^5$*, *Phys. Lett.* **B436**, 281 (1998) [hep-th/9806095].
11. I. Pesando, *A kappa gauge fixed type IIB superstring action on $AdS_5 \times S^5$*, *JHEP* **9811**, 002 (1998) [hep-th/9808020]; *All roads lead to Rome: Supersolvable and supercosets*, *Mod. Phys. Lett.* **A14**, 343 (1999) [hep-th/9808146].
12. R. Kallosh and J. Rahmfeld, *The GS string action on $AdS_5 \times S^5$*, *Phys. Lett.* **B443**, 143 (1998) [hep-th/9808038].
13. R. Kallosh and A.A. Tseytlin, *Simplifying superstring action on $AdS_5 \times S^5$*, *JHEP* **9810**, 016 (1998) [hep-th/9808088].
14. R.R. Metsaev and A.A. Tseytlin, *Superstring action in $AdS_5 \times S^5$: κ-symmetry light cone gauge*, hep-th/0007036.
15. R.R. Metsaev, C.B. Thorn and A.A. Tseytlin, *Light-cone Superstring in AdS Spacetime*, hep-th/0009171.
16. S. Forste, D. Ghoshal and S. Theisen, *Stringy corrections to the Wilson loop in $N = 4$ super Yang-Mills theory*, *JHEP* **9908**, 013 (1999) [hep-th/9903042].
17. N. Drukker, D.J. Gross and A.A. Tseytlin, *Green-Schwarz string in $AdS_5 \times S^5$: Semiclassical partition function*, *JHEP* **0004**, 021 (2000) [hep-th/0001204].
18. J. Maldacena, *Wilson loops in large N field theories*, *Phys. Rev. Lett.* **80**, 4859 (1998) [hep-th/9803002]; S.-J. Rey and J. Yee, *Macroscopic Strings as Heavy Quarks of Large N Gauge Theory and Anti-de Sitter Supergravity*, hep-th/9803001.
19. M.B. Green and J.H. Schwarz, *Supersymmetric String Theories*, *Phys. Lett.* **B109**, 444 (1982).
20. N. Berkovits, *Covariant quantization of the superstring*, hep-th/0008145; N. Berkovits and O. Chandia, *Superstring Vertex Operators in an $AdS_5 \times S^5$ Background*, hep-th/0009168.
21. I. Pesando, *On the fixing of the kappa gauge symmetry on AdS and flat background: The lightcone action for the type IIB string on $AdS_5 \times S^5$* , *Phys. Lett.* **B485**, 246 (2000) [hep-th/9912284].
22. L. Brink, O. Lindgren and B.E. Nilsson, *N=4 Yang-Mills Theory On The Light Cone*, *Nucl. Phys.* **B212**, 401 (1983); S. Mandelstam, *Light Cone Superspace And The Ultraviolet Finiteness Of The N=4 Model*, *Nucl. Phys.* **B213**, 149 (1983).
23. R.R. Metsaev, *Light cone formalism in AdS spacetime*, in *Proc. XIV Int. Workshop on High Energy Physics and Quantum Field Theory*, hep-th/9911016.
24. R.R. Metsaev, *Light cone form of field dynamics in anti-de Sitter spacetime and AdS/CFT correspondence*, *Nucl. Phys.* **B563**, 295 (1999) [hep-th/9906217].
25. R.R. Metsaev, *Light cone gauge formulation of IIB supergravity in $AdS_5 \times S^5$ background and AdS/CFT correspondence*, *Phys. Lett.* **B468**, 65 (1999) [hep-th/9908114].
26. E.S. Fradkin and A.A. Tseytlin, *Effective Action Approach To Superstring Theory*, *Phys. Lett.* **B160**, 69 (1985).
27. P. Candelas, G.T. Horowitz, A. Strominger and E. Witten, *Vacuum Configurations For Superstrings*, *Nucl. Phys.* **B258**, 46 (1985).
28. R.E. Rudd, *Light cone gauge quantization of 2-D sigma models*, *Nucl. Phys.* **B427**, 81 (1994) [hep-th/9402106].
29. A.M. Polyakov, private communication (Jan. 2000).
30. P. Goddard, J. Goldstone, C. Rebbi and C.B. Thorn, *Quantum Dynamics Of A Massless Relativistic String*, *Nucl. Phys.* **B56**, 109 (1973).
31. P. Claus, M. Gunaydin, R. Kallosh, J. Rahmfeld and Y. Zunger, *Supertwistors as quarks of $SU(2,2|4)$*, *JHEP* **9905**, 019 (1999) [hep-th/9905112]. I. Bandos, J. Lukierski, C. Preitschopf and D. Sorokin, *OSp supergroup manifolds, superparticles and supertwistors*, *Phys. Rev.* **D61**, 065009 (2000) [hep-th/9907113].

COSMOLOGICAL BREAKING OF SUPERSYMMETRY?

T. BANKS*
Department of Physics and Astronomy
Rutgers University, Piscataway, NJ 08855–0849, USA
E-mail: banks@physics.rutgers.edu

This is the text of my talk at the Strings 2000 Conference at the University of Michigan. It consists of three interlocking conjectures about the nature of SUSY breaking and the Cosmological Constant, as well as evidence from a toy model that demonstrates the validity of the claim that systems with a UV/IR connection can exhibit much larger SUSY breaking than local field theory.

1. Introduction - the Entropy of DeSitter Space

The purpose of this talk is to present three radical conjectures about the nature of M-theory, SUSY breaking, and the cosmological constant, as well as tangential evidence that the most implausible of the conjectures might indeed be true. The text follows the talk I gave at Strings 2000 at the University of Michigan, and covers material that was presented in Refs. 1 and 2, as well as an as yet unpublished manuscript.[3]

The first conjecture has to do with the well known "entropy of DeSitter space," which arises because DeSitter space has a hot cosmological horizon. A number of explanations of this entropy have appeared[4] but I do not believe that they are fully satisfactory. I propose instead, that just as there are quantum mechanical systems which describe M-theory[a] in asymptotically flat[b] and asymptotically AdS[6] limits, there is a quantum system with *a finite number of states, N*, which describes asymptotically DeSitter (AsDS) spacetimes. The number N is given in terms of the cosmological constant by

$$N = e^{12\pi M_P^4/\Lambda}. \tag{1}$$

(This equation is valid in 4 dimensions. From now on we will set $M_P = 1$.)

The phrase AsDS space refers to a spacetime which begins with a Big Bang. That is, there is a spacelike surface lying at a finite proper time in the past of any geodesic observer, on which the universe begins. In particular, DeSitter space itself is not AsDS. The DS universe is a sphere which shrinks down from infinite size in the infinite past to a minimum radius, and then reexpands to infinite radius. If the cosmological constant is smaller than the Planck scale, then semiclassical

---

*On leave from U.C. Santa Cruz.

[a]As always, I use this terminology to describe the, as yet incompletely understood, theory whose various weakly coupled string and low energy SUGRA limits we have been studying since 1984.

[b]The present author may be excused for assuming that Matrix Theory[5] provides a proof that asymptotically flat M-theory is described by a quantum mechanical system.

methods are valid in the remote past, and they assure us that we can encode an infinite amount of information in the initial conditions for DS space. The problem is that once back reaction is taken into account, most of these initial conditions will not evolve asymptotically to DS space in the future. Instead, singularities will be encountered at a finite time. The cosmological entropy bounds of Bousso, Fischler and Susskind, assure us that all regions of an AsDS universe can, for a wide range of initial data, evolve in a singularity free manner.

The primary motivation for this conjecture is that N is certainly a bound on the number of states in an AsDS space. Indeed, we know that the most efficient way to pack entropy into a region of spacetime is to form a black hole. But the black hole spectrum in AsDS space is bounded. The metric of a Schwarzchild-Desitter black hole is

$$ds^2 = -h(r)dt^2 + dr^2/h(r) + r^2 d\Omega^2, \tag{2}$$

where,

$$h(r) = 1 - 2M/r - r^2/R^2. \tag{3}$$

The timelike Killing vector of this metric becomes null on two surfaces, given by the two positive zeroes of $h(r)$. The smaller of these is the black hole horizon and the larger is the cosmological horizon. This metric makes sense only up to a maximal value of M (for fixed R), where the two zeroes of h coincide (though the geodesic distance between them remains finite). The sum of the entropies of both horizons of this maximal solution is only 2/3 of the total entropy of the DeSitter Horizon in the absence of black holes.

Experience in matching microscopic entropies computed in string theory with the Bekenstein-Hawking formula leads us to expect that the bounds are always saturated. This is the origin of the conjecture formulated in (1). One might object that the above analysis refers only to the entropy as seen by a single observer in DS space, while the manifold seems to have a much larger volume and indeed eventually grows to infinite size. I would claim that this is an illusion. In the holographic interpretation of the black hole information puzzle, we have learned that a view of the world from outside the black hole horizon is complete. We make a mistake if we ascribe independent degrees of freedom to the region inside the horizon. It is hard to see how this principle could work for black hole physics and be inapplicable to the DS horizon.

There is a more formal argument which leads to the same conclusion. While Euclidean functional integrals over metrics are certainly not the correct formulation of M-theory, the semiclassical approximation to such integrals does seem to provide us with a reservoir of correct results. From a Euclidean functional integral point of view, the global group of isometries of the DeSitter metric plays a very different role than the Minkowski or Anti DeSitter symmetry groups. Indeed, the latter are best viewed as equivalence classes of diffeomorphisms with nontrivial behavior on the boundary of asymptotically flat or AdS spacetimes. Euclidean DeSitter space is a sphere and has no boundary. Its group of isometries should be integrated over.

Indeed, in world sheet string theory, a situation in which functional integrals over metrics can be rigorously defined, this is precisely what we do with the $SL(2, C)$ symmetries of the two sphere. They are gauge transformations and we mod out by them.

Now consider the semiclassical approximation to M-theory in DeSitter space (states of M-theory whose low energy physics is described by N=1 four dimensional SUGRA may well have such DeSitter background solutions and this is likely to be the only situation in which DeSitter space arises in M-theory). It is described by quantum fields propagating on the Euclidean 4 sphere. If we analytically continue to Minkowski signature, we obtain quantum fields propagating on the static patch of DeSitter space, in a thermal state at the Hawking-DeSitter temperature.[7] We can obtain the Green's functions in other regions of DeSitter space by applying elements of the DeSitter group, but, as emphasized above, these should be thought of as gauge transformations. Thus, the degrees of freedom outside the cosmological horizon are gauge copies of those inside and should not be multiply counted in evaluating the entropy.

It is important to contrast this discussion of AsDS spaces with the temporary DeSitter eras of inflationary cosmology. These spacetimes have no cosmological horizon. If, for example, an inflationary universe evolves into a matter or radiation dominated FRW cosmology, which expands forever, then the Bousso-Fischler-Susskind[8] cosmological entropy bounds do not imply that the entropy is bounded by the area of the inflationary "horizon." The proper holographic screens are at null infinity. Indeed, it is clear that black holes of arbitrary size can be embedded in such a spacetime, so that there can be no finite bound on the entropy.

The conjecture embodied in (1) has several important consequences. It implies that the cosmological constant is not a calculable parameter in a low energy effective field theory. Rather, it is a fundamental input which determines the number of states in the Hilbert space which describes AsDS universes.[c] In conventional quantum mechanics, the number of physical states is a boundary condition of the problem, not a dynamically determined parameter.

Our disappointment at finding that we will not be able to calculate the cosmological constant is tempered by the fact that the above conjecture, solves, at one stroke, the most serious phenomenological problem of M-theory: *Why No SUSY ?*. For years we have imagined that we lived in a universe which was approaching an asymptotically flat vacuum state of M-theory. Indeed, we knew of candidates for entire families of such vacua, and over the years the evidence for their existence and stability as true quantum solutions of the theory has been mounting. The problem is that all of our candidates are exactly supersymmetric. Even if one fantasizes about the existence of a stable, nonsupersymmetric vacuum with vanishing cosmological constant, one is left with the conundrum of why we are not living in one of the supersymmetric ones. But all supersymmetric vacua are asymptotically flat or

[c]It is likely that recollapsing universes can also be described in the same Hilbert space, depending on initial conditions.

asymptotically AdS, and thus have an infinite number of physical states. They are inaccessible in a system with a finite number of states.

The hypothesis of a finite number of states for an AsDS universe thus resolves the puzzle of the DeSitter entropy and simultaneously cures the vacuum degeneracy problem of asymptotically flat M-theory. I find it the most plausible, as well as the most attractive, of the three conjectures I will present in this lecture.

2. The Importance of Being SUSic

The declared goal of most research in string theory since 1984 has been to find an asymptotically flat vacuum state of the theory that describes the microphysics of the real world. This was supposed to be a task separable from the more difficult problem of describing the real real world, which is evidently cosmological. The logic was that of the particle physicist: physics is local and microphysics is largely independent of cosmology at the present era. This logic was also driven by a practical complication which in modern terms can be seen to be a consequence of the holographic nature of M-theory. The only things we knew how to calculate in string theory were S-matrix elements, and there is no evident analog of the S-matrix in a general cosmology.[d]

My second conjecture is that the problem we have been posing for ourselves in insoluble, because all asymptotically flat, Poincare invariant ground states of M-theory must be exactly supersymmetric. Although I have no proof of this statement, I believe the proof will involve consideration of the spectrum of black holes. Black hole physics provides the ultimate example of the UV/IR connection in M-theory. Black holes are extremely high energy states of the theory, but they are large and have low curvature. As a consequence, many of their properties may be read reliably from the ostensible *low energy* limit of the theory, which is SUGRA. In particular, SUGRA gives us a computation of the degeneracy of black holes states at a given energy:

$$N(E) \sim e^{E^{\frac{(d-2)}{(d-3)}}} \tag{4}$$

This grows even faster than the Hagedorn degeneracy of perturbative string theory. It implies that correlation functions of generic operators in this quantum theory do no even exist as tempered distributions in the time variable.[9]

Many years ago, Kutasov and Seiberg[10] showed that the Hagedorn degeneracy of perturbative string theory generally led to tachyon instabilities. Only in theories where the perturbative boson and fermion spectra differed by at most the level density of a 1+1 dimensional field theory, could tachyons be avoided. My contention then is that there is a nonperturbative analog of the Kutasov-Seiberg result, which is stronger than that result precisely because the black hole spectrum is so much more divergent. Furthermore, because the UV/IR connection allows us such a revealing picture of the nature of the high energy states (at least on a coarse grained level)

[d]On the deepest level, what is at issue is the question of how one makes a choice of holographic screen. In general, I would claim this is a gauge choice, though in spacetimes with timelike or null boundaries, a single screen on the boundary may suffice.

we can almost argue that the difference in bosonic and fermionic energy densities in a hypothetical SUSY violating flat vacuum is in fact exponential. Indeed, although one could postulate exact discrete chiral symmetries, it seems plausible that the lightest fermion in a SUSY violating vacuum will have nonzero mass. We can transform bosonic and fermionic black holes into each other by dropping single minimal mass fermions in from infinity, and this gives us an easy computation of the difference in level densities.

The extant evidence for and against this conjecture weighs heavily on the positive side. The only evidence against it is that there are finely tuned SUGRA effective Lagrangians with broken SUSY and vanishing cosmological constant. It is obvious that this analysis is insensitive to the questions raised above about the black hole density of states.

Evidence for the conjecture is easier to find. SUSY breaking in perturbative string theory leads to a potential for the dilaton and other moduli. If this potential is positive then a controlled approximate description of the system seems to be available. One can construct solutions of the effective equations of motion in which the coupling starts at zero in the infinite past, rises to a finite but small value and returns to zero. The asymptotic Planck scale is infinite and there are no finite energy black hole solutions. The spectrum of asymptotic states is that of the free SUSY violating string theory, and satisfies the Kutasov Seiberg bound. The ten dimensional $O(16)^2$ heterotic string is an example of this behavior.

If instead, as in Scherk-Schwarz-Rohm compactifications, the potential is negative, the system is drawn into the strongly coupled regime with curvatures of order the string scale. We have no control over the behavior of the system but it is unlikely to describe an asymptotically flat vacuum state. Indeed, Fabinger and Horava[11] have argued that all such systems "tunnel to nothing." A Matrix Theory analysis of the same systems[12] also suggests strongly that they do not have asymptotically flat vacua. Indeed, the crucial role that exact SUSY plays in the definition of asymptotic states in Matrix Theory[5] is another argument in favor of our conjecture.

Finally, one might have tried to claim that a generalized AdS/CFT correspondence would give us a SUSY violating AdS vacuum of string theory for any nonsupersymmetric conformal field theory. A major problem with this idea is that it is difficult to present an example in which the putative AdS radius is much larger than the string scale. There must be a large and tuneable gap between the dimension of the stress tensor (which would correspond to gravitons) and other primary operators of the CFT. SUSY violating CFTs do not usually have lines of fixed points so the only sort of parameter which might provide the required gap in the spectrum of dimensions is the dimension of field space. But within the conventional regime of large N scaling, all known fixed points will have couplings of order one or smaller and there is no way to explain the gap.

Thus, within the bounds of our ability to calculate, all solid evidence points to the absence of asymptotically flat SUSY violating vacua. Our refusal to give up the search for one is based on our evident desire to make contact between M-theory and the real world, rather than on hard evidence that such vacua can exist.

3. SUSY and the Giant Black Holes

If SUSY is unbroken in asymptotically flat spacetimes, and M-theory is related to the real world, it follows that the origin of SUSY breaking is cosmological. This is the content of our third conjecture: the observed splitting between supermultiplets must be explained in terms of cosmology. We have long known that DeSitter space violates SUSY, but the magnitude of the observed SUSY breaking seems to be too large to be explained by the tiny cosmological constant of the universe.

A classical low energy SUGRA Lagrangian can exhibit spontaneous breaking of SUSY. If we eschew fine tuning, the relation between the SUSY breaking scale and the cosmological constant is

$$M_{SUSY} \sim \Lambda^{1/4}. \tag{5}$$

However, quantum corrections to this are generically large, and within the quantum field theory approximation, we find

$$M_{SUSY} \sim \Lambda^{1/2}. \tag{6}$$

Neither of these relations is compatible with the experimental lower bound on M_{SUSY}, given the experimental upper bound on, or perhaps observation of, the cosmological constant. The drastic difference between these formulae does point up the ultraviolet sensitivity of the relation between these two parameters.

Conventional, Wilsonian, renormalization theory would suggest that physics at energies higher than those described by field theory, cannot modify these conclusions for the better. In field theory, or any conventional regularization of it, high energies correspond to short distances. Short wavelength excitations are not sensitive to the value of the cosmological constant as long as it is much smaller than the energy scale in question. However, we have argued above that the dominant states in M-theory at energies above the Planck scale are large black holes. The properties of these states, at the very highest energy, are sensitive to Λ and a finite value of Λ acts as a UV cutoff on their spectrum.

Thus, quantum field theoretic calculations up to the Planck scale indicate an ultraviolet sensitivity of the calculation of M_{SUSY} for fixed cosmological constant, and the nature of the high energy spectrum above the Planck scale is such as to suggest that their contribution to the mass renormalization may depend strongly on the cosmological constant. We have only to convince ourselves that this "virtual black hole" contribution to the mass renormalization is sizable.

It is often stated that string theory is cutoff at high energies. The evidence for this comes from calculations of exclusive processes involving a fixed number of particles at high energies and momentum transfers. However, expectations from classical general relativity tell us that this is not the full story. Indeed, most relativists believe that a collision of two particles with center of mass energy much greater that the Planck scale, and impact parameter less than the Schwarzchild radius corresponding to that energy, will result, with probability one in the formation of a black hole.[13] If the black hole mass is large, the gross properties of the final state can be understand in terms of Hawking's calculation of black hole evaporation. That is the final state is a typical member of a thermal ensemble with very low temperature. The average energy of final state particles decreases with the center of mass energy and their average number increases. Thus, although it is true that exclusive processes with a small number of hard final state particles are highly suppressed, there is no sense in which the S-matrix becomes 1 at high energies.

It is of course an unjustified leap of faith to conclude from this that the virtual black hole contributions to mass renormalization of light particles is large. Our only guide to a general relation between real and virtual processes is quantum field theory, and it is surely inapplicable to the problem at hand. One reason to believe that the renormalizations are large, and indeed explain the correct scale of SUSY breaking, is the independent conjecture of the last section. It implies that the old paradigm of SUSY breaking in flat space will not work, and there seems little choice but to assume that cosmological effects are responsible for SUSY breaking.

The net result of these conjectures can be summarized as follows. The cosmological constant Λ is (in Planck units) a dimensionless parameter of M-theory which determines the number of physical states in the Hilbert space. As Λ goes to zero we have a sort of critical phenomenon in which the number of physical states goes to infinity. Even if we try to control this spectrum by grading it by the energy, the high energy density of states in the limiting theory blows up faster than in any known quantum theory. The example of the gentler Hagedorn spectrum in string theory shows us that SUSY may be required to guarantee stability in such a situation. The SUSY breaking scale will then go to zero with Λ and as in all critical phenomena, we may expect power law scaling $M_{SUSY} \sim \Lambda^{\alpha}$. The classical value of the critical exponent α is $1/2$. Field theoretic calculation shows that this classical approximation breaks down badly, but the field theory calculation itself breaks down at energies of order the Planck scale. We have no solid argument that the contribution of energies far above the Planck scale is small. If the correct value of the critical exponent turns out to be $\alpha = 1/8$ then, plugging in the (apparent) observed value of the cosmological constant we get a SUSY breaking scale in the TeV range.

The story is an attractive one, and it is really only the last of the conjectures that seems hard to swallow. In the next section we will present some calculations in a toy model. It is a noncommutative field theory which exhibits a UV/IR coupling vaguely reminiscent of that in quantum gravity. The supersymmetric version of the model is very well behaved. However, when we introduce what would have been a small and rather innocuous breaking of SUSY via the infrared boundary conditions, the quantum corrections to SUSY breaking observables are infinite in perturbation theory. I will also present a preliminary report on the analysis of a large N version of the model, which appears to resolve the problem by having a nontranslation invariant ground state.

4. Scherk-Schwarz SUSY Breaking in Noncommutative Field Theory

In quantum field theory, SUSY breaking by twisted boundary conditions is a mild, purely infrared breaking of the symmetry. In the UV, SUSY is restored, and non-renormalization theorems are respected. If the boundary conditions are imposed on a circle of radius R, SUSY breaking order parameters are finite and vanish like inverse powers of R when R goes to infinity. Consider for example the four dimensional Wess-Zumino models, the most general perturbatively renormalizable theories containing only four dimensional chiral superfields. In perturbation theory, these models have only logarithmic ultraviolet divergences and obey strong non-renormalization theorems. All divergences may be absorbed into a wave function renormalization matrix for the chiral fields. The vacuum energy vanishes identically. When such models are compactified on a circle of radius R with Scherk-Schwarz boundary conditions, the UV renormalizations are independent of R and of the boundary conditions. The vacuum energy is finite and of order $1/R^4$ as $R \to \infty$. The three dimensional masses of the Kaluza-Klein modes of the chiral fields have SUSY splittings of order $1/R$. Nonperturbatively, the models are UV trivial, because the coupling constants are renormalized only by wave function renormalization. However, if we use a SUSY cutoff, this problem only shows up at momenta of order $e^{-c/\lambda}$ where λ is the coefficient of a typical cubic term in the superpotential. Thus, if all the couplings are reasonably small, the UV triviality has little effect on low energy physics.

The noncommutative version of these models has been studied by a number of authors.[14] They retain the good behavior of their commutative partners to all orders in perturbation theory. In particular, the conversion of UV divergences into power law infrared singularities that was observed by the authors of Ref. 15 in generic noncommutative field theories, is absent in the perturbation series of the NCWZ models. It is easy to verify that these statements are not affected by compactification of the theory on a circle in a SUSY preserving manner.

In Ref. 2 we studied the perturbation series of the simplest NCWZ model in the presence of Scherk-Schwarz (SS) SUSY breaking. This is a model with a single chiral superfield and superpotential $W = m\Phi^2 + g\Phi^3$. Noncommutativity is introduced in the 23 plane, and the 3 direction is compactified by restricting p_3 to the values n/R for bosons and $(n + \frac{1}{2})/R$ for fermions. We found that at two loop order for the vacuum energy the nonplanar contributions are infinite. Similarly, one loop nonplanar contributions to the self energies of the modes (considered as three dimensional fields) are infinite. A similar phenomenon is observed when noncommutative $\mathcal{N} = 4$ Super Yang-Mills theory is SS compactified.

The mechanism which produces the divergences is easy to understand. Individual component graphs in the theory have quadratic UV divergences, which are converted into IR divergences in the presence of the noncommutativity parameter B (we choose this nonstandard symbol to avoid confusion with the supercoordinates).

If the model is exactly supersymmetric these IR divergences cancel between bosonic and fermionic graphs. But in the presence of SS SUSY breaking, fermions have no zero momentum modes along the SS circle, while bosons do. The divergences no longer cancel. We note that the resummation procedure of Ref. 15 does not remove these divergences. This is a consequence of the discreteness of toroidal momentum space. In a continuum SUSY violating noncommutative field theory the MSVR resummation produces an integrable singularity in the two loop expression for the vacuum energy. Here the singularity occurs at a discrete point, and gives an infinite contribution.

To obtain a systematic nonperturbative handle on these divergences, we are performing a study[3] of a large N vector version of NCWZ, with superpotential $W = \mu\Sigma^2 + (m + g\Sigma)\Phi^i\Phi^i$. The collective fields, which are classical in the large N limit, are the bosonic components of Σ. It turns out that for constant collective fields the gap equations are independent of the parameter B. The stationary solution is $\sigma = F_\sigma = 0$, up to small corrections of order $1/R$ coming from SS symmetry breaking.

The real action begins when one studies small fluctuations around these vacua. Straightforward large N perturbation theory will reproduce the divergences found in ordinary perturbation theory. However, although our analysis is not yet complete, the momentum dependence of the propagators of the collective fields seems to indicate an instability of the constant field background. One can lower the energy by looking at collective fields with nonzero momentum along the noncommutative directions. We believe that this indicates that the true vacuum of the model spontaneously breaks translation invariance. The prevalence of such striped vacua in noncommutative field theories was pointed out by Gubser and Sondhi.[16] Our proposed state differs from those studied by these authors because it violates translation invariance without breaking the internal $O(N)$ symmetry.

Rigorous investigation of these large N, nontranslation invariant vacua would require numerical work, because it is impossible to evaluate the collective field action analytically if the fields are not constant. However, we believe that we have been able to estimate the order of magnitude of the resulting SUSY breaking. It is larger by powers of R than that in the commutative version of the theory. These results will be reported in Ref. 3. We also believe that the large N expansion around the striped vacuum will not contain IR divergences because the internal lines all carry nonzero momentum in the circle direction.

5. Conclusions

The three conjectures that make up the core of this talk are interdependent, but may be studied independently. The claim that only a finite number of physical states are necessary to the description of AsDS spaces is a straightforward extrapolation of current ideas about the holographic principle and black holes. It is difficult for me to imagine how the latter ideas could be correct, but not the former. Further investigation of this claim would certainly be fruitful.

The claim that SUSY violating vacua with asymptotically flat spacetime and constant value of G_{NEWTON} at infinity do not exist is supported by myriad examples. The notion that the problem with such vacua is the rapidly increasing spectral density of black hole states is new, but fits with the old result of Kutasov and Seiberg[10] about the necessity of SUSY for the stability of perturbative string vacua. Note that the examples of controllable SUSY breaking in string theory all have G_N vanishing at infinity, either in space or time. This dramatically changes the spectrum of stable black holes in the theory. In linear dilaton vacua, the black hole spectrum is Hagedorn, while in vacua where there is a potential that pushes the dilaton to weak coupling, the stable state in the asymptotic past and future has no black hole excitations at all. It is also interesting to note that the most plausible candidates for SUSY violating asymptotically flat vacua with constant dilaton[17] have exact boson-fermion degeneracy for an infinite spectrum of states. While this is perhaps a reason to suspect that these states of string theory do not behave as they are advertised to (how is the degeneracy preserved without a symmetry?) it shows that if these models really exist, they are likely to exhibit cancellations similar to those in exactly supersymmetric models.

The most likely avenue for progress in proving this conjecture is to find a non-perturbative version of the Kutasov-Seiberg argument. The first step in such an argument would be to exhibit an analog of the torus partition function, and the modular transformation which links the high energy density of states to the stability of the ground state. The search for counterexamples to this conjecture should be pursued with vigor as well.

If one accepts the validity of the second conjecture, then something like the third appears to be the only alternative to a declaration that M-theory does not describe the world. This is the psychological reason that we have resisted the evidence for the second conjecture for so long. Of course, desperation is not an argument for the truth of the conjecture. The toy models provided by noncommutative field theory show us that our expectations for the size of SUSY breaking induced by global changes in the large scale structure of spacetime are colored by too long an acquaintance with local quantum field theory. Despite the existence of explicit proposals for holographic models of spacetime[5,6] we still have little understanding of how field theory arises as a limiting description, and thus of the ways in which the field theory rules break down.

While the toy models provide evidence that the third conjecture is not completely crazy, they do not provide a methodology for calculating the critical exponent relating SUSY breaking to the cosmological constant. It seems to me that the most promising way in to this question is via the study of the second conjecture. By understanding why SUSY is necessary to control the black hole spectrum in an asymptotically flat spacetime, we are likely to be able to discover how SUSY is broken when the asymptotically flat geometry is deformed.

Acknowledgments

This work was supported in part by the Department of Energy, grant DOE DE-FG02-96ER40559. The author thanks L.Susskind, R. Bousso, E. Martinec, and especially E.Witten, for challenging and informative discussions. Special thanks go to W.Fischler for collaboration and discussion throughout the course of the work reported here.

References

1. T. Banks, *Cosmological breaking of supersymmetry?*, hep-th/0007146.
2. T. Banks, W. Fischler, *Scherk-Schwarz SUSY breaking in noncommutative field theory*, hep-th/0007186.
3. T. Banks, M. Dine, W. Fischler, D. Pahel, *Scherk-Schwarz SUSY breaking in a large N noncommutative field theory*, Manuscript in preparation.
4. S.W. Hawking, J. Maldacena, A. Strominger, *DeSitter entropy, quantum entanglement and AdS/CFT*, hep-th/0002145; J. Maldacena, A. Strominger, *Statistical entropy of DeSitter space*, *JHEP* **9802**, 014 (1998) [gr-qc/9801096].
5. T. Banks, W. Fischler, S. Shenker, L. Susskind, *M-theory as a matrix model: a conjecture*, *Phys. Rev.* **D55**, 5112 (1997) [hep-th/9610043].
6. J. Maldacena, *The large N limit of superconformal theories and supergravity*, *Adv. Theor. Math. Phys.* **2**, 231 (1998) [hep-th/9711200]; S.S. Gubser, I.R. Klebanov, A.M. Polyakov, *Gauge theory correlators from noncritical string theory*, *Phys. Lett.* **B428**, 105 (1998) [hep-th/9802109]; E. Witten *Anti-de Sitter space and holography*, *Adv. Theor. Math. Phys.* **2**, 253 (1998) [hep-th/9802150].
7. R. Figari, R. Hoegh-Krohn and C. R. Nappi, *Interacting relativistic boson fields in the De Sitter universe with two space-time dimensions*, *Comm. Math. Phys.* **44**, 265 (1975).
8. W. Fischler and L. Susskind, *Holography and cosmology*, hep-th/9806039; R. Bousso, *A covariant entropy conjecture*, *JHEP* **9907**, 004 (1999) [hep-th/9905177]; *Holography in general space-times*, *JHEP* **9906**, 028 (1999) [hep-th/9906022]; *The holographic principle for general backgrounds*, *Class. Quant. Grav.* **17**, 997 (2000) [hep-th/9911002]; E.E. Flanagan, D. Marolf and R.M. Wald, *Proof of classical versions of the Bousso entropy bound and of the generalized second law*, *Phys. Rev.* **D62**, 084035 (2000) [hep-th/9908070].
9. O. Aharony, T. Banks, *Note on the quantum mechanics of M-theory*, *JHEP* **9903**, 016 (1999) [hep-th/9812237].
10. D. Kutasov, N. Seiberg, *Number of degrees of freedom, density of states, and tachyons in string theory and CFT*, *Nucl. Phys.* **B358**, 600 (1991).
11. M. Fabinger, P. Hořava, *Casimir effect between world branes in heterotic M-theory*, *Nucl. Phys.* **B580**, 243 (2000) [hep-th/0002073].
12. *A nonsupersymmetric matrix orbifold*, *JHEP* **0003**, 027 (2000) [hep-th/9910164].
13. H. Matschull, *Black hole creation in 2+1-dimensions*, *Class. Quant. Grav.* **16**, 1069 (1999) [gr-qc/9809087]; D. Amati, M. Ciafaloni and G. Veneziano, *Superstring collisions at planckian energies*, *Phys. Lett.* **B197**, 81 (1987); *Can Space-Time Be Probed Below The String Size?*, *Phys. Lett.* **B216**, 41 (1989); *Planckian scattering beyond the semiclassical approximation*, *Phys. Lett.* **B289**, 87 (1992); *Effective action and all order gravitational eikonal at Planckian energies*, *Nucl. Phys.* **B403**, 707 (1993); T. Banks, W. Fischler, *A model for high energy scattering in quantum gravity*, hep-th/9906038.
14. H.O. Girotti, M. Gomes, V.O. Rivelles and A.J. da Silva, *A consistent noncommutative*

field theory: The Wess-Zumino model, *Nucl. Phys.* **B587**, 299 (2000) [hep-th/0005272]; A.A. Bichl, J.M. Grimstrup, H. Grosse, L. Popp, M. Schweda and R. Wulkenhaar, *The superfield formalism applied to the noncommutative Wess-Zumino model*, *JHEP* **0010**, 046 (2000) [hep-th/0007050]; C. Chu and F. Zamora, *Manifest supersymmetry in non-commutative geometry*, *JHEP* **0002**, 022 (2000) [hep-th/9912153].

15. S. Minwalla, M. Van Raamsdonk and N. Seiberg, *Noncommutative perturbative dynamics*, hep-th/9912072.
16. S.S. Gubser, S. Sondhi, *Phase structure of noncommutative scalar field theories*, hep-th/0006119.
17. S. Kachru, J. Kumar, E. Silverstein, *Vacuum energy cancellation in a nonsupersymmetric string*, *Phys. Rev.* **D50**, 104004 (1999) [hep-th/9807076].

1+1 DIMENSIONAL NCOS AND ITS $U(N)$ GAUGE THEORY DUAL

IGOR R. KLEBANOV
Joseph Henry Laboratories, Princeton University
Princeton, NJ 08544, USA

JUAN MALDACENA
Lyman Laboratories, Harvard University
Cambridge, MA 02138, USA

We study some aspects of open string theories on D-branes with critical electric fields. We show that the massless open string modes that move in the direction of the electric field decouple. In the 1+1 dimensional case the dual theory is $U(N)$ SYM with electric flux, and the decoupling of massless open strings is dual to the decoupling of the $U(1)$ degrees of freedom. We also show that, if the direction along the electric field is compact, then there are finite energy winding closed string modes. They are dual to Higgs branch excitations of the SYM theory, and their energetics works accordingly. These properties provide new non-trivial evidence for the duality.

1. Introduction

We study some aspects of non-commutative open string theories. These are open string theories on branes in the presence of a critical electric field. More precisely, they are defined through a scaling limit where the electric field is taken to its critical value[1,2] (for earlier work, see Ref. 3, and for other recent work Refs. 4,5,6,7,8). The resulting open string theory is the same as the usual open string theory but with a new product which includes Moyal phases in the directions of the electric field, which we take to be the 01 directions. The Moyal phase is determined in terms of the string scale α'_e.

This open string field theory can be defined in various dimensionalities. We concentrate our discussion on the 1+1 dimensional case. In this case the theory on a D1 brane with N units of electric flux is related to a $U(N)$ SYM theory in a sector with one unit of electric flux.[9,3,10] N is related to the open string coupling through the formula $G_0^2 = \frac{1}{N}$. Thus, G_0^2 cannot exceed 1 and at this maximum value the NCOS actually becomes free.[11] To achieve weak string coupling, we need to take N to be large. In this way, the $1 + 1$ dimensional NCOS provides a new connection between large N gauge theory and weakly coupled strings. This duality has very interesting predictions for the open string field theory.

- The gauge theory dual to the NCOS has gauge group $U(N) = U(1) \times SU(N)/Z_N$ theory. This implies that we have a *free* $U(1)$ theory. We stress that it is free at all energies, not just low energies. The $SU(N)/Z_N$ theory with one unit of

Z_N flux has a mass gap.[12] We show in this paper that the massless open strings in 1+1 dimensional NCOS decouple and give the free $U(1)$ predicted by the duality. All other open string modes are massive: they describe excitations in the $SU(N)/Z_N$ theory.

- The field theory has a Higgs branch which has a positive energy density. This implies that if we compactify the field theory on a circle then we should observe a continuum above a certain calculable threshold corresponding to breaking $U(N) \to U(N-1) \times U(1)$ by a Higgs vev. The low energy excitations should now reflect the presence of massless $U(1)$ modes. We show that the NCOS theory couples to closed strings that wind around the circle with only one orientation. The spectrum of these closed strings is *precisely* as expected from the duality.
- Let us break $U(N) \to U(N-K) \times U(K)$ by a Higgs vev, keeping the electric flux in the $U(N-K)$ part. We expect that the low energy dynamics of $U(K)$, for a large Higgs vev, should be given by the Matrix strings,[13,14,15] i.e. by an orbifold $Sym(R^8)^K$ perturbed by a dimension 3 operator which depends on the Yang Mills coupling but not on N or K. We check that the low energy interactions of the closed string modes that couple to NCOS theory have precisely the correct coupling.
- We could consider the gauge theory with M units of electric flux. This corresponds to having M D-branes in NCOS theory. We check that the energetics are as expected.

After completing these checks of the duality, we make some remarks on the relation of this NCOS theory and the Matrix theory, where the $U(1)$ flux is related to D0 branes. In fact we show that NCOS theory in 1+1 dimensions is related to DLCQ of IIA string theory in the presence of D0 branes. We also make some remarks on the high-energy and finite temperature behaviors of the NCOS theory.

2. The Duality

In this section we summarize some aspects of the duality between $1+1$ dimensional NCOS theory and the $U(N)$ SYM.[10]

Non commutative open string theory (NCOS) is the theory that results from introducing a Moyal phase $\theta = 2\pi\alpha'_e$ in the usual open string field theory with parameters α'_e, G_0^2.[1,2] It is useful to view this theory as arising from a decoupling limit of usual string theories. These decoupling limits always involve $g_s \to \infty$ so that to properly analyze them we should go to S-dual descriptions. From this dual description we find that in the 1+1 dimensional case the decoupling limit reduces to the usual decoupling limit for N D-branes in the presence of a weak electric field[10] which gives a $U(N)$ super Yang Mills theory with one unit of electric flux. Here N is the number of units of the electric flux on the D1 brane in the picture where we define the NCOS limit. The subscript on α'_e is introduced to distinguish it from the

α' of the theory that has been used to define a decoupling limit: $\alpha' = \alpha'_e(1-F^2)$, where $F = F_{01}$ is the electric field on the original D1-brane.

Now let us find the relation between the parameters α'_e, G_0^2 of the NCOS and the parameters N, g_{YM}^2 of the gauge theory. In order to do this it is convenient to go to a parent string theory where the N D-branes live and take then the limit $\alpha' \to 0$. If we analyze this limit in the S-dual theory we find the NCOS limit. In Ref. 10 this limit was discussed. We note here that we are interested in defining a decoupling limit of the theory of a D1 brane in the presence of N units of $U(1)$ flux.

The relation between α'_e and the Yang-Mills coupling can be found following the steps in the duality to be (note that our convention for g_{YM}^2 differs by a factor of 2π from the standard one):

$$g_{YM}^2 = \frac{N^2}{\alpha'_e} \,. \tag{1}$$

The square of the open string coupling becomes[3,10]

$$G_0^2 = \frac{1}{N} \,. \tag{2}$$

More generally we could consider M coincident D-branes in NCOS theory. Now the relation between the electric field on M D1-branes and the number N of flux unit is modified compared to the $M = 1$ case.[16,11] The relation between parameters is found to be

$$g_{YM}^2 = \frac{N^2}{M^2\alpha'_e} \,, \qquad G_0^2 = \frac{M}{N} \,. \tag{3}$$

where the dual theory is $U(N)$ SYM with M units of flux. It is interesting that now G_0^2 may be an arbitrary fraction. Properties of the theory should depend sensitively on the value of the numerator and denominator: they should vary discontinuously as a function of G_0^2.

In the rest of the paper we check some of the predictions of this duality.

3. Free the U(1) !

The $U(1)$ modes of the gauge theory are dual to the massless modes in NCOS: the scalars in the vector representation of $SO(8)$ and their superpartners. One of the striking predictions of the duality is that the massless open string modes should decouple. This prediction is even more striking when we combine it with the statement that all the tree level amplitudes in the NCOS theory are the same as in the usual open string theory except for some Moyal phases. Thus, one might conclude that there must be a contradiction in the above statements since the tree level amplitudes involving massless string states are certainly non-vanishing in the usual string theory. What is important to realize, however, is that the statement about the Moyal phases is a statement about *ordered* correlation functions of vertex operators which correspond to specific Feynman diagrams. The complete amplitudes are given by summing over all orderings and thus the phases introduce the crucial difference.

Consider the simple case of a four point function. The relevant disk diagram decomposes into 6 possible orderings. There are 2 orderings for 3 of the vertex operators which are fixed at $0, 1, \infty$ and the fourth can be integrated over $(-\infty, 0)$, $(0,1)$ or $(1, +\infty)$. These three integrals are nonzero individually but their sum turns out to vanish! We perform these calculations explicitly in Appendix A.

In the rest of this section we will present a general argument for why these amplitudes vanish. This is due to the precise value of the Moyal phase in NCOS and it would not be true away from the $F = F_{cr}$ limit. In this limit the correlation function of world sheet fields at the boundary is given by

$$\langle X^+(t)X^-(0)\rangle = 4\alpha'_e(\log|t| + i\frac{\pi}{2}\text{sign}(t)) \tag{4}$$

where $X^\pm = X^0 \pm X^1$. The crucial observation is that we can think of the right hand side of (4), up to an irrelevant constant, as $\log(t)$, where $\log(t)$ is a holomorphic function of t such that when we go from $t > 0$ to $t < 0$ it produces the phase difference $i\pi$ that we see in (4). There is of course a question of branch cuts and we have to choose them in a definite way. Let us suppose that we are integrating a given massless vertex operator over the boundary of a disk. Since it is massless the on-shell condition implies $p_+p_- = 0$. Let us assume that $p_- = 0$. Then we choose all branch cuts in the log in (4) so that they are outside the disk. Then we can deform the contour of integration for t from the boundary of the disk to the interior and make it disappear, so that the resulting amplitude is zero. Notice that we did not have to assume that other vertex operators were also massless: the only assumption was that at least one is massless and we integrate that one first. Notice that this implies that three point amplitudes are also zero once we add over all orderings since we can take one of the vertex operators and integrate it dividing by the volume of integration, which is finite if we integrate only one. Alternatively we can argue that massless 3-point functions vanish by factorizing the 4-point functions.

For higher loop diagrams it seems that one needs to sum over the insertion of the massless state on all boundaries of the diagram in order to show that the amplitudes vanish. It would be interesting to analyze it more precisely.

Just in case the reader fears that all amplitudes would be zero we compute a four tachyon amplitude in bosonic NCOS theory in Appendix A, and we show that it is non-vanishing. In general, purely massive correlators do not vanish, which agrees with the expectation that $SU(N)/Z_N$ gauge theory has non-trivial dynamics. However, as we show in Appendix A, all massive correlators fall off faster at high energies than in the commutative open string. The $2 \to 2$ scattering cross-section depends on the energy as E^{-4} away from the poles. This is reminiscent of how it works in the gauge theory: $\sigma \sim g_{YM}^4/E^4$.

It should be noted that our argument about the decoupling of massless modes may be extended to $p+1$ dimensional NCOS theory with $p > 1$. The statement is that the $U(1)$ modes with momentum purely in the 1 direction (i.e. the direction of space-time non-commutativity) decouple from the theory. In the $p = 3$ case, analyzed in Ref. 2, the NCOS is dual to a space-space non-commutative Yang-Mills theory with $B_{23} \neq 0$. In general, the $U(1)$ modes do not decouple in a non-commutative theory, but if we have a mode that does not depend on x^2 and x^3, then it does decouple because the $U(1)$ theory in the remaining directions is commutative. Therefore, our argument translates into non-trivial checks of other dualities in Ref. 10 as well.

4. Higgs Branch and Closed Strings

In this section we analyze the second prediction of the duality: the existence of a Higgs branch with the right energetics. It is convenient for this discussion to compactify the Yang-Mills theory, and also the NCOS theory, on a circle. So we compactify the direction 1 on a circle of radius R. The $SU(N)$ with one unit of Z_N flux may be Higgsed to an $SU(N-K) \times U(1)^K$ with one unit of Z_{N-K} flux in the first factor. The energy cost for this can be calculated from the BPS mass formula for (p,q) strings in the gauge theory limit. The relevant formula says that in a $U(N)$ theory with M units of electric flux we have a ground state energy of the form

$$E(M,N) = 2\pi R \frac{g_{YM}^2}{4\pi} \frac{M^2}{N} \ . \tag{5}$$

Notice that this gives the energy above the $U(N)$ Yang Mills ground state with no flux. The energy of the Higgs branch we considered above, for which $M=1$, is

$$E_K = E(1, N-K) = \frac{Rg_{YM}^2}{2} \frac{1}{N-K} \sim \frac{Rg_{YM}^2}{2} \left(\frac{1}{N} + \frac{K}{N^2} + \frac{K^2}{N^3} + \cdots \right) \ . \tag{6}$$

On this Higgs branch we also expect to have a continuum of excitations, corresponding to configurations where the Higgs field has some momentum in the transverse directions. There are also massless excitations propagating along the $U(1)$s. Let us consider first the case of $K=1$: just a single $U(1)$. Then we expect to find an energy spectrum for this $U(1)$ which to leading order in N would be

$$E_1 = \frac{R}{2\alpha'_e} + \frac{1}{R}(N_L + N_R + \alpha'_e k^2/2) \tag{7}$$

where we have used the relation (1). Notice that the particular dependence of (7) on $N_{R,L}$ is just stating that we have massless free fields. There is also a level matching condition that $N_L - N_R$ should be an integer. Notice that in (7) we have used only the second term in the expansion 6. The first term represents the mass of the electric flux line itself:

$$(2\pi R)\frac{N}{4\pi\alpha'_e} = (2\pi R)\frac{1}{4\pi\alpha'_e G_0^2} \ . \tag{8}$$

From the dependence of the tension on G_0^2 we conclude that the flux line is dual to a D-string in the NCOS (this is not surprising because this is precisely the S-duality relation). Note, however, that (8) is a factor of two smaller than the naive expectation for the D1-brane tension in NCOS. The naive expectation is what we get by just stating that the mass of brane is given by the lowest term of the non-commutative analog of the Born-Infeld action.[17,18] On the other hand, if we just stick to the standard string field theory methods and we determine the mass of the brane in the the manner described below then we get the right value. In order to determine the mass of the brane, imagine that we do not take the complete decoupling limit but that we still keep the original α' finite. Then one way of computing the mass of the brane would be to consider the string field ϕ that parametrizes the position of the brane and to write its kinetic term as $S = \int dtC(\dot{\phi})^2$. Then we relate ϕ to the physical displacement r of the brane by considering a configuration with two D-branes and relating the mass of the open string computed in string field theory, which is linear in ϕ, to the mass we expect if we think we have a string stretching between the branes. Finally we write the action as $S = \int dtM\dot{r}^2/2$ and we read off the mass. The only difference in this procedure between the usual case and the case with the Moyal phases is that the mass of the stretched string as a function of r is different. In usual open string theory it is $m = \frac{r}{2\pi\alpha'}$ while in this case it is $m = \frac{r}{2\pi\alpha'_{eff}\sqrt{1-F^2}}$. Computing the mass using this method we get a mass that diverges since it represents the mass of the $(N, 1)$ string in the parent string theory before taking the decoupling limit. If we subtract from this mass the mass of the N D1 branes then we indeed get (8). It would be nice to derive the D-brane mass formula directly in the NCOS theory. We believe that the arguments relating the cubic string field theory action to the Born Infeld action should fail in the NCOS theory. In fact, we saw that the massless states are free, which is incompatible with a Born-Infeld action.

4.1. *Closed strings*

The factor of 2 discussed above is not the only puzzling factor in (6). The tension of the Higgs branch excitation calculated in (7) is $1/(4\pi\alpha'_e)$. Its scaling allows us to identify it with a closed string wound around the compact circle, except for a factor of 2. Thus, the duality predicts that, even though the masses of open strings are given by the usual formula, $m^2 = N/\alpha'_e$ (N is the integer oscillator level), the compactified NCOS theory contains wound closed strings of tension $1/(4\pi\alpha'_e)$, instead of the usual closed strings of tension $1/(2\pi\alpha'_e)$. In the following we confirm this prediction.

The presence of wound closed strings in NCOS is hardly a surprise: once we compactify NCOS then an open string can stretch around the circle and the two ends can join with a finite probability, since the open string coupling constant is finite. Therefore, the NCOS theory can emit wound closed strings into the bulk. Moreover the argument leading to (6) predicts that these strings can only wind with one orientation: the winding number K cannot be negative. We also expect this from the NCOS theory since the energy necessary to stretch the string in one direction is finite while it is infinite in the other direction in the scaling limit of Refs. 1,2.

In order to show directly that these closed strings are produced we sketch the calculation of the non-planar one loop diagram. The compactness of the 1 direction introduces the crucial modification that allows closed string poles, and also explains why the tension is twice smaller than in the commutative case (in Ref. 2 it was shown that there are no closed string poles in the non-compact case).

We consider a one-loop diagram, which is a cylinder. We introduce some vertex operators on one boundary and some on the other boundary so that we have some net momentum p flowing along the cylinder from one boundary to the other. As usual, p^1 is quantized: $p^1 = n/R$. We are interested in poles arising from the large s region, where s is the proper time along the closed string channel. These poles must be associated with on-shell closed strings. In presence of the NS-NS field B_{01}, the on-shell condition for closed strings with n units of momentum and m units of winding can be derived with standard methods (see, for example, Ref. 19):

$$-\alpha'(p_0)^2 - 2p_0 B_{01} mR + \frac{(mR)^2}{\alpha'}(1-B_{01}^2) + \alpha'(n/R)^2 + 2(N_L+N_R) + \alpha'_e k^2 = 0\ , \quad (9)$$

where k is the transverse momentum. These strings do not acquire infinite energy in the scaling limit $\alpha' = \alpha'_e(1 - B_{01}^2) \to 0$ where α'_e is held fixed. Terms quadratic in p_0 and n/R vanish, and we obtain a mass formula of non-relativistic type:

$$p_0 = \frac{mR}{2\alpha'_e} + \frac{\alpha'_e}{2mR}k^2 + \frac{N_L + N_R}{mR} \quad (10)$$

together with the standard level-matching condition $N_L - N_R = mn$. This formula is in *precise* agreement with the prediction of the duality (7). Especially striking is the fact that that the tension of the wound strings is indeed $1/(4\pi\alpha'_e)$. It is crucial that the quadratic dependence on p_0 cancels since it implies that $m > 0$ for positive energy states, so that strings can wind only in one sense. Notice also the appearance of a continuum associated to the momentum k in the transverse directions.

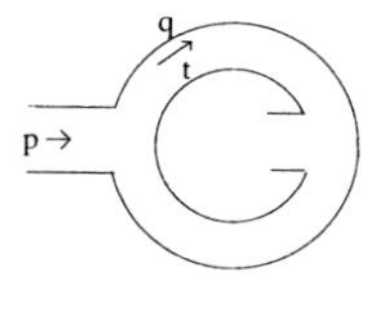

Fig. 1. Nonplanar open string diagram. In open string field theory we would build it from the cubic vertex and we would consider states carrying momentum q and $q + p$ along the loop.

We can also see the existence of the closed string poles directly in the open string interpretation of this diagram, in the spirit of the explanation in Ref. 2. The idea is to think about the non-planar diagram in open string field theory where one would have to integrate over the momentum propagating along the loop (see Figure 1). Then the only difference between the NCOS diagram and the usual diagram is the presence of the phase $e^{2ip\wedge q}$ where $p \wedge q = \pi\alpha'_e(p_0 q_1 - p_1 q_0)$, p is the momentum flowing through the closed string channel and q is the momentum running along the loop. But now, since momenta are quantized we get that $q_1 = n/R$. So we integrate over q_0 and sum over n. The rest of the diagram has the same value as in the usual case. For simplicity we set $p_1 = 0$. The propagators can be exponentiated and we have to compute something of the form

$$\int dq_0 \sum_n e^{i2\pi\alpha'_e p_0 n/R} e^{-2\pi\alpha'_e(-q_0^2+(n/R)^2)t+\beta p_0 q_0} . \tag{11}$$

The integral over q_0 is of the usual form and will produce the usual $\exp(+s\frac{\alpha'_e}{2}p_0^2)$ in the closed string channel (after analytic continuation). The sum over n is the interesting part. By Poisson resummation we will get

$$\sum_m e^{-s\frac{\alpha'_e}{2}|mR/\alpha'_e - p_0|^2} . \tag{12}$$

We now have to multiply this times the result of the integral over q_0 times the rest of the string diagram which for small t or large s has the standard expansion in terms of closed strings coming from modular transforming the small t diagram. Putting it all together we have

$$\sum_m \exp\left\{-s\left(\frac{\alpha'_e}{2}(|mR/\alpha'_e - p_0|^2 - p_0^2 + k^2) + 2N\right)\right\} . \tag{13}$$

We see that the quadratic part in p_0 cancels and then there is a pole at

$$p_0 = \frac{mR}{2\alpha'_e} + \frac{\alpha'_e}{2mR}k^2 + \frac{2N}{mR} \tag{14}$$

where $N = N_L = N_R$ which are equal because we assumed $p_1 = 0$. In the case of non-vanishing p_1 the on-shell condition generalizes to (10).

It is interesting to consider the interaction of these closed strings. From the gauge theory we expect that their low energy interactions are described by a orbifold theory $Sym(R^8)^K$ deformed by a dimension 3 operator with a coefficient of the order of $\frac{1}{g_{YM}}$.[15] From the string theory description we expect the same but with a coefficient $G_0^2\sqrt{\alpha'_e}$. Fortunately, using (1), (2) we find that the two coincide.

Now let us turn to the interpretation of the third term in (6). The NCOS action will contain a term of the form

$$S = \frac{R}{\alpha'_e G_0^2}\frac{1}{2}(2\pi\alpha'_e)^2 \dot{A}_1^2 . \tag{15}$$

For configurations where $A_1 \sim A_1 + 1/R$ depends only on time, the Moyal phases do not appear. As in the usual discussion, if we consider a configuration with K units of momentum conjugate to A_1, then the energy, expanded to quadratic order, will be

$$E = \frac{R}{\alpha'_e}\frac{G_0^2 K^2}{2} . \tag{16}$$

We see from (1), (2) that we reproduce the third term of (6). Since the total winding number for fundamental strings is being kept fixed we find that this A_1 field will have to carry momentum precisely equal to the number of strings that have been emitted into the bulk.

4.2. *Finite temperature*

Instead of compactifying the x^1 direction we could consider the Euclidean theory and compactify x^0 on a circle of radius R. This describes NCOS at temperature $1/(2\pi R)$. Considering closed strings wound around the Euclidean time direction, we find the dispersion relation

$$\alpha'(p_1)^2 + 2ip_1 B_{01} mR + \frac{(mR)^2}{\alpha'}(1 - B_{01}^2) + \alpha'(n/R)^2 + 2(N_L + N_R - 1) + \alpha'_e k^2 = 0\,. \tag{17}$$

Notice the tachyonic shift now present in the oscillator energies.[20] In the scaling limit we have

$$-ip_1 = \frac{mR}{2\alpha'_e} + \frac{\alpha'_e}{2mR}k^2 + \frac{N_L + N_R - 1}{mR}\,. \tag{18}$$

As the temperature is increased, a massless closed string state appears at $T = 1/(\pi\sqrt{8\alpha'_e})$. This is precisely the Hagedorn temperature expected from studying the open string density of states. It would be interesting to study the connection between the closed strings and the NCOS Hagedorn transition in more detail.

4.3. *Coincident D-branes*

Here we consider the case of M coincident D-branes in NCOS theory. The $U(1)$ part of the massless degrees of freedom is free as before. On the other hand we have other nearly massless excitations. These excitations might not be truly massless once we include the effects of the interactions. This should be dual to $U(N)$ Yang-Mills theory with M units of flux. In the Yang Mills theory we expect the energy of this configuration to be proportional to

$$Rg_{YM}^2 \frac{M^2}{2N}\,. \tag{19}$$

Using (3) we find that this agrees with the expected mass of M D-branes in NCOS theory, up to a factor of two mentioned earlier.

A simple observation is that the energy cost to separate them in NCOS theory is zero to leading order in $M/N = G_0^2$. If N and M are coprime we expect that this marginal direction should be completely lifted. In this case we expect from the field theory that the energy cost to separate them is of order $1/N^3 \sim G_0^2/\alpha'_e$ (up to factors of order M). This is an effect that should arise at one loop. Naively the one loop diagram in NCOS is the same as in the usual string theory, so this effect would vanish. What should probably happen is that the mechanism that ensures that G_0^2 is quantized will in this case produce a F flux on the branes that would give rise to the energy difference. It would be interesting to understand the physics quantizing the coupling.

5. Conclusions

In this paper we have carried out some consistency checks on the duality conjectures involving non-commutative open string theory (NCOS). We have mainly focused on the 1+1 dimensional case which is the only one where the non-commutativity is consistent with the Lorentz invariance (although the parity invariance is obviously broken). In this case the NCOS is obtained in the limit of critical electric field along a D-string, and it is believed to be dual to the maximally supersymmetric 1+1 dimensional $U(N)$ gauge theory with a single flux quantum. Thus, 1+1 dimensional NCOS is a new example of gauge field/string duality.[21,22]

In the SYM theory the $SU(N)/Z_N$ dynamics has a mass gap while the $U(1)$ part decouples at all energy scales. Perhaps our most striking result is that the dual NCOS respects this decoupling through the vanishing of all amplitudes involving massless particles.

We have also considered the NCOS with a compact spatial dimension along the electric field, and showed that in this case open strings couple to winding modes of the closed string sector. The mass formula for such closed strings is in accord with the energetics on the Higgs branch of the gauge theory: the bulk closed strings are dual to individual D-strings which have become liberated from the $(1, N)$ bound state.

After some number of such closed strings have been liberated, their assembly gives rise to a Matrix string theory set-up.[13,14,15] Indeed, 1+1 dimensional maximally supersymmetric SYM theory appeared in the Matrix description of type IIA string theory.[13,14,15] It was observed there that one unit of flux in the Yang Mills theory corresponded to a D0 brane of type IIA in the infinite momentum frame. The large N simplifications that we are seeing both in the open string theory and closed string theory are the simplifications that happen for the DLCQ limit of string theory. In particular we are analyzing a D0 brane moving at almost the speed of light. Indeed after T-dualizing along 1 direction, critical electric field becomes the velocity equal to the speed of light.[16,1]

It is also interesting to notice an analogy between NCOS and the SL(2,R) WZW model. In the SL(2,R) WZW model, one has short strings and long strings.[23] The short strings are analogous to the open strings of NCOS and the long strings, which can wind only in one direction, are analogous to the closed strings in NCOS. In both cases the coupling constant is quantized, at least in the supersymmetric context. This is not a coincidence since both theories are describing strings in critical electric fields or H fields.

It would be interesting to consider unstable branes in the presence of electric fields. There is an important difference between magnetic field and electric fields on brane-antibrane pair. In the case of magnetic fields the induced $D(p-2)$ charge in the anti-Dp brane is the opposite to that of the Dp brane. So the statement that the tachyon condensation is the same in the open string theory with Moyal phases as defined in Ref. 24 as in the usual string theory is just a manifestation of Sen's argument[25] that brane antibrane anihillation is universal. In the case that we have a constant electric field on the brane the system has a net fundamental string charge. This implies that after tachyon condensation we should be left with closed strings.[25] It is interesting to notice that we can define the NCOS limit also for unstable branes like Dp-antiDp branes in string theory, or D-branes in bosonic string theory. This will give us an NCOS string theory which has a tachyon. Tachyon condensation in NCOS theory is the same as in the usual string theory since the tachyon condensate is constant. In particular we could consider the case of a D1-anti-D1 pair in the NCOS limit. This theory is dual to a $U(N)$ theory broken to $U(N/2)\times U(N/2)$ with opposite $U(1)$ fluxes in each factor. Then we have a relation between parameters as in (3) with $M = 2$. If we naively compute the action at the minimum it seems to be off by a factor of two from the expected answer. This factor of two is related to the factor of two present before in the mass of the D-brane. It would be nice to understand this more precisely.

Acknowledgments

We are grateful to C. Callan, A. Hashimoto, C. Herzog, N. Itzhaki, S. Minwalla, A. Polyakov, A. Strominger, L. Susskind and N. Toumbas for useful discussions. The research of I.R.K. was supported in part by the NSF grant PHY-9802484, and the James S. McDonnell Foundation grant No. 91-48. The research of J.M. was supported in part by DOE grant DE-FGO2-91ER40654, NSF grant PHY-9513835, the Sloan Foundation and the David and Lucile Packard Foundation.

References

1. N. Seiberg, L. Susskind and N. Toumbas, *Strings in background electric field, space/time non-commutativity and a new non-critical string theory*, *JHEP* **0006**, 021 (2000) [hep-th/0005040].
2. R. Gopakumar, J. Maldacena, S. Minwalla and A. Strominger, *S-duality and noncommutative gauge theory*, *JHEP* **0006**, 036 (2000) [hep-th/0005048].
3. S. Gukov, I.R. Klebanov and A.M. Polyakov, *Phys. Lett.* **B423**, 64 (1998) [hep-th/9711112].
4. N. Seiberg, L. Susskind and N. Toumbas, *Space/time noncommutativity and causality*, *JHEP* **0006**, 044 (2000) [hep-th/0005015].
5. O. Ganor, G. Rajesh and S. Sethi, *Duality and noncommutative gauge theory*, hep-th/0005046.
6. J. Barbon and E. Rabinovici, *Stringy fuzziness as the custodian of time-space noncommutativity*, *Phys. Lett.* **B486**, 202 (2000) [hep-th/0005073].
7. J. Gomis and T. Mehen, *Space-time noncommutative field theories and unitarity*, *Nucl. Phys.* **B591**, 265 (2000) [hep-th/0005129].
8. T. Harmark, *Supergravity and space-time non-commutative open string theory*, *JHEP* **0007**, 043 (2000) [hep-th/0006023].
9. H. Verlinde, *Matrix String Interpretation of the Large N Loop Equation*, hep-th/9705029.

10. R. Gopakumar, S. Minwalla, N. Seiberg and A. Strominger, *(OM) Theory in Diverse Dimensions*, *JHEP* **0008**, 008 (2000) [hep-th/0006062].
11. I.R. Klebanov, Talk delivered at Lennyfest, May 2000.
12. E. Witten, *Bound States Of Strings And p-Branes*, *Nucl. Phys.* **B460**, 335 (1996) [hep-th/9510135].
13. L. Motl, *Proposals on nonperturbative superstring interactions*, hep-th/9701025.
14. T. Banks and N. Seiberg, *Strings from matrices*, *Nucl. Phys.* **B497**, 41 (1997) [hep-th/9702187].
15. R. Dijkgraaf, E. Verlinde and H. Verlinde, *Matrix String Theory*, *Nucl. Phys.* **B500**, 43 (1997).
16. C.G. Callan, I.R. Klebanov, *D-Brane Boundary State Dynamics*, *Nucl. Phys* **B465**, 473 (1996) [hep-th/9511173].
17. E.S. Fradkin and A.A. Tseytlin, *Nonlinear Electrodynamics From Quantized Strings*, *Phys. Lett.* **B163**, 123 (1985).
18. A. Abouelsaoud, C. Callan, C. Nappi and S. Yost, *Open Strings In Background Gauge Fields*, *Nucl. Phys.* **B280**, 135 (1987).
19. J. Polchinski, *String Theory*, vol. 1, sec. 8.4.
20. J.J. Atick and E. Witten, *The Hagedorn transition and the number of degrees of freedom of string theory*, *Nucl. Phys.* **B310**, 291 (1988); I. Kogan, *Vortices on the world sheet and string's critical dynamics*, *JETP Lett.* **45**, 709 (1987); B. Sathiapalan, *Vortices on the string world sheet and constraints on toral compactification*, *Phys. Rev.* **D35**, 3277 (1987).
21. G. 't Hooft, *A planar diagram theory for strong interactions*, *Nucl. Phys.* **B72**, 461 (1974).
22. A.M. Polyakov, *String Theory and Quark Confinement*, *Nucl. Phys. Proc. Suppl.*, **68**, 1 (1998) [hep-th/9711002].
23. J. Maldacena and H. Ooguri, *Strings in AdS(3) and SL(2,R) WZW model. I*, hep-th/0001053.
24. N. Seiberg and E. Witten, *String theory and noncommutative geometry*, hep-th/9912072.
25. A. Sen, *Universality of the tachyon potential*, *JHEP* **9912**, 027 (1999) [hep-th/9911116].

Appendix A

In this appendix we compute some scattering amplitudes on D1-branes in the presence of electric fields, following Ref. 3. We compute them for a general noncommutativity parameter θ and then we set $\theta = 2\pi\alpha'_e$ which corresponds to the NCOS limit.

The simplest nontrivial scattering calculation is the four-point amplitude for massless NS open strings:[3]

$$A_4(\xi_1, p_1; \xi_2, p_2; \xi_3, p_3; \xi_4, p_4) \sim \int \{d\sigma_1 d\sigma_2 d\sigma_3 d\sigma_4\}$$
$$\langle \xi_1 \cdot V_0(p_1, \sigma_1)\, \xi_2 \cdot V_0(p_2, \sigma_2)\, \xi_3 \cdot V_{-1}(p_3, \sigma_3)\, \xi_4 \cdot V_{-1}(p_4, \sigma_4)\rangle$$

The subindex indicates the picture. The vertex operators for scalar particles (the massless transverse modes of the string) have the form:

$$V^j_{-1}(p_1, z) = e^{-\phi(z)}\, \psi^j(z)\, e^{ip_1 \cdot X(z)} \tag{A.1}$$

$$V_0^j(p_2, z) = \left(\partial X^j(z) + ip_2 \cdot \psi(z)\psi^j(z)\right) \, e^{ip_2 \cdot X(z)}$$

where $j = 2, \ldots, 9$, while the momenta p^α are longitudinal, $\alpha = 0, 1$. z must lie on the real axis.

The Green function on the boundary:

$$\langle X^\alpha(\sigma_1) X^\beta(\sigma_2) \rangle = -2\alpha'_e \eta^{\alpha\beta} \ln |\sigma_1 - \sigma_2| + i\frac{\theta}{2}\varepsilon^{\alpha\beta} \text{sign}(\sigma_1 - \sigma_2) \,. \tag{A.2}$$

The second term gives rise to phases that depend on the ordering of the operators.

The $1 + 1$ dimensional massless kinematics is very restrictive. In the center of mass frame the momenta are

$$p_1^\alpha = \begin{pmatrix} p \\ p \end{pmatrix} , \quad p_2^\alpha = \begin{pmatrix} p \\ -p \end{pmatrix} , \quad p_3^\alpha = \begin{pmatrix} -p \\ -p \end{pmatrix} , \quad p_4^\alpha = \begin{pmatrix} -p \\ p \end{pmatrix} .$$

Let us position V_2 and V_4 at 0 and 1, V_3 at ∞, and integrate over the position of V_1 from 0 to 1. This gives the s-channel contribution to the amplitude that was calculated in:[3]

$$A^s = -\frac{G_0^2}{\alpha'_e} \delta^2(\sum_i p_i) \pi s \frac{\cos(s\frac{\theta}{2\alpha'_e})}{\sin(\pi s)} \xi_1 \cdot \xi_3 \xi_2 \cdot \xi_4 \tag{A.3}$$

with s given by (this corrects a numerical factor in (18) of Ref. 3)

$$s = -t = 4p^2 \alpha'_e \tag{A.4}$$

So far we have examined the s-channel amplitude, but it turns out that the other channels give comparable contributions. They are not affected by the phases, and their form is

$$A^{t,u} \sim s^2 \xi_1 \cdot \xi_3 \xi_2 \cdot \xi_4 \Gamma(-u) \left[\frac{\Gamma(-s)}{\Gamma(1+t)} + \frac{\Gamma(-t)}{\Gamma(1+s)} \right] \tag{A.5}$$

In computing this amplitude we imagine that we artificially add some momentum in the transverse directions as if we were on a higher dimensional brane, so that $u \neq 0$ and then we take the $u \to 0$ limit. The poles in u cancel and we obtain

$$A^{t,u} = \frac{G_0^2}{\alpha'_e} \delta^2(\sum_i p_i) \pi s \frac{\cos(\pi s)}{\sin(\pi s)} \xi_1 \cdot \xi_3 \xi_2 \cdot \xi_4 \tag{A.6}$$

Adding this to the s-channel, we have the complete 4-point amplitude

$$A^4 = \frac{G_0^2}{\alpha'_e} \delta^2(\sum_i p_i) \pi s \frac{\cos(\pi s) - \cos(s\frac{\theta}{2\alpha'_e})}{\sin(\pi s)} \xi_1 \cdot \xi_3 \xi_2 \cdot \xi_4 \tag{A.7}$$

For $\theta = 0$ this has the correct pole structure, with the t and u channels canceling half of the poles in the s-channel amplitude: they cancel the poles with the residue of the unphysical sign. Remarkably, for $\theta = 2\pi\alpha'_e$, the amplitude vanishes identically!

We have also calculated some bosonic string 4-point functions involving the massless scalar particles. The amplitude for 4 massless scalars turns out to have exactly the same form as (A.7). For the forward scattering of a scalar off a tachyon, we find that the complete amplitude is

$$\frac{G_0^2}{\alpha'_e}\delta^2(\sum_i p_i)\pi(s+1)\frac{\cos[\pi(s+1)]-\cos[(s+1)\frac{\theta}{2\alpha'_e}]}{\sin[\pi(s+1)]}\xi_1\cdot\xi_2 \ , \tag{A.8}$$

which again vanishes in the NCOS limit, $\theta = 2\pi\alpha'_e$.

If all operators are massive, then the amplitude no longer vanishes in general. To check this we have calculated the complete 4-tachyon amplitude on the bosonic D1-brane:

$$A = -\frac{G_0^2}{\alpha'_e}\delta^2(\sum_i p_i)\pi(s+2)\frac{\cos(\frac{\theta}{2\alpha'_e}\sqrt{s(s+4)})-\cos(\pi s)}{\sin(\pi s)} \tag{A.9}$$

Obviously, this no longer vanishes for $\theta = 2\pi\alpha'_e$. However, in the NCOS limit the amplitude has much softer UV behavior than the $\theta = 0$ amplitude. Indeed, for large s we may approximate

$$\cos(\pi\sqrt{s(s+4)}) \approx \cos[\pi(s+2) - 2\pi/(s+2)] \ .$$

Therefore, for large s,

$$A \sim \frac{G_0^2}{\alpha'_e}\delta^2(\sum_i p_i)\frac{\sin[\pi s - \pi/(s+2)]}{\sin[\pi s]} \ ,$$

which implies that the residues of the poles at $s = n$ fall off as $1/(n+2)$ for large n. Away from the poles, however, the amplitude is energy independent at high energies. This should be contrasted with the commutative open string where this amplitude grows as E^2.

This softened UV behavior of NCOS massive amplitudes is general. If we consider 4 particles of mass m, then the amplitude has a factor

$$\cos(\pi\sqrt{s(s-4m^2\alpha'_e)}) - \cos(\pi s) \ ,$$

which leads to a cancellation for large s.

It is interesting that this behavior of 4-point amplitudes is similar to the energy dependence in the perturbative Yang-Mills theory in 1+1 dimensions. Away from the poles the NCOS amplitude is of order $\frac{G_0^2}{\alpha'_e}$ in the high-energy limit, so that the $2 \to 2$ scattering cross-section is

$$\sigma \sim \frac{G_0^4}{(\alpha'_e E^2)^2} \ . \tag{A.10}$$

The perturbative amplitude for 4 W-bosons is energy-independent and of order g_{YM}^2, so that the scattering cross-section is $\sim (g_{YM}/E)^4$.

SOME COMMENTS ON BRANES, G-FLUX, AND K-THEORY

GREGORY MOORE
Department of Physics, Rutgers University
Piscataway, NJ 08855, USA

Summary of a talk explaining three ways in which string theory and M-theory are related to the mathematics of K-theory.

1. Introduction

This talk was divided into 3 parts. In each part we describe how considerations related to D-branes and RR fields of type II string theory naturally lead to mathematics related to K-theory. Part one reviews some work done with D.-E. Diaconescu and E. Witten, and reported in Refs. 1,2. Here we review the way K-theory theta functions are related to M-theory partition functions. Part two reports on some work in progress with G. Segal. Details will appear in a future publication.[3] Here we show how simple considerations in topological field theory lead to a picture of D-branes being classified by K-theory. Part three announced material which was done with J. Harvey, and has been described in detail in Ref. 4. Here we explain how the K-theory of C^* algebras fits in naturally with recent developments on tachyon condensation in the presence of B-fields.

2. Partition Functions: From M to K

The partition function Z_{IIA} of type IIA superstring theory on a smooth compact spin 10-manifold X should be related to the partition function Z_M of M-theory on the associated manifold $Y = X \times S^1$. Nevertheless, showing that this is the case when X is topologically complicated turns out to be a very subtle problem. This is the problem which was addressed and solved in detail in Refs. 1,2.

In brief, we consider a long-distance/weak-coupling limit. In the IIA string we take $g_{\mu\nu} = tg^0_{\mu\nu}$, where $g^0_{\mu\nu}$ is some fixed metric and we let $t \to +\infty, g_{\text{string}} \to 0$. On the M-theory side we take the long-distance limit (except for the M-theory circle!). In both theories the partition functions reduce to the form:

$$Z = (\text{simple} \quad \text{factors}) \times \sum_{\text{G}-\text{flux}} e^{-S(G)} \tag{1}$$

Formulating precisely both the sum and the measure in Eq. 1 is tricky.

2.1. *Sum in type IIA theory*

In this case $G = G_0 + G_2 + \cdots + G_{10}$ is the sum of all fluxes. In formulating the sum there are two basic inputs one must take into account. First, even in the IIA theory, one should consider G to be a *selfdual field*: $G = *G$. Second, there is a quantization condition on the allowed fluxes. In Refs. 5,6 it is argued that the quantization condition is

$$[\frac{G}{2\pi}] = \text{ch}(x + \theta/2)(\hat{A}(TX))^{1/2} \tag{2}$$

where $[G]$ is the DeRham cohomology class of the G-field (a real differential form) while x, θ are K-theory classes. θ is a quantum shift, and *the topological sectors for RR fluxes are labelled by a K-theory class $x \in K^0(X)$.* (For a recent discussion see Ref. 7.) The definition of the partition function of a self-dual scalar field in 2 dimensions can be generalized to higher dimensions. This was done in Ref. 8 for self-dual 3-forms in the 5-brane, and generalized in Ref. 9 to the case of 10-dimensional RR fields. The procedure amounts to the quantization of a certain principally polarized abelian variety. The underlying torus is $K^0(X) \otimes R/(K^0(X)/K^0(X)_{tors})$ which turns out to have a natural metric, complex structure, and positive integral $(1,1)$ form. In this way one arrives at the type IIA K-theory theta function

$$\Theta_{IIA} = \sum e^{-KE(G)} e^{i\Phi(G)} \tag{3}$$

Here the sum runs over all cohomological values of G_0, G_2, G_4 consistent with the *existence* of a K-theory class x such that Eq. 2 can hold. The existence of such an x puts constraints on the possible cohomology class for G_4 (as described in section 2.3 below). The kinetic energy is

$$KE(G) = t^5 \parallel G_0 \parallel^2 + t^3 \parallel G_2 \parallel^2 + t \parallel G_4 \parallel^2 \tag{4}$$

and corresponds to the standard supergravity action. Here $\| \cdot \|^2$ is the standard Hodge norm in the metric $g^0_{\mu\nu}$. On the other hand, the phase is rather subtle, and follows from the considerations of Refs. 9,1,2. It has the structure:

$$e^{i\Phi(G)} = \Omega(x) \exp[2\pi i \int_X (-\frac{1}{15} G_2^5 + \frac{1}{6} G_2^3 G_4 + \cdots)] \tag{5}$$

The terms in the exponential are new topological phases which must be included in the 10-dimensional supergravity action in order to match properly to M-theory on a circle. The function $\Omega(x)$ takes values ± 1, and is based on a mod-two index. It can be thought of as $(-1)^{N(x)}$ where $N(x)$ is the number of Ramond fermion zeromodes on a IIB brane with K-theory charge x. One of the most striking aspects of $\Omega(x)$ is that there is no local formula for the mod two index![10] This is related to the well-known difficulties in formulating the action for a chiral field, and constitutes a significant departure from standard Lagrangian formulations of field theory.

2.2. *M-theory partition function*

Now we define precisely the M-theory partition function on a compact smooth spin 11-manifold Y in the long distance limit. As with the RR partition function, there is a subtle quantization condition and a phase. Both were analyzed by Witten a few years ago in Ref. 11. The quantization condition depends on a *cohomology* class $a \in H^4(Y, Z)$, and is $[\frac{G(a)}{2\pi}] = a - \lambda/2$ where λ is the characteristic class for the spin structure on Y. It satisfies $2\lambda = p_1(TY)$. The phase is, roughly speaking $\Omega_M(C) = \exp[2\pi i \int_Y (CGG + CX_8)]$, where X_8 is the famous R^4 correction to 11-dimensional supergravity.[12,13] However, if $a \neq 0$ then C is not globally defined, and one must formulate this phase carefully. One approach (adopted in Ref. 11) is to choose a bounding 12-manifold Z, $\partial Z = Y$ on which G extends, and set $\Omega_M(C) = \exp[2\pi i \int_Z (\alpha G^3 + \beta G X_8)]$, where α, β are certain numerical coefficients. (That such extensions even exist is a nontrivial result of Stong.) While this is perfectly correct, it can be difficult to work with. Moreover, it is not manifestly well-defined because there can be different choices for Z. The best way to understand why the phase is well-defined, as pointed out by Witten in Ref. 11, is to give the formulation of the phase in terms of E_8 gauge theory in 12 dimensions.

Here we will give a slightly different (but equivalent) definition of the M-theory phase based on work in progress with Emanuel Diaconescu.[14] An M-theory 3-form "C-field" can be defined to be a quadruple: (V, A, G, c) where $V \to X$ is an E_8 vector bundle and A is a connection on V. $G \in \Omega^4(X, R)$ is a real differential form, and $c \in \Omega^3(X, R)/\Omega^3_Z(X)$, where $\Omega^3_Z(X)$ are 3-forms with integral periods (they are necessarily closed). These data must satisfy

$$\frac{G}{2\pi} = \frac{1}{60}\mathrm{Tr}_{248}\frac{F^2}{8\pi^2} + \frac{1}{32\pi^2}\mathrm{Tr}R^2 + d\frac{c}{2\pi} \tag{6}$$

and are subject to an equivalence relation $(V_1, A_1, G_1, c_1) \sim (V_2, A_2, G_2, c_2)$ if $G_2 = G_1$, and $CS(A_1, A_2) = c_2 - c_1$, where CS is the Chern-Simons form associated to the above characteristic class. It should be stressed that this definition is equivalent to the statement that the M-theory 3-form is a Cheeger-Simons differential character $C \in \widehat{H}^3(Y, U(1))$. The proof of this fact relies on the remarkable result that E_8 bundles in less than 15 dimensions are classified by a single characteristic class $a \in H^4(Y, Z)$.[15] Let us denote the corresponding bundle by $V(a)$.

In terms of these data the M-theory phase is given by

$$\Omega_M(C) = \exp\left[2\pi i\left(\frac{\eta(D_{V(a)}) + h(D_{V(a)})}{4} + \frac{\eta(D_{RS}) + h(D_{RS})}{8}\right)\right] \cdot \omega(c) \tag{7}$$

Here D_V is the Dirac operator coupled to the connection A on the bundle $V(a)$, D_{RS} is the Rarita-Schwinger operator, $h(D) =$ number of zeromodes of the operator D on Y, and $\eta(D)$ is the eta invariant of Atiyah-Patodi-Singer. The phase $\omega(c)$ is given by

$$\omega(c) = \exp\left[2\pi i \int_Y \left(c(\bar{G}^2 + X_8) + cdc\bar{G} + \frac{1}{3}c(dc)^2\right)\right] \tag{8}$$

where $\bar{G} = \frac{1}{60}\mathrm{Tr}_{248}\frac{F^2}{8\pi^2} + \frac{1}{32\pi^2}\mathrm{Tr}R^2$. The generalization of this formula to 11-manifolds with boundary is not trivial, and will be reported in Ref. 14.[a]

2.3. *Equality of the partition functions*

In the case when $Y = X \times S^1$, with the supersymmetric spin structure on the circle S^1, and a "C-field" pulled back from X we can compare the partition functions at leading order, $\mathcal{O}(e^{-t})$ in the large distance expansion. Such contributions come solely from G_4. In Refs. 1,2 it is shown that both partition functions reduce to determinants times the theta function

$$\Theta = \sum e^{-\|G(a)\|^2}(-1)^{f(a)} \tag{9}$$

where the sum is over cohomology classes $a \in H^4(X,Z)$ such that $Sq^3(a) = 0$, where Sq^3 is the Steenrod squaring operation. On the IIA side, this is precisely the condition for the existence of a K-theory class x with $\mathrm{ch}(x) = a + \cdots$. On the M-theory side, this arises because the subtle M-theory phase is sensitive to torsion information in the cohomology class a of the M-theory 4-form. The sum over the torsion classes projects onto a's such that $Sq^3(a) = 0$. It is in this somewhat indirect way that the classification of IIA RR fluxes via K-theory turns out to be compatible with the classification of M-theory 4-flux by cohomology. The $Z/2Z$-valued function $f(a)$ is a mod-two index for the Dirac operator coupled to an E_8 connection in 10-dimensions, and again follows, nontrivially, from the definition of the phase in both M-theory and in IIA theory.

2.4. *Further results*

The basic computation sketched above can be extended in many directions. First, as shown in Refs. 1,2 the result extends to the next subleading order in the large t expansion at order e^{-t^3}. This involves summing over nontrivial circle bundles $Y \to X$ on the M-theory side, and correspondingly over nontrivial configurations for G_2 as well as for G_4 on the IIA side. There are many directions for further work including computation of one-loop determinants, extension to the type I setting, and inclusion of topologically nontrivial B-fields.[b] Work in progress with E. Diaconescu[14] is extending the result to include "instanton" amplitudes where membranes and fivebranes are inserted in the partition function.

Among the many open problems in the subject two are outstanding, but probably very difficult. First, E_8 gauge theory has thus far proved to be a mathematical convenience. Whether or not there is a deeper physical significance to the presence of 11 and 12-dimensional E_8 gauge theories is a fascinating open issue. Second, it

[a]The formula for $\omega(c)$ reported in the talk at Strings2000 omitted the second and third terms above.

[b]The considerations of section 2.1 extend straightforwardly to the case of flat B-fields $B \in H^2(X, U(1))$, if $[H] \in H^3(X,Z)$ is torsion. They reproduce rather nicely the couplings of the RR sector to the B-field.

is clear that the relation of K-theory to RR charges should have a profound generalization to include fundamental string and NS 5-brane charges. One focal point for finding this generalization is the issue of how the relation of RR charges and K-theory can be compatible with the S-duality of IIB string theory. (This was, in fact, the central motivating question behind Refs. 1,2.) For some provocative recent results on this subject see Ref. 16.

3. Sewing Constraints and D-branes

The previous section focused on a long-distance approach to the emergence of K-theory. It is of interest to understand how the relevance of K-theory can be seen from a more "microscopic" or fundamental approach, for example, from considerations of conformal field theory, or of string field theory. Such considerations lead one directly to think about the K-theory of algebras, a subject that was very much "in the air" at the Strings2000 meeting. In this section we report on one way of understanding the connection between K-theory and D-branes based on simple considerations of sewing and topological field theory. This is based on on-going work with Graeme Segal. We recommend Segal's Stanford lectures[17] as useful background.

The axiomatic formulation of open and closed strings was considered some time ago by P. Horava.[18] Nevertheless, we think it is interesting to reformulate it in the light of the connection between K-theory and D-branes. After submission of this manuscript to the editors, but before putting it on hep-th, a paper appeared with some overlapping results for this section.[19]

Given a closed string background one can ask: "What are the possible D-branes?" Similarly, given a closed conformal field theory $\mathcal{C}$ one can ask: "What are the possible associated boundary conformal field theories?" These questions are too hard to answer at present. Nevertheless, it turns out that if one replaces a conformal field theory by a 2d topological field theory then the question is solvable, yet not entirely trivial.

Recall the ancient folktheorem that 2D topological field theories are in 1-1 correspondence with commutative Frobenius algebras: In/out circles map to in/out Hilbert spaces, and an oriented surface maps to a linear operator between in and out spaces. The multiplication in the Frobenius algebra is defined by the pants diagram, the trace and unit are defined by the operator corresponding to the disk, with opposite orientations. The axioms of a commutative associative algebra are equivalent to the consistency of sewing arbitrary surfaces from these three components. Let us now allow both open and closed strings. Then, the 2 dimensional surfaces have two kinds of boundaries. There are in- and out-going intervals corresponding to the in- and out-going open strings, and there are the "free boundaries" corresponding to the ends of the open string "moving along a D-brane." Free boundaries carry boundary condition labels $a, b, \ldots$ which should be thought of as objects in a linear category.[17] The open string sectors $\mathcal{O}_{ab}$ are the morphisms between the objects a and b in the linear category. Our question: "Given a closed 2d topological field theory, what are the possible boundary conditions?" should be asked as two questions: First, "What are the algebraic conditions that encode consistency of open and closed string sewing?" and second, "What linear categories are consistent with these conditions?" It is useful to focus first on the case of a single boundary condition on both ends of the open string, that is, to let $\mathcal{O} = \mathcal{O}_{aa}$. Then we have

Theorem 1: To give an open and closed topological field theory is to give

1. A commutative Frobenius algebra $(\mathcal{C}, \theta_{\mathcal{C}}, 1_{\mathcal{C}})$.
2. A not necessarily commutative Frobenius algebra $(\mathcal{O}, \theta_{\mathcal{O}}, 1_{\mathcal{O}})$.
3. A homomorphism $\iota_* : \mathcal{C} \to Z(\mathcal{O})$, where $Z(\mathcal{O})$ is the center of $\mathcal{O}$, such that $\iota_*(1_{\mathcal{O}}) = 1_{\mathcal{C}}$, and such that:

$$\pi = \iota_* \iota^* \tag{10}$$

Here ι^* is the adjoint to ι_*, defined by $\theta_{\mathcal{O}}(\psi \iota_*(\phi)) = \theta_{\mathcal{C}}(\iota^*(\psi)\phi)$, while $\pi : \mathcal{O} \to \mathcal{O}$ is an operator defined by the double-twist diagram.

The condition Eq. 10 is closely related to the "Cardy condition" of boundary conformal field theory. The operator π is defined using only the open string data $(\mathcal{O}, \theta_{\mathcal{O}}, 1_{\mathcal{O}})$. Indeed, if ψ_μ is a basis for $\mathcal{O}$ and ψ^μ is a dual basis relative to $\theta_{\mathcal{O}}$ then $\pi(\psi) = \sum_\mu \psi_\mu \psi \psi^\mu$. In pictures, Eq. 10 follows because the double twist diagram can be viewed both as an open string diagram, and also as a closed string channel diagram. Indeed ι^* is the open to closed string transition, while ι_* is the closed to open transition.

The claim of the theorem is that in the case of open and closed strings with a single type of boundary condition on all free boundaries, the above conditions are equivalent to the consistency conditions for sewing. This statement is not trivial to prove. The conditions are roughly the same as those found in Ref. 20 but are not precisely the same.

Let us now consider the solutions to the above conditions. In general this is a difficult problem, but if the "fusion rules" (i.e. the regular representation matrices) of $\mathcal{C}$ are diagonalizable then we can classify the $\mathcal{O}$'s. When the fusion rules are diagonalizable we say $\mathcal{C}$ is "semisimple." In this case we may introduce basic idempotents such that $\epsilon_i \epsilon_j = \delta_{ij} \epsilon_i$ and $\mathcal{C} = \oplus_i C \cdot \epsilon_i$. Indeed, given any basis ϕ_μ for $\mathcal{C}$ one can diagonalize the fusion rules with a matrix S^i_μ. Letting $\mu = 0$ correspond to the identity element we can write: $\epsilon_i = \sum_\mu S^i_0 (S^{-1})^\mu_i \phi_\mu$.

Theorem 2: If $\mathcal{C}$ is semisimple then $\mathcal{O} = \mathrm{End}_{\mathcal{C}}(M)$ for M a finitely generated projective $\mathcal{C}$-module.

The theorem states that the possible $\mathcal{O}$'s are simply sections of a vector bundle over the "spacetime" Spec($\mathcal{C}$). One should think of ϵ_i as corresponding to the spacetime points in Spec($\mathcal{C}$). Recall that in the Gelfand-Naimark theorem one defines Spec($\mathcal{C}$) as the space of characters of the algebra $\mathcal{C}$. Indeed, the character corresponding to ϵ_i is $\chi(\phi) = \theta_{\mathcal{C}}(\epsilon_i \phi)/\theta_i$ where $\theta_i = \theta_{\mathcal{C}}(\epsilon_i)$. These traces are part of the invariants of a Frobenius algebra. Then $\mathcal{O} = \oplus_i \mathrm{End}(W_i)$ for some collection of finite dimensional vector spaces W_i.

In this formalism one can easily work out the formula for the "boundary state," defined by $B = \iota^*(1_{\mathcal{O}})$. This object is an element of $\mathcal{C}$ and has the property that insertions of n factors of B in $\theta_{\mathcal{C}}(B^n \cdots)$ corresponds geometrically to adding n holes to the surface. In formulae $B = \sum_i \dim(W_i)\epsilon_i/\sqrt{\theta_i}$, The squareroot in this formula is significant and is related to the standard fact that the closed string coupling is the square of the open string coupling. Moreover, if we consider a *family* of Frobenius algebras, then, if the subfamily of semisimple algebras has nontrivial fundamental group, transport around nontrivial loops can result in monodromy such that $\epsilon_i \to \epsilon_{\sigma(i)}$ where σ is a permutation, while $\sqrt{\theta_i} \to \pm\sqrt{\theta_{\sigma(i)}}$. Because of this one should generalize vector spaces to virtual vector spaces above to allow for negative integers in $\dim(W_i)$.

Now, let us consider multiple boundary conditions. The open string spaces $\mathcal{O}_{ab}$ for distinct boundary conditions a, b is a bimodule for $\mathcal{O}_{aa} \otimes \mathcal{O}_{bb}$. The Cardy condition generalizes in the obvious way $\pi_a^b = \iota_a \iota^b$, where ι^b is the open to closed transition for boundary conditions of type bb, and ι_a is the closed to open transition for boundary conditions of type aa. This condition can be shown to imply that $\mathcal{O}_{ab} = \mathrm{Hom}(W_a, W_b)$ where W_a, W_b are projective modules for $\mathcal{C}$. The conclusion then is that *the linear category classifying the boundary conditions for the closed 2d topological field theory is the K-theory $K_0(\mathcal{C})$ for the commutative closed string Frobenius algebra.*

Many interesting examples of the above theorems applied to families of Frobenius algebras can be given. Moreover, the discussion can be generalized to "orbifolds" that is, to the equivariant case. Analogs of theorems 1 and 2 above can be stated, at least in the semisimple case. It is possible to introduce a "B-field" even in this topological setting. The "untwisted sector" of the theory defines an ordinary commutative Frobenius algebra $\mathcal{C}_1$, and $\mathrm{Spec}(\mathcal{C}_1)$ will be a finite G-space X, where G is the orbifold group. The B-field is valued in G-equivariant cohomology $H_G^2(X; U(1))$ and has a corresponding fieldstrength $h \in H_G^3(X; Z)$. The linear category is then given by the K-theory $K_{G,h}(X)$ of G-equivariant, twisted bundles of algebras over the "spacetime" X, in accord with previous works. Details will be described in Ref. 3.

The above results might have some bearing on the subject of topological open strings and D-branes in Calabi-Yau manifolds. For recent work on this subject see Refs. 21,22,23.

4. Noncommutative Tachyons and K-theory

Part three of the talk announced results that have since been published in Ref. 4. We will therefore be very brief here.

There has been much recent progress on understanding tachyon condensation by using the technology of noncommutative field theory.[24,25,26,27,28] Consider the bosonic or the type II string. The general picture is to consider spacetime to be a product of a commutative and a noncommutative manifold $X_c \times X_{nc}$. A field on a noncommutative manifold is equivalent to an operator on Hilbert space. Therefore, if an unstable D25, or D9 brane wraps this spacetime then its tachyon field will be described by a field on X_c, valued in operators on Hilbert space. At large noncommutativity parameter the equations of motion for the tachyon field say that it is a projection operator (in the bosonic string) or a partial isometry (in the type II string). This fits in perfectly with the K-theory classification of D-branes. Indeed, the isomorphism classes of complex rank n vector bundles on X_c are in one-one correspondence with homotopy classes $[X_c, BU(n)]$ where $BU(n)$ is the space of rank n projection operators on Hilbert space. Similarly, the Atiyah-Janich model for K-theory shows that we can identify $K^0(X_c)$ with the homotopy classes $[X, \mathcal{F}]$ where $\mathcal{F}$ is the space of Fredholm operators on Hilbert space. By polar decomposition one can restrict to the space of partial isometries.

In the type II string a very special class of partial isometries can be written corresponding to the "noncommutative Atiyah-Bott-Shapiro construction." One takes $[x^i, x^j] = -i\theta^{ij}$ with θ of maximal rank and $i = 1, \ldots, 2p$. Letting Γ_i be chiral gamma matrices of rank 2^{p-1} we can form the partial isometry in the polar decomposition of $\Gamma_i x^i$. This is the tachyon field corresponding to condensing a D9 brane to a D(9-2p) brane transverse to a noncommutative plane of dimension 2p. The fact that the D-brane charge is unchanged by turning on a B-field is equivalent to the index theorem identifying topological and analytical indices. For further details, and further developments of these ideas, see Ref. 4, and references therein.

Recently, in collaboration with E. Martinec we have generalized these ideas to include D-branes on orbifolds. The algebra of functions on X_{nc} is replaced by a crossed-product algebra. Fractional branes and discrete torsion are very naturally and easily incorporated into the formalism. In addition, the formalism gives an interesting perspective on the formulation of the theory of D-branes in asymmetric orbifolds. Details will appear in Ref. 29.

Acknowledgments

I would like to thank the organizers for the opportunity to present this talk at Strings2000. The above work is all collaborative and would never have been possible without the insights and contributions of D.-E. Diaconescu, J. Harvey, E. Martinec, G. Segal, and E. Witten. This work is supported by DOE grant DE-FG02-96ER40949.

References

1. E. Diaconescu, G. Moore and E. Witten, *E_8 Gauge Theory, and a Derivation of K-Theory from M-theory*, hep-th/0005090.
2. D.-E. Diaconescu, G. Moore and E. Witten, *A Derivation of K-theory from M-theory*, hep-th/0005091.
3. G. Moore and G. Segal, Work in progress.
4. J. Harvey and G. Moore, *Noncommutative Tachyons and K-theory*, hep-th/0009030.
5. G. Moore and E. Witten, *Self-duality, RR fields and K-Theory*, *JHEP* **0005**, 032

(2000) [hep-th/9912279].

6. D.S. Freed and M.J. Hopkins, *On Ramond-Ramond Fields and K-Theory*, *JHEP* **0005**, 044 (2000) [hep-th/0002027].
7. D.S. Freed, *Dirac Charge Quantization and Generalized Differential Cohomology*, hep-th/0011220.
8. E. Witten, *Five-Brane Effective Action In M-Theory*, *J. Geom. Phys.* **22**, 103 (1997) [hep-th/9610234].
9. E. Witten, *Duality Relations among Topological Effects in String Theory*, *JHEP* **0005**, 031 (2000) [hep-th/9912086].
10. M.F. Atiyah and I.M. Singer, *The index of elliptic operators: V*, *Ann. Math.* **93**, 139 (1971).
11. E. Witten, *On Flux Quantization in M-Theory and the Effective Action*, *J. Geom. Phys.* **22**, 1 (1997) [hep-th/9609122].
12. M.J. Duff, J.T. Liu and R. Minasian, *Eleven dimensional origin of string/string duality: a one loop test*, *Nucl. Phys.* **B452**, 261 (1995) [hep-th/9506126].
13. C. Vafa and E. Witten, *A One-loop Test of String Duality*, *Nucl. Phys.* **B447**, 261 (1995) [hep-th/9505053].
14. E. Diaconescu and G. Moore, Work in progress.
15. E. Witten, *Topological Tools In Ten-Dimensional Physics*, in *Unified String Theories*, 1985 Santa Barbara Proceedings, eds. M. Green and D. Gross (World Scientific, 1986).
16. P.S. Aspinwall and M.R. Plesser, *D-branes, Discrete Torsion and the McKay Correspondence*, hep-th/0009042.
17. See ITP lectures by G. Segal, http://doug-pc.itp.ucsb.edu/online/geom99/.
18. P. Horava, *Equivariant Topological Sigma Models*, *Nucl. Phys.* **B418**, 571 (1994) [hep-th/9309124].
19. C.I. Lazaroiu, *On the structure of open-closed topological field theory in two-dimensions*, hep-th/0010269.
20. D. Lewellen, *Sewing constraints for conformal field theories on surfaces with boundaries*, *Nucl. Phys.* **B372**, 654 (1992).
21. K. Hori, A. Iqbal and C. Vafa, *D-Branes and Mirror Symmetry*, hep-th/0005247.
22. M.R. Douglas, *D-branes, Categories, and N=1 supersymmetry*, hep-th/0011017.
23. C.I. Lazaroiu, *Instanton amplitudes in open-closed topological field theory*, hep-th/0011257.
24. R. Gopakumar, S. Minwalla and A. Strominger, *Noncommutative Solitons*, *JHEP* **0005**, 020 (2000) [hep-th/0003160].
25. J. Harvey, P. Kraus, F. Larsen and E. Martinec, *D-branes and Strings as Non-commutative Solitons*, *JHEP* **0007**, 042 (2000) [hep-th/0005031].
26. K. Dasgupta, S. Mukhi and G. Rajesh, *Noncommutative Tachyons*, *JHEP* **0006**, 022 (2000) [hep-th/0005006].
27. E. Witten, *Noncommutative tachyons and string field theory*, hep-th/0006071.
28. E. Witten, *Overview of K-theory applied to strings*, hep-th/0007175.
29. E. Martinec and G. Moore, *Noncommutative solitons on orbifolds*, in preparation.

SPACE-TIME UNCERTAINTY AND NONCOMMUTATIVITY IN STRING THEORY

TAMIAKI YONEYA
Institute of Physics, University of Tokyo
Komaba, Tokyo 153–8902, Japan

We analyze the nature of space-time nonlocality in string theory. After giving a brief overview on the conjecture of the space-time uncertainty principle, a (semi-classical) reformulation of string quantum mechanics, in which the dynamics is represented by the noncommutativity between temporal and spatial coordinates, is outlined. The formalism is then compared to the space-time noncommutative field theories associated with nonzero electric B-fields.

1. Motivations

What is string theory? This is a question we have been continually asking ourselves in exploring string theory as a hint towards the ultimate unified theory of all interactions including quantum gravity.

One of the most characteristic features of string theory is the existence of a fundamental constant, string length $\ell_s \sim \sqrt{2\pi\alpha'}$, which sets a natural cutoff scale for the ultraviolet part of quantum fluctuations for particle fields associated with the spectrum of string states. This implies that string theory must necessarily exhibit some nonlocality and/or certain fuzziness with respect to the short distance structure of space-time.

From this point of view, it is quite remarkable that string theory gives a completely well-defined analytic S-matrix which essentially satisfies all the axioms for physically acceptable theory satisfying, at least perturbatively, Lorentz invariance, (macro) causality and unitarity. In contrast to this, various past attempts toward nonlocal field theories failed to give sensible results, because of lack of self-consistency or of suitable guiding principles for constructing nontrivially interacting theories. It thus seems an important task in uncovering its underlying principles to characterize the nonlocality of string theory. Our attitude is that, given string theory, we should learn how to formulate the idea of fundamental length from the structure of string theory, rather than postulating some arbitrary principles from scratch.

Recent development on the connection of string theory with the external B-field to non-commutative geometric field theories might provide a good hint for pursuits in this direction. However, we should keep in mind that the nature of nonlocality originated from the B-field is nothing to do with the extendedness of strings. In the present article, I would like to, first, review briefly the old proposal of a 'spacetime uncertainty principle' as a possible general characterization of the space-time structure of string theory at short distances, and then to discuss some ideas toward a reformulation of string theory in such a way that the noncommutativity between space and time is manifest. I hope that the comparison of the nature of the latter noncommutativity to the one of the typical noncommutative field theories which have the algebra of space-time coordinates of the Moyal-type product associated with the electric B-field is useful for deepening our understanding of string theory.

2. Space-Time Uncertainty

The main idea for proposing the space-time uncertainty relation[1,2] comes from a simple analogy concerning the nature of string quantum mechanics. The crucial requirement of the ordinary string perturbation theory is the world-sheet conformal invariance. Indeed, most of the important merits of string theory as a possible unified theory are due to the conformal invariance. In particular, the elimination of the ultraviolet of divergence in the presence of gravity is essentially due to the modular invariance, which is the part of the conformal symmetry. From the viewpoint of generic two-dimensional field theory, the conformal invariance forces us to choose a very particular class of all possible two-dimensional field theories, corresponding to the fixed points of Wilsonian renormalization group. This is quite analogous to the imposition of Bohr-Sommerfeld quantization conditions to classical mechanics, in which the adiabatic invariance of action variables to be quantized can be regarded as a characterization of the quantization condition. In the final formulation of quantum mechanics, the quantization condition was replaced by the more universal framework such as Hilbert space and operator algebra acting in it. This analogy suggests us the importance of reinterpreting the conformal invariance requirement by elevating it to a more universal form, which may ultimately be formulated in a way that does not depend on perturbative methods.

One of the crucial properties related to modular invariance is expressed as the 'reciprocity relation' of the 'extremal length'. The extremal length is a conformally invariant notion of length associated with families of curves on general Riemann surfaces. If we consider some finite region Ω and a set Γ of arcs on Ω, the extremal length of Γ is defined by $\lambda_\Omega(\Gamma) = \sup_\rho \frac{L(\Gamma,\rho)^2}{A(\Omega,\rho)}$ with $L(\Gamma,\rho) = \inf_{\gamma\in\Gamma} L(\gamma,\rho)$, $A(\Omega,\rho) = \int_\Omega \rho^2 dz d\bar{z}$ where ρ is the possible metric function giving the length $L(\gamma,\rho) \equiv \int_\gamma \rho|dz|$ on Ω of a curve in Γ in the conformal gauge. Since any Riemann surface can be composed of a set of quadrilaterals pasted along the boundaries (with some twisting operations, in general), it is sufficient to consider the extremal length for an arbitrary quadrilateral segment Ω. Let the two pairs of opposite sides of Ω be α, α' and β, β'. Take Γ be the set of all connected set of arcs joining α and α'. The set of arcs joining β and β' is called the conjugate set of arcs, denoted by Γ^*. We then have two extremal lengths, $\lambda_\Omega(\Gamma)$ and $\lambda_\Omega(\Gamma^*)$. Then the reciprocity relation is that

$$\lambda_\Omega(\Gamma)\lambda_\Omega(\Gamma^*) = 1. \tag{1}$$

The simplest example is just the rectangle with the Euclidean sizes a and b in the Gauss plane. In this case, we can easily prove that $\lambda(\Gamma) = a/b, \quad \lambda(\Gamma^*) = b/a$. For details, we refer the reader to a more extensive review[3] and the mathematical references cited there.

To see how the reciprocity of the extremal length reflects to target space-time, let us consider the Polyakov amplitude for the mapping from the rectangle on a Riemann surface to a rectangular region in space-time with the side lengths A, B with the boundary condition $(0 \le \xi_1 \le a, 0 \le \xi_2 \le 1)$ $x^\mu(0,\xi_2) = x^\mu(a,\xi_2) = \delta^{\mu 2} B\xi_2/b, x^\mu(\xi_1, 0) = x^\mu(\xi_1, b) = \delta^{\mu 1} A\xi_1/a$. Then the amplitude contains the factor

$$\exp\Big[-\frac{1}{\ell_s^2}\Big(\frac{A^2}{\lambda(\Gamma)} + \frac{B^2}{\lambda(\Gamma^*)}\Big)\Big], \tag{2}$$

multiplied by a power-behaved prefactor. Thus the fluctuations of two space-time lengths A and B satisfy an 'uncertainty relation' $\Delta A \Delta B \sim \ell_s^2$. For general and more complicated boundary conditions, it is not easy to establish a simple relation such as above between the extremal lengths and the space-time lengths, since there are various ambiguities in defining space-time lengths in terms of string variables. After all *only* legitimate observables allowed in string theory is the on-shell *S-matrix*. However, it seems natural to conjecture that the above relation sets a limitation, in some averaged sense, on the measurability of the lengths in space-time in string theory, since conformal invariance must be valid to all orders of string perturbation theory and the random nature of boundaries generally contributes to further fuzziness on the space-time lengths. Note that this reciprocity relation exhibits one of the most important duality relations in string amplitudes between ultraviolet and infrared structures. Since in the Minkowski metric one of the lengths is always dominantly time-like, we propose the following uncertainty relation on the space-time lengths

$$\Delta T \Delta X \gtrsim \ell_s^2 \tag{3}$$

as a universal characterization of the short-distance space-time structure of string theory. This relation was originally proposed by the present author[1] in 1987 independently of other proposals of similar nature, for example, the notion of 'minimal distance.[4]'

The consistency of this 'space-time uncertainty' relation with the high-energy behaviors of the perturbative string amplitudes was analyzed in Ref. 3 to which I refer the reader for details and relevant references. We find that generically there are many instances where the above uncertainty relation is far from being saturated. However, so far all the known results seem to be consistent with the validity of the space-time uncertainty relation as an *inequality*. In particular, in the high-energy and high-momentum-transfer limit, both the temporal and spatial uncertainties increase linearly with respect to energy for fixed-genus amplitudes. However, the proportional constant decreases for higher genera and the well-known behavior $|A_{\text{resum}}(s,\phi)| \sim \exp\big(-\sqrt{6\pi^2 s f(\phi)/\log s}\big)$ of the Borel-summed amplitude[5] is consistent with the saturation of the equality in (3) up to some possible logarithmic corrections that perhaps depend on how to precisely define the space-time uncertainties. This may be an indication that the relation (3) is indeed valid independently of string coupling g_s.

A further support for the validity of the relation is its effectiveness for D-branes. For example, the effective Yang-Mills theories for the low-velocity D-p-branes predict that the characteristic spatial (*transverse* to D-p-branes) and temporal scales of D-p-brane scattering oppositely scale with respect to the string coupling, namely as $\Delta X \sim g_s^{-1/(3-p)}\ell_s$ and $\Delta T \sim g_s^{1/(3-p)}$ for $p \geq 0$ and for $p \neq 3$. Although the case $p = 3$ is special in that the effective Yang-Mills theory is conformally invariant, the conformal transformation property is actually consistent with the space-time uncertainty relation as discussed in Ref. 6. We can also derive these characteristic scales directly without recourse to the effective Yang-Mills theory. For example, the characteristic scales $\Delta X \sim g_s^{1/3}\ell_s, \Delta T \sim g_s^{-1/3}\ell_s$ of D-particle-D-particle scattering are a direct consequence[7] of the space-time uncertainty relation and the ordinary quantum mechanical Heisenberg relation, given the fact that the mass of the D-particle is proportional to $1/g_s$. All these properties are natural from the viewpoint of open string theories where the relation (3) must be valid.

Finally, let us discuss one important question. Since the relation (3) is independent of string coupling g_s, it seems at first sight that it does not take into account gravity. So what is its relation to the Planck scale which is the characteristic scale of quantum gravity? In string theory, the existence of gravity[8,9] can also be regarded as an important consequence of the world sheet conformal invariance. This is due to the possibility of deforming the background space-time by a linearized gravitational wave. However, in perturbation theory, the coupling strength of the gravitational wave is an independent parameter determined by the vacuum expectation value of dilaton. In this sense, the string coupling can not be a fundamental constant which appears in the universal nonperturbative property of string theory. Thus in oder to take into account the Planck length for the space-time uncertainty relation, we have to put that information by hand. Now we shall show that by combining the Planck scale with the space-time uncertainly relation, we can derive the M-theory scale without invoking D-branes or membranes.

For that purpose, it is useful first to reinterpret the meaning of the Planck length using a similar language of the stringy space-time uncertainties, by considering the limitation of the notion of classical space-time as the background against the possible formation of virtual black holes in the short distance regime. If we want to probe the space-time structure in time direction to order δT, the quantum mechanical uncertainty relation tells us that the uncertainty with respect to the energy of order $\delta E \sim 1/\delta T$ is necessarily induced. If we further require that the structure of the background space-time is not influenced appreciably by this amount of fluctuation, the spatial scale δX to be probed can not be smaller than the Schwarzschild radius associated with the energy fluctuation. Hence, $\delta X \gtrsim (G_D \delta E)^{1/(D-3)}$ in D-space-time dimensions. This sets the relation for the characteristic gravitational scales in

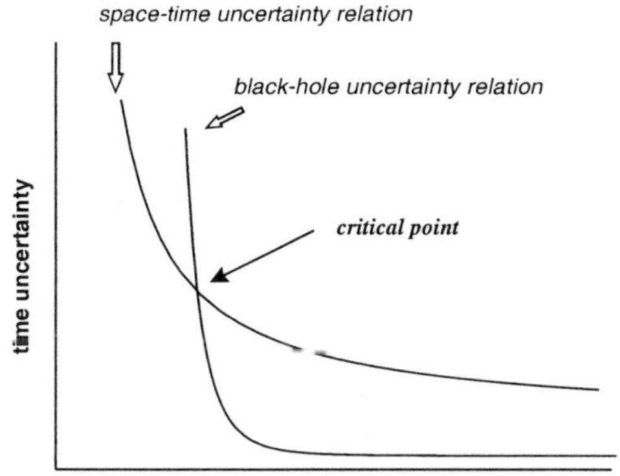

Fig. 1. This diagram schematically shows the structure of the space-time uncertainty relation and the uncertainty relation associated with the Planck scale. The critical point is where the two relations meet.

the form

$$\delta T(\delta X)^7 \gtrsim g_s^2 \ell_s^8. \tag{4}$$

in D =10 dimensional string theory. This may be called the 'black-hole uncertainty' relation. Note, however, that the nature of the scales $\delta T, \delta X$ is different from those in the relation (3). The uncertainties in (4) only express limitations, for observers at asymptotic infinity, with respect to spatial and temporal resolutions, below which the naive classical space-time picture without the formation of microscopic black holes can no longer be applied. In contrast to this, the space-time uncertainty relation sets the more fundamental limitation below which the space-time geometry itself loses its meaning. Then, the most important characteristic scale associated with the existence of gravitation in string theory corresponds to the point of their crossover. The critical crossover scales are then determined as

$$\Delta X_c \sim \delta X_c \sim g_s^{1/3} \ell_s, \quad \Delta T_c \sim \delta T_c \sim g_s^{-1/3} \ell_s.$$

It is quite remarkable that the spatial critical scale ΔX_c coincides with the M-theory scale. Let us look at the Fig. 1 to appreciate the meaning of these scales. For $\Delta T < \Delta T_c$ there is no region where the fluctuation containing the microscopic black hole associated with quantum fluctuations is important in string theory, while, for $\Delta T > \Delta T_c$, there is a region where $(\Delta T)^{-1}\ell_s^2 < \Delta X < \Delta X_c$ is satisfied, and hence black hole formation at the microscopic level becomes appreciable in string theory. The importance of this region increases as the string coupling grows larger.

3. Space-Time Noncommutativity

The validity of the fundamental uncertainty relation of the form (3) suggests the existence of certain noncommutative space-time structure that underlies string theory. Indeed, the expression (2) is strongly reminiscent of the Wigner representation

$\rho(p,q) \sim \exp\big[-\big((p/\Delta p)^2 + (q/\Delta q)^2\big)\big]$ of the density matrix corresponding to the gaussian wave packet in particle quantum mechanics, suggesting that the space-time of string theory is something analogous to the classical phase space in particle quantum mechanics. However, the usual quantum mechanics of strings does not directly show any such noncommutativity between space and time. This raises a question: Is there any alternative formulation of string quantum mechanics in which the space-time noncommutativity is manifest? In such a formulation,[3] we expect that the world-sheet conformal symmetry would be translated into a quite different language. There might be, hopefully, a chance of providing a hint toward some nonperturbative formulation of string theory.

Let us start from the following version of the Nambu-Goto-Schild action:

$$S_{ngs} = -\int_\Sigma d^2\xi \Big\{\frac{1}{e}\Big[-\frac{1}{2(4\pi\alpha')^2}(\epsilon^{ab}\partial_a X^\mu \partial_b X^\nu)^2\Big] + e\Big\}. \tag{5}$$

We only consider the case of bosonic strings, expecting that the extension to superstrings will not cause any fundamental difficulty. By eliminating nonpropagating auxiliary field $e(\xi)$, the action reduces to the ordinary Nambu-Goto action. In this form, the conformal invariance of string theory is buried in the existence of the standard Virasoro constraint $\mathcal{P}^2 + \frac{1}{4\pi\alpha'}\acute{X}^2 = 0, \quad \mathcal{P}\cdot\acute{X} = 0$ which does *not* explicitly involve the world sheet auxiliary field and the world-sheet metric. It is important for later interpretation that the Hamiltonian constraint comes from the equation

$$\frac{1}{4\pi\alpha'}\sqrt{-\frac{1}{2}(\epsilon^{ab}\partial_a X^\mu \partial_b X^\nu)^2} = e \tag{6}$$

for the auxiliary field e. Recall also that causality of string theory is embodied in the time-like nature of the area element $\epsilon^{ab}\partial_a X^\mu \partial_b X^\nu$ in this formalism. Now in order to rewrite the action such that it becomes quadratic in the space-time coordinates, we introduce another auxiliary field $b_{\mu\nu}(\xi)$ which transforms as a world-sheet scalar and simultaneously as an antisymmetric tensor with respect to the space-time indices:

$$S_{ngs2} = -\int_\Sigma d^2\xi \Big\{\frac{1}{4\pi\alpha'}\epsilon^{ab}\partial_a X^\mu \partial_b X^\nu b_{\mu\nu} + e\Big(\frac{1}{2}b_{\mu\nu}^2 + 1\Big)\Big\}. \tag{7}$$

Now the constraint (6) is replaced by that for the new auxiliary field $b_{\mu\nu}$,

$$\frac{1}{2}b_{\mu\nu}^2 = -1,. \tag{8}$$

The first auxiliary field e only plays the role of Lagrange multiplier for this condition. Namely, the requirement of conformal invariance is essentially reinterpreted as the condition that the world-sheet b field is time-like or 'electric'.

Let us consider the quantization of this action by regarding the b-field as an external field. Since the action is then first order with respect to the world-sheet time (τ) derivative, the system has second class constraints $P_\mu = b_{\mu\nu}\partial_\sigma X^\nu/2\pi\alpha'$, relating the components of the generalized coordinates and momenta directly. The Dirac bracket taking into account this constraint is given as

$$\left\{X^\mu(\sigma_1), \partial_\sigma b_{\nu\alpha}(\sigma_2)X^\alpha(\sigma_2)\right\}_D = 2\pi\alpha'\delta^\mu_\nu\delta(\sigma_1 - \sigma_2). \tag{9}$$

Remembering that the b-field is dominantly time-like by (8), we see that the space $X^i(\xi)$ and the time $X^0(\xi)$ become indeed noncommutative. In particular, the center-of-mass time $T \equiv (1/2\pi)\int_0^{2\pi} d\sigma X^0$ and the spatial extension X defined by

$$X \equiv -\int d\sigma\, b_{0i}(\sigma)\partial_\sigma X^i \tag{10}$$

satisfy $(2\pi\alpha' \to \ell_s^2)$

$$\{T, X\}_D = \ell_s^2. \tag{11}$$

Here for simplicity we have assumed a closed string. That the expression (10) can be adopted justifiably as the measure of spatial (longitudinal) extension of strings can be seen by remembering that in the semi-classical approximation the b-field is just proportional to the area element of the world sheet of strings, $b^{\mu\nu} = -\frac{1}{4\pi\alpha'}\epsilon^{ab}\partial_a X^\mu \partial_b X^\nu/e$, which is derived by taking the variation of the action with respect to the b-field. Note that in this approximation the auxiliary e-field is determined by the normalization condition (6) through (8). In this way, we have now reformulated the string mechanics, at least in the semi-classical approximation, in such a way that the noncommutativity between spatial extension and time is manifest. This naturally conforms to the general property (3) of the space-time uncertainties, derived on the basis of the world-sheet conformal symmetry in the previous section, as it should be. For example, the constraint (8) indeed replaces the role of conformal invariance.

Remarks

1. The nature of the noncommutativity discussed above is close, at least formally, to that associated with the antisymmetric space-time external $B_{\mu\nu}(X)$ field in the presence of D-branes. Note, however, an obvious difference that the present noncommutativity is intrinsic to the extendedness of strings and is nothing to do with the choice of the background of string theory. If we add the B-field background in considering D-branes, there arises an additional contribution to the noncommutativity, since the first term in the action is deformed as $b_{\mu\nu} \to b_{\mu\nu} + 2\pi\alpha' B_{\mu\nu}$. Of course, when $B_{\mu\nu}$ is constant, it affects only at the end points of open strings. Let us briefly treat the open string boundary in the present formalism. If we allow free variations for δX^μ at the boundary without B-field along the D-brane world volume, the boundary condition is $\partial_\tau X^\nu b_{\mu\nu} = 0$. In the semi-classical approximation we are using, we can set $b^{\mu\nu} = -\frac{1}{4\pi\alpha'}\epsilon^{ab}\partial_a X^\mu \partial_b X^\nu/e$. It is convenient here to choose the orthonormal world-sheet coordinate satisfying $\dot{X} \cdot X' = 0$ and $\dot{X}^2 + X'^2 = 0$ where the Lorentz contractions are done only over the directions along the D-brane world volume. Then, using (6) the boundary condition becomes the usual Neumann

condition $X'^{\mu} = 0$. Now suppose we add a constant space-time B-field which corresponds to the additional boundary term $(1/2)\int_{\partial\Sigma} d\tau X^{\mu}(\tau)\partial_\tau X^{\nu}(\tau)B_{\mu\nu}$. Then the boundary condition is $X'^{\mu} + 2\pi\alpha' B^{\mu\nu}\dot{X}_{\nu} = 0$. If the B-field is magnetic, it does not affect the time-like nature of the string coordinate at the boundary. However, if that is electric and the magnitude exceeds a critical value $1/2\pi\alpha'$ such that $b_{0i} + 2\pi\alpha' B_{0i}$ can vanish, the time-like nature of the open string boundary would be lost, leading to a violation of causality in the dynamics of open strings and D-branes. This can be seen as follows. Without losing generality we can assume that the only nonzero component of the B-field is $B_{01} = B$, providing the direction 1 is along the D-brane world volume. The boundary condition together with the coordinate condition leads to the relation

$$\dot{X}^2 = -X'^2 = (2\pi\alpha' B)^2((\dot{X}^1)^2 - (\dot{X}^0)^2).$$

When $(2\pi\alpha' B)^2 > 1$, this implies that the vector $\dot{X}^{\mu}$ is space-like at the open-string boundary. Thus it is impossible[10] to decouple the string scale from that of the noncommutativity associated to electric B-field. Namely, we can not define sensible field theory limits satisfying causality and unitarity using electric B field[11] without open-string degrees of freedom. This seems to indicate that the feasibility of the space-time noncommutativity is inextricably connected to stringy nonlocality.

2. It is perhaps instructive to make a further comparison of the present formalism with the naive field theory model in which the noncommutativity between space and time is introduced explicitly. A scalar field theory in a noncommutative space-time with space-time commutation relation $[x, t] = i\theta$ can be constructed by assuming the product of the fields are defined by the Moyal product

$$\phi(x) * \phi(x) = e^{i\frac{\theta}{2}(\partial_{x_1}\partial_{t_2} - \partial_{t_1}\partial_{x_2})}\phi(x_1)\phi(x_2)\Big|_{x_1=x_2=x, t_1=t_2=t}.$$

For notational clarity, we consider only (1+1)-dimensional part of space-time. Then a 3-point interaction vertex takes the following form in the coordinate representation

$$\begin{aligned}&\int dxdt\,(\phi_1 * \phi_2 * \phi_3)(x,t)\\ &= (\pi\theta)^{-2}\Big(\prod_{i=1}^{3}\int dx_i dt_i\Big)\exp i\Big[\frac{2}{\theta}\sum_{cyclic}(x_i t_j - t_i x_j)\Big]\prod_{i=1}^{3}\phi_i(x_i,t_i).\end{aligned} \tag{12}$$

The exponent $\frac{2}{\theta}\sum_{cyclic}(x_i t_j - t_i x_j)$ in this expression is the formal analogue of the first term of the string action (7) with the identification $\theta \sim \alpha'$. Actually, there is a crucial difference in that the exponential factor in field theory case directly leads to non-causal shifts of the time coordinates in proportion to the momenta of external lines as $t_1 - t_2 \sim \theta p_3, \ldots$. Hence the sign of the time shifts depends on the direction of momenta. In the case of strings, the connection between the external momenta and the shifts, if any, of the time is not so direct as in the case of the simple Moyal product. For example, the center-of-mass spatial coordinates do not directly contribute to the noncommutativity in (11). The dynamics of strings is completely local at each point of world sheet and hence the time-like nature of the area element ensures causality in the evolution of the system. As discussed above, causality is preserved as long as the external space-time electric B field does not exceed the critical value.

However, the above formal analogy prompts us to speculate a possibility of formulating the space-time uncertainty and noncommutativity as a certain kind of 'deformation' from classical space-time geometry to quantum and stringy geometry. It is a major challenge to find some unique characterization of such stringy deformation of space-time geometry.

3. Another remark which might be useful in understanding the nature of the present reformulation is that the counterpart in particle theory of what we have done above is simply the momentum representation of the particle propagator. We can start from the familiar particle action

$$L = -\frac{1}{2}\int d\tau\Big(-\frac{1}{e}(\frac{dx^\mu}{d\tau})^2 + em^2\Big).$$

By introducing another auxiliary field p_μ which is now a space-time vector corresponding to the line element of the world line, we can rewrite it as

$$L_p = \int d\tau\Big(p_\mu \frac{dx^\mu}{d\tau} - \frac{1}{2}e(p^2+m^2)\Big).$$

Obviously, the role of the Virasoro condition in strings is now played by the mass-shell condition $p^2+m^2=0$ requiring that the momentum is time-like (or light-like when $m=0$). In the particle case, the action $\int d\tau\, p_\mu \frac{dx^\mu}{d\tau}$ defines a usual Poisson structure. In analogy with this, the string action $\frac{1}{4\pi\alpha'}\int d^2\xi\, \epsilon^{ab}\partial_a X^\mu \partial_b X^\nu b_{\mu\nu}$ can be regarded as defining a generalized Poisson structure which is appropriate to strings.[a] One natural possibility along this line would be to regard the auxiliary b-field as a sort of momentum variable corresponding to the area element of string world sheet. Since, comparing to the particle case, the string case has one additional dimension, this line of thought leads to the use of the so-called Nambu bracket. Such a possibility was indeed suggested in Ref. 12. Unfortunately, there seems to be no appropriate quantization procedure based on this interpretation. The interpretation we have given in the present article by means of the ordinary Dirac bracket quantization seems to be the only viable possibility toward quantization. Our discussion, however, remains still at a very formal level. It is an open question whether the above formalism can lead to a new exactly calculable scheme in full-fledged quantum theory. It is tempting to speculate a possibility of some tractable integral representation of string amplitudes where the ordinary moduli parameters of Riemann surfaces are integrated over. Instead of the moduli parameters, we should have some different integration variables corresponding to the 'area momenta'. I leave such a possibility for the reader as an interesting new direction in exploring string theory.

[a] After writing the previous review,[3] the present author came to know by reading reference 12 that a suggestion which is closely related to the present remark was first made by Nambu in Ref.13.

4. Another relevant question related to the more precise formulation of the present approach is to discuss the curved background. A natural way of including the space-time metric is by introducing the vielbein field $e_\mu^A(X)$ where A is the local Lorentz index. We can assume that the auxiliary field $b_{AB}(\xi)$ now transforms as an antisymmetric tensor at each local Lorentz frame on the space-time point $X^\mu(\xi)$ with the constraint $(1/2)b_{AB}^2 = -1$. The area element is then written as $\epsilon^{ab}\partial_a X^\mu \partial_b X^\nu e_\mu^A(x(\xi))e_\nu^B(x(\xi))b_{AB}(\xi)$. The Dirac bracket relation is more complicated than the flat space, with the right hand side depending on the space-time coordinates. We expect that the requirement of consistent quantization would lead to the condition for the space-time background which should be equivalent to the familiar β-function condition of renormalization group. It is an important problem to work this out. Of course, once we could arrive at a satisfiable characterization of the deformed geometry as suggested above, such a property would be an evident consequence of the general formalism.

Even apart from further clarification and refinements of the ideas discussed here, there are innumerably many other remaining questions, such as the relevance of the space-time uncertainties and noncommutativity for black-hole physics, the interpretation from the viewpoint of 11 dimensional M-theory, consistency with S- and T- dualities, interpretation of possible other scales than the M-theory scale, the role of supersymmetry, and so on. Some of these questions have been treated partially in Ref. 3. Real answers to most of the deeper questions, however, must be left to the future.

Acknowledgments

I would like to thank the organizers of Strings 2000 conference for inviting me. The present work is supported in part by Grant-in-Aid for Scientific Research (No. 12440060) from the Ministry of Education, Science and Culture.

References

1. T. Yoneya, in *Wandering in the Fields*, eds. K. Kawarabayashi and A. Ukawa (World Scientific, 1987), p. 419.
2. T. Yoneya, *Mod. Phys. Lett.* **A4**, 1587 (1989).
3. T. Yoneya, *Prog. Theor. Phys.* **103**, 1081 (2000) [hep-th/0004074].
4. D. Gross, talk at the Munich conference in 1987, See Proc. XXIV Int. Conf. High Energy Physics, Munich, eds. R. Lotthaus and J. Kühn (Springer Verlag, 1989); D. Gross and P. Mende, *Nucl. Phys.* **B303**, 407 (1988).
5. P. Mende and H. Ooguri, *Nucl. Phys.* **B303**, 407 (1988).
6. A. Jevicki and T. Yoneya, *Nucl. Phys.* **B535**, 335 (1998).
7. M. Li and T. Yoneya, *Phys. Rev. Lett.* **78**, 1219 (1997); M. Li and T. Yoneya, *Chaos, Solitons and Fractals* **10**, 423 (1999) [hep-th/9806240].
8. T. Yoneya, *Lett. Nuovo. Cim.* **8**, 951 (1973); *Prog. Theor. Phys.* **51**, 1907 (1974); **56**, 1310 (1976).
9. J. Scherk and J. Schwarz, *Nucl. Phys.* **B81**, 118 (1974); *Phys. Lett.* **57B**, 463 (1975).
10. N. Seiberg, L. Susskind and N. Toumbas, *JHEP* **0006**, 044 (2000) [hep-th/0005015];

R. Gopakumar, J. Maldacena, S. Minwalla and A. Strominger, *JHEP* **0006**, 036 (2000) [hep-th/0005048]; J.L.F. Barbón and E. Ravinovici, *Phys. Lett.* **B486**, 202 (2000) [hep-th/0005073].
11. J. Gomis and T. Mehren, *Nucl. Phys.* **B591**, 265 (2000) [hep-th/0005129].
12. I.V. Kanatchikov, *Rept. Math. Phys.* **40**, 225 (1997) [hep-th/9710069].
13. Y. Nambu, *Phys. Lett.* **92B**, 327 (1980).

NONABELIAN D-BRANES AND NONCOMMUTATIVE GEOMETRY

ROBERT MYERS
Department of Physics, McGill University, 3600 University Street
Montréal, Québec, H3A 2T8, Canada
E-mail: rcm@hep.physics.mcgill.ca

We discuss the nonabelian world-volume action which governs the dynamics of N coincident Dp-branes. In this theory, the branes' transverse displacements are described by matrix-valued scalar fields, and so this is a natural physical framework for the appearance of noncommutative geometry. One example is the dielectric effect by which Dp-branes may be polarized into a noncommutative geometry by external fields. Another example is the appearance of noncommutative geometries in the description of intersecting D-branes of differing dimensions, such as D-strings ending on a D3- or D5-brane.

1. Introduction

The idea that noncommutative geometry should play a role in physical theories is an old one.[1,2] In the past few years, it has also become a topic of increasing interest to string theorists. From one point of view, the essential step in realizing a noncommutative geometry is replacing the spacetime coordinates by noncommuting operators: $x^\mu \to \hat{x}^\mu$. In this replacement, however, there remains a great deal of freedom in defining the nontrivial commutation relations which the operators $\hat{x}^\mu$ satisfy. Some explicit choices that have appeared in physical problems are as follows:

(i) Canonical commutation relations:

$$[\hat{x}^\mu, \hat{x}^\nu] = i\theta^{\mu\nu} \qquad \theta^{\mu\nu} \in \boldsymbol{C}$$

Such algebras have appeared in the Matrix theory description of planar D-branes[3] — for a review, see Ref. 4. The latter stimulated an ongoing investigation by string theorists of noncommutative field theories which arise in the low energy limit of planar D-branes carrying constant B-field fluxes.[5]

(ii) Quantum space relations:

$$\hat{x}^\mu\, \hat{x}^\nu = q^{-1}\, R^{\mu\nu}{}_{\rho\tau}\, \hat{x}^\rho\, \hat{x}^\tau \qquad R^{\mu\nu}{}_{\rho\tau} \in \boldsymbol{C}$$

These algebras received some attention from physicists in the early 1990's — see, *e.g.*, Ref. 6 — and have appeared more recently in the geometry of the moduli space of N=4 super-Yang-Mills theory.[7]

(iii) Lie algebra relations:

$$[\hat{x}^\mu, \hat{x}^\nu] = i f^{\mu\nu}{}_\rho\, \hat{x}^\rho \qquad f^{\mu\nu}{}_\rho \in \boldsymbol{C}$$

Such algebras naturally arise in the description of fuzzy spheres as was discovered in early attempts to quantize the supermembrane.[8] These noncommutative geometries have been applied in Matrix theory to describe spherical branes.[9,10]

The latter noncommutative descriptions of spheres also arise in various contexts in the physics of D-branes, as will be discussed in the following. We begin in section 2 with a discussion of the nonabelian D-brane action. Section 3 presents an outline of the dielectric effect for D-branes, and section 4 gives a discussion of how noncommutative geometries arise in the description of intersecting branes. Sections 2 and 3 are essentially a summary of the material appearing in Ref. 11. Section 4 describes that for Refs. 12 and 13. We direct the interested reader to these papers for more detailed discussions.

2. Nonabelian D-brane Action

One remarkable aspect of the D-brane story[14] is that the $U(1)$ gauge symmetry of an individual D-brane is enhanced to a nonabelian U(N) symmetry for N coincident Dp-branes.[15] This effect lies at the heart of such recent developments as the entropy counting of near-BPS black holes[16] and the AdS/CFT correspondence.[17] Refs. 11 and 18 recently made progress in constructing the world-volume action describing the dynamics of nonabelian D-branes. Just as for the abelian theory, the nonabelian action has two distinct pieces. First, one has the Born-Infeld action

$$-T_p \int d^{p+1}\sigma \,\mathrm{STr}\left\{e^{-\phi}\sqrt{-\det\left(P\left[E_{ab}+E_{ai}(Q^{-1}-\delta)^{ij}E_{jb}\right]+\lambda\, F_{ab}\right)\det(Q^i{}_j)}\right\} \tag{1}$$

with $E_{\mu\nu} = G_{\mu\nu} + B_{\mu\nu}$ and $Q^i{}_j \equiv \delta^i{}_j + i\lambda\,[\Phi^i, \Phi^k]\,E_{kj}$. Here, T_p is the Dp-brane tension and λ denotes the inverse of the (fundamental) string tension, *i.e.*, $\lambda = 2\pi\ell_s^2$. This Born-Infeld action describes the couplings of the Dp-brane to the massless Neveu-Schwarz (NS) fields of the bulk closed string theory, *i.e.*, the metric, dilaton and Kalb-Ramond two-form. The interactions with the massless Ramond-Ramond (RR) fields are incorporated in the second part of the action, the Wess-Zumino term

$$\mu_p \int \mathrm{STr}\left\{P\left[e^{i\lambda \mathrm{i}_\Phi \mathrm{i}_\Phi}(\sum C^{(n)}\, e^B)\right] e^{\lambda\, F}\right\}\ . \tag{2}$$

Here $C^{(n)}$ denote the n-form RR potentials. Eq. (2) shows that a Dp-brane is naturally charged under the (p+1)-form RR potential with charge μ_p.

The transverse displacements of the branes are parameterized by the world-volume scalar fields, Φ^i. In the nonabelian theory, these scalars are in the adjoint representation of the U(N) gauge symmetry. Hence the physics of nonabelian D-branes becomes a natural setting for noncommutative geometry given that the displacements are encoded in matrix-valued fields. The world-volume scalars appear in the action in three ways: First, there is the explicit appearance in the first exponential in Eq. (2). Here, i_Φ denotes the interior product by Φ^i regarded as a vector in the transverse space, *e.g.*, acting on an n-form $C^{(n)} = \frac{1}{n!}C^{(n)}_{\mu_1\cdots\mu_n}dx^{\mu_1}\cdots dx^{\mu_n}$,

$$\mathrm{i}_\Phi\mathrm{i}_\Phi C^{(n)} = \frac{1}{2(n-2)!}[\Phi^i,\Phi^j]\,C^{(n)}_{ji\mu_3\cdots\mu_n}dx^{\mu_3}\cdots dx^{\mu_n}\,. \tag{3}$$

It is essential that the scalars are matrix-valued since for ordinary vectors, $\mathrm{i}_v\mathrm{i}_v C^{(n)} = 0$. Next, both actions involve the pull-back (denoted by $P[\cdots]$) of various spacetime tensors to the world-volume. This involves covariant derivatives of the scalars, *e.g.*,

$$P[E]_{ab} = E_{ab} + \lambda\,E_{ai}\,D_b\Phi^i + \lambda\,E_{ib}\,D_a\Phi^i + \lambda^2\,E_{ij}D_a\Phi^iD_b\Phi^j\,. \tag{4}$$

Finally, the bulk supergravity fields are in general functions of all of the spacetime coordinates, and so in the world-volume action, they are implicitly functionals of the nonabelian scalars. For example, the metric functional appearing in the D-brane action would be given by a nonabelian Taylor expansion

$$G_{\mu\nu} = \sum_{n=0}^{\infty}\frac{\lambda^n}{n!}\,\Phi^{i_1}\cdots\Phi^{i_n}\,(\partial_{x^{i_1}}\cdots\partial_{x^{i_n}})G_{\mu\nu}(\sigma^a,x^i)|_{x^i=0}\,. \tag{5}$$

In both parts of the action, there is a single symmetrized gauge trace which encompasses all of the scalars appearing in these various ways (and the gauge fields).

The implicit appearance of the nonabelian scalars in the functional dependence[19] and pull-back[20] of the background fields was originally argued on general grounds. The interesting commutator couplings to the bulk fields were first discovered in the construction of the nonabelian action in Refs. 11 and 18. There, the action was deduced by demanding that the nonabelian theory must be consistent with T-duality. All of these different types of interactions have been confirmed to leading order by examining string scattering amplitudes[21] using the methods of Ref. 22.

An individual Dp-brane carrying a flux of $B+F$ acts as a source for RR potentials with a lower form degree,[23] *i.e.*, $C^{(n)}$ with $n = p-1, p-3, \ldots$. In the nonabelian action (2), the operator $\mathrm{i}_\Phi\mathrm{i}_\Phi$ has form degree −2, and so world-volume interactions also appear involving the higher RR forms. That is in the nonabelian theory, a Dp-brane can also couple to the RR potentials with $n = p+3, p+5, \ldots$ through the additional commutator interactions. For example, the D0-brane action includes a linear coupling to C^3, which is the potential corresponding to D2-brane charge,

$$\begin{aligned} &i\lambda\,\mu_0\int \mathrm{Tr}\,P\left[\mathrm{i}_\Phi\mathrm{i}_\Phi C^{(3)}\right] \qquad\qquad (6)\\ &\quad = i\frac{\lambda}{2}\mu_0\int dt\ \mathrm{Tr}\left(C^{(3)}_{tjk}(\Phi,t)\,[\Phi^k,\Phi^j] + \lambda C^{(3)}_{ijk}(\Phi,t)\,D_t\Phi^k\,[\Phi^k,\Phi^j]\right)\,. \end{aligned}$$

Note that the first term on the right hand side has the form of a source for D2-brane charge. Essentially this interaction appears in the construction of D2-branes in Matrix theory taking the large N limit.[3] Here with finite N, this term would vanish upon taking the trace if $C^{(3)}_{tjk}$ was simply a function of the world-volume coordinate t. However, in general these three-form components are functionals of Φ^i. Hence, while there would be no monopole coupling to D2-brane charge, nontrivial expectation values of the scalars can give rise to higher multipole couplings.

3. Dielectric Branes

In flat space, the nonabelian scalar potential arises from the Born-Infeld action:

$$V = T_p \,\mathrm{Tr}\sqrt{det(Q^i{}_j)} = \mathrm{N}T_p - \frac{T_p\lambda^2}{4}\mathrm{Tr}([\Phi^i,\Phi^j]\,[\Phi^i,\Phi^j]) + \dots \tag{7}$$

A nontrivial set of extrema of this potential is given by taking the scalars as constant commuting matrices, *i.e.*,

$$[\Phi^i,\Phi^j] = 0 \tag{8}$$

for all i and j. Since they are commuting, the Φ^i may be simultaneously diagonalized and the eigenvalues are interpreted as the separated positions of N fundamental Dp-branes in the transverse space.

From the results described section 2, it is clear that in going from flat space to general background fields, the scalar potential is modified by new interactions and so one should reconsider the analysis of the extrema. It turns out that this yields an interesting physical effect that is a precise analog for D-branes of the dielectric effect in ordinary electromagnetism. That is when Dp-branes are placed in a nontrivial background field for which the Dp-branes would normally be regarded as neutral, new terms will be induced in the scalar potential, and one expects that generically there will be new extrema beyond those found in flat space. In particular, there can be nontrivial extrema with noncommuting expectation values of the Φ^i. This would correspond to the external fields polarizing the Dp-branes to expand into a (higher dimensional) noncommutative world-volume geometry. This is the analog of the familiar process where an electric field can induce a separation of charges in neutral materials. In this case, the polarized material then carries an electric dipole (and possibly higher multipoles). The latter is also seen in the D-brane analog. At a noncommutative extremum, the gauge traces of products of scalars will be nonvanishing in various interactions involving the supergravity fields. Hence, the Dp-branes act as a source for the latter bulk fields.

To make these ideas explicit, we illustrate the process with a simple example. We consider N D0-branes in a constant background RR field $F^{(4)}$, *i.e.*, the field strength associated with D2-brane charge. We find that the D0-branes expand into a noncommutative two-sphere which represents a spherical bound state of a D2-brane and N D0-branes.

Consider a background with the RR four-form field strength:

$$F^{(4)}_{tijk} = -2f\varepsilon_{ijk} \qquad \text{for } i,j,k \in \{1,2,3\} \tag{9}$$

where f is a constant. Since $F^{(4)} = dC^{(3)}$, we must consider the coupling of the D0-branes to the RR three-form potential, in Eq. (6). If one explicitly introduces the nonabelian Taylor expansion (5), one finds the leading order interaction may be written as

$$\frac{i}{3}\lambda^2\mu_0\int dt\,\mathrm{Tr}\left(\Phi^i\Phi^j\Phi^k\right)F^{(4)}_{tijk}(t)\ . \tag{10}$$

Combining Eq. (10) with the leading order Born-Infeld potential yields

$$V(\Phi) = \mathrm{N}T_0 - \frac{\lambda^2 T_0}{4}\mathrm{Tr}([\Phi^i,\Phi^j]^2) - \frac{i}{3}\lambda^2\mu_0\mathrm{Tr}\left(\Phi^i\Phi^j\Phi^k\right)F^{(4)}_{tijk}(t)\ . \tag{11}$$

Substituting in the background field (9) and $\mu_0 = T_0$, $\delta V(\Phi)/\delta\Phi^i = 0$ yields

$$0 = [[\Phi^i,\Phi^j],\Phi^j] + i\,f\varepsilon_{ijk}[\Phi^j,\Phi^k]\ . \tag{12}$$

Note that commuting matrices (8) describing separated D0-branes still solve this equation. The value of the potential for these solutions is simply $V_0 = NT_0$, the mass of N D0-branes. Another interesting solution of Eq. (12) is: $\Phi^i = f/2\,\alpha^i$ where α^i are an N×N matrix representation of the SU(2) algebra

$$[\alpha^i,\alpha^j] = 2i\,\varepsilon_{ijk}\,\alpha^k\ . \tag{13}$$

Now evaluating the value of the potential (11) for this new solution yields

$$V = \mathrm{N}T_0 - \frac{\pi^2\ell_s^3 f^4}{6g}\mathrm{N}^3\left(1-\frac{1}{\mathrm{N}^2}\right) \tag{14}$$

using $T_0 = 1/(g\ell_s)$ and the irreducible representation of the SU(2) algebra. Hence the noncommutative solution has lower energy than a solution of commuting matrices, and the latter configuration of separated D0-branes is unstable towards condensing out into this noncommutative solution. The irreducible representation appears to describe the ground state as it yields a lower energy than any reducible representation of the SU(2) algebra.

Geometrically, one can recognize the SU(2) algebra as that corresponding to the noncommutative or fuzzy two-sphere.[8] The radius of the fuzzy two-sphere is

$$R = \frac{\lambda}{\sqrt{\mathrm{N}}}\left(\sum\mathrm{Tr}[(\Phi^i)^2]\right)^{1/2} = \pi\ell_s^2 f\mathrm{N}\left(1-\frac{1}{\mathrm{N}^2}\right)^{1/2} \tag{15}$$

in the ground state solution. From the matrix model construction of Kabat and Taylor,[10] one can infer this ground state is not simply a spherical arrangement of D0-branes rather the noncommutative solution actually represents a spherical D2-brane with N D0-branes bound to it. In the present context, the latter can be verified by seeing that this configuration has a dipole coupling to the RR four-form. The precise form of this coupling is calculated by substituting the noncommutative scalar solution into the world-volume interaction (10), which yields

$$-\frac{R^3}{3\pi g\ell_s^3}\left(1-\frac{1}{\mathrm{N}^2}\right)^{-1/2}\int dt\,F^{(4)}_{t123}\ . \tag{16}$$

for the ground state solution. Physically this $F^{(4)}$-dipole moment arises because antipodal surface elements on the sphere have the opposite orientation and so form small pairs of separated membranes and anti-membranes.

Given that the noncommutative ground state solution corresponds to a bound state of a spherical D2-brane and N D0-branes, this system can be analyzed from the point of view of the (abelian) world-volume theory of a D2-brane. One would consider a spherical D2-brane carrying a flux of the U(1) gauge field strength representing the N bound D0-branes, and at the same time, sitting in the RR four-form background (9). In fact, the results match those calculated in the framework of the D0-branes remarkably well. The results for the energy, radius and dipole coupling are the same as in Eqs. (14), (15) and (16), respectively, except that the factors of $(1 - 1/N^2)$ are absent.[11] This agreement is surprising since it appears there is no common domain where the two pictures can both produce reliable results. One expects that the D2-brane calculations would be valid when $R \gg \ell_s$ while naively the D0-brane calculations would be valid when $R \ll \ell_s$. However, a more careful consideration of the latter shows that the D0-brane calculations are reliable for $R \ll \sqrt{N}\ell_s$. This inequality ensures that the scalar field commutators appearing in the full nonabelian potential (7) are small so that the Taylor expansion of the square root converges rapidly. Hence for large N, there is a large domain of overlap where both of the dual pictures are reliable.

The example considered above must be considered simply a toy calculation demonstrating the essential features of the dielectric effect for D-branes. A more complete calculation would require analyzing the D0-branes in a consistent supergravity background. For example, the present case could be extended to consider the asymptotic supergravity fields of a D2-brane, where the RR four-form would be slowly varying but the metric and dilaton fields would also be nontrivial. Alternatively, one can find solutions with a constant background $F^{(4)}$ in M-theory, namely the $AdS_4 \times S^7$ and $AdS_7 \times S^4$ backgrounds. In lifting the D0-branes to M-theory, they become gravitons carrying momentum in the internal space. Hence the expanded D2-D0 system considered here correspond to the giant gravitons of Ref. 24. The analog of the D2-D0 bound state in a constant background $F^{(4)}$ corresponds to M2-branes[25] with internal momentum expanding into AdS_4, while that in a constant H field corresponds to the M2-branes[24] expanding on S^4. Alternatively, the dielectric effect has been found to play a role in other string theory contexts, for example, in the resolution of certain singularities in the AdS/CFT correspondence,[26] or in describing D-branes in the spacetime background of a WZW model.[27] Further, one can consider background field configurations which through the dielectric effect generate more complicated noncommutative geometries.[28]

4. Intersecting Branes

One interesting aspect of the abelian Born-Infeld action is that it supports solitonic configurations describing lower-dimensional branes protruding from the original D-brane.[29] For example, in the case of a D3-brane, one finds spike or bion solutions corresponding to fundamental strings and/or D-strings extending out of the D3-brane. In these configurations, both the world-volume gauge fields and transverse scalar fields are excited. The gauge field corresponds to that of a point charge arising from the end-point of the attached string, *i.e.*, an electric charge for a fundamental string and a magnetic monopole charge for a D-string. The scalar field describes the deformation of the D3-brane geometry caused by attaching the strings. These solutions seem to have a surprisingly wide range of validity. In fact, one can show that the electric spike corresponding to a fundamental string is a solution of the full string theory equations of motion.[30] Further the dynamics of these solutions, as probed through small fluctuations, agrees with the expected string behavior.[31]

For the system of N D-strings ending on a D3-brane, there is also a dual description in terms of the nonabelian world-volume theory of the N D-strings. There one finds solutions which have an interpretation in terms of noncommutative geometry as describing the D-strings expanding out in a funnel to become an orthogonal D3-brane. In fact, there is an extensive discussion of this system in the literature — see, *e.g.*, Ref. 32 or 33 — where the emphasis was on the close connection[32] of the D-string equations to the Nahm equations for BPS monopoles.[34] In Ref. 12, our emphasis was on the interpretation of these solutions in terms of noncommutative geometry and the remarkable agreement that one finds with the D3-brane spikes in the large N limit.

For N D-strings in flat space, the dynamics is determined completely by the Born-Infeld action (1) which reduces to[11,36]

$$S = -T_1 \int d^2\sigma \, \mathrm{STr}\sqrt{-\det\left(\eta_{ab} + \lambda^2 \partial_a \Phi^i Q^{-1}_{ij} \partial_b \Phi^j\right) \, \det\left(Q^{ij}\right)} \,, \tag{17}$$

where $Q^{ij} = \delta^{ij} + i\lambda[\Phi^i, \Phi^j]$. Implicitly here, the world-volume gauge field has been set to zero, which is a consistent truncation for the configurations considered below. To find the dual description of the bion solutions of the D3-brane theory,[29] one needs a static solution which represents the D-strings expanding into a D3-brane. The corresponding geometry would be a long funnel where the cross-section at fixed σ has the topology of a two-sphere. In this context, the latter cross-section naturally arises as a fuzzy two-sphere[8] if the scalars have values in an N$\times$N matrix representation of the SU(2) algebra (13). Hence one is lead to consider the ansatz

$$\Phi^i = \frac{R(\sigma)}{\lambda\sqrt{\mathrm{N}^2-1}}\,\alpha^i, \quad i = 1,2,3, \tag{18}$$

where we will focus on case where the α^i are the irreducible $N \times N$ SU(2) matrices. Then with the normalization in Eq. (18), the function $|R(\sigma)|$ corresponds precisely to the radius of the fuzzy two-sphere

$$R(\sigma)^2 = \frac{\lambda^2}{\mathrm{N}} \sum \mathrm{Tr}[\Phi^i(\sigma)^2] \,. \tag{19}$$

Substituting the ansatz (18) into the matrix equations of motion derived from (17), one finds a simple class of solutions of the form:

$$R(\sigma) = \pm\frac{\mathrm{N}\pi\ell_s^2}{\sigma}\left(1-\frac{1}{\mathrm{N}^2}\right)^{1/2} \tag{20}$$

The geometry of the solution, Eqs. (18) and (20), certainly has the desired funnel shape. The fuzzy two-sphere shrinks to zero size as $\sigma \to \infty$ and opens up to fill the $x^{1,2,3}$ hypersurface at $\sigma = 0$. By examining the nonabelian Wess-Zumino action (2), one can show that the noncommutative solution induces a coupling to RR four-form potential $C^{(4)}_{t123}$. This calculation confirms then that, with the minus (plus) sign in Eq. (20), the D-strings expand into a(n anti-)D3-brane which fills the $x^{1,2,3}$ directions.[12] Given that the funnel solution of the D-string theory and the bion spike of the D3-brane theory are both BPS, one might expect that there will be a good agreement between these two dual descriptions. The formula for the height of D3-brane spike above the $x^{1,2,3}$ hyperplane is[29]

$$\sigma = \frac{\mathrm{N}\pi\ell_s^2}{R}\ . \tag{21}$$

Comparing to Eq. (20), one finds that for large N the two descriptions are describing the same geometry up to $1/\mathrm{N}^2$ corrections. One finds similar quantitative agreement for large N in calculating the energy, the RR couplings and the low energy dynamics in the two dual descriptions.[12] As in the discussion of the dielectric effect, one can argue that the D3-brane description is valid for $R \gg \ell_s$ while the D-string description is reliable for $R \ll \sqrt{\mathrm{N}}\ell_s$. Hence one can understand the good agreement between these dual approaches for large N since there is a large domain of overlap where both are reliable.

Note that in the configurations considered in this section, there are no nontrivial supergravity fields in the ambient spacetime. Hence the appearance of the noncommutative geometry in these solutions is quite distinct from that in the dielectric effect, where the external fields drive the D-branes into a certain geometry in the ground state. In the funnel solutions, the noncommutative geometry was introduced in the ansatz (18) by hand. An interesting extension of these solutions is then to replace the SU(2) generators by those corresponding to some other noncommutative geometry, *i.e.*, to replace Eq. (18) by

$$\Phi^i = \frac{R(\sigma)}{\lambda\sqrt{C}}\,G^i \tag{22}$$

where the G^i are new $\mathrm{N}\times\mathrm{N}$ constant matrices satisfying $\sum(G^i)^2 = N\,C$.

One particular example that we have examined in detail[13] is the case where G^i are the generators describing a fuzzy four-sphere — these may be found in, *e.g.*, Ref. 10. In this case, the funnel describes the D-strings expanding into a D5-brane. Interestingly for small R, one finds $\sigma \propto \ell_s^2/R$ as in Eq. (21). In fact, this behavior is universal for all funnels on the D-string in any situation, and the only difference is in the overall coefficient. This universal behavior arises because the leading order (in λ) equations of motion always have the same form. However, for large R, nonlinearities in the Born-Infeld action (17) become important, and this behavior is modified. For the case of the fuzzy four-sphere, one has $\sigma \propto \ell_s^4/R^3$ at large R. The same kind of behavior is also found for the corresponding solutions in the dual D5-brane world-volume theory, although of course in this description, the nonlinearities of the Born-Infeld action are important for small R. An interesting feature of the D5-brane spike is that it is also nonabelian in character. Charge conservation arguments indicate that the D-string acts as a source of second Chern class in the world-volume of the D5-brane.[37] More precisely, if N D-strings end on a D5-brane, then

$$\frac{1}{8\pi^2}\int_{S^4} \mathrm{Tr}(F \wedge F) = \mathrm{N}\ , \tag{23}$$

for any four-sphere surrounding the D-string endpoint. Hence both of the dual descriptions have a noncommutative character. Again, we find that the dual constructions seem to agree at large N, however, the details of the solutions are more complex. In part, the latter must be due to the fact that the $D5 \perp D1$ system is not supersymmetric.

Acknowledgments

This research was supported by NSERC of Canada and Fonds FCAR du Québec. I would like to thank Neil Constable and Oyvind Tafjord for collaborations in the research presented in Refs. 12 and 13.

References

1. H.S. Snyder, *Phys. Rev.* **71**, 38 (1947); *Phys. Rev.* **72**, 68 (1947).
2. A. Connes, *Noncommutative Geometry* (Academic Press, 1994).
3. T. Banks, W. Fischler, S.H. Shenker and L. Susskind, *Phys. Rev.* **D55**, 5112 (1997) [hep-th/9610043]; T. Banks, N. Seiberg and S. Shenker, *Nucl. Phys.* **B490**, 91 (1997) [hep-th/9612157].
4. W. Taylor, *The M(atrix) model of M-theory*, lectures given at NATO Advanced Study Institute on Quantum Geometry, Akureyri, Iceland, 10-20 Aug 1999 [hep-th/0002016].
5. A. Connes, M.R. Douglas and A. Schwarz, *JHEP* **9802**, 003 (1998) [hep-th/9711162]; N. Seiberg and E. Witten, *JHEP* **9909**, 032 (1999) [hep-th/9908142].
6. J. Wess and B. Zumino, *Nucl. Phys. Proc. Suppl.* **18B**, 302 (1991); P. Schupp, P. Watts and B. Zumino, *Lett. Math. Phys.* **25**, 139 (1992) [hep-th/9206029].
7. D. Berenstein, V. Jejjala and R.G. Leigh, *Nucl. Phys.* **B589**, 196 (2000) [hep-th/0005087]; hep-th/0006168.
8. J. Hoppe, *Quantum Theory of a Massless Relativistic Surface and a Two-Dimensional Bound State Problem*, MIT Ph.D. thesis, 1982 (available at http://www.aei-potsdam.mpg.de/~hoppe); B. de Wit, J. Hoppe and H. Nicolai, *Nucl. Phys.* **B305**, 545 (1988).
9. D. Kabat and W. Taylor, *Adv. Theor. Math. Phys.* **2**, 181 (1998) [hep-th/9711078].
10. J. Castelino, S. Lee and W. Taylor, *Nucl. Phys.* **B526**, 334 (1998) [hep-th/9712105].
11. R.C. Myers, *JHEP* **9912**, 022 (1999) [hep-th/9910053].
12. N.R. Constable, R.C. Myers and O. Tafjord, *Phys. Rev.* **D61**, 106009 (2000) [hep-th/9911136].

13. N.R. Constable, R.C. Myers and O. Tafjord, in preparation.
14. J. Polchinski, *TASI lectures on D-branes*, hep-th/9611050; C.V. Johnson, *D-brane primer*, hep-th/0007170.
15. E. Witten, *Nucl. Phys.* **B460**, 335 (1996) [hep-th/9510135].
16. A.W. Peet, *TASI lectures on black holes in string theory*, hep-th/0008241.
17. O. Aharony, S.S. Gubser, J. Maldacena, H. Ooguri and Y. Oz, *Phys. Rept.* **323**, 183 (2000) [hep-th/9905111].
18. W. Taylor and M. Van Raamsdonk, *Nucl. Phys.* **B573**, 703 (2000) [hep-th/9910052].
19. M.R. Douglas, *Nucl. Phys. Proc. Suppl.* **68**, 381 (1998) [hep-th/9707228]; *Adv. Theor. Math. Phys.* **1**, 198 (1998) [hep-th/9703056].
20. C.M. Hull, *JHEP* **9810**, 011 (1998) [hep-th/9711179]; H. Dorn, *Nucl. Phys.* **B494**, 105 (1997) [hep-th/9612120].
21. M.R. Garousi and R.C. Myers, *Nucl. Phys.* **B542**, 73 (1999) [hep-th/9809100]; *World-volume Potentials on D-branes*, hep-th/0010122.
22. M.R. Garousi and R.C. Myers, *Nucl. Phys.* **B475**, 193 (1996) [hep-th/9603194]; A. Hashimoto and I.R. Klebanov, *Nucl. Phys. Proc. Suppl.* **55B**, 118 (1997) [hep-th/9611214].
23. M.R. Douglas, hep-th/9512077.
24. J. McGreevy, L. Susskind and N. Toumbas, *JHEP* **0006**, 008 (2000) [hep-th/0003075].
25. M.T. Grisaru, R.C. Myers and O. Tafjord, *JHEP* **0008**, 040 (2000) [hep-th/0008015]; A. Hashimoto, S. Hirano and N. Itzhaki, *JHEP* **0008**, 051 (2000) [hep-th/0008016].
26. J. Polchinski and M.J. Strassler, hep-th/0003136.
27. C. Bachas, M. Douglas and C. Schweigert, *JHEP* **0005**, 048 (2000) [hep-th/0003037]; M. Li, hep-th/0003173.
28. S.P. Trivedi and S. Vaidya, *JHEP* **0009**, 041 (2000) [hep-th/0007011].
29. C.G. Callan and J.M. Maldacena, *Nucl. Phys.* **B513**, 198 (1998) [hep-th/9708147]; G.W. Gibbons, *Nucl. Phys.* **B514**, 603 (1998) [hep-th/9709027]; P.S. Howe, N.D. Lambert and P.C. West, *Nucl. Phys.* **B515**, 203 (1998) [hep-th/9709014].
30. L. Thorlacius, *Phys. Rev. Lett.* **80**, 1588 (1998) [hep-th/9710181].
31. S. Lee, A. Peet and L. Thorlacius, *Nucl. Phys.* **B514**, 161 (1998) [hep-th/9710097]; D. Bak, J. Lee and H. Min, *Phys. Rev.* **D59**, 045011 (1999) [hep-th/9806149]; K.G. Savvidy and G.K. Savvidy, *Nucl. Phys.* **B561**, 117 (1999) [hep-th/9902023]; D. Kastor and J. Traschen, *Phys. Rev.* **D61**, 024034 (2000) [hep-th/9906237].
32. D.-E. Diaconescu, *Nucl. Phys.* **B503**, 220 (1997) [hep-th/9608163].
33. J.P. Gauntlett, J. Gomis, P.K. Townsend, *JHEP* **9801**, 003 (1998) [hep-th/9711205]; D. Brecher, *Phys. Lett.* **B442**, 117 (1998) [hep-th/9804180]; A. Giveon and D. Kutasov, *Rev. Mod. Phys.* **71**, 983 (1999) [hep-th/9802067]; A. Kapustin and S. Sethi, *Adv. Theor. Math. Phys.* **2**, 571 (1998) [hep-th/9804027]; D. Tsimpis, *Phys. Lett.* **B433**, 287 (1998) [hep-th/9804081]; K. Hashimoto, *Prog. Theor. Phys.* **101**, 1353 (1999) [hep-th/9808185].
34. W. Nahm, *Phys. Lett.* **B90**, 413 (1980); *The construction of all self-dual multi-monopoles by the ADHM method*, in *Monopoles in quantum field theory*, ed. Craigie *et al.* (World Scientific, Singapore, 1982).
35. A. Hashimoto, *Phys. Rev.* **D57**, 6441 (1998) [hep-th/9711097].
36. A.A. Tseytlin, hep-th/9908105.
37. G.W. Semenoff and K. Zarembo, *Nucl. Phys.* **B556**, 247 (1999) [hep-th/9903140].

STABLE NON-BPS STATES AND THEIR HOLOGRAPHIC DUALS

SUNIL MUKHI and NEMANI V. SURYANARAYANA
Tata Institute of Fundamental Research
Homi Bhabha Road, Mumbai 400 005, India

Stable non-BPS states can be constructed and studied in a variety of contexts in string theory. Here we review some interesting constructions that arise from suspended and wrapped branes. We also exhibit some stable non-BPS states that have holographic duals.

1. Introduction

Type II string theory has stable, BPS Dp-branes with $p = 0, 2, 4, 6, 8$ in type IIA, and $p = -1, 1, 3, 5, 7, 9$ in type IIB. For the other values of p one finds unstable, non-BPS branes: $p = -1, 1, 3, 5, 7, 9$ in type IIA and $p = 0, 2, 4, 6, 8$ in type IIB theory.[1,2] The spectrum on an unstable D-brane in superstring theory is the spectrum of a single open string, but without GSO projection. Hence there is a real tachyon.

The BPS branes are of course stable, while the non-BPS branes can decay, via tachyon condensation, into the vacuum, or into lower (BPS or non-BPS) branes. A pair of a BPS brane and its antibrane is also unstable and can decay similarly.

This is quite a general paradigm. In flat backgrounds, type IIA branes are either BPS and stable, or non-BPS and unstable. It is interesting to look for backgrounds which admit non-BPS but stable branes. In this situation, masses are not protected by BPS formulae. We can hope to disentangle effects of duality from effects of supersymmetry.

If the backgrounds are themselves non-supersymmetric then things rapidly become difficult. The most accessible situations are those where the backgrounds are supersymmetric, but the states that we study are not. Some examples are orbifolds, orientifolds and Calabi-Yau compactifications. Another class of examples is provided by suspended brane constructions.[3,4] These all have lower supersymmetry than flat space, which helps to find stable non-BPS states.

In the following, we first investigate brane-antibrane configurations in the flat-space background of type II superstring theory and identify some stable non-BPS states. Next, we turn to AdS-type backgrounds and their holographically dual gauge theories.[5] Here, We analyze stable, non-BPS configurations of branes wrapped over cycles in the $AdS_5 \times T^{1,1}$ background that is dual to 3-branes at a conifold.[6] In the course of the discussion we will make extensive use of the conifold singularity and its brane-construction dual.[7] ALE spaces will also play an auxiliary role.

2. Singularities, Brane Duals and Fractional Branes

Let us start with type IIB on a Z_2 ALE singularity along the (6789) directions. Via T-duality along x^6, the ALE singularity turns into a pair of parallel NS5-branes in type IIA string theory, extending along the (12345) directions and located at different points along the x^6 direction.[8]

The ALE singularity hides a 2-cycle Σ of zero size, which can be resolved to get an Eguchi-Hanson space. But at the orbifold point, the NS-NS B-field has a flux of $\frac{1}{2}$ through this 2-cycle.[9] In the brane dual, the NS5-branes are symmetrically located along the x^6 circle. This duality extends beyond the orbifold point. Varying the B-flux in the ALE corresponds to varying the relative x^6 separations of the NS5-branes.[10]

If we bring a D3-brane into the plane of an ALE singularity, it can split into a pair of fractional D3-branes $f3, f3'$ of charge and tension α and $1-\alpha$ where $\alpha = \int_\Sigma B$ is the B-flux.[11] The fractional branes are interpreted as:

$$\begin{array}{ll} f3: & \text{D5 wrapped on } \Sigma \\ f3': & \overline{\text{D}}5 \text{ wrapped on } \Sigma, \quad \int_\Sigma F = 1 \end{array}$$

In the dual brane construction, a D4-brane wrapped on x^6 can be brought in to touch the NS5-branes, where it can break into two pieces (Fig.1(a)). The gauge

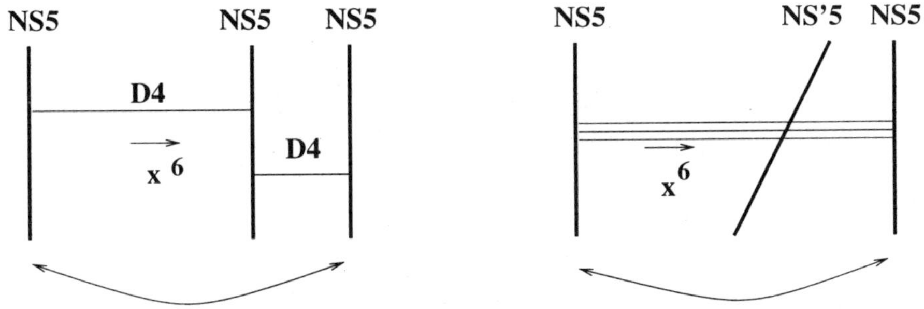

Fig. 1. (a) T-dual of D3-brane at an ALE singularity. (b) T-dual of D3-branes at a conifold.

group $U(1) \times U(1)$ and the presence of bi-fundamental matter is also evident from the brane construction.

An analogous relation holds for the conifold singularity along the (456789) directions. It is dual to a similar brane construction but with rotated NS5-branes (Fig.1(b)).[7] This model too has bi-fundamental matter, but also a quartic superpotential as long as there is more than one D4-brane as shown in the figure.[12]

3. Fractional Branes and a Stable Non-BPS Configuration

An interesting class of non-BPS brane configurations is obtained from the system of an adjacent brane-antibrane pair.[3] In some cases, this can be analyzed using perturbative string theory, via duality to ALE or conifold singularities.

The configuration of interest contains a pair of parallel NS5-branes oriented as was just discussed. In the two intervals between the NS5-branes, we place a D4-brane and a $\overline{\text{D}}$4-brane (Fig.2(a)). The NS5-brane configuration is T-dual to an

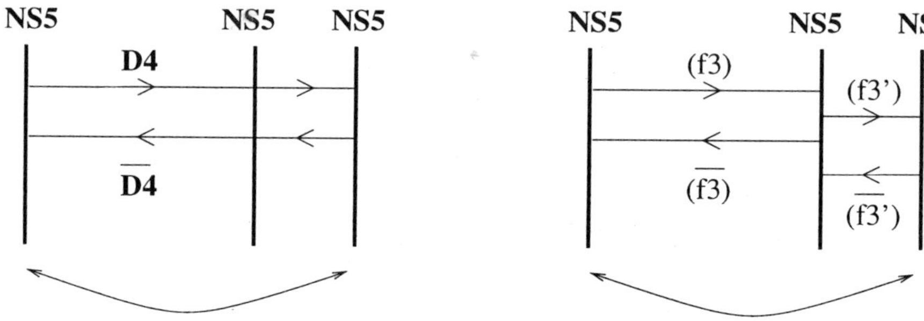

Fig. 2. (a) Brane-antibrane pair at an ALE. (b) The four types of (T-dual) fractional branes.

ALE singularity. The D4 and $\overline{\text{D}}$4-brane in the intervals T-dualize into a fractional D3-brane and a fractional antibrane. Let us try to understand this correspondence in more detail.

A D3 $-$ $\overline{\text{D}}$3 pair at a Z_2 ALE singularity splits into 4 distinct types of fractional branes, which we call $f3, f3', \overline{f3}, \overline{f3}'$. These are interpreted as follows:

$$\begin{array}{rll} f3: & \text{D5 wrapped on } \Sigma, & \int_\Sigma F = 0 \\ f3': & \overline{\text{D}}\text{5 wrapped on } \Sigma, & \int_\Sigma F = 1 \\ \overline{f3}: & \overline{\text{D}}\text{5 wrapped on } \Sigma, & \int_\Sigma F = 0 \\ \overline{f3}': & \text{D5 wrapped on } \Sigma, & \int_\Sigma F = 1 \end{array} \tag{1}$$

Introducing a D4 $-$ $\overline{\text{D}}$4 pair in the brane construction, we see that it too can break into four distinct pieces: This is the Coulomb branch, and we can identify the four fractional branes as in Fig.2(b).

Since we are interested in studying an adjacent D4 $-$ $\overline{\text{D}}$4 pair, we see that the dual fractional branes are $f3$ and $\overline{f3}'$. This system has a net D5-brane charge of $+2$, and a net D3-brane charge of $2\alpha - 1$.

The open strings connecting adjacent branes correspond in the ALE dual to the following Chan-Paton factors:

$$\begin{array}{rcl} f3 - \overline{f3}' & : & \frac{1}{2}(\sigma_3 + i\sigma_2) \otimes (\sigma_1 + i\sigma_2) \\ \overline{f3}' - f3 & : & \frac{1}{2}(\sigma_3 - i\sigma_2) \otimes (\sigma_1 - i\sigma_2) \\ f3' - \overline{f3} & : & \frac{1}{2}(\sigma_3 - i\sigma_2) \otimes (\sigma_1 + i\sigma_2) \\ \overline{f3} - f3' & : & \frac{1}{2}(\sigma_3 + i\sigma_2) \otimes (\sigma_1 - i\sigma_2) \end{array} \tag{2}$$

These are all odd under the ALE projection. Therefore the strings connecting $f3$ to $\overline{f3}'$ have no tachyonic or massless bosonic states. In fact, these strings only give massless fermions.

Next we construct the boundary states corresponding to the fractional D3-branes, and use them to compute the force between the adjacent pair of interest. There are four independent consistent boundary states for D3, $\overline{\mathrm{D}}3$, which can be identified with the four fractional branes $f3, f3', \overline{f3}', \overline{f3}$.[13]

$$\begin{aligned}
|\mathrm{D}3,+\rangle &= \tfrac{1}{2}\Big(|U\rangle_{NSNS} + |U\rangle_{RR} + |T\rangle_{NSNS} + |T\rangle_{RR}\Big) && : f3 \\
|\mathrm{D}3,-\rangle &= \tfrac{1}{2}\Big(|U\rangle_{NSNS} + |U\rangle_{RR} - |T\rangle_{NSNS} - |T\rangle_{RR}\Big) && : f3' \\
|\overline{\mathrm{D}}3,+\rangle &= \tfrac{1}{2}\Big(|U\rangle_{NSNS} - |U\rangle_{RR} - |T\rangle_{NSNS} + |T\rangle_{RR}\Big) && : \overline{f3}' \\
|\overline{\mathrm{D}}3,-\rangle &= \tfrac{1}{2}\Big(|U\rangle_{NSNS} - |U\rangle_{RR} + |T\rangle_{NSNS} - |T\rangle_{RR}\Big) && : \overline{f3}
\end{aligned} \tag{3}$$

The amplitude of interest is:

$$\begin{aligned}
&\int_0^\infty dl \langle \overline{\mathrm{D}}3,+|e^{-lH_c}|\mathrm{D}3,+\rangle = \int_0^\infty \tfrac{dt}{2t}\,\mathrm{tr}\;_{NS-R}\left(\tfrac{1-(-1)^F}{2}\,\tfrac{1-R}{2}\,e^{-2tH_0}\right) \\
&= \tfrac{v^{(4)}}{32(2\pi)^4}\int_0^\infty \tfrac{dt}{t^3}\left\{\tfrac{f_3(\widetilde{q})^8+f_4(\widetilde{q})^8-f_2(\widetilde{q})^8}{f_1(\widetilde{q})^8} - 4\tfrac{f_4(\widetilde{q})^4 f_3(\widetilde{q})^4+f_4(\widetilde{q})^4 f_3(\widetilde{q})^4}{f_1(\widetilde{q})^4 f_2(\widetilde{q})^4}\right\}
\end{aligned} \tag{4}$$

This simplifies to:

$$\frac{v^{(4)}}{16(2\pi)^4}\int_0^\infty \frac{dt}{t^3}\frac{f_4(\widetilde{q})^8}{f_1(\widetilde{q})^8}\left[1 - 4\frac{f_1(\widetilde{q})^4 f_3(\widetilde{q})^4}{f_2(\widetilde{q})^4 f_4(\widetilde{q})^4}\right] \tag{5}$$

The integrand is strictly negative, implying that the force between the $f3$ and $\overline{f3}'$ is repulsive. Thus we find that the force between an adjacent suspended brane-antibrane pair is repulsive.

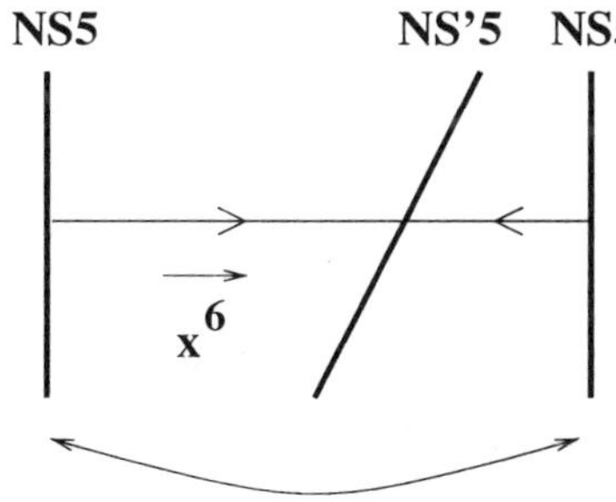

Fig. 3. Adjacent brane-antibrane pair at a T-dual conifold.

Now consider a "twist" on the configuration of adjacent brane-antibrane pairs that we discussed earlier.[4] We rotate one NS5-brane as in Fig.3. Thus we now have an NS5 and an NS5'-brane, making up the brane dual of the conifold. The adjacent D4-brane-antibrane pair is dual to fractional D3-branes at a conifold.

Physically, we expect a repulsive force between the adjacent brane and antibrane, as was shown earlier in the unrotated model. But there is also a classical attraction since the branes cannot separate without being stretched. This leads to a possibility of stable equilibrium at finite displacement. In fact we get a more complicated and interesting result exhibiting a phase transition as a function of the radius r of the compact x^6 direction.

The energy of the stretched D4-brane is

$$\mathcal{V} T_4 \sqrt{L^2 + 2r^2} \tag{6}$$

where $\mathcal{V}$ is an (infinite) volume factor, T_4 is the tension of a BPS D4-brane, and L is the separation between the NS5 and NS5'-branes.

We assume that the repulsion is as for the ALE (unrotated) case, since it comes from strings connecting the $D4 - \overline{D}4$ pair across each NS5-brane. After a calculation,[4] we find that the shape of the potential depends on the separation parameter L as shown in Figs.4(a) and 4(b) for small L and large L respectively. Here $y \sim r$ with a constant rescaling.

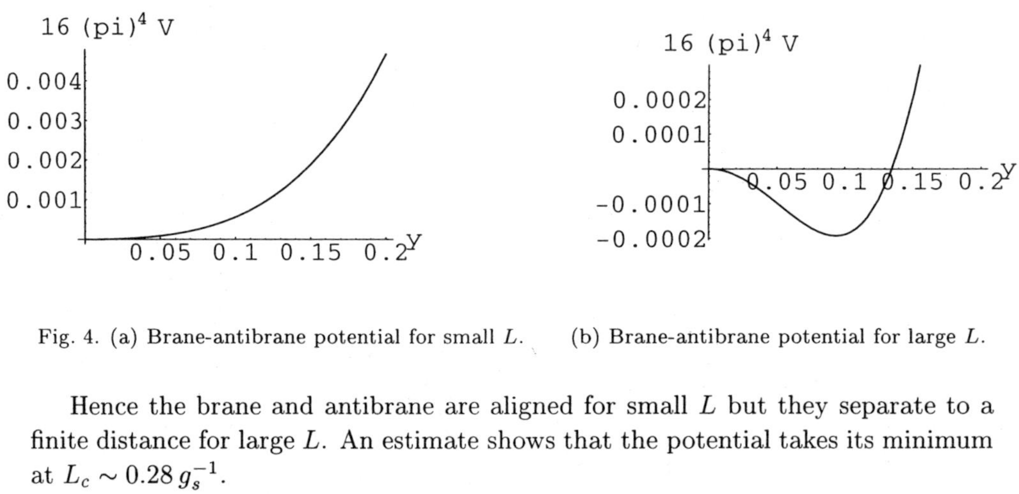

Fig. 4. (a) Brane-antibrane potential for small L. (b) Brane-antibrane potential for large L.

Hence the brane and antibrane are aligned for small L but they separate to a finite distance for large L. An estimate shows that the potential takes its minimum at $L_c \sim 0.28\, g_s^{-1}$.

4. Branes at a Conifold and Non-BPS States in AdS_5

If we bring N D3-branes to a conifold singularity and take the large-N limit, we end up with a $\frac{1}{4}$-supersymmetric background of type IIB: $AdS_5 \times T^{1,1}$ where $T^{1,1}$ is a particular Einstein 5-manifold.[6] If we T-dualize the conifold we get a model of rotated NS5-branes. N D3-branes at the conifold become N D4-branes wrapped round the x^6 circle, as described above.[7]

The adjacent brane-antibrane model that we have described above does not have an AdS dual. If we add N D4-branes to it, then the $\overline{D}4$ will annihilate against a fractional D4-brane, leaving $N-1$ whole D4-branes plus two fractional D4-branes, as shown in Fig.5. Let us now describe a stable non-BPS brane construction that,

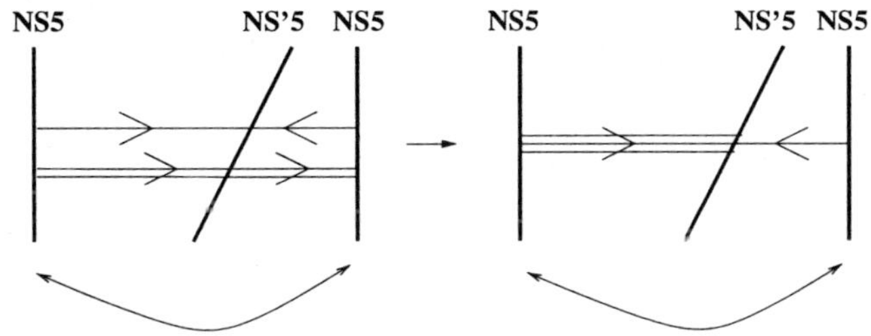

Fig. 5. Adjacent brane-antibrane pair at a T-dual conifold.

instead, does have an AdS dual.

Take N D4-branes as before and introduce a D2-brane in the first interval, as in Fig.6. In the conifold geometry, this corresponds to the introduction of a

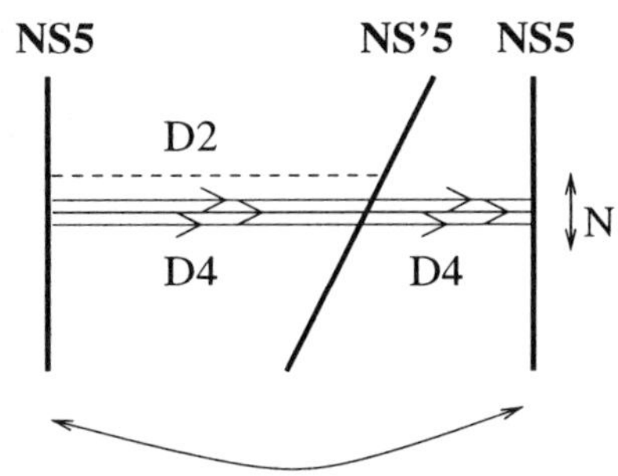

Fig. 6. T-dual conifold with D2 and D4-branes.

fractional D-string in the plane of the singularity. This configuration is clearly non-supersymmetric. For example, the strings joining a D2-brane and N D4-branes in the interval will be tachyonic. This part of the configuration will decay into a stable bound state of the D4-branes and the D2-brane. While this is BPS by itself, the neighbouring interval still has only D4-branes, as in Fig.7. The $(D2, D4)$ bound state and the D4-branes preserve incompatible supersymmetries. Hence the whole system is non-BPS, much as for an adjacent brane-antibrane pair. In the conifold geometry, we have a fractional D-string bound to N $f3$-branes and coincident with N $f3'$ branes.

This system is stable, and we can take the large N limit. In this limit, the conifold geometry is replaced by its 5-manifold base, the Einstein space $T^{1,1}$. Topologically,

$$T^{1,1} \sim S^2 \times S^3$$

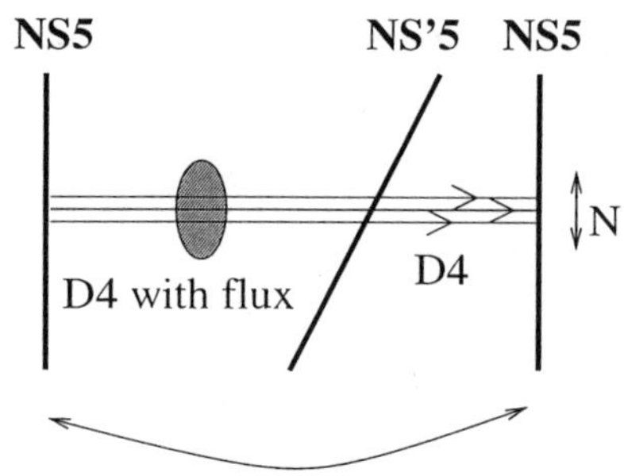

Fig. 7. T-dual conifold with D4-branes and flux.

The S^2 is the same 2-cycle that was of vanishing size before taking the large-N limit. The fractional D-string was actually a D3-brane wrapped on this S^2.[14,10] Hence, in the large N limit, the fractional D-string can be identified with a "fat string" obtained by wrapping a D3-brane on S^2. We will analyze this fat string further in the following section.

At this point, it is instructive to list all the unwrapped and wrapped branes of this model:

Dimension	Unwrapped	S^2	S^3	$S^2 \times S^3$
-1	$D(-1)$	$D1$	$UD2$	$UD4$
0	$UD0$	$UD2$	$D3$	$D5$
1	$D1$	$D3$	$UD4$	$UD6$
2	$UD2$	$UD4$	$D5$	$D7$
3	$D3$	$D5$	$UD6$	$UD8$
4	$UD4$	$UD6$	$D7$	$D9$

In this table, the prefix "U" indicates an unstable brane. The remaining branes are stable. Some of these wrapped branes have been studied previously.[14,10] For example, the D5 wrapped on S^2 is known to be a domain wall that augments the gauge group: $U(N) \times U(N) \to U(N+1) \times U(N)$, while the D3 wrapped on S^2 is our fat string. We would like to understand its holographic dual description. The Euclidean D-string wrapped on S^2 gives rise to an instanton, while the (unstable) $UD2$ on S^2 is an unstable D0-brane.[a] We will comment on their holographic duals too.

5. Properties of the Fat String

The nature of the fat string depends on the B-flux through S^2. In general we have

$$\int_{S^2} B_{NS,NS} = \alpha, \qquad \int_{S^2} B_{RR} = \beta$$

[a]These are distinct from the standard type IIB D-instantons and D0-branes, whose holographic duals have been studied previously.[15,16]

The $U(N) \times U(N)$ gauge theory on the 3-branes has couplings and θ-angles given by:[17]

$$\begin{aligned} \tau_1 &= \beta + \alpha\tau_s \\ \tau_2 &= -\beta + (1-\alpha)\tau_s \end{aligned} \tag{7}$$

where $\tau_s = \frac{\chi_{RR}}{2\pi} + \frac{i}{g_s}$.

The fat string carries D-string charge α and F-string charge β, by virtue of the Chern-Simons coupling

$$\int B_{NS,NS} \wedge B_{RR} \rightarrow \alpha \int B_{RR} + \beta \int B_{NS,NS}$$

on a D3-brane. It is convenient to choose $\beta = 0$. The tension of the fat string can be estimated from integrating the DBI action of a D3-brane over S^2:

$$T_{\text{fat}} \sim T_3 \int_{S^2} \sqrt{\det g + (B_{NS,NS})^2}$$

In the flat space limit, the S^2 is of zero size and this becomes

$$T_{\text{fat}} \sim T_3\, \alpha$$

which shows that it is BPS. On the other hand at large N the dominant contribution comes from

$$T_{\text{fat}} \sim T_3 \int_{S^2} \sqrt{g} \sim \frac{N}{(g_s N)^{\frac{1}{2}} \alpha'}$$

As with fractional branes, there are really two complementary fat strings, the second one being an anti D3-brane wrapped over S^2 and having a magnetic flux $\int F = 1$ over the cycle. We call this a fat$'$ string. It has a D-string charge $(1-\alpha)$. The non-BPS nature of fat strings, and their charges, imply that a fat string and a fat$'$ string can annihilate with loss of energy into a D-string.

Recall how a D-string is understood in holography.[18] In $AdS_5 \times S^5$, a D-string parallel to the boundary corresponds to a magnetic flux tube. As the string falls towards the horizon, the flux tube fattens and in the limit becomes a constant flux. The same result holds for a D-string in $AdS_5 \times T^{1,1}$, but the flux is in the diagonal of the $U(N) \times U(N)$ gauge group.

The fat string is similarly a flux tube in the boundary theory, but this time the flux is only in one $U(N)$ factor. This is consistent with its non-BPS nature. On a 3-brane we have nonlinearly realized supersymmetry that acts on the gauginos as:

$$\delta^* \lambda_\alpha^{(1)} = \frac{1}{4\pi\alpha'} \eta_\alpha^*, \quad \delta^* \lambda_\alpha^{(2)} = \frac{1}{4\pi\alpha'} \eta_\alpha^*$$

and linearly realized supersymmetry:

$$\delta \lambda_\alpha^{(1)} = F_{23}^{(1)} \sigma_\alpha^{23\,\beta}\, \eta_\beta, \quad \delta \lambda_\alpha^{(2)} = F_{23}^{(2)} \sigma_\alpha^{23\,\beta}\, \eta_\beta$$

If and only if the fluxes are diagonal: $F^{(1)} = F^{(2)} = F$, there is a surviving set of linearly realized supersymmetries, described by choosing

$$\eta^*_\alpha = -4\pi\alpha' F_{23}\sigma^{23\,\beta}_{\alpha}\,\eta_\beta$$

For non-diagonal fluxes, no supersymmetry is preserved.

One can also compute the potential experienced by the non-BPS fat string, and also study Wilson/'t Hooft loops in the *AdS* context.[19]

Let us comment briefly on some of the other wrapped branes. The euclidean D1 wrapped on S^2 is a new "D-instanton". It is expected to be dual to a Yang-Mills instanton in the first factor of $U(N) \times U(N)$. It has its own associated sphaleron, the D2-brane of type IIB wrapped on S^2. The relation between the two is parallel to the one between unwrapped D-instantons and D0-branes, studied recently.[20,16]

6. Conclusions

The stable brane-antibrane construction that we have exhibited should describe an interesting non-SUSY model field theory. Microscopically it has a pair of branes separated by a finite calculable distance. Such constructions might be useful in making brane-world type models. It would be interesting to understand the spectrum and interactions of the effective low-energy field theory on these branes, which have so far not been worked out in complete detail.

Recall that the conventional BPS brane constructions are most useful when we can use S-duality (in type IIB) or the duality with M-theory (in type IIA). What do we learn from these dualities about brane-antibrane constructions? A stable non-BPS brane configuration in type IIA theory, such as the one we have exhibited, must have a well-defined M-theory limit as an M5-brane wrapping a 2-cycle. Because the configuration is not BPS, the 2-cycle will not be holomorphic, so a novel approach would be required to determine it.

In the present discussion, "fat" objects were associated to one $U(N)$ factor while "thin" objects are diagonal in $U(N) \times U(N)$. This is quite general. The study of branes at more complicated singularities gives rise to product gauge groups involving many factors, and one should then be able to construct large numbers of non-BPS objects associated to one or more of these factors. Each one will be a "fat" object, related to a fractional brane at the generalized singularity.

The stable non-BPS configurations described here are particularly suitable for investigation using the AdS/CFT correspondence. This should help us to generalize many of the notions of holography to situations without supersymmetry, nevertheless retaining some control over the dynamics.

References

1. A. Sen *Non-BPS States and Branes in String Theory*, hep-th/9904207.
2. O. Bergman and M. Gaberdiel, *Non-BPS Dirichlet Branes*, *Class. Quant. Grav.* **17**, 961 (2000) [hep-th/9908126].

3. S. Mukhi, N.V. Suryanarayana and D. Tong, *Brane-Antibrane Constructions*, *JHEP* **0003**, 015 (2000) [hep-th/0001066].
4. S. Mukhi and N.V. Suryanarayana, *A Stable Non-BPS Configuration from Intersecting Branes and Antibranes*, *JHEP* **0006**, 001 (2000) [hep-th/0003219].
5. J. Maldacena, *The Large N Limit of Superconformal Field Theories and Supergravity*, *Adv. Theor. Math. Phys.* **2**, 231 (1998) [hep-th/9711200].
6. I. Klebanov and E. Witten, *Superconformal Field Theory on Three-branes at a Calabi-Yau Singularity*, *Nucl. Phys.* **B536**, 199 (1998) [hep-th/9807080].
7. A. Uranga, *Brane Configurations for Branes at Conifolds*, *JHEP* **9901**, 022 (1999) [hep-th/9811004];
K. Dasgupta and S. Mukhi, *Brane Constructions, Conifolds and M-Theory*, *Nucl. Phys.* **B551**, 204 (1999) [hep-th/9811139].
8. A. Karch, D. Lüst and D. Smith, *Equivalence of Geometric Engineering and Hanany-Witten via Fractional Branes*, *Nucl. Phys.* **B533**, 348 (1998) [hep-th/9803232].
9. P.S. Aspinwall, *Enhanced Gauge Symmetries and K3 Surfaces*, *Phys. Lett.* **B357**, 329 (1995) [hep-th/9507012].
10. K. Dasgupta and S. Mukhi, *Brane Constructions, Fractional Branes and Anti-deSitter Domain Walls*, *JHEP* **9907**, 008 (1999) [hep-th/9904131].
11. M.R. Douglas and G. Moore, *D-Branes, Quivers, and ALE Instantons*, hep-th/9603167;
J. Polchinski, *Tensors from K3 Orientifolds*, *Phys. Rev.* **D55**, 6423 (1997) [hep-th/9606165];
M.R. Douglas, *Enhanced Gauge Symmetry in M(atrix) Theory*, *JHEP* **9707**, 004 (1997) [hep-th/9612126];
D.-E. Diaconescu, M.R. Douglas and J. Gomis, *Fractional Branes and Wrapped Branes*, *JHEP* **9802**, 013 (1998) [hep-th/9712230].
12. O. Aharony and A. Hanany, *Branes, Superpotentials and Superconformal Fixed Points*, *Nucl. Phys.* **B504**, 239 (1997) [hep-th/9704170].
13. M. Gaberdiel and B. Stefanski, *Dirichlet Branes on Orbifolds*, *Nucl. Phys.* **B578**, 58 (2000) [hep-th/9910109].
14. S. Gubser and I. Klebanov, *Baryons and Domain Walls in an $N = 1$ Superconformal Gauge Theory*, *Phys. Rev.* **D58**, 125025 (1998) [hep-th/9808075].
15. C.-S. Chu, P.-M. Ho and Y.-Y. Wu, *D Instanton in AdS(5) and Instanton in SYM(4)*, *Nucl. Phys.* **B541**, 179 (1999) [hep-th/9806103];
M. Bianchi, M.B. Green, S. Kovacs and G. Rossi, *Instantons in Supersymmetric Yang-Mills and D Instantons in IIB Superstring Theory*, *JHEP* **9808**, 013 (1998) [hep-th/9807033].
16. N. Drukker, D.J. Gross and N. Itzhaki, *Sphalerons, Merons and Unstable Branes in ADS*, *Phys. Rev.* **D62**, 086007 (2000) [hep-th/0004131].
17. A. Lawrence, N. Nekrasov, C. Vafa, *On Conformal Field Theories in Four-Dimensions*, *Nucl. Phys.* **B533**, 199 (1998) [hep-th/9803015].
18. V. Balasubramanian, P. Kraus, A. Lawrence and S. Trivedi, *Holographic Probes of Anti-deSitter Space-times*, *Phys. Rev.* **D59**, 104021 (1999) [hep-th/9808017].
19. S. Mukhi and N.V. Suryanarayana, *Holographic Duals of Stable Non-BPS configurations*, to appear.
20. J.A. Harvey, P. Horava and P. Kraus, *D Sphalerons and the Topology of String Configuration Space*, *JHEP* **0003**, 021 (2000) [hep-th/0001143].

REPRESENTATIONS OF SUPERCONFORMAL ALGEBRAS IN THE $\mathrm{AdS}_{7/4}/\mathrm{CFT}_{6/3}$ CORRESPONDENCE

SERGIO FERRARA
CERN Theoretical Division
CH 1211 Geneva 23, Switzerland
and Laboratori Nazionali di Frsacati, INFN, Italy

EMERY SOKATCHEV
Laboratoire d'Annecy-le-Vieux de Physique Théorique LAPTH - BP 110*
F-74941 Annecy-le-Vieux Cedex, France

We perform a general analysis of representations of the superconformal algebras $\mathrm{OSp}(8/4,\mathbb{R})$ and $\mathrm{OSp}(8^*/2N)$ in harmonic superspace. We present a construction of their highest-weight UIR's by multiplication of the different types of massless conformal superfields ("supersingletons"). In particular, all "short multiplets" are classified. Representations undergoing shortening have "protected dimension" and may correspond to BPS states in the dual supergravity theory in anti-de Sitter space. These results are relevant for the classification of multitrace operators in boundary conformally invariant theories as well as for the classification of AdS black holes preserving different fractions of supersymmetry.

1. Introduction

Superconformal algebras and their representations play a crucial rôle in the AdS/CFT correspondence because of their dual rôle of describing the gauge symmetries of the AdS bulk supergravity theory and the global symmetries of the boundary conformal field theory.[1,2,3]

A special class of configurations which are particularly relevant are the so-called BPS states, i.e. dynamical objects corresponding to representations which undergo "shortening".

These representations can only occur when the conformal dimension of a (super)primary operator is "quantized" in terms of the R symmetry quantum numbers and they are at the basis of the so-called "non-renormalization" theorems of supersymmetric quantum theories.[4]

There exist different methods of classifying the UIR's of superconformal algebras. One is the so-called oscillator construction of the Hilbert space in which a given UIR acts.[5] Another one, more appropriate to describe field theories, is the realization of such representations on superfields defined in superspaces.[6,7] The latter are "supermanifolds" which can be regarded as the quotient of the conformal supergroup by some of its subgroups.

In the case of ordinary superspace the subgroup in question is the supergroup

*UMR 5108 associée à l'Université de Savoie

obtained by exponentiating a non-semisimple superalgebra which is the semidirect product of a super-Poincaré graded Lie algebra with dilatation ($SO(1,1)$) and the R symmetry algebra. This is the superspace appropriate for non-BPS states. Such states correspond to bulk massive states which can have "continuous spectrum" of the AdS mass (or, equivalently, of the conformal dimension of the primary fields).

BPS states are naturally associated to superspaces with lower number of "odd" coordinates and, in most cases, with some internal coordinates of a coset space G/H. Here G is the R symmetry group of the superconformal algebra, i.e. the subalgebra of the even part which commutes with the conformal algebra of space-time and H is some subgroup of G having the same rank as G.

Such superspaces are called "harmonic"[8] and they are characterized by having a subset of the initial odd coordinates θ. The complementary number of odd variables determines the fraction of supersymmetry preserved by the BPS state. If a BPS state preserves K supersymmetries then the θ's of the associated harmonic superspace will transform under some UIR of H_K.

For 1/2 BPS states, i.e. states with maximal supersymmetry, the superspace involves the minimal number of odd coordinates (half of the original one) and H_K is then a maximal subgroup of G. On the other hand, for states with the minimal fraction of supersymmetry H_K reduces to the "maximal torus" whose Lie algebra is the Cartan subalgebra of G.

It is the aim of the present paper to give a comprehensive treatment of BPS states related to "short representations" of superconformal algebras for the cases which are most relevant in the context of the AdS/CFT correspondence, i.e. the $d = 3$ ($N = 8$) and $d = 6$ ($N = (2,0)$). The underlying conformal field theories correspond to world-volume theories of N_c copies of M_2, M_5 and D_3 branes in the large N_c limit[9,10,11,12,13] which are "dual" to AdS supergravities describing the horizon geometry of the branes.[14]

The present contribution summarizes the results which have already appeared elsewhere.[15,16,17] We first carry out an abstract analysis of the conditions for Grassmann (G-)analyticity[18] (the generalization of the familiar concept of chirality[7]) in a superconformal context. We find the constraints on the conformal dimension and R symmetry quantum numbers of a superfield following from the requirement that it do not depend on one or more Grassmann variables. Introducing G-analyticity in a traditional superspace cannot be done without breaking the R symmetry. The latter can be restored by extending the superspace by harmonic variables[19,8,20,21,22,23,24] parametrizing the coset G/H_K. We also consider the massless UIR's ("supersingleton" multiplets),[25,26] first as constrained superfields in ordinary superspace[27,28,29] and then, for a part of them, as G-analytic harmonic superfields.[8,24,29] Next we use supersingleton multiplication to construct UIR's of $OSp(8^*/2N)$ and $OSp(8/4,\mathbb{R})$. We show that in this way one can reproduce the complete classification of UIR's of Ref. 30. We also discuss different kinds of shortening which certain superfields (not of the BPS type) may undergo. We conclude the paper by listing the various BPS states in the physically relevant cases of M_2 and M_5 branes horizon geometry where only one type of supersingletons appears.

Massive towers corresponding to 1/2 BPS states are the K-K modes coming from compactification of M-theory on $AdS_{7/4} \times S_{4/7}$.[31,9] Short representations of superconformal algebras also play a special rôle in determining N-point functions from OPE.[32,33]

Another area of interest is the classification of AdS black holes,[34,35,36,37] according to the fraction of supersymmetry preserved by the black hole background.

In a parallel analysis with black holes in asymptotically flat background,[38] the AdS/CFT correspondence predicts that such BPS states should be dual to superconformal states undergoing "shortening" of the type discussed here.

2. The Six-Dimensional Case

In this section we describe highest-weight UIR's of the superconformal algebras $OSp(8^*/2N)$ in six dimensions. Although the physical applications refer to $N = 1$ and $N = 2$, it is worthwhile to carry out the analysis for general N, along the same lines as in the four-dimensional case.[39,40] We first examine the consequences of G-analyticity and conformal supersymmetry and find out the relation to BPS states. Then we will construct UIR's of $OSp(8^*/2N)$ by multiplying supersingletons. The results exactly match the general classification of UIR's of $OSp(8^*/2N)$ of Ref. 30.

2.1. *The conformal superalgebra $OSp(8^*/2N)$ and Grassmann analyticity*

The standard realization of the conformal superalgebra $OSp(8^*/2N)$ makes use of the superspace

$$\mathbb{R}^{6|8N} = \frac{OSp(8^*/2N)}{\{K, S, M, D, T\}} = (x^\mu, \theta^{\alpha\, i}) \tag{2.1}$$

where $\theta^{\alpha\, i}$ is a left-handed spinor carrying an index $i = 1, 2, \ldots, 2N$ of the fundamental representation of the R symmetry group $USp(2N)$. Unlike the four-dimensional case, here chirality is not an option but is already built in. The only way to obtain smaller superspaces is through Grassmann analyticity. We begin by imposing a single condition of G-analyticity on the superfield defined in (2.1):

$$q^1_\alpha \Phi(x, \theta) = 0 \tag{2.2}$$

which amounts to considering the coset

$$\mathbb{A}^{6|4(2N-1)} = \frac{OSp(8^*/2N)}{\{K, S, M, D, T, Q^1\}} = (x^\mu, \theta^{\alpha\, 1,2,\ldots,2N-1}) \tag{2.3}$$

From the algebra of $OSp(8^*/2N)$ we obtain

$$m_{\mu\nu} = 0\,, \tag{2.4}$$

$$t^{11} = t^{12} = \ldots = t^{1\, 2N-1} = 0\,, \tag{2.5}$$

$$4t^{1\, 2N} + \ell = 0\,. \tag{2.6}$$

Eq. (2.4) implies that the superfield Φ must be a Lorentz scalar. In order to interpret Eqs. (2.5), (2.6), we need to split the generators of USp($2N$) into raising operators (corresponding to the positive roots), $T^{k\,2N-l}$, $k = 1,\ldots,N$, $l = k,\ldots,2N - k$ (simple if $l = k$), $[\mathrm{U}(1)]^N$ charges $H_k = -2T^{k\,2N-k+1}$, $k = 1,\ldots,N$ and lowering operators. The Dynkin labels a_k of a USp($2N$) irrep are defined as follows:

$$a_k = H_k - H_{k+1}\ ,\quad k = 1,\ldots,N-1\ ,\quad a_N = H_N\ , \tag{2.7}$$

so that, for instance, the generator Q^1 is the HWS of the fundamental irrep $(1,0,\ldots,0)$

Now it becomes clear that (2.5) is part of the USp($2N$) irreducibility conditions whereas (2.6) relates the conformal dimension to the sum of the Dynkin labels:

$$\ell = 2\sum_{k=1}^{N} a_k\ . \tag{2.8}$$

Let us denote the highest-weight UIR's of the OSp($8^*/2N$) algebra by

$$\mathcal{D}(\ell; J_1, J_2, J_3; a_1,\ldots,a_N)$$

where ℓ is the conformal dimension, J_1, J_2, J_3 are the SU*(4) Dynkin labels and a_k are the USp($2N$) Dynkin labels of the first component. Then the G-analytic superfields defined above are of the type

$$\Phi(\theta^{1,2,\ldots,2N-1}) \ \Leftrightarrow\ \mathcal{D}(2\sum_{k=1}^{N} a_k; 0,0,0; a_1,\ldots,a_N)\ . \tag{2.9}$$

From the superconformal algebra it is clear that we can go on in the same manner until we remove half of the θ's, namely $\theta^{N+1},\ldots,\theta^{2N}$. Each time we have to set a new Dynkin label to zero. We can summarize by saying that the superconformal algebra OSp($8^*/2N$) admits the following short UIR's corresponding to BPS states:

$$\frac{p}{2N}\ \mathrm{BPS}:\quad \mathcal{D}(2\sum_{k=p}^{N} a_k; 0,0,0; 0,\ldots,0,a_p,\ldots,a_N)\ ,\quad p = 1,\ldots,N\ . \tag{2.10}$$

2.2. *Supersingletons*

There exist three types of massless multiplets in six dimensions corresponding to ultrashort UIR's (supersingletons) of OSp($8^*/2N$) (see, e.g., Ref. 41 for the case $N = 2$). All of them can be formulated in terms of constrained superfields as follows.

(i) The first type is described by a superfield $W^{\{i_1\ldots i_n\}}(x,\theta)$, $1 \le n \le N$, which is antisymmetric and traceless in the external USp($2N$) indices (for even n one can impose a reality condition). It satisfies the constraint (see Refs. 27 and 42)

$$D^{(k}_{\alpha} W^{\{i_1)i_2\ldots i_n\}} = 0 \qquad \Rightarrow \mathcal{D}(2; 0,0,0; 0,\ldots,0,a_n = 1,0,\ldots,0) \tag{2.11}$$

The components of this superfield are massless fields. In the case $N = n = 1$ this is the on-shell $(1,0)$ hypermultiplet and for $N = n = 2$ it is the on-shell $(2,0)$ tensor multiplet.[27,28]

(ii) The second type is described by a (real) superfield without external indices, $w(x,\theta)$ obeying the constraint

$$D^{(i}_{[\alpha} D^{j)}_{\beta]} w = 0 \qquad \Rightarrow \mathcal{D}(2;0,0,0;0,\dots,0)\ . \tag{2.12}$$

(iii) Finally, there exists an infinite series of multiplets described by superfields with n totally symmetrized external Lorentz spinor indices, $w_{(\alpha_1\dots\alpha_n)}(x,\theta)$ (they can be made real in the case of even n) obeying the constraint

$$D^i_{[\beta} w_{(\alpha_1]\dots\alpha_n)} = 0 \qquad \Rightarrow \mathcal{D}(2+n/2;n,0,0;0,\dots,0)\ . \tag{2.13}$$

As shown in Ref. 16, the six-dimensional massless conformal fields only carry reps $(J_1,0)$ of the little group $\mathrm{SU}(2)\times\mathrm{SU}(2)$ of a light-like particle momentum. This result is related to the analysis of conformal fields in d dimensions.[43,44] This fact implies that massless superconformal multiplets are classified by a single SU(2) and USp($2N$) R-symmetry and are therefore identical to massless super-Poincaré multiplets in five dimensions. Some physical implication of the above circumstance have recently been discussed in Ref. 45 where it was suggested that certain strongly coupled $d = 5$ theories effectively become six-dimensional.

2.3. *Harmonic superspace*

The massless multiplets *(i)*, *(ii)* admit an alternative formulation in harmonic superspace (see Refs. 46,47,29 for $N = 1,2$). The advantage of this formulation is that the constraints (2.11) become conditions for G-analyticity. We introduce harmonic variables describing the coset $\mathrm{USp}(2N)/[\mathrm{U}(1)]^N$:

$$u \in \mathrm{USp}(2N): \qquad u^I_i u^i_J = \delta^I_J\ ,\quad u^I_i \Omega^{ij} u^J_j = \Omega^{IJ}\ ,\quad u^I_i = (u^i_I)^*\ . \tag{2.14}$$

Here the indices i,j belong to the fundamental representation of USp($2N$) and I,J are labels corresponding to the $[\mathrm{U}(1)]^N$ projections. The harmonic derivatives

$$D^{IJ} = \Omega^{K(I} u^{J)}_i \frac{\partial}{\partial u^K_i} \tag{2.15}$$

form the algebra of $\mathrm{USp}(2N)_R$ realized on the indices I,J of the harmonics.

Let us now project the defining constraint (2.11) with the harmonics $u^K_k u^1_{i_1}\dots u^n_{i_n}$, $K = 1,\dots,n$:

$$D^1_\alpha W^{12\dots n} = D^2_\alpha W^{12\dots n} = \dots = D^n_\alpha W^{12\dots n} = 0 \tag{2.16}$$

where $D^K_\alpha = D^i_\alpha u^K_i$ and $W^{12\dots n} = W^{\{i_1\dots i_n\}} u^1_{i_1}\dots u^n_{i_n}$. Indeed, the constraint (2.11) now takes the form of a G-analyticity condition. In the appropriate basis in superspace the solution to (2.16) is a short superfield depending on part of the odd coordinates:

$$W^{12\dots n}(x_A, \theta^1, \theta^2, \dots, \theta^{2N-n}, u)\ . \tag{2.17}$$

In addition to (2.16), the projected superfield $W^{12\dots n}$ automatically satisfies the USp($2N$) harmonic irreducibility conditions

$$D^{K\ 2N-K} W^{12} = 0\ , \quad K = 1, \dots, N \tag{2.18}$$

(only the simple roots of USp($2N$) are shown). The equivalence between the two forms of the constraint follows from the obvious properties of the harmonic products $u^K_{[k} u^K_{i]} = 0$ and $\Omega^{ij} u^K_i u^L_j = 0$ for $1 \le K < L \le n$. The harmonic constraints (2.18) make the superfield ultrashort.

Finally, in case (ii), projecting the constraint (2.12) with $u^I_i u^I_j$ where $I = 1, \dots, N$ (no summation), we obtain the condition

$$D^I_\alpha D^I_\beta w = 0\ . \tag{2.19}$$

It implies that the superfield w is *linear* in each projection $\theta^{\alpha I}$.

2.4. *Series of UIR's of OSp*$(8^*/2N)$ *and shortening*

It is now clear that we can realize the BPS series of UIR's (2.10) as products of the different G-analytic superfields (supersingletons) (2.16).[a] BPS shortening is obtained by setting the first $p-1$ USp($2N$) Dynkin labels to zero:

$$\frac{p}{2N}\ \text{BPS}: \quad W^{[0,\dots,0,a_p,\dots,a_N]}(\theta^1, \theta^2, \dots, \theta^{2N-p}) = (W^{1\dots p})^{a_p} \dots (W^{1\dots N})^{a_N} \tag{2.20}$$

(note that even if $a_1 \neq 0$ we still have $1/2N$ shortening).

We remark that our harmonic coset USp($2N$)/$[\mathrm{U}(1)]^N$ is effectively reduced to

$$\frac{\mathrm{USp}(2N)}{\mathrm{U}(p) \times [\mathrm{U}(1)]^{N-p}} \tag{2.21}$$

in the case of $p/2N$ BPS shortening (just as it happened in four dimensions). Such a smaller harmonic space was used in Ref. 29 to formulate the $(2,0)$ tensor multiplet.

A study of the most general UIR's of OSp($8^*/2N$) (similar to the one of Ref. 48 for the case of SU($2,2/N$)) is presented in Ref. 30. We can construct these UIR's by multiplying the three types of supersingletons above:

$$w_{\alpha_1\dots\alpha_{m_1}} w_{\beta_1\dots\beta_{m_2}} w_{\gamma_1\dots\gamma_{m_3}}\ w^k\ W^{[a_1,\dots,a_N]} \tag{2.22}$$

where $m_1 \ge m_2 \ge m_3$ and the spinor indices are arranged so that they form an SU*(4) UIR with Young tableau (m_1, m_2, m_3) or Dynkin labels $J_1 = m_1 - m_2, J_2 =$

[a]As a bonus, we also prove the unitarity of these series, since they are obtained by multiplying massless unitary multiplets.

$m_2 - m_3, J_3 = m_3$. Thus we obtain four distinct series:

$$\text{A)} \qquad \ell \geq 6 + \frac{1}{2}(J_1 + 2J_2 + 3J_3) + 2\sum_{k=1}^{N} a_k \ ;$$

$$\text{B)} \qquad J_3 = 0 \ , \qquad \ell \geq 4 + \frac{1}{2}(J_1 + 2J_2) + 2\sum_{k=1}^{N} a_k \ ;$$

$$\text{C)} \qquad J_2 = J_3 = 0 \ , \qquad \ell \geq 2 + \frac{1}{2}J_1 + 2\sum_{k=1}^{N} a_k \ ;$$

$$\text{D)} \qquad J_1 = J_2 = J_3 = 0 \ , \qquad \ell = 2\sum_{k=1}^{N} a_k \ . \tag{2.23}$$

The superconformal bound is saturated when $k = 0$ in (2.22). Note that the values of the conformal dimension we can obtain are "quantized" since the factor w^k has $\ell = 2k$ and k must be a non-negative integer to ensure unitarity. With this restriction Eq. (2.23) reproduces the results of Ref. 30. However, we cannot comment on the existence of a "window" of dimensions $2+\frac{1}{2}J_1+2\sum_{k=1}^{N} a_k \leq \ell \leq 4+\frac{1}{2}J_1+2\sum_{k=1}^{N} a_k$ conjectured in Ref. 30.[b]

In the generic case the multiplet (2.22) is "long", but for certain special values of the dimension some shortening can take place.[30]

3. The Three-Dimensional Case

In this section we carry out the analysis of the $d = 3$ $N = 8$ superconformal algebra $\mathrm{OSp}(8/4,\mathbb{R})$ in a way similar to the above. Some of the results have already been presented in Ref. 15. As in the previous cases, our results could easily be extended to $\mathrm{OSp}(N/4,\mathbb{R})$ superalgebras with arbitrary N. The $N = 2$ and $N = 3$ cases were considered in Ref. 50.

3.1. *The conformal superalgebra OSp*$(8/4,\mathbb{R})$ *and Grassmann analyticity*

The standard realization of the conformal superalgebra $\mathrm{OSp}(8/4,\mathbb{R})$ makes use of the superspace

$$\mathbb{R}^{3|16} = \frac{\mathrm{OSp}(8/4,\mathbb{R})}{\{K,S,M,D,T\}} = (x^\mu, \theta^{\alpha\, i}) \ . \tag{3.1}$$

In order to study G-analyticity we need to decompose the generators Q^i_α under $[\mathrm{U}(1)]^4 \subset \mathrm{SO}(8)$. Besides the vector representation 8_v of SO(8) we are also going to use the spinor ones, 8_s and 8_c. In this context we find it convenient to introduce the four subgroups U(1) by successive reductions: $\mathrm{SO}(8) \rightarrow \mathrm{SO}(2)\times\mathrm{SO}(6) \sim \mathrm{U}(1)\times\mathrm{SU}(4) \rightarrow [\mathrm{SO}(2)]^2\times\mathrm{SO}(4) \sim [\mathrm{U}(1)]^2\times\mathrm{SU}(2)\times\mathrm{SU}(2) \rightarrow [\mathrm{SO}(2)]^4 \sim [\mathrm{U}(1)]^4$.

[b]In a recent paper[49] the UIR's of the six-dimensional conformal algebra SO(2, 6) have been classified. Note that the superconformal bound in case A (with all $a_i = 0$) is stronger that the purely conformal unitarity bounds found in Ref. 49.

Denoting the four U(1) charges by $\pm$, $(\pm)$, $[\pm]$ and $\{\pm\}$, we decompose the three 8-dimensional representations as follows:

$$8_v: \quad Q^i \quad \rightarrow \quad Q^{\pm\pm},\ Q^{(\pm\pm)},\ Q^{[\pm]\{\pm\}}, \tag{3.2}$$

$$8_s: \quad \phi^a \quad \rightarrow \quad \phi^{+(+)[\pm]},\ \phi^{-(-)[\pm]},\ \phi^{+(-)\{\pm\}},\ \phi^{-(+)\{\pm\}} \tag{3.3}$$

$$8_c: \quad \sigma^{\dot a} \quad \rightarrow \quad \sigma^{+(+)\{\pm\}},\ \sigma^{-(-)\{\pm\}},\ \sigma^{+(-)[\pm]},\ \sigma^{-(+)[\pm]} \tag{3.4}$$

Let us denote a quasi primary superconformal field of the OSp(8/4, $\mathbb{R}$) algebra by the quantum numbers of its HWS:

$$\mathcal{D}(\ell; J; a_1, a_2, a_3, a_4) \tag{3.5}$$

where ℓ is the conformal dimension, J is the Lorentz spin and a_i are the Dynkin labels (see, e.g., Ref. 51) of the SO(8) R symmetry.

In order to build G-analytic superspaces we have to add one or more projections of Q^i_α to the coset denominator. In choosing the subset of projections we have to make sure that: i) they anticommute among themselves; ii) the subset is closed under the action of the raising operators of SO(8). Then we have to examine the consistency of the vanishing of the chosen projections with the conformal superalgebra. Thus we find the following sequence of G-analytic superspaces corresponding to BPS states:

$$\frac{1}{8}\ \text{BPS}: \quad \left\{ \begin{array}{l} q^{++}_\alpha \Phi = 0 \ \rightarrow \\ \Phi(\theta^{++}, \theta^{(\pm\pm)}, \theta^{[\pm]\{\pm\}}) \\ \mathcal{D}(a_1 + a_2 + \frac{1}{2}(a_3 + a_4); 0; a_1, a_2, a_3, a_4) \end{array} \right. \tag{3.6}$$

$$\frac{1}{4}\ \text{BPS}: \quad \left\{ \begin{array}{l} q^{++}_\alpha \Phi = q^{(++)}_\alpha \Phi = 0 \ \rightarrow \\ \Phi(\theta^{++}, \theta^{(++)}, \theta^{[\pm]\{\pm\}}) \\ \mathcal{D}(a_2 + \frac{1}{2}(a_3 + a_4); 0; 0, a_2, a_3, a_4) \end{array} \right. \tag{3.7}$$

$$\frac{3}{8}\ \text{BPS}: \quad \left\{ \begin{array}{l} q^{++}_\alpha \Phi = q^{(++)}_\alpha \Phi = q^{[+]\{+\}}_\alpha \Phi = 0 \ \rightarrow \\ \Phi(\theta^{++}, \theta^{(++)}, \theta^{[+]\{\pm\}}, \theta^{[-]\{+\}}) \\ \mathcal{D}(\frac{1}{2}(a_3 + a_4); 0; 0, 0, a_3, a_4) \end{array} \right. \tag{3.8}$$

$$\frac{1}{2}\ \text{BPS (type I)}: \quad \left\{ \begin{array}{l} q^{++}_\alpha \Phi = q^{(++)}_\alpha \Phi = q^{[+]\{\pm\}}_\alpha \Phi = 0 \ \rightarrow \\ \Phi(\theta^{++}, \theta^{(++)}, \theta^{[+]\{\pm\}}) \\ \mathcal{D}(\frac{1}{2}a_3; 0; 0, 0, a_3, 0) \end{array} \right. \tag{3.9}$$

$$\frac{1}{2}\ \text{BPS (type II)}: \quad \left\{ \begin{array}{l} q^{++}_\alpha \Phi = q^{(++)}_\alpha \Phi = q^{[\pm]\{+\}}_\alpha \Phi = 0 \ \rightarrow \\ \Phi(\theta^{++}, \theta^{(++)}, \theta^{[\pm]\{+\}}) \\ \mathcal{D}(\frac{1}{2}a_4; 0; 0, 0, 0, a_4) \end{array} \right. \tag{3.10}$$

Note the existence of two types of 1/2 BPS states due to the two possible subsets of projections of q^i closed under the raising operators of SO(8).

3.2. *Supersingletons and harmonic superspace*

The supersingletons are the simplest OSp(8/4, $\mathbb{R}$) representations of the type (3.9) or (3.10) and correspond to $\mathcal{D}(1/2; 0; 0, 0, 1, 0)$ or $\mathcal{D}(1/2; 0; 0, 0, 0, 1)$. The existence of two distinct types of $d = 3$ $N = 8$ supersingletons has first been noted in Ref. 52. Each of them is just a collection of eight Dirac supermultiplets[26] made out of "Di" and "Rac" singletons.[25]

In order to realize the supersingletons in superspace we note that the HWS in the two supermultiplets above has spin 0 and the Dynkin labels of the 8_s or 8_c of SO(8), correspondingly. Therefore we take a scalar superfield $\Phi_a(x^\mu, \theta^\alpha_i)$ (or $\Sigma_{\dot a}(x^\mu, \theta^\alpha_i)$) carrying an external 8_s index a (or an 8_c index $\dot a$). These superfields are subject to the following on-shell constraints:[c]

$$\text{type I:} \qquad D^i_\alpha \Phi_a = \frac{1}{8}\gamma^i_{a\dot b}\tilde\gamma^j_{\dot b c} D^j_\alpha \Phi_c \; ; \tag{3.11}$$

$$\text{type II:} \qquad D^i_\alpha \Sigma_{\dot a} = \frac{1}{8}\tilde\gamma^i_{\dot a b}\gamma^j_{b\dot c} D^j_\alpha \Sigma_{\dot c} \; . \tag{3.12}$$

The two multiplets consist of a massless scalar in the 8_s (8_c) and spinor in the 8_c (8_s).

The harmonic superspace description of these supersingletons can be realized by taking the harmonic coset[d]

$$\frac{\mathrm{SO}(8)}{[\mathrm{SO}(2)]^4} \sim \frac{\mathrm{Spin}(8)}{[\mathrm{U}(1)]^4} \; . \tag{3.13}$$

Since $\mathrm{SO}(8) \sim \mathrm{Spin}(8)$ has three inequivalent fundamental representations, $8_s, 8_c, 8_v$, following Ref. 57 we introduce three sets of harmonic variables:

$$u^A_a \; , \; w^{\dot A}_{\dot a} \; , \; v^I_i \tag{3.14}$$

where A, $\dot A$ and I denote the decompositions of an 8_s, 8_c and 8_v index, correspondingly, into sets of four U(1) charges (see (3.2)-(3.4)). Each of the 8×8 real matrices (3.14) belongs to the corresponding representation of $\mathrm{SO}(8) \sim \mathrm{Spin}(8)$. This implies that they are orthogonal matrices (this is a peculiarity of SO(8) due to triality):

$$u^A_a u^B_a = \delta^{AB} \; , \quad w^{\dot A}_{\dot a} w^{\dot B}_{\dot a} = \delta^{\dot A\dot B} \; , \quad v^I_i v^J_i = \delta^{IJ} \; . \tag{3.15}$$

Further, we introduce harmonic derivatives (the covariant derivatives on the coset (3.13)):

$$D^{IJ} = u^A_a (\gamma^{IJ})^{AB} \frac{\partial}{\partial u^B_a} + w^{\dot A}_{\dot a} (\gamma^{IJ})^{\dot A\dot B} \frac{\partial}{\partial w^{\dot B}_{\dot a}} + v^{[I}_i \frac{\partial}{\partial v^{J]}_i} \; . \tag{3.16}$$

They respect the algebraic relations (3.15) among the harmonic variables and form the algebra of SO(8) realized on the indices $A, \dot A, I$ of the harmonics.

---

[c] See also Ref. 29 for the description of a supersingleton related to ours by SO(8) triality. Superfield representations of other $OSp(N/4)$ superalgebras have been considered in Refs. 53,54.

[d] A formulation of the above multiplet in harmonic superspace has been proposed in Ref. 29 (see also Refs. 55 and 56 for a general discussion of three-dimensional harmonic superspaces). The harmonic coset used in Ref. 29 is Spin(8)/U(4). Although the supersingleton itself does indeed live in this smaller coset (see Section), its residual symmetry $U(4)$ would not allow us to multiply different realizations of the supersingleton. For this reason we prefer from the very beginning to use the coset (3.13) with a minimal residual symmetry.

We now use the harmonic variables for projecting the supersingleton defining constraints (3.11), (3.12). It is easy to show that the projections $\Phi^{+(+)[+]}$ and $\Sigma^{+(+)\{+\}}$ satisfy the following G-analyticity constraints:

$$D^{++}\Phi^{+(+)[+]} = D^{(++)}\Phi^{+(+)[+]} = D^{[+]\{\pm\}}\Phi^{+(+)[+]} = 0 \,, \tag{3.17}$$

$$D^{++}\Sigma^{+(+)\{+\}} = D^{(++)}\Sigma^{+(+)\{+\}} = D^{[+]\{\pm\}}\Sigma^{+(+)\{+\}} = 0 \tag{3.18}$$

where $D^I_\alpha = v^I_i D^i_\alpha$, $\Phi^A = u^A_a \Phi_a$ and $\Sigma^{\dot A} = w^{\dot A}_{\dot a}\Sigma_{\dot a}$. This is the superspace realization of the 1/2 BPS shortening conditions (3.9), (3.10). In the appropriate basis in superspace $\Phi^{+(+)[+]}$ and $\Sigma^{+(+)\{+\}}$ depend on different halves of the odd variables as well as on the harmonic variables:

$$\text{type I}: \qquad \Phi^{+(+)[+]}(x_A, \theta^{++}, \theta^{(++)}, \theta^{[+]\{\pm\}}, u, w) \,, \tag{3.19}$$

$$\text{type II}: \qquad \Sigma^{+(+)\{+\}}(x_A, \theta^{++}, \theta^{(++)}, \theta^{[\pm]\{+\}}, u, w) \,. \tag{3.20}$$

In addition to the G-analyticity constraints (3.17), (3.18), the on-shell superfields $\Phi^{+(+)[+]}$, $\Sigma^{+(+)\{+\}}$ are subject to the SO(8) irreducibility harmonic conditions obtained by replacing the SO(8) raising operators by the corresponding harmonic derivatives. The combination of the latter with Eq. (3.17) is equivalent to the original constraint (3.11).

3.3. *OSp*(8/4, ℝ) *supersingleton composites*

One way to obtain short multiplets of OSp(8/4, ℝ) is to multiply different analytic superfields describing the type I supersingleton. The point is that above we chose a particular projection of, e.g., the defining constraint (3.11) which lead to the analytic superfield $\Phi^{+(+)[+]}$. In fact, we could have done this in a variety of ways, each time obtaining superfields depending on different halves of the total number of odd variables. Leaving out the 8_v lowest weight θ^{--}, we can have four distinct but equivalent analytic descriptions of the type I supersingleton:

$$\begin{aligned} &\Phi^{+(+)[+]}(\theta^{++}, \theta^{(++)}, \theta^{[+]\{+\}}, \theta^{[+]\{-\}}) \,, \\ &\Phi^{+(+)[-]}(\theta^{++}, \theta^{(++)}, \theta^{[-]\{+\}}, \theta^{[-]\{-\}}) \,, \\ &\Phi^{+(-)\{+\}}(\theta^{++}, \theta^{(--)}, \theta^{[+]\{+\}}, \theta^{[-]\{+\}}) \,, \\ &\Phi^{+(-)\{-\}}(\theta^{++}, \theta^{(--)}, \theta^{[+]\{-\}}, \theta^{[-]\{-\}}) \,. \end{aligned} \tag{3.21}$$

Then we can multiply them in the following way:

$$(\Phi^{+(+)[+]})^{p+q+r+s}(\Phi^{+(+)[-]})^{q+r+s}(\Phi^{+(-)\{+\}})^{r+s}(\Phi^{+(-)\{-\}})^{s} \tag{3.22}$$

thus obtaining three series of OSp(8/4, ℝ) UIR's exhibiting 1/8, 1/4 or 1/2 BPS shortening:

$$\begin{aligned} &\frac{1}{8}\ \text{BPS:} \quad \mathcal{D}(a_1 + a_2 + \frac{1}{2}(a_3 + a_4), 0; a_1, a_2, a_3, a_4) \,, \quad a_1 - a_4 = 2s \geq 0 \,; \\ &\frac{1}{4}\ \text{BPS:} \quad \mathcal{D}(a_2 + \frac{1}{2}a_3, 0; 0, a_2, a_3, 0) \,; \\ &\frac{1}{2}\ \text{BPS:} \quad \mathcal{D}(\frac{1}{2}a_3, 0; 0, 0, a_3, 0) \end{aligned} \tag{3.23}$$

where

$$a_1 = r + 2s \ , \quad a_2 = q \ , \quad a_3 = p \ , \quad a_4 = r \ . \tag{3.24}$$

We see that multiplying only one type of supersingletons cannot reproduce the general result of Section for all possible short multiplets. Most notably, in (3.23) there is no 3/8 series. The latter can be obtained by mixing the two types of supersingletons:

$$[\Phi^{+(+)[+]}(\theta^{++},\theta^{(++)},\theta^{[+]\{\pm\}})]^{a_3}[\Sigma^{+(+)\{+\}}(\theta^{++},\theta^{(++)},\theta^{[\pm]\{+\}})]^{a_4} \tag{3.25}$$

(or the same with Φ and Σ exchanged). Counting the charges and the dimension, we find exact matching with the series (3.8):

$$\frac{3}{8}\ \text{BPS:}\quad \mathcal{D}(\frac{1}{2}(a_3+a_4);0;0,0,a_3,a_4)\ . \tag{3.26}$$

Further, mixing two realizations of type I and one of type II supersingletons, we can construct the 1/4 series

$$[\Phi^{+(+)[+]}]^{a_2+a_3}[\Phi^{+(+)[-]}]^{a_2}[\Sigma^{+(+)\{+\}}]^{a_4} \tag{3.27}$$

which corresponds to (3.7):

$$\frac{1}{4}\ \text{BPS:}\quad \mathcal{D}(a_2+\frac{1}{2}(a_3+a_4);0;0,a_2,a_3,a_4)\ . \tag{3.28}$$

Finally, the full 1/8 series (3.6) (i.e., without the restriction $a_1 - a_4 = 2s \geq 0$ in (3.23)) can be obtained in a variety of ways.

3.4. *BPS states of OSp*(8/4, ℝ)

Here we give a summary of all possible OSp(8/4, ℝ) BPS multiplets. Denoting the UIR's by

$$\mathcal{D}(\ell;J;a_1,a_2,a_3,a_4) \tag{3.29}$$

where ℓ is the conformal dimension, J is the spin and a_1, a_2, a_3, a_4 are the SO(8) Dynkin labels, we find four BPS conditions:

3.4.1.

$$\frac{1}{8}\ \text{BPS}:\qquad q^{++}_\alpha = 0\ . \tag{3.30}$$

The corresponding UIR's are:

$$\mathcal{D}(a_1+a_2+\frac{1}{2}(a_3+a_4);0;a_1,a_2,a_3,a_4) \tag{3.31}$$

and the harmonic coset is

$$\frac{\text{Spin}(8)}{[\text{U}(1)]^4}\ . \tag{3.32}$$

If $a_2 = a_3 = a_4 = 0$ this coset becomes Spin(8)/U(4).

3.4.2.

$$\frac{1}{4}\ \text{BPS}: \qquad q_\alpha^{++} = q_\alpha^{(++)} = 0\ . \tag{3.33}$$

The corresponding UIR's are:

$$\mathcal{D}(a_2 + \frac{1}{2}(a_3 + a_4); 0; 0, a_2, a_3, a_4) \tag{3.34}$$

and the harmonic coset is

$$\frac{\text{Spin}(8)}{[\text{U}(1)]^2 \times \text{U}(2)}\ . \tag{3.35}$$

If $a_3 = a_4 = 0$ this coset becomes $\text{Spin}(8)/\text{U}(1) \times [\text{SU}(2)]^3$.

3.4.3.

$$\frac{3}{8}\ \text{BPS}: \qquad q_\alpha^{++} = q_\alpha^{(++)} = q_\alpha^{[+]\{+\}} = 0\ . \tag{3.36}$$

The corresponding UIR's are:

$$\mathcal{D}(\frac{1}{2}(a_3 + a_4); 0; 0, 0, a_3, a_4) \tag{3.37}$$

and the harmonic coset is

$$\frac{\text{Spin}(8)}{\text{U}(1) \times \text{U}(3)}\ . \tag{3.38}$$

3.4.4.

$$\frac{1}{2}\ \text{BPS (type I)}: \qquad q_\alpha^{++} = q_\alpha^{(++)} = q_\alpha^{[+]\{+\}} = q_\alpha^{[+]\{\pm\}} = 0\ ; \tag{3.39}$$

$$\frac{1}{2}\ \text{BPS (type II)}: \qquad q_\alpha^{++} = q_\alpha^{(++)} = q_\alpha^{[+]\{+\}} = q_\alpha^{[\pm]\{+\}} = 0\ . \tag{3.40}$$

The corresponding UIR's are:

$$\frac{1}{2}\ \text{BPS (type I)}: \qquad \mathcal{D}(\frac{1}{2}a_3; 0; 0, 0, a_3, 0)\ ; \tag{3.41}$$

$$\frac{1}{2}\ \text{BPS (type II)}: \qquad \mathcal{D}(\frac{1}{2}a_4; 0; 0, 0, 0, a_4)\ . \tag{3.42}$$

and the harmonic coset is

$$\frac{\text{Spin}(8)}{\text{U}(4)}\ . \tag{3.43}$$

Acknowledgments

E.S. is grateful to the TH Division of CERN for its kind hospitality. The work of S.F. has been supported in part by the European Commission TMR programme ERBFMRX-CT96-0045 (Laboratori Nazionali di Frascati, INFN) and by DOE grant DE-FG03-91ER40662, Task C.

References

1. J.M. Maldacena, *Adv. Theor. Math. Phys.* **2**, 231 (1998) 231; *Int. J. Theor. Phys.* **38**, 1113 (1999).
2. S.S. Gubser, I.R. Klebanov and A.M. Polyakov, *Phys. Lett.* **B428**, 105 (1998) [hep-th/9802109].
3. E. Witten, *Adv. Theor. Math. Phys.* **2**, 253 (1998) [hep-th/9802150].
4. J. Wess and B. Zumino, *Phys. Lett.* **B49**, 52 (1974); J. Iliopoulos and B. Zumino, *Nucl. Phys.* **B76**, 310 (1974); S. Ferrara, J. Iliopoulos and B. Zumino, *Nucl. Phys.* **B77**, 413 (1974).
5. I. Bars and M. Günaydin, *Comm. Math. Phys.* **91**, 31 (1983); M. Günaydin and S.J. Hyun, *J. Math. Phys.* **29**, 2367 (1988); M. Günaydin and C. Saclioglu, *Comm. Math. Phys.* **87**, 159 (1982); *Phys. Lett.* **B108**, 180 (1982).
6. A. Salam and J. Strathdee, *Nucl. Phys.* **B76**, 477 (1974).
7. S. Ferrara, J. Wess and B. Zumino, *Phys. Lett.* **B51**, 239 (1974).
8. A. Galperin, E. Ivanov, S. Kalitzin, V. Ogievetsky and E. Sokatchev, *Class. Quant. Grav.* **1**, 469 (1984).
9. O. Aharony, Y. Oz and Z. Yin, *Phys. Lett.* **B430**, 87 (1998) [hep-th/9803051].
10. S. Minwalla, *JHEP* **10**, 002 (1998) [hep-th/9803053].
11. K.G. Leigh and M. Rozsli, *Phys. Lett.* **B431**, 311 (1998) [hep-th/9803068].
12. E. Halyo, *JHEP* **04**, 011 (1998) [hep-th/9803077].
13. P. Claus, R. Kallosh and A. Van Proeyen, *Nucl. Phys.* **B518**, 117 (1998).
14. O. Aharony, S.S. Gubser, J. Maldacena, H. Ooguri and Y. Oz, *Phys. Rept.* **323**, 183 (2000) [hep-th/9905111].
15. S. Ferrara and E. Sokatchev, *Conformal primaries of* $OSp(8/4,\mathbb{R})$ *and BPS states in* AdS_4, *JHEP* **0005**, 038 (2000) [hep-th/0003051].
16. S. Ferrara and E. Sokatchev, *Representations of (1,0) and (2,0) superconformal algebras in six dimensions: massless and short superfields, Lett. Math. Phys.* **51**, 55 (2000) [hep-th/0001178].
17. S. Ferrara and E. Sokatchev, *Superconformal interpretation of BPS states in AdS geometries*, hep-th/0005151.
18. A. Galperin, E. Ivanov and V.I. Ogievetsky, *JETP Lett.* **33**, 168 (1981).
19. A.A. Rosly, *Constraints in supersymmetric Yang-Mills theory as integrability conditions*, in *Theor. Group Methods in Physics*, eds. M. Markov, V. Man'ko, Nauka, Moskwa, (1983) 263; *Class. Quant. Grav.* **2**, 693 (1985).
20. E. Ivanov, S. Kalitsyn, Nguyen Ai Viet and V. Ogievetsky, *J. Phys.* **A18**, 3433 (1985).
21. A. Galperin, E. Ivanov, S. Kalitzin, V. Ogievetsky and E. Sokatchev, *Class. Quant. Grav.* **2**, 155 (1985); R. Kallosh, *Pis'ma ZHETF* **41**, 172 (1985); A.A. Rosly and A.S. Schwarz, *Comm. Math. Phys.* **105**, 645 (1986).
22. A. Galperin, E. Ivanov and V. Ogievetsky, *Sov. J. Nucl. Phys.* **46**, 543 (1987).
23. I. Bandos, *Theor. Math. Phys.* **76**, 783 (1988).
24. G.G. Hartwell and P.S. Howe, *Int. J. Mod. Phys.* **A10**, 3901 (1995) [hep-th/9412147]; *Class. Quant. Grav.* **12**, 1823 (1995).
25. M. Flato and C. Fronsdal, *Lett. Math. Phys.* **2**, 421 (1978); *Phys. Lett.* **97B**, 236 (1980); *J. Math. Phys.* **22**, 1100 (1981); *Phys. Lett.* **B172**, 412 (1986).
26. C. Fronsdal, *Phys. Rev.* **D26**, 1988 (1982).
27. P.S. Howe, G. Sierra and P.K. Townsend, *Nucl. Phys.* **B221**, 331 (1983).
28. E. Bergshoeff, E. Sezgin and A. Van Proeyen, *Nucl. Phys.* **B264**, 653 (1986).
29. P.S. Howe, in *Supersymmetries and Quantum Symmetries*, eds. J. Wess and E.A. Ivanov, Lecture Notes in Physics (Springer, 1999), pp. 68-78.
30. S. Minwalla, *Adv. Theor. Math. Phys.* **2**, 781 (1998).

31. M. Günaydin, P. van Nieuwenhuizen and N. Warner, *Nucl. Phys.* **B255**, 63 (1985); M. Pernici, K. Pilch and P. Van Nieuwenhuizen, *Phys. Lett.* **B143**, 103 (1984).
32. E. D'Hoker and B. Pioline, *Near-Extremal Correlators and Generalized Consistent Truncation for $AdS_{4/7} \times S^{4/7}$*, *JHEP* **0007**, 021 (2000) [hep-th/0006103].
33. R. Corrado, B. Florea and R. McNees, *Correlation Functions of Operators and Wilson Surfaces in the d=6, (0,2) Theory in the Large N Limit*, *Phys. Rev.* **D60**, 085011 (1999) [hep-th/9902153]; F. Bastianelli, S. Frolov and A.A. Tseytlin, *Three point correlators of Stress Tensors in Maximally Supersymmetric Conformal Theories in D=3 and D=6*, *Nucl. Phys.* **B578**, 139 (2000) [hep-th/9911135]; F. Bastianelli and R. Zucchini, *Three point functions for a Class of Chiral Operators in Maximally Supersymmetric CFT at Large N*, *Nucl. Phys.* **B574**, 107 (2000) [hep-th/9909179]; F. Bastianelli and R. Zucchini, *Phys. Lett.* **B467**, 61 (1999).
34. S.W. Hawking, C.J. Hunter and M.M. Taylor-Robinson, *Phys. Rev.* **D59**, 064005 (1999).
35. A. Chamblin, R. Emparan, C.V. Johnson and R.C. Myers, *Phys. Rev.* **D60**, 064018 (1999) [hep-th/9902170].
36. K. Behrndt, M. Cvetic and W. Sabra, *Nucl. Phys.* **B553**, 317 (1999) [hep-th/9810227].
37. M.J. Duff and J.T. Liu, *Nucl. Phys.* **B554**, 237 (1999) [hep-th/9901149].
38. S. Ferrara and J. Maldacena, *Class. Quant. Grav.* **15**, 749 (1998); S. Ferrara and M. Günaydin, *Int. J. Mod. Phys.* **A13**, 2075 (1998).
39. L. Andrianopoli, S. Ferrara, E. Sokatchev and B. Zupnik, *Shortening of primary operators in N-extended $SCFT_4$ and harmonic-superspace analyticity*, hep-th/9912007.
40. S. Ferrara and E. Sokatchev, *Short representations of $SU(2, 2/N)$ and harmonic superspace analyticity*, *Lett. Math. Phys.* **52**, 247 (2000) [hep-th/9912168].
41. M. Günaydin and S. Takemae, *Unitary supermultiplets of OSp(8/4) and the AdS7/CFT6 duality*, *Nucl. Phys.* **B578**, 405 (2000) [hep-th/9910110].
42. J.-H. Park, *Nucl. Phys.* **B539**, 599 (1999).
43. W. Siegel, *Int. J. Mod. Phys.* **A4**, 2015 (1989).
44. E. Angelopoulos and M. Laoues, *Rev. Math. Phys.* **10**, 271 (1998).
45. C. Hull, *Strongly coupled gravity and duality*, *Nucl. Phys.* **B583**, 237 (2000) [hep-th/0004195].
46. P.S. Howe, K. Stelle and P. West, *Class. Quant. Grav.* **2**, 815 (1985).
47. B.M. Zupnik, *Sov. J. Nucl. Phys.* **44**, 512 (1986); Proceedings of the 1985 Yurmala conference "Theoretical-group methods in physics", v. 1, p.52-59, Nauka, Moscow, 1986.
48. V. K. Dobrev and V. B. Petkova, *Phys. Lett.* **B162**, 127 (1985); *Fortschr. Phys.* **35**, 7 (1987); **35**, 537 (1987).
49. S. Ferrara and C. Fronsdal, *Conformal fields in higher dimensions*, hep-th/0006009.
50. D. Fabbri, P. Fré, L. Gualtieri and P. Termonia, *Class. Quant. Grav.* **17**, 55 (2000); P. Fré, L. Gualtieri and P. Termonia, *Phys. Lett.* **B471**, 27 (1999).
51. L. Frappat, A. Sciarrino and P. Sorba, *Dictionary on Lie algebras and superalgebras* (Academic Press, 2000).
52. M. Günaydin, B. Nilsson, G. Sierra and P. Townsend, *Phys. Lett.* **B176**, 45 (1986).
53. E.A. Ivanov and A. Sorin, *J. Phys. A* **13**, 1159 (1980).
54. D. Fabbri, P. Fre, L. Gualtieri, P. Termonia, *Class. Quant. Grav.* **17**, 55 (2000).
55. B.M. Zupnik and D.V. Khetselius, *Sov. J. Nucl. Phys.* **47**, 730 (1988).
56. P.S. Howe and M.I. Leeming, *Class. Quant. Grav.* **11**, 2843 (1994).
57. A. Galperin, P.S. Howe and K.S. Stelle, *Nucl. Phys.* **B368**, 248 (1992).

THE ENHANÇON, MULTIMONOPOLES AND FUZZY GEOMETRY

CLIFFORD V. JOHNSON
Department of Mathematical Sciences
University of Durham, South Road
Durham DH1 3LE, United Kingdom
E-mail: c.v.johnson@durham.ac.uk

The presentation at Strings 2000 was intended to be in two main parts, but there was only time for part one. However both parts appeared on the online proceedings, and are also included in this document. The first part concerns an exploration of the connection between the physics of the "enhançon" geometry arising from wrapping N D6–branes on the K3 manifold in Type IIA string theory and that of a charge N BPS multi–monopole. This also relates to the physics of 2+1 dimensional $SU(N)$ gauge theory with eight supercharges. The main results uncovered by this exploration are: *a)* better insight into the non–perturbative geometry of the enhançon; *b)* the structure of the moduli space geometry, and its characterisation in terms of a family of Atiyah–Hitchin–like manifolds; *c)* the use of Nahm data to describe aspects of the geometry, showing that the enhançon locus itself has a description as a fuzzy sphere. Part two discusses the addition of extra D2–branes into the geometry. Two probe computations show the difference between the geometry as seen by D2–branes and that seen by wrapped D6–branes, and the accompanying gauge theory interpretations are discussed.

1. Wrapping Branes

In the context of superstring theory, one of the simplest ways to realize a supersymmetric $SU(N)$ gauge theory with no flavours and only eight supercharges is *via* the low energy limit of N D–branes wrapped on manifolds which break half of the supersymmetry of the parent string theory. The brane breaks the other half.

For $D = 4$, $\mathcal{N} = 2$, or $D = 3$, $\mathcal{N} = 4$, the appropriate thing to do is wrap D7–branes or (respectively) D6–branes, on the manifold $K3$. Let us focus on the latter. We would like to understand as much as we can about the gauge theory by studying the physics of these wrapped branes. Perhaps new insights can be gained at large N, since in that case the back reaction on the supergravity fields is large, and some of the physics of the branes —and hence the gauge theory — is encoded in a non–trivial supergravity solution.

The characteristic scale of the curvature of the supergravity solution is $L = g_s\ell_s N$, (where g_s is the string coupling and $\ell_s^2 = \alpha' = 1/(2\pi T)$, with T the string tension) while the $SU(N)$ Yang–Mills coupling is:

$$g_{\rm YM}^2 = (2\pi)^4\ell_s^3 g_s V^{-1} \; . \tag{1}$$

Here, V is the volume of the $K3$ upon which we wrap the branes.

With N large we can stay at weak string coupling and low energy ($g_s, \ell_s \ll 1$), while sensibly discussing supergravity (without quantum and stringy effects), since L

can be large. Asking that V be small enough (for a suitable wrapping to the resulting 2+1 dimensional theory) keeps us in the infra–red (IR) of the gauge theory, where $g_{\rm YM}$ is large. Therefore, if we are going to learn anything at all from the gravity solution, it will concern the gauge theory at both large N and strong coupling.

While all of the above is true, it does not guarantee that we will succeed in our quest, since *a*) We need to make sure that we find a smooth and sensible supergravity geometry and *b*) We must understand it well enough to deduce anything from it, and *c*) There is no guarantee that there is necessarily a clean separation between the supergravity physics and that of the gauge theory in a way which achieves a complete duality between the two as in the case with the AdS/CFT correspondence.[1]

Issues *a*) and *b*) were addressed in Ref. 2, where many aspects of the solution were uncovered. In particular, the main naive geometry one would write for the $K3$ wrapped D6 is singular, and Ref. 2 showed how this naive geometry (containing a "repulson" singularity[3]) is replaced by a non–singular one. Essentially, the correct geometry contains a spherical shell of smeared D6–branes which has radius

$$r_{\rm e} = \frac{V^*}{V - V^*} g_s N \ell_s \ ; \tag{2}$$

where $V^* \equiv (2\pi\ell_s)^4$ is a critical volume which we shall discuss later.[a]

Issue *c*) is a subtle one which requires further work. As pointed out in Ref. 2, there is *no* clear separation between supergravity and the gauge theory, and so either the only recourse is to study the entire type IIA string theory in this background, or to seek some other effective theory in the limit, which might capture the full dynamics at large N. Some conjectural remarks in the latter direction are contained in Ref. 5 and we will not pursue this further here.

Let us instead focus on learning more about the low energy dynamics of the gauge theory and large N, and attempt to see what we can learn.

2. Probe Results

The supergravity solution for the N wrapped D6–branes is naively:

$$ds^2 = (Z_2 Z_6)^{-\frac{1}{2}}(-dt^2 + dx_1^2 + dx_2^2) + (Z_2 Z_6)^{\frac{1}{2}}(dr^2 + r^2 d\Omega_2^2) + \left(V\frac{Z_2}{Z_6}\right)^{\frac{1}{2}} ds_{K3}^2 \ , \tag{3}$$

with ds_{K3}^2 the metric of $K3$, normalised to unit volume, and the harmonic functions are

$$Z_2 \ = \ 1 - \frac{(2\pi)^4 g_s N \alpha'^{5/2}}{2Vr} \ , \quad Z_6 = 1 + \frac{g_s N \alpha'^{1/2}}{2r} \ , \tag{4}$$

and $d\Omega_2^2 = d\theta^2 + \sin^2\theta\, d\phi^2$. The true volume of $K3$ varies with radius and is $V(r) = VZ_2(r)/Z_6(r)$. The repulson singularity is at $V(r) = 0$.

[a]Note that versions of this story for other gauge groups have been worked out in Ref. 4.

One of the telling results of the analysis of Ref. 2 was the result that the metric seen by a single wrapped D6–brane probe in the fields produced by all the others is given by:

$$ds^2 = F(r)\left(dr^2 + r^2 d\Omega_2^2\right) + F(r)^{-1}\left(ds/2 - \mu_2 C_\phi d\phi/2\right)^2 , \tag{5}$$

where

$$F(r) = \frac{Z_6}{2g_s}\left(\mu_6 V(r) - \mu_2\right) , \tag{6}$$

and $C_\phi = -(r_6/g_s)\cos\theta$. The basic D6– and D2–brane charges are[6] $\mu_6 = (2\pi)^{-6}\alpha'^{-7/2}$ and $\mu_2 = (2\pi)^{-2}\alpha'^{-3/2}$, respectively.[b]

This moduli space metric shows that the kinetic term of the brane vanishes at the enhançon radius r_e, and so it cannot proceed inwards of that radius. So in equations (3), the geometry for $r < r_e$ cannot be correct, since it is that associated with branes at $r = 0$. The idea of Ref. 2, bolstered by other facts we will mention shortly, ($e.g$ $V(r_e) = V^*$) is that to a first approximation, the geometry in the interior should be a flat, sourceless geometry, matched onto the exterior solution in Eqn. (3) at $r = r_e$.[c]

The limit where we isolate the massless fields —looking for the gauge theory— is to take $\alpha' \to 0$, holding the gauge coupling (1) and $U = r/\alpha'$ finite.[1] (The latter should be thought of as the characteristic energy scale at which the $SU(N)$ breaks to $SU(N-1) \times U(1)$ representing the separating of the probe.) In this case, the metric becomes

$$ds^2 = f(U)\left(dU^2 + U^2 d\Omega_2^2\right) + f(U)^{-1}\left(d\sigma - \frac{N}{8\pi^2}A_\phi d\phi\right)^2 ,$$

where

$$f(U) = \frac{1}{8\pi^2 g_{\rm YM}^2}\left(1 - \frac{g_{\rm YM}^2 N}{U}\right) , \tag{7}$$

the $U(1)$ monopole potential is $A_\phi = -\cos\theta$, and $\sigma = s\alpha'/2$. This metric is meaningful only for $U > g_{\rm YM}^2 N$. It is the Euclidean Taub–NUT metric, with a negative mass. It is a hyperKähler manifold, because $\nabla f = \nabla \times A$, where $A = (N/8\pi^2)A_\phi d\phi$.

There are many facts about the geometry and the gauge theory which we can understand in this metric. First, for geometry, the radius $U_e = g_{\rm YM}^2 N$ is the place where the spacetime should be treated carefully. In fact, it is the scaled version of the "enhançon" radius r_e, (see Eqn. (2)). Here the volume of the $K3$ becomes equal to V^*, and a generic $U(1)$ in the six dimensional supergravity is joined by massless W–bosons to achieve an enhanced $SU(2)$ gauge theory. The wrapped branes are BPS monopoles of that $U(1)$, and as such cease to be pointlike and smear out at symmetry enhancement. Thus the geometry of spacetime is different from that which one would write for the case when the wrapped branes are pointlike.

[b] Strictly speaking, the spirit of our discussion says that we should have replaced N by $N-1$ in the probe result, since we have separated off the probe. This is not really essential, and so we will not do it.

[c] See Ref. 7 for a supergravity discussion of the physics of this sort of excision process.

Concerning gauge theory, the factor $f(U)$ represents the one loop running of the Yang–Mills coupling (see figure 1). This is because U is the vacuum expectation value (vev) of the complex scalar ϕ in the $\mathcal{N} = 2$ gauge multiplet which breaks $SU(N) \to SU(N-1) \times U(1)$. In the low energy effective theory, this vev controls the effective coupling, since the model becomes a sigma model whose kinetic term is $f(U)\partial_\mu U \partial^\mu U$.

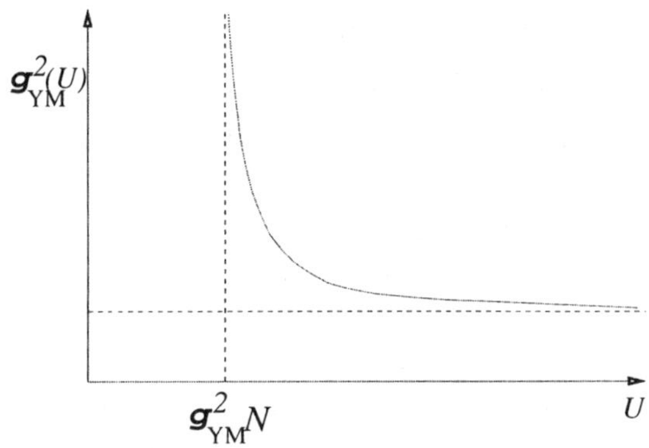

Fig. 1. The running gauge coupling.

The coupling diverges at the scale $U = g^2_{\rm YM}N$. This should be the scale at which the $SU(N)$ is restored, but in fact it is not, since the divergence happens at $U \neq 0$. In other words, the branes are not coincident, and so (extending the discussion to any of the constituent branes), the generic gauge group is $U(1)^{N-1}$: we are out on the Coulomb branch of the gauge theory, and the origin is modified from being an $SU(N)$ point by quantum corrections.

3. Moduli Space Metrics

What we looked at just now was a four dimensional submanifold of the full $4N-4$ dimensional moduli space of the $SU(N)$ gauge theory. This full moduli space, which is hyperKähler, is proposed to be isomorphic to the relative moduli space[8] of N BPS monopoles of 4D $SU(2)$ gauge theory spontaneously broken to $U(1)$.[9,10,11] Here, the wrapped D6–branes play the role of the monopoles, since they couple magnetically to the $U(1)$ which gets enhanced to an $SU(2)$ when $V(r) = V^*$. (There are 23 other $U(1)$'s which will not concern us here.) In the directions perpendicular to the wrapped brane, with coordinates (r, θ, ϕ), they appear as monopoles in the $3+1$ dimensional problem.

So let us learn both about the gauge theory and the monopole moduli space. First let us consider the case of two monopoles ($N = 2$). In that case, it is known that the metric on the gauge theory moduli space is the Atiyah–Hitchin manifold, (as for the two monopole moduli space) which is given as follows:[12,13]

$$\begin{aligned} ds^2_{\rm AH} &= f^2 d\rho^2 + a^2\sigma_1^2 + b^2\sigma_2^2 + c^2\sigma_3^2 \ , \quad \text{where} \\ \sigma_1 &= -\sin\psi d\theta + \cos\psi \sin\theta d\phi \ ; \\ \sigma_2 &= \cos\psi d\theta + \sin\psi\sin\theta d\phi \ ; \\ \sigma_3 &= d\psi + \cos\theta d\phi \ ; \\ \frac{2bc}{f}\frac{da}{d\rho} &= (b-c)^2 - a^2 \ , \text{ and cyclic perms.}; \quad \rho = 2K(k) \,, \end{aligned} \tag{8}$$

and $K(k)$ is the elliptic integral of the first kind:

$$K(k) = \int_0^{\frac{\pi}{2}} (1 - k^2 \sin^2\tau)^{\frac{1}{2}} d\tau \ . \tag{9}$$

Also, k, the "modulus", runs from 0 to 1, so $\pi \le \rho \le \infty$.

In our moduli space metric (7), if we rescale $\rho = U/g^2_{\rm YM}$ and $\psi = 4\pi^2\sigma$, then we get:

$$\begin{aligned} ds^2 &= \frac{g^2_{\rm YM}}{8\pi^2} ds^2_{\rm TN} \ , \quad \text{with} \\ ds^2_{\rm TN} &= \left(1 - \frac{2}{\rho}\right)(d\rho^2 + \rho^2 d\Omega_2^2) + 4\left(1-\frac{2}{\rho}\right)^{-1}(d\psi + \cos\theta d\phi)^2 \ . \end{aligned} \tag{10}$$

This is precisely the form of the metric that one gets by expanding the Atiyah–Hitchin metric in large ρ. The difference between the Atiyah–Hitchin manifold and this "negative mass" ($m = -1$ in standard units, see *e.g.* Ref. 14) Taub–NUT metric is a family of exponential corrections of the form $e^{-\rho}$, which are non–perturbative in a large ρ expansion.

From the point of view of the gauge theory, these corrections are of order $e^{-U/g^2_{\rm YM}}$, and so are instanton corrections. These corrections render the complete manifold smooth —the $\rho = 2$ singularity is just an artifact of the large ρ expansion— which is consistent with gauge theory expectations.

The combination $\psi = 4\pi^2\sigma = \tilde{\sigma}$ is the dual scalar to the gauge theory photon and has period 2π. The fact that ψ is 2π periodic and not 4π gives us an $SO(3)$ isometry for the Atiyah–Hitchin manifold instead of the naive $SU(2)$ of the Taub–NUT metric. In fact, it can be proven that the *unique* smooth hyperKähler manifold with this isometry and the above asymptotic behaviour is the Atiyah–Hitchin manifold, essentially proving that it is the correct result for the $SU(2)$ gauge theory moduli space.[9,15]

What about N different from 2? Well, the key observation[5] is that the rescalings $\rho = 2U/g^2_{\rm YM}N$ and $\psi = \sigma 8\pi^2/N$, give

$$ds^2 = \frac{g^2_{\rm YM}N^2}{32\pi^2} ds^2_{\rm TN} \ , \tag{11}$$

generalising our previous result (11) connecting the moduli space metric to the same negative mass Taub–NUT spacetime. It is tempting to conclude that the manifold controlling our non–perturbative corrections for all N is again the Atiyah–Hitchin manifold, but we should be careful.[d] Note that changing N will not change the period of the probe $U(1)$. $\tilde{\sigma} = 4\pi^2\sigma$ should still have period 2π. Therefore the period of ψ in the metric above is now $4\pi/N$. Furthermore, the instanton corrections in the gauge theory will still be typically $e^{-U/g^2_{\rm YM}}$, and so this translates into $e^{-N\rho/2}$ for the corrections to the $m = -1$ Taub–NUT metric.

Let us then summarise what we have learned for our non–perturbative metrics for any N. They are Atiyah–Hitchin–like in the sense that:

a) They differ from negative mass Taub–NUT by non–perturbative corrections of the form $e^{-N\rho/2}$.

b) They have only a *local* $SU(2)$ action. Globally, it is broken by instanton corrections to $SU(2)/\mathbf{Z}_N$.

Some remarks are in order. One might worry that the manifolds that we seek might be singular, especially given property *b*). This would be an awkward feature from the gauge theory perspective, since there is nothing like a Higgs branch available which might be opening up at the singularity, which is typically what happens in situations like this. (Classically there is no Higgs branch, and it seems imprudent to conjecture that one might be generated quantum mechanically.)

However, it is natural that there might be a harmless singularity in the moduli space following from the fact that we are studying a four dimensional subspace of the full $4N-4$ dimensional moduli space, which is known to be smooth. It is natural to expect that there are therefore $4N-8$ adjustable parameters in the full metric, which can resolve the singularity. The worry that the parameters do not appear in the metric written above is alleviated by the fact that these are non–perturbative parameters. They are monopole positions and phases: parameters of different vacua from the point of view of the one we have focussed on.

From the point of view of the monopole problem, or from the enhançon picture, this simply corresponds to the other monopoles which make up the internal structure of the $N-1$ monopole having to move around somewhat in order to accommodate the merging with the single monopole probe.

Property *a*) also deserves some remarks. While the instanton corrections thus identified will smooth the singularity (together with the possible need for a resolution by adjusting $4N-8$ parameters) in the moduli space, it is clear that in the strict large N limit, the corrections are extremely small because of the N in the exponential. So in fact, the strict supergravity result for the location and nature of the enhançon is quite accurate. (See also Ref. 16.)

It is worth comparing this to the two monopole case ($N = 2$). There, the naive singularity of the Taub–NUT metric is at $\rho = 2$. It is in fact completely removed by non–perturbative effects, and the radius $\rho = \pi$ is the smallest possible value. This corresponds to the axisymmetric case of the two monopoles being exactly on top of

---

[d]We confess to being less than careful in the original version of Ref. 5, and also in the Strings 2000 talk and the transparencies displayed at `http://feynman.physics.lsa.umich.edu/strings2000/`

one another.[17] (This is naively a singularity of the complete metric but is in fact a "bolt" singularity, which is harmless.)

It is clear that there is a natural generalisation of this for each N. For very large N, the coincident monopole case is the enhançon itself, at $\rho = 2$ of $U = g_{\rm YM}^2 N$.

Another point worth making about large N here is that while ρ is a relative coordinate in the case of the two–monopole case, and the Taub–NUT and Atiyah–Hitchin manifolds have nothing to do with spacetime geometry, this is not true here: For large N, ρ is a rescaled centre of mass coordinate of the $N-1$ monopole system being probed by a single one. But since N is large, the centre of mass is essentially at the centre of the $N-1$ monopole, and so serves as a good spacetime coordinate, with the large monopole at the origin.

4. Fuzzy Geometry

Another appearance of the enhançon geometry is when a large number, N, of D3–branes are stretched between a pair of NS5–branes, as in figure 2(a). This situation

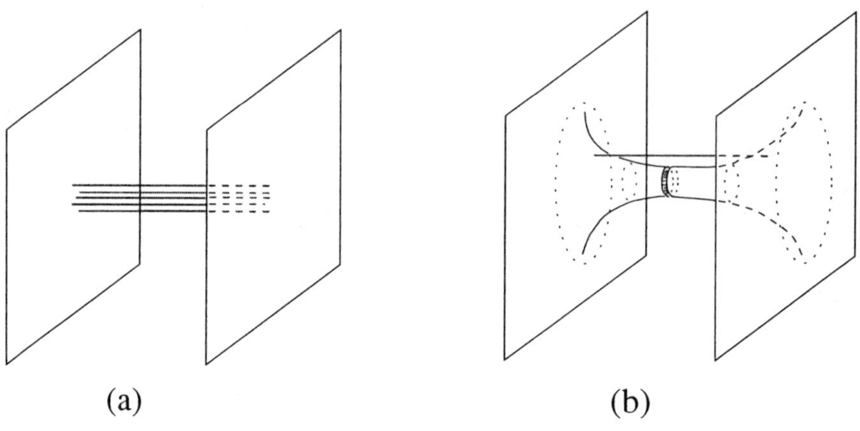

Fig. 2. (a) A slice of the configuration of D3–branes stretching between NS5–branes. (b) A slice of the resulting "double trumpet" shape of the NS5–branes at large N. (A separated probe is also shown.) This system has a natural description in terms of the Nahm equations as explained in the text. The enhançon is the place (an S^2) where the NS5–branes touch.

is related to the $K3$–wrapped D6–brane system by a chain of dualities.[2] The D3–branes pull the NS5–branes into a shape such that at large N they meet in an S^2 geometry, as in figure 2(b). This S^2 is the enhançon, since when two NS5–branes in type IIB string theory meet, there is an enhanced $SU(2)$ gauge symmetry describing their relative collective geometry. The endpoints of the D3–branes act as monopoles of the spontaneously broken $SU(2)$, where the Higgs vev is the asymptotic distance between the NS5–branes.

It is intuitively clear that in this and the wrapped D6–brane picture, the enhançon geometry is only a smooth sphere at large N. The geometry is made of N

smeared objects, the monopoles.

To find a quantitative description of this geometry needs new variables. In this D3/NS5 system, the variables are intuitive: N D3–branes have 3 $N \times N$ matrix valued coordinates, Φ^i $(i = 1,2,3)$, representing their transverse motions within the NS5–brane worldvolumes. These coordinates are governed by Nahm's equations in the BPS case:[18,19]

$$\frac{d\Phi^i}{d\sigma} = \frac{1}{2}\epsilon_{ijk}[\Phi^j, \Phi^k] , \tag{12}$$

where $\sigma \in [-1,1]$, (where we have placed the NS5–branes at ± 1) and the appropriate solutions are those for which the $\Phi^i(\sigma)$ have a pole at the ends of the interval, with residues Σ^i, which form the $N \times N$ irreducible representation of $SU(2)$:[20,21,22]

$$[\Sigma^i, \Sigma^j] = 2i\epsilon^{ijk}\Sigma^k . \tag{13}$$

It is clear from this description that the D3–brane geometry has a non–commutative geometry underlying its description, the general solution taking one $SU(2)$ representation and twisting it into another while traversing the interval. A slice through this geometry, such as the enhançon itself at $\sigma = 0$, is clearly a non–commutative or "fuzzy" sphere.[23] As a simple attempt at creating an approximate solution, a symmetric "double trumpet" configuration can be made by gluing two sections of the infinite funnel solution of Ref. 24:

$$\Phi^i = f(\sigma)\Sigma^i , \qquad f(\sigma) = \frac{g_s}{\sigma \pm 1} . \tag{14}$$

A complete solution would of course connect smoothly through the middle, but this crude solution serves to illustrate the point:[e] A cross section of this is a fuzzy sphere, with radius R given by:

$$R^2 = 4\pi^2\alpha' \sum_{i=1}^{3} \mathrm{Tr}(\Phi^i)^2 = 4\pi^2\alpha'(N^2 - 1)f(\sigma) , \tag{15}$$

giving a radius for the enhançon (at $\sigma = 0$) of $R_e \sim g_s \ell_s N$, which is we saw in the wrapped D6–brane description.

It is noteworthy that the geometry of this object made of expanding branes is described as a fuzzy sphere just as the brane expansion in a background R–R field produced by the dielectric brane effect.[26] One is led to speculate that there might be further couplings to be deduced, such as multipole couplings to *higher* rank R–R fields *via* curvature terms on the D–brane worldvolume.[f]

---

[e] The idea is to make a solution which is spherically symmetric. The only monopole solution which is spherically symmetric is the one–monopole.[25] But we expect that for large N, we can make a solution which approaches spherical symmetry on the sufficiently granular (supergravity) scale.

[f] Contrast this with the route taken in Ref. 2, where it was deduced indirectly that the branes smear, using the electric couplings[27] to *lower* rank R–R fields, producing a place where the tension will go to zero, W–bosons of an enhanced gauge symmetry, implying BPS monopoles, smearing, *etc.*

5. Other Branes and other Probes

Now although the wrapping of a D6–brane on $K3$ induces a negative D2–brane charge within the part of its worldvolume transverse to $K3$, it must be stressed that this is not an anti–D2–brane. It preserves the same supersymmetries that a D2–brane with that orientation would.

In fact, it is useful to consider adding extra D2–branes to the wrapped D6–brane arrangement with the same orientation as the unwrapped part of the D6–branes. Let us add M of them. The net D2–brane charge as measured at infinity is therefore $M-N$, and it is interesting to note that we can cancel away the net charge by setting $M=N$.

There is an $SU(M)$ gauge theory on the worldvolume of these new branes, and there are of course strings which can stretch between these branes and the other set. Since these ones are not wrapped, the 2+1 dimensional gauge theory coupling is simply

$$g^2_{\rm YM,2} = g_s/\ell_s \ . \tag{16}$$

This form will produce an interesting conflict with the gauge theory coupling (1) of the wrapped brane gauge theory when we take decoupling limits shortly.

A natural question to ask is what happens to the geometry with the addition of these extra branes? Well, the first thing to note is that the supergravity geometry is simply modified by changing the harmonic function (4) which yields the D2–brane charge:

$$Z_2 = 1 - \frac{(2\pi)^4 g_s N\alpha'^{5/2}}{2Vr} \longrightarrow \widetilde{Z}_2 = 1 + \frac{(2\pi)^4 g_s (M-N)\alpha'^{5/2}}{2Vr} \ . \tag{17}$$

Note that this modifies, for example, the result for the volume of $K3$, which is now $V(r) = V\widetilde{Z}_2/Z_6$.

The next thing to notice is that we can probe with wrapped D6–branes as before, but we also can probe with pure D2–branes. The result for probing with a wrapped D6–brane is much like that which we saw before. It is equation (5), but now with

$$F(r) = \frac{Z_6}{2g_s}\left(\mu_6 V(r) - \mu_2\right) = \frac{1}{2g_s}\left(\mu_6\widetilde{Z}_2 - \mu_2 Z_6\right) \ . \tag{18}$$

This means that everything we have seen is generically the same as before. There is an enhançon, which is still where $V(r) = V^*$, but for increasing M, this happens at smaller radius, given by:

$$\tilde{r}_{\rm e} = \frac{V^*}{V - V^*} g_s\ell_s(N-M) \ . \tag{19}$$

After taking a decoupling limit, we will get similar results with N replaced by $N-M$, and so, for example the scaled enhançon radius is $\widetilde{U}_{\rm e} = (N-M)g^2_{\rm YM}$. The result for probing with the pure D2–brane is easy to deduce. Take the previous results (5) and (18), set μ_6 to zero, and reverse the sign of μ_2. So we get in this case

$$F(r) = \frac{\mu_2 Z_6}{2g_s} \ . \tag{20}$$

So we see that the pure D2–branes completely ignore the D2–brane component, and we just get the usual result similar to that for D2–branes probing D6–branes in flat space. The effects of the wrapping on the transverse space simply do not show up. This is seen more clearly when one places this result for $F(r)$ into the metric (5) and see that it is essentially the Taub–NUT metric, now with a positive mass set by N. This will be precisely the case when we take a decoupling limit shortly.

So in fact the M D2–branes which make up the geometry can clearly live at $r = 0$, since there is no reason for them to be stuck at the enhançon radius. After excising the geometry interior to the radius r_e, instead of replacing it with a flat geometry as we did before,[2] we should replace it with a geometry produced by M D2–branes in the interior, sitting at $r = 0$, appropriately matched at $\tilde{r}_e$ to the exterior geometry.[g]

This is all made clearer when we take the decoupling limit of this result. Again we have a choice. We can take $\alpha' \to 0$, holding $U = r/\alpha'$ fixed, but we can choose to hold *a*) the Yang–Mills coupling, (1) of the wrapped D6–branes, fixed, or *b*) the coupling (16) on the pure D2–branes fixed. We cannot do both. The former stays with the wrapped D6–brane's $SU(N)$ gauge theory, and treats the D2–branes as merely supplying extra hypermultiplets.[h] The latter limit, instead, treats the D6–branes as hypermultiplets within the D2–brane $SU(M)$ gauge theory.

For the case *a*) we get the metric in equation (7) with scaled function

$$f(U) = \frac{N}{16\pi^2 U} \ , \tag{21}$$

a pure one loop contribution. Meanwhile, for case *b*) we have instead the scaled function

$$f(U) = \frac{1}{8\pi^2 g_{\mathrm{YM},2}^2}\left(1 + \frac{N g_{\mathrm{YM},2}^2}{2U}\right) \ . \tag{22}$$

(It is interesting that one limit allows the tree level part to survive —coming from the "1" in the metric— while the other does not. The second case is reminiscent of an AdS–like limit.) The latter is precisely what one should get for the moduli space result. There is a $U(1)$ on the D2–brane and $N/2$ massless flavours supplied by the D6–branes.

With unwrapped D6–branes, we would have the following to say about the location $U = 0$: The metric is smooth there for one flavour, as the NUT singularity is removable, but for higher numbers of flavours, the singularity is not removable (recall that the periodicity of σ is fixed by the gauge theory), but has the happy interpretation of being the origin of a Higgs branch. This is the branch where the D2–brane becomes a fat instanton inside the D6–brane $SU(N)$ gauge theory. (More can be found out about this in Refs. 29,30.)

Here, with wrapped branes we have a subtlety. We have asked that we excise the naive spacetime geometry inside $r = \widetilde{r}_e$ (or $U = \widetilde{U}_e$), and replace it with a

[g] See Ref. 28 for further discussion of this sort of procedure.

[h] We must be careful though, since the D2–branes are too small to make the gauge dynamics of their "flavour group", $SU(M)$, decouple. So they are probably only formally hypermultiplets.

geometry which has no D6–branes inside that radius, so in fact there is no way to form instantons, and hence no Higgs branch. This appears to fit the fact that the D6–branes are wrapped on a small $K3$: we would need $\mathbf{R}^4$ to form genuine instantons.

Acknowledgments

CVJ would like to thank the organisers of the Strings 2000 conference at the University of Michigan for an invitation to present this work, and for an extremely enjoyable conference. A comment at the conference by Alex Buchel and an innocent–seeming question from an anonymous referee of Ref. 5 are gratefully acknowledged.

References

1. J. Maldacena, *Adv. Theor. Math. Phys.* **2**, 231 (1998) [hep-th/9711200].
2. C.V. Johnson, A. Peet and J. Polchinski, *Phys. Rev.* **D61**, 086001 (2000) [hep-th/9911161].
3. K. Behrndt, *Nucl. Phys.* **B455**, 188 (1995) [hep-th/9506106];
   R. Kallosh and A. Linde, *Phys. Rev.* **D52**, 7137 (1995) [hep-th/9507022];
   M. Cvetič and D. Youm, *Phys. Lett.* **B359**, 87 (1995) [hep-th/9507160].
4. L. Järv and C.V. Johnson, hep-th/0002244, to appear in Phys. Rev. **D**.
5. C.V. Johnson, hep-th/0004068.
6. J. Polchinski, *Phys. Rev. Lett.* **75**, 4724 (1995) [hep-th/9510017].
7. C.V. Johnson, R.C. Myers, A.W. Peet and S.F. Ross, to appear.
8. E.J. Weinberg, *Phys. Rev.* **D20**, 936 (1979);
   W. Nahm, *Phys. Lett.* **B85**, 373 (1979).
9. N. Seiberg and E. Witten, in *Saclay 1996, The mathematical beauty of physics*, hep-th/9607163.
10. G. Chalmers and A. Hanany, *Nucl. Phys.* **B489**, 223 (1997) [hep-th/9608105].
11. A. Hanany and E. Witten, *Nucl. Phys.* **B492**, 152 (1997) [hep-th/9611230].
12. M.F. Atiyah and N.J. Hitchin, *The Geometry And Dynamics Of Magnetic Monopoles*, M.B. Porter Lectures (Princeton University Press, 1988).
13. G.W. Gibbons and N.S. Manton, *Nucl. Phys.* **B274**, 183 (1986).
14. T. Eguchi, P.B. Gilkey and A.J. Hanson, *Phys. Rept.* **66**, 213 (1980).
15. N. Dorey, V.V. Khoze, M.P. Mattis, D. Tong and S. Vandoren, *Nucl. Phys.* **B502**, 59 (1997) [hep-th/9703228].
16. M.R. Douglas and S.H. Shenker, *Nucl. Phys.* **B447**, 271 (1995) [hep-th/9503163].
17. R.S. Ward, *Comm. Math. Phys.* **79**, 317 (1981); *Phys. Lett.* **B102**, 136 (1981); *Comm. Math. Phys.* **83**, 563 (1982).
18. W. Nahm, *The Construction Of All Selfdual Multi-Monopoles By The ADHM Method* (Talk), in *Monopoles In Quantum Field Theory. Proceedings, Monopole Meeting, Trieste, Italy, December 11-15, 1981*, eds. N.S. Craigie, P. Goddard and W. Nahm (World Scientific, 1982).
19. D. Diaconescu, *Nucl. Phys.* **B503**, 220 (1997) [hep-th/9608163].
20. N. Hitchin, *Comm. Math. Phys.* **83**, 579 (1982); *Comm. Math. Phys.* **89**, 148 (1982).
21. H. Nakajima, *Monopoles And Nahm's Equations*, in *Sanda 1990, Proceedings, Einstein metrics and Yang-Mills connections*, pp. 193-211.
22. D. Tsimpis, *Phys. Lett.* **B433**, 287 (1998) [hep-th/9804081].
23. J. Madore, *Annals Phys.* **219**, 187 (1992); *Class. Quant. Grav.* **9**, 69 (1992); *Phys. Lett.* **B263**, 245 (1991).

24. N.R. Constable, R.C. Myers and O. Tafjord, *Phys. Rev.* **D61**, 106009 (2000).
25. E.B. Bogomolny, *Sov. J. Nucl. Phys.* **24**, 449 (1976);
M.K. Prasad and C.M. Sommerfield, *Phys. Rev. Lett.* **35**, 760 (1975).
26. R.C. Myers, *JHEP* **9912**, 022 (1999) [hep-th/9910053].
27. M. Bershadsky, C. Vafa and V. Sadov, *Nucl. Phys.* **B463**, 398 (1996) [hep-th/9510225].
28. C.V. Johnson and R.C. Myers, to appear.
29. N. Seiberg, *Phys. Lett.* **B384**, 81 (1996) [hep-th/9606017].
30. C.V. Johnson, *D-Brane Primer*, hep-th/0007170.

M-THEORY DUALITY AND BPS-EXTENDED SUPERGRAVITY

BERNARD DE WIT
Institute for Theoretical Physics and Spinoza Institute
Utrecht University, Utrecht, The Netherlands

We discuss toroidal compactifications of maximal supergravity coupled to an extended configuration of BPS states which transform consistently under the U-duality group. Under certain conditions this leads to theories that live in more than eleven spacetime dimensions, with maximal supersymmetry but only partial Lorentz invariance. We demonstrate certain features of this construction for the case of nine-dimensional $N = 2$ supergravity.

1. Introduction

Not too much is known about the underlying degrees of freedom of M-theory. From perturbative string theory one knows about the existence of massive and massless string states. The latter are captured in supergravity where one can explore solitonic solutions which correspond to the branes that arise in (nonperturbative) string theory. In spacetimes with compactified dimensions there exist winding and momentum states. These are BPS states, which can in principle be incorporated into supergravity as matter supermultiplets. As with all BPS states, one has a rather precise knowledge about their mass spectrum and multiplicity, which suffices for many applications. Massive string states decouple in the zero-slope limit; some of them are also BPS. More daring ideas about the underlying degrees of freedom of M-theory are based on supermembrane and matrix theories[1,2,3] (for recent reviews, see Refs. 4).

M-theory is subject to a large number of duality symmetries and equivalences. From the string perspective one knows about S- and T-duality which are conjectured[5] to be contained into the so-called U-duality group. This is an arithmetic subgroup of the group of nonlinearly realized symmetries (henceforth denoted by G) of the maximal supergravity theories.[6] Hence the U-dualities coincide with the geometrical symmetries of supergravity. They are conjectured to be exact symmetries of (toroidally compactified) M-theory and therefore act on the BPS states. For a comprehensive review, see Ref. 7. The group G is expected to be broken down to an integer-valued subgroup because it must respect the (perturbatively or nonperturbatively generated) charge lattice of the BPS states. It is unavoidable that the inclusion of BPS states leads us back (at least in special limits) to eleven-dimensional supergravity[8] or extensions thereof, for the simple reason that the BPS states include the Kaluza-Klein (KK) states needed to elevate the theory to eleven dimensions. Of course, a multitude of arguments has already been presented all pointing to the conclusion that eleven-dimensional supergravity has a role to play (in particular, see Refs. 5,9,10).

Clearly, an important distinction between toroidally compactified M-theory and toroidally compactified eleven-dimensional supergravity resides in the BPS states. The compactifications of M-theory on a torus T^n are conjectured to be invariant under the duality group $\mathrm{E}_{n(n)}(\mathbf{Z})$, which includes the $\mathrm{SL}(2,\mathbf{Z})$ S-duality group of the IIB superstring as a subgroup. This conjecture includes the BPS states which must transform according to the same duality group. In contrast with this, the toroidal compactifications of eleven-dimensional supergravity generate towers of KK states whose central charges are associated with momenta in the compactified dimensions. These states cannot transform under the duality symmetry, simply because the central charges that they carry are too restricted; central charges associated with the two- and five-brane charges in eleven dimensions vanish for the KK states. Obviously the massless states coincide for the two theories. But in view of the fact that M-theory contains more BPS states than the KK states associated with the momenta on the hyper-torus, one may wonder why the spacetime dimension should remain restricted to eleven. This is one of the questions that we will address here.

In principle, the central charges transform according to the automorphism group $\mathrm{H_R}$ of the supersymmetry algebra that rotates the supercharges and commutes with the D-dimensional Lorentz group, where $D = 11 - n$. In most cases, however, the central charges can also be assigned to the larger U-duality group. However, there are exceptions for spacetime dimensions $D = 11 - n \leq 5$ for the low-dimensional branes (for a discussion, see Refs. 7,11). On the other hand we note that the antisymmetric gauge fields of supergravity are assigned to representations of G and not of $\mathrm{H_R}$ (the only exceptions exist in $D = 4, 8$ where one has to rely on the Hodge duality rotations of the field strengths in order to realize the full duality group).

When the central charges, and thus the corresponding BPS states that carry those charges, constitute representations of the U-duality group, it may be possible to incorporate all these BPS states into a local supergravity field theory in a way that is U-duality invariant. We call such theories BPS-extended supergravities. The construction of these theories is motivated in part by some old intriguing result that the toroidal compactifications of eleven-dimensional supergravity, without truncation to the massless modes, exhibit traces of the hidden symmetries,[12,13] while one of the reasons for the lack of invariance of the untruncated theory is precisely the incompleteness of the massive KK states with respect to the hidden symmetries. Interestingly enough, by insisting on U-duality for the BPS states one usually goes beyond eleven-dimensional supergravity. To study the properties of these new theories we discuss the simplest version corresponding to a toroidal compactification to nine spacetime dimensions, where we couple a variety of BPS supermultiplets to the unique $N = 2$ supergravity theory, which is thus based on 32 supercharges. The supergravity theory has been studied previously,[14] but the main emphasis here is on the BPS states. We also make contact with earlier work[15] where it was argued that the BPS multiplets have a natural interpretation in terms of the momentum and wrapping states on the M-theory torus. The IIA/B T-duality and the IIA/M S-duality combines as a duality between IIB theory on $R^9 \times S^1$ and M-theory on $R^9 \times T^2$. As we were alluding to above, this theory can be interpreted as a theory living in twelve dimensions (although there is no twelve-dimensional Lorentz invariance). In special decompactification limits it coincides with eleven-dimensional or IIA/B supergravity.[16] In the next sections we discuss various aspects of this nine-dimensional BPS-extended theory.

2. Maximal Supersymmetry and BPS States in Nine Dimensions

Let us start by considering the BPS multiplets that are relevant in nine spacetime dimensions from the perspective of supergravity, string theory and (super)membranes. It is well known that the massive supermultiplets of IIA and IIB string theory coincide, whereas the massless states comprise inequivalent supermultiplets, for the simple reason that they transform according to different representations of the SO(8) helicity group. When compactifying the theory on a circle, massless IIA and IIB states in nine spacetime dimensions transform according to identical SO(7) representations of the helicity group and constitute equivalent supermultiplets. The corresponding interacting field theory is the unique $N = 2$ supergravity theory in nine spacetime dimensions. However, the supermultiplets of the BPS states, which carry momentum along the circle, remain inequivalent, as they remain assigned to the inequivalent representations of the group SO(8) which is now associated with the restframe (spin) rotations of the massive states. The momentum states of the IIA and the IIB theories will be denoted henceforth as KKA and KKB states, respectively. The fact that they constitute inequivalent supermultiplets, has implications for the winding states in order that T-duality will remain valid. This is discussed below.

In Ref. 16 this question was investigated in detail and considered in the context of $N = 2$ supersymmetry in nine spacetime dimensions with Lorentz-invariant central charges. These central charges are encoded in a two-by-two real symmetric matrix Z^{ij}, which can be decomposed as

$$Z^{ij} = b\,\delta^{ij} + a\,(\cos\theta\,\sigma_3 + \sin\theta\,\sigma_1)^{ij}\,. \tag{1}$$

Here $\sigma_{1,3}$ are the real symmetric Pauli matrices. We note that the central charge associated with the parameter a transforms as a doublet under the SO(2) group that rotates the two supercharge spinors, while the central charge proportional to the parameter b is SO(2) invariant. Subsequently one shows that BPS states that carry these charges must satisfy the mass formula,

$$M_{\rm BPS} = |a| + |b|\,. \tag{2}$$

Here one can distinguish three types of BPS supermultiplets. One type has central charges $b = 0$ and $a \neq 0$. These are 1/2-BPS multiplets, because they are annihilated by half of the supercharges. The KKA supermultiplets that comprise KK states of IIA supergravity are of this type. Another type of 1/2-BPS multiplets has central charges $a = 0$ and $b \neq 0$. The KKB supermultiplets that comprise the KK states of IIB supergravity are of this type. Finally there are 1/4-BPS multiplets (annihilated by one fourth of the supercharges) characterized by the fact that neither a nor b vanishes.

For type-II string theory one obtains these central charges in terms of the left- and right-moving momenta, $p_{\rm L}$, $p_{\rm R}$, that characterize winding and momentum along S^1. However, the result takes a different form for the IIA and the IIB theory,

$$Z^{ij} = \begin{cases} \frac{1}{2}(p_{\rm L} + p_{\rm R})\delta^{ij} + \frac{1}{2}(p_{\rm L} - p_{\rm R})\sigma_3^{ij}\,, & \text{(for IIB)} \\ \frac{1}{2}(p_{\rm L} - p_{\rm R})\delta^{ij} + \frac{1}{2}(p_{\rm L} + p_{\rm R})\sigma_3^{ij}\,. & \text{(for IIA)} \end{cases} \tag{3}$$

The corresponding BPS mass formula is equal to

$$M_{\rm BPS} = \tfrac{1}{2}|p_{\rm L} + p_{\rm R}| + \tfrac{1}{2}|p_{\rm L} - p_{\rm R}|\,. \tag{4}$$

For $p_{\rm L} = p_{\rm R}$ we confirm the original identification of the momentum states, namely that IIA momentum states constitute KKA supermultiplets, while IIB momentum states constitute KKB supermultiplets. For the winding states, where $p_{\rm L} = -p_{\rm R}$, one obtains the opposite result: IIA winding states constitute KKB supermultiplets, while IIB winding states constitute KKA supermultiplets. The 1/4-BPS multiplets arise for string states that have either right- or left-moving oscillator states, so that either $M_{\rm BPS} = |p_{\rm L}|$ or $|p_{\rm R}|$ with $p_{\rm L}^2 \neq p_{\rm R}^2$. All of this is entirely consistent with T-duality,[17,18] according to which there exists a IIA and a IIB perspective, with decompactification radii are that inversely proportional and with an interchange of winding and momentum states.

It is also possible to view the central charges from the perspective of the eleven-dimensional (super)membrane. Assuming that the two-brane charge takes values in the compact coordinates labeled by 9 and 10, which can be generated by wrapping the membrane over the corresponding T^2, one readily finds the expression,

$$Z^{ij} = Z_{9\,10}\,\delta^{ij} - (P_9\,\sigma_3^{ij} - P_{10}\,\sigma_1^{ij})\,. \tag{5}$$

When compactifying on a torus with modular parameter τ and area A, the BPS mass formula takes the form

$$\begin{aligned} M_{\rm BPS} &= \sqrt{P_9^2 + P_{10}^2} + |Z_{9\,10}| \\ &= \frac{1}{\sqrt{A\,\tau_2}}|q_1 + \tau\, q_2| + T_{\rm m}A\,|p|\,. \end{aligned} \tag{6}$$

Here $q_{1,2}$ denote the momentum numbers on the torus and p is the number of times the membrane is wrapped over the torus; $T_{\rm m}$ denotes the supermembrane tension. Clearly the KKA states correspond to the momentum modes on T^2 while the KKB states are associated with the wrapped membranes on the torus. Therefore there is a rather natural way to describe the IIA and IIB momentum and winding states starting from a (super)membrane in eleven spacetime dimensions. This point was first emphasized in Ref. 15.

3. BPS-Extended Supergravity

It is possible to consider $N = 2$ supergravity in nine spacetime dimensions and couple it to the simplest BPS supermultiplets corresponding to the KKA and KKB states. As shown in the previous section there are three central charges and nine-dimensional supergravity possesses precisely three gauge fields that couple to these charges. From the perspective of eleven-dimensional supergravity compactified on T^2, the KK states transform as KKA multiplets. Their charges transform obviously with respect to an SO(2) associated with rotations of the coordinates labeled by 9 and 10. Hence we have a double tower of these charges with corresponding KKA supermultiplets. On the other hand, from the perspective of IIB compactified on S^1, the KK states constitute KKB multiplets and their charge is SO(2) invariant. Here we have a single tower of KKB supermultiplets. However, from the perspective of nine-dimensional supergravity one is led to couple both towers of KKA and KKB supermultiplets simultaneously. In that case one obtains some dichotomic theory,[16] which we refer to as BPS-extended supergravity. In the case at hand this new theory describes the ten-dimensional IIA and IIB theories in certain decompactification limits, as well as eleven-dimensional supergravity. But the theory is in some sense truly twelve-dimensional with three compact coordinates, although there is no twelve-dimensional Lorentz invariance, not even in a uniform decompactification limit, as the fields never depend on all the twelve coordinates! Whether this kind of BPS-extended supergravity offers a viable scheme in a more general context than the one we discuss here, is not known. In fact, not much work has been done on incorporating the coupling of BPS multiplets into a supersymmetric field theory. Nevertheless, in the case at hand we know a lot about these couplings from our knowledge of the T^2 compactification of eleven-dimensional supergravity and the S^1 compactification of IIB supergravity.

The fields of nine-dimensional $N = 2$ supergravity are listed in table 1, where we also indicate their relation with the fields of eleven-dimensional and ten-dimensional IIA/B supergravity upon dimensional reduction. It is not necessary to work out all the nonlinear field redefinitions here, as the corresponding fields can be uniquely identified by their scaling weights under SO(1,1), a symmetry of the massless theory that emerges upon dimensional reduction and is associated with scalings of the internal vielbein. The scalar field σ is related to G_{99}, the IIB metric component in the compactified dimension, by $G_{99} = \exp(\sigma)$; likewise it is related to the determinant of the eleven-dimensional metric in the compactified dimensions, which is equal to $\exp(-\frac{4}{3}\sigma)$. The precise relationship follows from comparing the SO(1,1) weights through the dimensional reduction of IIB and eleven-dimensional supergravity. In nine dimensions supergravity has two more scalars, which are described by a nonlinear sigma model based on $\mathrm{SL}(2,\mathbf{R})/\mathrm{SO}(2)$. The coset is described by the complex doublet of fields ϕ^α, which satisfy a constraint $\phi^\alpha\phi_\alpha = 1$ and are subject to a lo-

Table 1. The bosonic fields of the eleven dimensional, type-IIA, nine-dimensional $N = 2$ and type-IIB supergravity theories. The eleven-dimensional and ten-dimensional indices, respectively, are split as $\hat{M} = (\mu, 9, 10)$ and $M = (\mu, 9)$, where $\mu = 0, 1, \ldots 8$. The last column lists the SO(1,1) scaling weights of the fields.

$D = 11$	IIA	$D = 9$	IIB	SO(1,1)
$\hat{G}_{\mu\nu}$	$G_{\mu\nu}$	$g_{\mu\nu}$	$G_{\mu\nu}$	0
$\hat{A}_{\mu\,9\,10}$	$C_{\mu\,9}$	B_μ	$G_{\mu\,9}$	-4
$\hat{G}_{\mu\,9}, \hat{G}_{\mu\,10}$	$G_{\mu\,9}\ , C_\mu$	A_μ^α	$A_{\mu\,9}^\alpha$	3
$\hat{A}_{\mu\nu\,9}, \hat{A}_{\mu\nu\,10}$	$C_{\mu\nu\,9}, C_{\mu\nu}$	$A_{\mu\nu}^\alpha$	$A_{\mu\nu}^\alpha$	-1
$\hat{A}_{\mu\nu\rho}$	$C_{\mu\nu\rho}$	$A_{\mu\nu\rho}$	$A_{\mu\nu\rho\sigma}$	2
$\hat{G}_{9\,10}, \hat{G}_{9\,9}, \hat{G}_{10\,10}$	$\phi, G_{9\,9}, C_9$	ϕ^α	ϕ^α	0
		$\exp(\sigma)$	$G_{9\,9}$	7

cal SO(2) invariance, so that they describe precisely two scalar degrees of freedom ($\alpha = 1, 2$). We expect that the local SO(2) invariance can be incorporated in the full BPS-extended supergravity theory and can be exploited in the construction of the couplings of the various BPS supermultiplets to supergravity.

We already mentioned the three abelian vector gauge fields which couple to the central charges. There are two vector fields A_μ^α, which are the KK photons from the T^2 reduction of eleven-dimensional supergravity and which couple therefore to the KKA states. From the IIA perspective these correspond to the KK states on S^1 and the D0 states. From the IIB side they originate from the tensor fields, which confirms that they couple to the IIB (elementary and D1) winding states. These two fields transform under SL(2), which can be understood from the perspective of the modular transformation on T^2 as well as from the S-duality transformations that rotate the elementary strings with the D1 strings. The third gauge field, denoted by B_μ, is a singlet under SL(2) and is the KK photon on the IIB side, so that it couples to the KKB states. On the IIA side it originates from the IIA tensor field, which is consistent with the fact that the IIA winding states constitute KKB supermultiplets.

From the perspective of the supermembrane, the KKA states are the momentum states on T^2, while the KKB states correspond to the membranes wrapped around the torus. While it is gratifying to see how all these correspondences work out, we stress that from the perspective of nine-dimensional $N = 2$ supergravity, coupled to the smallest 1/2-BPS states, the results follow entirely from supersymmetry.

The resulting BPS-extended theory incorporates eleven-dimensional supergravity and the two type-II supergravities in special decompactification limits. But, as we stressed above, we are dealing with a twelve-dimensional theory here, of which three coordinates are compact, except that no field can depend on all of the three compact coordinates. The theory has obviously two mass scales associated with the KKA and KKB states. We return to them in a moment. Both S- and T-duality are manifest, although the latter has become trivial as the theory is not based on a specific IIA or IIB perspective. One simply has the freedom to view the theory from a IIA or a IIB perspective and interpret it accordingly.

We should discuss the fate of the group $\mathrm{G} = \mathrm{SO}(1,1) \times \mathrm{SL}(2,\mathbf{R})$ of pure supergravity after coupling the theory to the BPS multiplets. The central charges of the BPS states form a discrete lattice, which is affected by this group. Hence, after coupling to the BPS states, we only have a discrete subgroup that leaves the charge lattice invariant. This is the group $\mathrm{SL}(2,\mathbf{Z})$.

The KKA and KKB states and their interactions with the massless theory can be understood from the perspective of compactified eleven-dimensional and IIB supergravity. In this way we are able to deduce the following BPS mass formula,

$$M_{\mathrm{BPS}}(q_1,q_2,p) = m_{\mathrm{KKA}}\,\mathrm{e}^{3\sigma/7}\,|q_\alpha \phi^\alpha| + m_{\mathrm{KKB}}\,\mathrm{e}^{-4\sigma/7}\,|p|\,, \tag{7}$$

where q_α and p refer to the integer-valued KKA and KKB charges, respectively, and m_{KKA} and m_{KKB} are two independent mass scales. This formula can be compared to the membrane BPS formula (6) in the eleven-dimensional frame. One then finds that

$$m_{\mathrm{KKA}}^2\, m_{\mathrm{KKB}} \propto T_{\mathrm{m}}\,, \tag{8}$$

with a numerical proportionality constant.

4. Supertraces and R^4-Terms

In this BPS-extended supergravity theory one can integrate out the BPS supermultiplets and study their contribution to the effective action. In this way one makes contact with the R^4-terms that were considered by Refs. 19,20,21. This amounts to a one-loop calculation in a nine-dimensional field theory. Since we evaluate the coefficient of the R^4 term at zero momentum, the relevant amplitude has the structure of a box diagram in massive φ^3 theory in nine spacetime dimensions (relying on the usual recombination with the non-box graphs that often appears in gauge theories). The contributions from the KKA and KKB states are equal to[22]

$$A_4^{\mathrm{KKA+KKB}} = \frac{1}{(2\pi)^9}\int \mathrm{d}^9k \sum_{q_1,q_2,p} \frac{1}{[k^2+M^2(q_1,q_2,p)]^4}\,. \tag{9}$$

The momentum integral is linearly divergent and it is multiplied by a so-called helicity supertrace B_n, with $n = 8$, which correctly counts all the virtual states belonging to the KKA and KKB supermultiplets. These helicity supertraces can be calculated straightforwardly and depend on the supermultiplet of virtual states in the loop. One finds that 1/2-BPS multiplets have a zero supertrace when $n < 8$, the 1/4-BPS multiplets have a zero supertrace when $n < 12$, and the generic supermultiplets have a zero supertrace for $n < 16$. Therefore only the KKA and KKB supermultiplets are expected to contribute to the R^4-terms (at one loop).

After a Poisson resummation this result can be written as follows (we use the BPS mass formula (6)),

$$\begin{aligned} A_4^{\rm KKA+KKB} &= \frac{2}{3}\,\frac{A^{-1/2}}{(4\pi)^6}\sum_{q_1',q_2'}\frac{\tau_2^{3/2}}{|q_1'+\tau q_2'|^3}+\frac{4}{3}\,\frac{T_{\rm m}\,A}{(4\pi)^6}\sum_{p'}\frac{1}{p'^2} \\ &= \frac{2}{3}\,\frac{1}{(4\pi)^6}\Big[A^{-1/2}\,f(\tau,\bar\tau)+\tfrac{2}{3}\pi^2\,T_{\rm m}\,A\Big]\,. \end{aligned} \tag{10}$$

In the last line we dropped the terms in the sum for $q_1' = q_2' = 0$ and $p' = 0$. These terms represent the ultraviolet divergences. The modular function $f(\tau,\bar\tau)$ is defined by[19]

$$f(\tau,\bar\tau) = \sum_{(q_1',q_2')\neq(0,0)}\frac{\tau_2^{3/2}}{|q_1'+\tau q_2'|^3}\,. \tag{11}$$

This result is invariant (as it should) under the IIB S-duality symmetry $\tau \to (a\tau + b)/(c\tau + d)$ with a, b, c and d integers satisfying $ac - bd = 1$. The contribution from the KKB states is such that the result is compatible with T-duality of type-II string theory.

It is a gratifying result that including the KKA and the KKB states in the context of a uniform regularization scheme leads directly to a result that is consistent with T-duality. Nevertheless, there remains a number of critical questions. The degree of divergence of the initial box graphs is not reduced by supersymmetry, because the relevant helicity supertraces for the KKA and the KKB multiplets, which represent the sum over the virtual states, yield identical multiplicative factors. The subtraction dependence also reflects itself in the fact that the KKA contributions disappear in the decompactification limit to eleven dimensions, $A \to \infty$, which is counterintuitive. Hence the result remains in principle subtraction dependent. In fact the uniform regularization implies that the sum over the KKA and the KKB states leads to double counting of the massless states.

5. Concluding Remarks

We have considered extensions of supergravity where one couples to a supersymmetric set of BPS states that is complete with respect to the U-duality group. We used maximal nine-dimensional supergravity as an example, coupled to both KKA and KKB supermultiplets. From the point of view of U-duality, this example is not quite satisfactory in demonstrating all relevant features, as the KKA and the KKB states transform separately under U-duality. In lower-dimensional spacetimes the situation is different and the corresponding BPS states transform irreducibly under U-duality. Nevertheless, in this way one obtains a dichotomic theory that can be regarded as a twelve-dimensional field theory, which, upon suitable decompactification limits, leads to eleven-dimensional supergravity or ten-dimensional IIA/B supergravity.

There are interesting questions regarding the field-theoretic coupling of the fields associated with BPS states. In the case at hand these can in principle be answered, because the couplings can be deduced from the coupling of the massive KK fields in the compactifications of eleven-dimensional and IIB supergravity. One such questions concerns the role of the local symmetry H = SO(2) that one uses in the description of the SL(2)/SO(2) coset space for the nonlinear sigma model. This symmetry has a (composite) gauge field, which does not describe additional degrees of freedom.

What is difficult to answer at present is the question whether such BPS-extended theories can be fully consistent as local field theories. It is known that the coupling of fields of higher spins can lead to minimal-coupling inconsistencies. As long as we consider terms that are at most quadratic in the massive fields, there seems no immediate discrepancy. However, once one includes the higher-order couplings between the BPS states, this is not so clear. Yet, it would be interesting to calculate these couplings as in this way one can make contact with the algebra of BPS states discussed in Ref. 23.

As we mentioned before, the extension with BPS states is interesting from the perspective of alternative formulations of eleven-dimensional supergravity which have been written down long ago.[12,13] There it was found that the eleven-dimensional theory does exhibit nontrivial traces of the hidden symmetry groups $E_{7(7)}$ and $E_{8(8)}$, although the invariance under these groups is only realized upon truncating to the massless states. However, to some extent the lack of invariance can be traced back to the incompleteness of the BPS states with respect to U-duality (for a recent discussion, see Ref. 11). Therefore it seems reasonable to expect that there do exist interacting field theories of supergravity coupled to BPS states which are consistent and respect U-duality and therefore are suitable starting points for learning more about M-theory. On the other hand, we do expect certain obstacles to this program for the case of toroidal compactifications to low spacetime dimensions. First of all, we already know of some inconsistencies in the U-duality assignments of the central charges, and secondly, the BPS charges will be mutually nonlocal. For instance, the pointlike charges in $D = 4$ spacetime dimensions correspond to electric and magnetic charges, which can not be incorporated in the framework of a local field theory.

Acknowledgments

I thank Ivan Herger, Dieter Lüst, Hermann Nicolai and Niels Obers for valuable discussions.

References

1. E. Bergshoeff, E. Sezgin and P.K. Townsend, *Phys. Lett.* **B189**, 75 (1987).
2. B. de Wit, J. Hoppe and H. Nicolai, *Nucl. Phys.* **B305**, 545 (1988).
3. T. Banks, W. Fischler, S.H. Shenker and L. Susskind, *Phys. Rev.* **D55**, 5112 (1997) [hep-th/9610043].
4. H. Nicolai and R. Helling, *Supermembranes and M(atrix) Theory*, in Proc. of the Trieste String School on non-perturbative aspects of String Theory and Supersymmetric

Gauge Theories, 1998, hep-th/9809103;

B. de Wit, *Supermembranes and Super Matrix Models*, in *Quantum Aspects of Gauge Theories, Supersymmetry and Unification*, Proc. Curfu, 1998, eds. A. Ceresole, C. Kounnas, D. Lüst and S. Theisen (Lecture notes in Physics 525, Springer, 1999), hep-th/9902051;

W. Taylor, *The M(atrix) Model of M Theory*, lectures given at the NATO Advanced Study Institute on Quantum Geometry, Iceland, 1999, hep-th/0002016.

5. C.M. Hull and P.K. Townsend, *Nucl. Phys.* **B438**, 109 (1995) [hep-th/9410167].
6. E. Cremmer and B. Julia, *Phys. Lett.* **76B**, 409 (1978).
7. N.A. Obers and B. Pioline, *Phys. Rept.* **318**, 113 (1999) [hep-th/9809039].
8. E. Cremmer, B. Julia and J. Scherk, *Phys. Lett.* **76B**, 409 (1978).
9. P.K. Townsend, *Phys. Lett.* **B350**, 184 (1995) [hep-th/9501068].
10. E. Witten, *Nucl. Phys.* **B443**, 85 (1995) [hep-th/9503124].
11. B. de Wit and H. Nicolai, *Hidden Symmetries, Central Charges and All That*, to be published in the Proceedings of the Gürsey Memorial Conference II, June 2000, Istanbul.
12. B. de Wit and H. Nicolai, *Phys. Lett.* **155B**, 47 (1985); *Nucl. Phys.* **B274**, 363 (1986).
13. H. Nicolai, *Phys. Lett.* **187B**, 363 (1987).
14. E. Bergshoeff, C. Hull and T. Ortin, *Nucl. Phys.* **B451**, 547 (1995) [hep-th/9504081].
15. J.H. Schwarz, *An $SL(2,Z)$ Multiplet of Type-II Superstrings*, *Phys. Lett.* **B360**, 13 (1995), Erratum *ibid*, **B364**, 252 (1995) [hep-th/9508143]; *Nucl. Phys. Proc. Suppl.* **49**, 183 (1996) [hep-th/9509148].
16. M. Abou-Zeid, B. de Wit, D. Lüst and H. Nicolai, *Phys. Lett.* **466B**, 144 (1999) [hep-th/9908169].
17. M. Dine, P. Huet and N. Seiberg, *Nucl. Phys.* **B322**, 301 (1989).
18. J. Dai, R.G. Leigh and J. Polchinski, *Mod. Phys. Lett.* **A4**, 2073 (1989).
19. M.B. Green, M. Gutperle and P. Vanhove, *Phys. Lett.* **B409**, 177 (1997) [hep-th/9706175].
20. M.B. Green and M. Gutperle, *Nucl. Phys.* **B498**, 195 (1997) [hep-th/9701093].
21. M.B. Green and P. Vanhove, *Phys. Lett.* **B408**, 122 (1997) [hep-th/9704145]; *Phys. Rev.* **D61**, 104011 (2000) [hep-th/9910056].
22. B. de Wit and D. Lüst, *Phys. Lett.* **B477**, 299 (2000) [hep-th/9912225].
23. J. Harvey and G. Moore, *Nucl. Phys.* **B463**, 315 (1996) [hep-th/9510182].

TOP TEN PROBLEMS IN FUNDAMENTAL PHYSICS*

M.J. DUFF
Michigan Center for Theoretical Physics
Randall Laboratory, Department of Physics, University of Michigan
Ann Arbor, MI 48109–1120, USA
E-mail: mduff@umich.edu

In 1900 the world-renowned mathematician David Hilbert presented twenty-three problems at the International Congress of Mathematicians in Paris. These problems have inspired mathematicians throughout the last century. As a piece of millennial madness, all participants of the Strings 2000 Conference were invited to help formulate the ten most important unsolved problems in fundamental physics. Each participant was allowed to submit one candidate problem for consideration. To qualify, the problem must not only have been important but also well-defined and stated in a clear way. The best ten problems were selected at the end of the conference by a selection panel consisting of David Gross, Edward Witten and myself.

1. Top Ten Unsolved Problems

The results, in no particular order, are:

(1) *Are all the (measurable) dimensionless parameters that characterize the physical universe calculable in principle or are some merely determined by historical or quantum mechanical accident and uncalculable?*

David Gross, Institute for Theoretical Physics, University of California, Santa Barbara.

(2) *How can quantum gravity help explain the origin of the universe?*

Edward Witten, California Institute of Technology and Institute for Advanced Study, Princeton.

(3) *What is the lifetime of the proton and how do we understand it?*

Steve Gubser, Princeton University and California Institute of Technology.

(4) *Is Nature supersymmetric, and if so, how is supersymmetry broken?*

Sergio Ferrara, CERN European Laboratory of Particle Physics;
Gordon Kane, University of Michigan.

*Research supported in part by DOE Grant DE-FG02-95ER40899.

(5) *Why does the universe appear to have one time and three space dimensions?*

Shamit Kachru, University of California, Berkeley;
Sunil Mukhi, Tata Institute of Fundamental Research;
Hiroshi Ooguri, California Institute of Technology.

(6) *Why does the cosmological constant have the value that it has, is it zero and is it really constant?*

Andrew Chamblin, Massachusetts Institute of Technology;
Renata Kallosh, Stanford University.

(7) *What are the fundamental degrees of freedom of M-theory (the theory whose low-energy limit is eleven-dimensional supergravity and which subsumes the five consistent superstring theories) and does the theory describe Nature?*

Louise Dolan, University of North Carolina, Chapel Hill;
Annamaria Sinkovics, Spinoza Institute;
Billy & Linda Rose, San Antonio College.

(8) *What is the resolution of the black hole information paradox?*

Tibra Ali, Department of Applied Mathematics and Theoretical Physics, Cambridge University;
Samir Mathur, Ohio State University.

(9) *What physics explains the enormous disparity between the gravitational scale and the typical mass scale of the elementary particles?*

Matt Strassler, Institute for Advanced Study, Princeton.

(10) *Can we quantitatively understand quark and gluon confinement in Quantum Chromodynamics and the existence of a mass gap?*

Igor Klebanov, Princeton University;
Oyvind Tafjord, McGill University.